Advances in Intelligent Systems and Computing

Volume 714

Series editor

Janusz Kacprzyk, Polish Academy of Sciences, Warsaw, Poland
e-mail: kacprzyk@ibspan.waw.pl

The series "Advances in Intelligent Systems and Computing" contains publications on theory, applications, and design methods of Intelligent Systems and Intelligent Computing. Virtually all disciplines such as engineering, natural sciences, computer and information science, ICT, economics, business, e-commerce, environment, healthcare, life science are covered. The list of topics spans all the areas of modern intelligent systems and computing such as: computational intelligence, soft computing including neural networks, fuzzy systems, evolutionary computing and the fusion of these paradigms, social intelligence, ambient intelligence, computational neuroscience, artificial life, virtual worlds and society, cognitive science and systems, Perception and Vision, DNA and immune based systems, self-organizing and adaptive systems, e-Learning and teaching, human-centered and human-centric computing, recommender systems, intelligent control, robotics and mechatronics including human-machine teaming, knowledge-based paradigms, learning paradigms, machine ethics, intelligent data analysis, knowledge management, intelligent agents, intelligent decision making and support, intelligent network security, trust management, interactive entertainment, Web intelligence and multimedia.

The publications within "Advances in Intelligent Systems and Computing" are primarily proceedings of important conferences, symposia and congresses. They cover significant recent developments in the field, both of a foundational and applicable character. An important characteristic feature of the series is the short publication time and world-wide distribution. This permits a rapid and broad dissemination of research results.

More information about this series at http://www.springer.com/series/11156

Chhabi Rani Panigrahi · Arun K. Pujari
Sudip Misra · Bibudhendu Pati
Kuan-Ching Li
Editors

Progress in Advanced Computing and Intelligent Engineering

Proceedings of ICACIE 2017, Volume 2

Springer

Editors
Chhabi Rani Panigrahi
Department of Computer Science
Rama Devi Women's University
Bhubaneswar, Odisha
India

Bibudhendu Pati
Department of Computer Science
Rama Devi Women's University
Bhubaneswar, Odisha
India

Arun K. Pujari
Department of Computer Science
Central University of Rajasthan
Ajmer, Rajasthan
India

Kuan-Ching Li
Department of Computer Science and Information Engineering (CSIE)
Providence University
Taichung
Taiwan

Sudip Misra
Department of Computer Science and Engineering
Indian Institute of Technology Kharagpur
Kharagpur
India

ISSN 2194-5357 ISSN 2194-5365 (electronic)
Advances in Intelligent Systems and Computing
ISBN 978-981-13-0223-7 ISBN 978-981-13-0224-4 (eBook)
https://doi.org/10.1007/978-981-13-0224-4

Library of Congress Control Number: 2018938653

Printed on acid-free paper

This Springer imprint is published by the registered company Springer Nature Singapore Pte Ltd.
The registered company address is: 152 Beach Road, #21-01/04 Gateway East, Singapore 189721, Singapore

Preface

This volume contains the papers presented at the 2nd International Conference on Advanced Computing and Intelligent Engineering (ICACIE) 2017: The 2nd International Conference on Advanced Computing and Intelligent Engineering (www.icacie.com) was held during November 23–25, 2017, at the Central University of Rajasthan, India. There were 618 submissions, and each qualified submission was reviewed by a minimum of two Technical Program Committee members using the criteria of relevance, originality, technical quality, and presentation. The committee accepted 109 full papers for oral presentation at the conference, and the overall acceptance rate is 18%.

ICACIE 2017 was an initiative taken by the organizers, which focuses on research and applications on the topics of advanced computing and intelligent engineering. The focus was also to present state-of-the-art scientific results, to disseminate modern technologies, and to promote collaborative research in the field of advanced computing and intelligent engineering.

Researchers presented their work in the conference and had an excellent opportunity to interact with eminent professors, scientists, and scholars in their area of research. All participants were benefitted from discussions that facilitated the emergence of innovative ideas and approaches. Many distinguished professors, well-known scholars, industry leaders, and young researchers were participated in making ICACIE 2017 an immense success.

We had also industry and academia panel discussion, and we invited people from software industries like TCS and Infosys, and from DRDO.

We thank all the Technical Program Committee members and all reviewers/ sub-reviewers for their timely and thorough participation during the review process.

We express our sincere gratitude to Honorable Vice-Chancellor and General Chair, Prof. Arun K. Pujari, Central University of Rajasthan, for allowing us to organize ICACIE 2017 on the campus and for his valuable moral and timely support. We also thank Prof. A. K. Gupta, Dean Research, for his valuable guidance. We appreciate the time and efforts put in by the members of the local organizing team at the Central University of Rajasthan, especially the faculty members of different departments, student volunteers, administrative and account section staff, guest house management, and hostel management staff, to make ICACIE 2017 successful. We thank Mr. Subhashis Das Mohapatra, System Analyst, C. V. Raman College of Engineering, Bhubaneswar, for designing and maintaining ICACIE 2017 Web site.

We are very grateful to all our sponsors, especially DRDO, for their generous support toward ICACIE 2017.

Bhubaneswar, India	Chhabi Rani Panigrahi
Ajmer, India	Arun K. Pujari
Kharagpur, India	Sudip Misra
Bhubaneswar, India	Bibudhendu Pati
Taichung, Taiwan	Kuan-Ching Li

About the Book

The book focuses on theory, practice, and applications in the broad areas of advanced computing techniques and intelligent engineering. This two-volume book includes 109 scholarly articles, which have been accepted for presentation from over 618 submissions in the 2nd International Conference on Advanced Computing and Intelligent Engineering held at the Central University of Rajasthan, India, during November 23–25, 2017. With a total of 109 papers, the first volume of this book consists of 55 numbers of papers and the second volume contains 54 papers. This book brings together academic scientists, professors, research scholars, and students to share and disseminate their knowledge and scientific research works related to advanced computing and intelligent engineering. It helps to provide a platform to the young researchers to find the practical challenges encountered in these areas of research and the solutions adopted. The book helps to disseminate the knowledge about some innovative and active research directions in the field of advanced computing techniques and intelligent engineering, along with some current issues and applications of related topics.

Contents

About the Editors

Dr. Chhabi Rani Panigrahi is Assistant Professor in the Department of Computer Science at Rama Devi Women's University, Bhubaneswar, India. She completed her Ph.D. in the Department of Computer Science and Engineering, Indian Institute of Technology Kharagpur, India. Her areas of research interests include Software Testing and Mobile Cloud Computing. She holds 17 years of teaching and research experience. She has published several international journals and conference papers. She is a Life Member of Indian Society for Technical Education (ISTE) and Member of IEEE and Computer Society of India (CSI). She also served as Guest Editor of IJCNDS and IJCSE journals. She was the Organizing Chair of ICACIE 2016, ICACIE 2017, and WiE Chair of IEEE ANTS 2017.

Prof. Arun K. Pujari is Faculty and Dean of the School of Computer and Information Sciences at the University of Hyderabad (UoH) and has been appointed as the Vice-Chancellor of the Central University of Rajasthan. He has earlier served as the Vice-Chancellor of Sambalpur University, Odisha, in 2008. He got his Ph.D. from IIT Kanpur in 1980. He has more than 15 years of experience as Dean and Head of the Department. He has served as a member of high-level bodies such as UGC, DST, DRDO, ISRO, and AICTE. He has over 100 publications to his credit and has wide exposure to national and international arenas. His two published books are *Data Mining Techniques* and *Database Management System.*

Dr. Sudip Misra is Professor in the Department of Computer Science and Engineering at the Indian Institute of Technology Kharagpur. Prior to this, he was associated with Cornell University, USA; Yale University, USA; Nortel Networks, Canada; and the Government of Ontario, Canada. He received his Ph.D. in Computer Science from Carleton University, Ottawa, Canada. He has several years of experience working in the academia, government, and the private sectors. His current research interests include Wireless Sensor Networks, Internet of Things

(IoT), Software Defined Networks, Cloud Computing, Big Data Networking, Computer Networks. He is the author of over 260 scholarly research papers, including more than 150 reputed journal papers. He has published nine books in the areas of advanced computer networks.

Dr. Bibudhendu Pati is Associate Professor in the Department of Computer Science at Rama Devi Women's University, Bhubaneswar, India. He has around 21 years of experience in teaching and research. His areas of research interests include Wireless Sensor Networks, Cloud Computing, Big Data, Internet of Things, and Advanced Network Technologies. He completed his Ph.D. from IIT Kharagpur. He is a Life Member of Indian Society for Technical Education, Computer Society of India and Senior Member of IEEE. He has got several papers published in reputed journals, conference proceedings, and books of international repute. He also served as Guest Editor of IJCNDS and IJCSE journals. He was the General Chair of ICACIE 2016, and IEEE ANTS 2017 international conference.

Prof. Kuan-Ching Li is a Professor of Computer Science and Engineering at Providence University, Taiwan. He received guest and distinguished chair professorships from universities in China and other countries, and a recipient of awards and funding support from several agencies and industrial companies. He has been actively involved in many conferences and workshops in program/general/steering conference chairman positions and has organized numerous conferences related to high-performance computing and computational science and engineering. Besides the publication of research papers, he is co-author/co-editor of more than 15 technical professional books published by CRC Press, Springer, McGraw-Hill and IGI Global. He is a Fellow of IET, a life member of TACC, a senior member of the IEEE and a member of the AAAS, and Editor-in-Chief of International Journal of Computational Science and Engineering (IJCSE), International Journal of Embedded Systems (IJES), and International Journal of High-Performance Computing and Networking (IJHPCN), published by Inderscience. His research interests include GPU/many-core computing, Big Data and Cloud.

Part I
Cloud Computing Security, Distributed Systems and Software Engineering

Secure Cloud-Based Federation for EHR Using Multi-authority ABE

Siddhesh Mhatre and Anant V. Nimkar

Abstract Cloud computing is developed as the most influential perfect models in the IT businesses starting late. Because of the progress implied in cloud computing, it will help data innovation in the healthcare industry. In existing healthcare model, outsourcing storage or accessing record from untrusted cloud servers become a challenging issue for security and privacy of data. An access control model is a productive approach that guarantees the information security in the cloud-based framework. In this work, we present a framework to provide expressive, proficient and revocable healthcare access control for a federation-based model using multi-Authority Ciphertext-Policy based Encryption (CP-ABE) scheme. The existing CP-ABE scheme is not able to fulfil all security need to protect healthcare records and control of privilege revocation problem. This research paper proposes the federation-based multi-Authority CP-ABE (F-CPABE) scheme for healthcare system with its subordinate strategies to outline design to healthcare records in federation-based access control scheme. The attribute revocation technique in this scheme helps to resolve both forward and backward security challenges. It has reduced attribute management overhead from a centralized system and also reduces time complexity.

Keywords Multi-authority · CP-ABE · Access control model
Federation · Electronic health record

1 Introduction

The healthcare organizations have made more than amazing progress from simple paper-based health record to Electronic Medical Records (EMR), from manual surgeries process to robotic surgeries, remote observations for patient and Hospital

S. Mhatre (✉) · A. V. Nimkar
Department of Computer Engineering, Sardar Patel Institute of Technology,
University of Mumbai, Mumbai, India
e-mail: siddhesh_mhatre@spit.ac.in

A. V. Nimkar
e-mail: anantvnimkar@spit.ac.in

C. R. Panigrahi et al. (eds.), *Progress in Advanced Computing and Intelligent Engineering*, Advances in Intelligent Systems and Computing 714,
https://doi.org/10.1007/978-981-13-0224-4_1

Information Systems (HIS), after the involvement of IT in the Healthcare industry [4]. The healthcare records are stored electronically at better places such as with Patient, Medical Practitioners (MPs), Care Delivery Organizations (CDOs) and they are called Patient Healthcare Records (PHR), Electronic Medical Records (EMR), and Electronic Health Records (EHR) respectively [5]. Healthcare organizations are attempting to deal with a different set of the health record [8].

Access control model is a way which provides a guarantee that the records are securely stored at cloud storage with proper access protection [2]. It analyses, evaluates and configures healthcare cloud and services for the secure exchange of EHR. This scheme permits information outsourcing to various cloud suppliers for storing data with the help of access control model in the distributed storage. Ciphertext-Policy based Encryption (CP-ABE) is one of the secure ways for immediate and controlled access to the stored information [7–9]. As the outcome of this research, the proposed framework not just permits EHR information storage but also permits incorporation of various associations and security of records. In existing ABE schemes, as a number of attributes under policy increases size of cipher text becomes very large and overhead on the system increases. The proposed model will help to reduce attribute management from the point of a centralized system and cipher text complexity.

The main contributions of this research can be summarized as follows:

- The proposed model speaks with a various structure of EHRs, for example, drug store, patients, care conveyance association, facility lab, etc.
- This model is used to make unified cloud storage framework in which patients and healthcare members can store and look into own records. They can share records to enhance patient's health by consideration with other healthcare organization.
- This scheme enhances the proficiency of the CP-ABE method by converting the technique into multi-authority CP-ABE. In multi-authority CP-ABE distributed storage framework, client's attributes can be changed intensely on their solicitation. This model provides facilities to a client such as new policies or changes some current ascribes to redesign his consent of information access.

In this research paper, we proposed a new federation-based multi-authority access control model using CP-ABE scheme with detailed construction of the model. Then, we compare proposed the model with existing CP-ABE schemes and results proved that proposed scheme is more efficient than existing CP-ABE schemes. We also mention and resolve both forward and backward security challenges.

The rest of the paper is organized as follows. Section 2 of a literature survey contains a summary and an overview of ABE related work. Section 3 presents our proposed federation-based cloud system with the detailed construction of scheme and Sect. 4 describes framework and assumptions for proposed scheme. Whereas Sect. 5 is focused on implementation and performance analysis of proposed scheme with computation and time complexity. This proposed scheme concluded with a remark in Sect. 6.

2 Literature Survey

There is a vast amount of research done on Access Control models and Attribute-Based Access Control [6]. Yanli et al. [14] describes the framework for encryption and re-encryption using users attributes and attribute group keys in a new scheme called Secure Personal Electronic Medical Record (SPEMR) scheme. It is privileged separation under the multi-owner settings with fine-grained access control. Samydurai et al. [7] describe an access control framework to deal with multi-authority systems with an efficient encryption scheme. It is scalable and dynamic multi-authority scheme. In this scheme, it is difficult to select one access control method to satisfy federation needs [5]. The access control mechanism used in healthcare need to satisfy all participant requirements, i.e. doctors, medical practitioners, patients and medical authorities, etc. to provide secure access to the records. Every member needs to get to specific fields of the health record keeping in mind the end goal to do his employment. To address the above-mentioned issues of access control and compliance management, we present a secure EHRs sharing framework based on multi-authority ABE. This will safely deal with the entrance for composite EHRs coordinated from different medicinal services suppliers at various granularity levels. The proposed scheme also supports Health Insurance Portability and Accountability Act (HIPAA) compliance management to ensure that it satisfies all compliant with HIPAA regulations in clouds.

Attribute-Based Encryption (ABE) is the most common and proficient used cryptographic technique in industry. In this literature, we are concerned about cryptographic enforcement mechanisms for CP-ABE. Sahai et al. [2] introduce the concept of ABE in which an encrypted cipher text is associated with a set of attributes and the private key of the user for an access policy over attributes. The user can only decrypt information if it satisfies the attributes. Sahai et al. [2] present the idea of ABE in which an encrypted information with selected attributes related to the owner and the private key of the owner to create decryption policy for a set of allowed attributes. The user can just unscramble data on the off chance that if it fulfils the properties of policy. Joseph Akinyele et al. [1] improved it by including non-monotonic formula and Goyel et al. [3] improved impressibility of ABE which supports any monotonic formula. To overcome limitations of existing ABE, Wang et al. [11] proposed a multi-authority ABE-based scheme called Multi-Authority based Attribute-Based Encryption (MA-ABE). This is a cloud-based scheme which supports multi authorities to provide access control mechanism with efficient encryption and decryption. In this paper, we work on Yang et al. CP-ABE with multi-authority access control mechanism for the healthcare organization with revocation of authority at any time in the system.

3 Proposed Method for EMR Cloud Federation System Model

The proposed EHR federation manages distinctive elements in various clouds. All health record are made and stored by their own distributed storage in healthcare organization. Cloud Exchange is in charge of the safe trade of the health record inside the cloud federation. There are numerous substances which are needed in health record for the investigation of health status. EHR cloud federation needs access control system which manages retrieval of the health record safely. In the cloud-based storage frameworks, Ciphertext-Policy Attribute-Based Encryption (CP-ABE) is used for protecting data from unauthorized access with a high level on access security and benefits of the information proprietors more straightforward access strategies [10, 13] (Fig. 1).

The proposed F-CPABE model consist of five elements in the framework: Federation certificate authority (FCA), Attribute authorities (AAs), Cloud Exchange system, Data owners and Data consumers.

- Federation Certificate authority (FCA): This is a globally trusted authority in the federated framework. The FCA initializes the system and permit clients and Attribute authorities (AAs) in the framework. It provides the unique identity to all the elements involved in the federated system. A unique identity is allotted by the Federation certificate authority (FCA) to every single legitimate client in the framework and furthermore, creates the global public key for the clients.

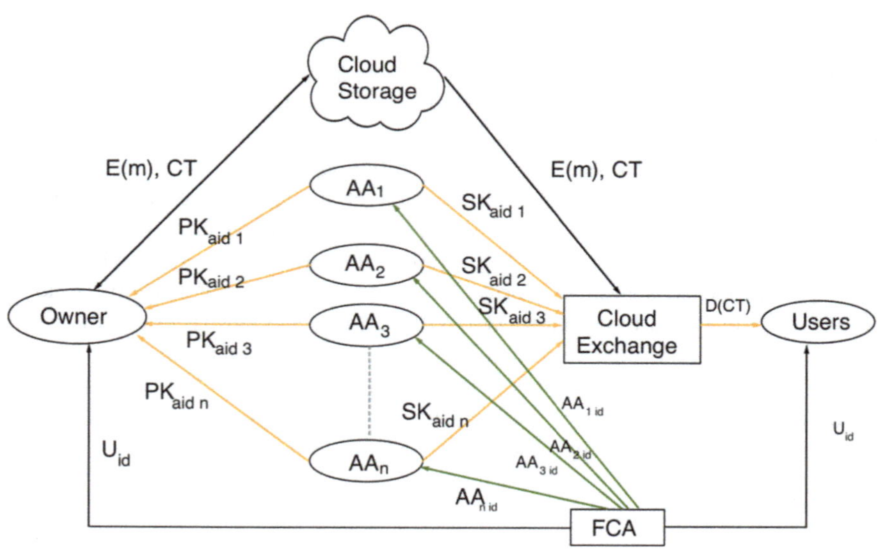

Fig. 1 Federation-based cloud storage system model

FCA also registers Attribute Authorities (AAs) in federation system and generate master public key and master secret key for particular AAs. FCA is not part of in any other quality administration and, the production of client master keys are connected with characteristics of administration.

- Attribute Authorities (AA): It attributes management authority which contains healthcare industries, CDOs, etc. It is responsible for management of users attributes at any time in the system according to their identity or role in its domain. It can add or revoke users attributes at any time in the system. In proposed EHR federation, each attribute assigned by attribute authorities is associated with its respective AA. The same user can have a different set of attributes assign by various AAs. The user does not store attributes as it is taken care by AAs. All the elements come under AA, are managed by attribute authorities, i.e. a number of users, doctors, hospitals, etc. Every AA in EHR federation has full control over the user structure, roles, and semantics of its attributes in the system. Each AAs in EMR federation system generates a local attribute for each user as well as manages revocation and change in the role of the user with reflecting attributes.
- Data Owners: In EHR federation, Data owners can be divided into several components according to the creation of the record. In the health ecosystem, medical practitioner creates EMR record for the patient where medical practitioner with nurse has full access to the health record of the patient. In general, the owner of a health record defines the access policy using attributes from AA or can request attributes from different AAs and encrypt using policies. In EHR federation, we provide multiple ownership for the data which can make changes in records or change in access policy for the record.
- Cloud Storage: Health records made by various participants in the EHR federation for sending encrypted information and receiving information from the server. The cloud server cannot be fully trusted for the information stock-piling and managing its control. This model gives CP-ABE access control system where only the clients who fulfil the attributes are permitted to get the information from EHR federation. All AAs can have their own cloud storage or can share same storage space.
- Cloud exchange: It provides secure record sharing between different AAs without accessing their cloud storage. If AA requires record from other AA, then it can put a request for record sharing to cloud exchange. Cloud exchange gets the data from the AA having the record and gives it to the respective AA.

3.1 Federation Access Control Scheme

To outline the EHR Federation, the most important security issue is to provide attribute revocation for federation-based access control scheme and provide the secure station to information trade in the combined cloud system. However, CP-ABE plan cannot be specifically connected to the federated cloud due to the numerous security issues. In existing scheme, all the encryption and decryption are managed by the central authority where in proposed scheme encryption and decryption are managed by the AAs. Existing healthcare system does not support the revocation of the access to provide security to the health records. A federation-based multi-authority access control model with revocable attributes is been proposed for controlling unauthorized access to healthcare records. This scheme improves the conventional role based model with the new federation-based access control model to work with multiple authorities. It distinguishes the functions of EMR Federation certificate authority (FCA) to the Attribute authorities (AAs). In this FCA acknowledge all the clients in the framework and assign global unique identity (UID) to the client and AAs deal with every property for the client and secret key. In Federation-based access control scheme, secret key allocated by various AAs can be tied together by cloud exchange for decryption of record. Each AAs is associated with an attribute authority identity number (AAID), so in any case, If FCA generates the same attribute for AAs it can be distinguishable.

Whenever EMR cloud-based Federation attribute revocation happens, client or any medical professional and CDOs may upgrade security rule. EHR Federation handles attribute revocation in the framework by controlling every attribute in the system. When revocation happens only components which are associated with the access policy and ciphertext are updated without informing revoked user. All the ciphertexts updating process handled by the server are done in the background and the user need not have to manually update all the ciphertext.

3.2 *Attribute Revocation Method*

In healthcare system revoking user access is a challenging security issue. The proposed method provides attribute revocation to a limit and stops unauthorized access to the healthcare data. It can be divided into two types:

1. Backward security: The user whose privilege revoked cannot be able to decrypt the updated cipher text.
2. Forward security: Changes in the privilege or new user with sufficient privilege can able to decrypt the previous cipher text using its public attributes.

Access revocation is used for taking away access from selected attribute in EMR federation cloud data. The participant who does not require the access of records any longer can be revoked using this method. At some point in healthcare ecosystem, need for revocation of the access is required. For example, if a patient P_1 visit a hospital H_1 for treatment where he has received treatment from medical practitioners M_1 and M_2, then the health record will be cooperatively created by P_1

and H_1, M_1 and M_2 but after some time if medical practitioners M_2 is not required for the treatment then P_1, H_1 or M_1 can revoke access to medical practitioners M_2. The removed user (whose attribute is revoked) will not be able to decrypt new cipher text because all the attributes policies by them are updated (Backward security).

A new user with sufficient attributes in the system then authorized user or the existing user can also add new user to access control list if he has sufficient permission to add a member. For example, EMR which is encrypted under the policy for H_1 healthcare AND (M_1 Doctor OR N_1 Nurse), which means under the policy M_1 OR N_1 from H1 can be able to decrypt data and they can also add any other doctor to the access control list (forward security).

4 Framework

The Framework setup for the federation-based access control model is as follows:

1. **System Initialization**: It is an initial phase of the system consisting of FCA setup and AA setup. The FCA registers new Attribute Authority (AA) and initializes it. The FCA setup is kept running in the federated framework. It takes attribute information from the user and AAs. For every client and attribute authority, it produces authentication id or user id (UID) as well as creates the unique master public key (MPK) and unique master secret key (MSK) for the certificate. It utilizes the elliptic curve with bilinear maps (or pairings). To introduce a group in the elliptic curve (EC), Two cyclic group of G and G1 created by FCA by using p and q of similar prime ordered group. It additionally picks hash function as shown in Eq. (1).

$$H\colon \{0,1\}^{*} \rightarrow G \text{ and } a, b \in Zp \tag{1}$$

For EC with bilinear maps (or pairings), it utilizes symmetric bend with a 512-piece base field. Then FCA generates two random numbers from a set of prime numbers Zp, to generate keys for the registered authority. (1) User Registration to FCA: All the user registrations are performed under the federation rules which is managed by FCA. Every User needs to enrol with own attributes to the FCA for the unique user id. In the enrolment event, if the user is legitimate in the federation framework, the FCA doles out a comprehensively extraordinary unique personality uid to the user. The FCA produces global secret key (SK_{uid}) and the users global public key (PK_{uid}) belongs to every user uid as

$$\text{UserSetup}(U_A) \rightarrow ((SK_{uid}, PK_{uid}), \text{Certificate}(uid)) \tag{2}$$

The FCA additionally produces an endorsement Certificate (Certificate(uid)) for the user uid. At that point, the FCA sends user public key and secret key for the certificate. All keys assign to the user are managed by itself. (2) AA Registration to FCA: Each AA enrols itself to the FCA amid the framework. In this model AAs have a legitimate power in this framework, the FCA first assign a global unique AA identity AA_{aid}. The FCA sends the master secret and public key (MSK_{aid}, MPK_{aid}) to that AA.

$$\text{AASetup}(U_A) \rightarrow ((MSK_{uid}, MPK_{uid}), \text{Certificate}((AA_{uid})) \quad (3)$$

2. **Data Encryption**: The encryption algorithm takes inputs health record (m) data which is hosted on the cloud of different AAs. Data encryption performed for each AAs in the system for all the enrolled user's data. AA uses its public keys MPK_{aid_k} and access policy P of all the involved attributes. Cipher text is denoted as the CT. Encryption is done as in Eq. (4)

$$\text{Ecrypt}(m, MPK_{aid_k}, P) \rightarrow (CT) \quad (4)$$

It isolates the information into a few information segments as $m = \{m_1, m_2, m_3, \ldots, m_4\}$ as indicated by the rational granularities. It scrambles information segments with the policy given to the encryption by utilizing symmetric encryption strategies. It then characterizes structure m_i for every attribute by running the encryption algorithm with policy.

3. **Data Decryption**: All the legitimate clients in the framework can uninhibitedly question any intrigued encoded information. After getting the information from the cloud exchange server, the client uses this algorithm to decode the ciphertext CT by utilizing its global secret keys (SK_{uid}). Once the properties of the client have fulfilled the policy of the cipher text then the only client are liable to decrypt the record.

$$\text{DecryptPHR}(CT, SK_{uid}, A_{uid}) \rightarrow (m) \quad (5)$$

For the EHR records, decryption algorithms run by AAs provides access to the EHR record. AAs take attributes of the user to decrypt data. AAs use its public keys MPK_{aid_k} and access policy P of all the involved attributes. A set of attributes from all the AAs of involved users in the encryption set I_A and Ciphertext (CT). The Decryption algorithm

$$\text{DecryptEHR}(CT, MSK_{aid}, \{A_{aid_k}\}_{aid_k} \in I_A) \rightarrow (m) \quad (6)$$

4. **Cloud Exchange**: Cloud exchange helps to share data from different AAs. Cloud exchange creates the virtual record which will help to share data among

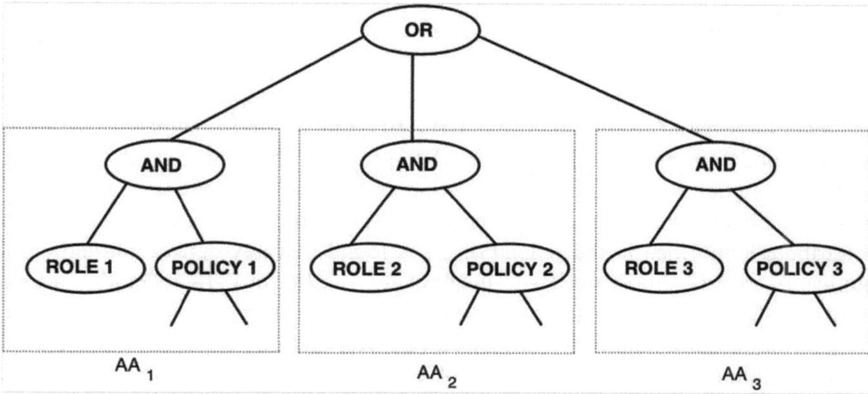

Fig. 2 Access structure for attributes in federation-based cloud system

the different AAs users. If hospital H_1 want to share some records with hospital H_2 then hospital H_1 encrypts data using owners attribute and request for the attributes from the hospital H_2 under AA_2. After getting attributes from AA_2, hospital H_1 of AA_1 encrypts data using OR GATE for with access structure.

Figure 2 demonstrates the multi-authority attribute-based encryption access structure for the share data. In this access structure or Access Tree (AT), four necessities for the model are characterized

- The tree root must be an OR entryway.
- Each offspring of the tree root must be an AND entryway. This must be twofold (binary) gates.
- For every sub-tree of the level 1 AND gates, the right child must be an attribute access policy tree (Policy (K)). This may likewise be an unfilled tree. This tree is named the attribute access policy tree connected with Role K.
- For every sub-tree of the level 1 AND gates, the left child must be a Role quality (Role (K)). This is a leaf. It might be an unfilled tree if and just if the right child of the level 1 AND entryway is avoided tree too. Once access structure is defined its attributes can be used for the encryption of data. All the AA accesses the structure which is defined by the owner of the record to provide access to the other users. Access structure model has two principle points of interest that suit our necessities for the federation-based cloud system which is not optimal in Attribute-Centric ABE model. This model provides auditing efficiency and Policy design flexibility. Other Attribute-Centric ABE Model treats role as a standard property (it is not required to be obligatory). Along with this, there may be an approach that does exclude a role quality, and the reviewing effectiveness of the federation base cloud model is lost.

5 Performance Analysis

To demonstrate the feasibility of proposed approach, we implemented prototype of the federation-based multi-authority CP-ABE model and compared with the existing model of multi-authority CP-ABE proposed by Yang et al. [12]. The F-CPABE model is compared with existing in terms of computation and performance efficiency for encryption and decryption time with consideration of a number of attribute authorities in the federation. It is found that key generation and communication cost is almost similar to the existing models. In EHR federation access control system, only attribute authorities need to store attributes of the user who enrol to the attribute authority. In this model, user and any other authority except attribute authority (AA) store the involved attributes in the system, therefore, storage overhead of the central authority and a user is reduced as compare to the existing model. Attribute authority (AA) is not responsible for managing and storing the public key (PK) or secret key (SK) assign to the owner or users. It will help to reduce storage overhead on Attribute authority. In proposed model, there is no central authority for managing all the attributes and this helps to reduce attribute collusion in the system. In the F-CPABE system, AAs do not communicate with each other. All the communication takes place between AA and cloud exchange, therefore, the communication cost of our access control model is lesser than other models. In this model, communication for the revocation of access and newly added user for authorization is much less because cipher text and policy updates occur only in AAs side unlike other models with the central authority. This helps to reduce overhead from the server to communicate with other AAs. EHR Federation cloud exchange manages all the data sharing from all the AAs so the communication is only with the cloud exchange system which avoids direct access to the cloud storage of other AAs. In other models, for data sharing, all AAs have to share their secret key with each other which can be used for unauthorized access of data.

We achieved federation-based cloud model with a multi-authority CP-ABE scheme on amazon cloud with 3 instances of Ubuntu server with 1 GB RAM and CPU of 2.50 GHz. To implement F-CPABE cloud system, cpabe toolkit [2] with help of other libraries Pairing-Based Cryptography (PBC) to perform mathematical operation of Pairing and charm crypto library for cryptographic operations in the

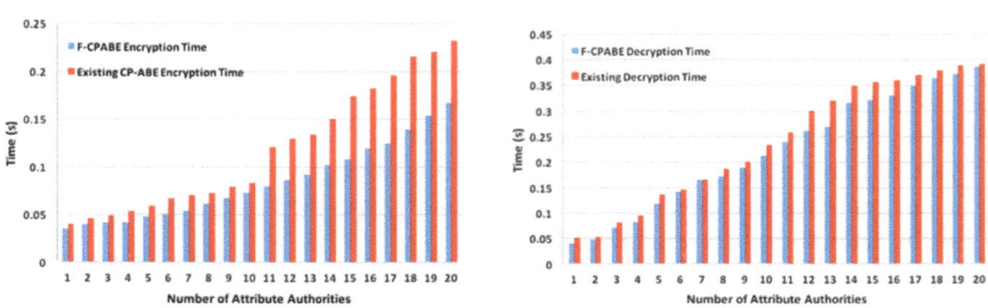

Fig. 3 Performance comparison of F-CPABE with existing multi-authority us CP-ABE

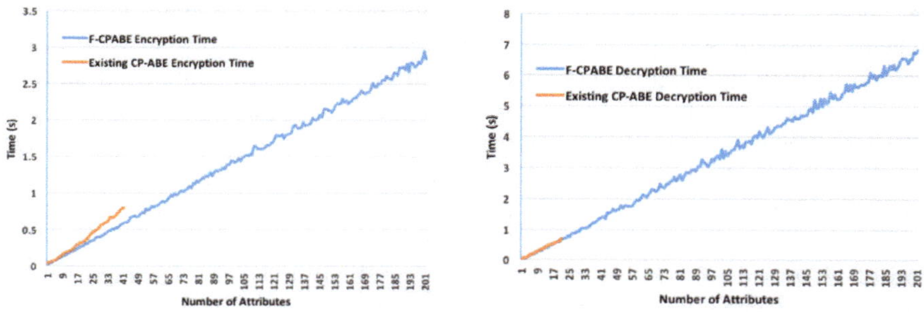

Fig. 4 Time comparison of the of the F-CPABE with existing model

federation base access control model is used. We measure simulation results timings for existing CP-ABE and F-CPABE for the mean of 20 trials for 20 AAs (AA_1, AA_2, …, AA_{20}) with all of the 1 MB file. Figure 3 shows encryption time and decryption time of our model with respect to existing model. It shows that encryption time and decryption time of our model is less than the existing CP-ABE because all encryption and decryption of record are done at AA's side and does not include communication with the centralized server. In this simulation, the encryption and decryption for 200 user attributes (a_1, a_2, …, a_{200}) are carried out and compare with existing scheme. It shows that encryption and decryption performance of F-CPABE is linear with respect to the number of attributes but existing schemes not consistent and it's hard to process when more than 50 attributes are used in existing schemes as shown in Fig. 4 with time utilization. The execution of our scheme is fairly more fascinating. It is marginally harder to gauge without an exact application since simulation timing depends on access policies used in cryptographic operations. The policy tree is randomly generated which is described in Sect. 4 with changing attributes and the size of the policy tree. The trees were produced with beginning from just a root node (OR), then more than once adding a child node to an arbitrarily chosen from different attribute authorities involved in the model. For encryption of data we took random attributes from the attribute authorities involved or have access to health record. Similarly, every running of decryption algorithm we randomly took attributes from attribute authorities who want to access the data and measured the time for the attribute who fulfil the policy structure for the data and excluding the attributes that did not fulfil it.

6 Conclusion

This research used to express two critical security and protection issues in federation-based health care in distributed computing environments: access control on the composite EHRs and HIPAA compliance administration. To address those two issues, a novel framework based on access control policy and logical techniques has been presented. All the more particularly, an EHR information schema approach is proposed to produce composite EHR sharing. In view of the composite EHR information schema, conveyed EHR cases from different healthcare areas can be accumulated into a composite EHR example. Also proposed and illustrated federation-based cloud storage for a healthcare organization by utilizing multi-authority CP-ABE scheme as access control mechanism. The proposed Multi-Authority Attribute-Based Encryption scheme provides an efficient, effective and expressive solution to the health records security problems in sharing health records. It provides revocable solution to the store EHR in the cloud storage. This is more reliable model under some hard cryptographic assumption. This is compared with existing in terms of computation and performance efficiency for encryption and decryption time with consideration of a number of attribute authorities in the federation. We indicated great improvement in encryption and decryption time. It was found that key generation and communication cost is almost similar to the existing models. Our future plan is to implement a model, based on multi-authority attribute-based encryption, to provide the security solution for healthcare data with the real-time application support. The secure trade of information from one cloud supplier to other and furthermore give backing to the Internet of things (IOT).

References

1. Joseph A Akinyele et al. “Securing electronic medical records using attribute-based encryption on mobile devices”. In: Proceedings of the 1st ACM workshop on Security and privacy in smartphones and mobile devices. ACM. 2011, pp. 75–86.
2. John Bethencourt, Amit Sahai, and Brent Waters. “Ciphertext-policy attribute based encryption”. In: Security and Privacy, 2007. SP’07. IEEE Symposium on. IEEE. 2007, pp. 321–334.
3. Vipul Goyal et al. “Bounded ciphertext policy attribute based encryption”. In: International Colloquium on Automata, Languages, and Programming. Springer. 2008, pp. 579–591.
4. Nimmy John and SanathShenoy. “Health cloud-Healthcare as a service (HaaS)”. In: Advances in Computing, Communications and Informatics (ICACCI, 2014 International Conference on. IEEE. 2014, pp. 1963–1966.
5. Anant V Nimkar and Soumya K Ghosh. “An access control model for cloud-based emr federation”. In: International Journal of Trust Management in Computing and Communications 2.4 (2014), pp. 330–352.
6. Sushmita Ruj, Amiya Nayak, and Ivan Stojmenovic. “DACC: Distributed access control in clouds”. In: Trust, Security and Privacy in Computing and Communications (TrustCom), 2011 IEEE 10th International Conference on. IEEE. 2011, pp. 91–98.
7. A Samydurai et al. “Secured Health Care Information exchange on cloud using attribute based encryption”. In: Signal Processing, Communication and Networking (ICSCN), 2015 3rd International Conference on. IEEE. 2015, pp. 1–5.
8. Danilo FS Santos, Angelo Perkusich, and Hyggo O Almeida. “Standard-based and distributed health information sharing for mHealth IoT systems”. In: e-Health Networking, Applications and Services (Healthcom), IEEE 16th International Conference on. IEEE. 2014, pp. 94–98.
9. Vijayaraghavan Varadharajan, Alon Amid, and Sudhanshu Rai. “Policy based Role Centric Attribute Based Access Control model Policy RC-ABAC”. In: Computing and Network Communications (CoCoNet), 2015 International Conference on. IEEE. 2015, pp. 427–432.

10. Chang Ji Wang et al. “An efficient cloud-based personal health records system using attribute-based encryption and anonymous multi-receiver identity-based encryption”. In: P2P, Parallel, Grid, Cloud and Internet Computing (3PGCIC), 2014 Ninth International Conference on. IEEE. 2014, pp. 74–81.
11. Yun Wang, Dalei Zhang, and Hong Zhong. “Multi-authority based weighted attribute encryption scheme in cloud computing”. In: Natural Computation (ICNC), 2014 10th International Conference on. IEEE. 2014, pp. 1033–1038.
12. Chao-Tung Yang et al. “Implementation of a medical image file accessing system on cloud computing”. In: Computational Science and Engineering (CSE), 2010 IEEE 13th International Conference on. IEEE. 2010, pp. 321–326.
13. Kan Yang and Xiaohua Jia. “Expressive, efficient, and revocable data access control for multi-authority cloud storage”. In: IEEE transactions on parallel and distributed systems 25.7 (2014), pp. 1735–1744.
14. Chen Yanli, Song Lingling, and Yang Geng. “Attribute-based access control for multi-authority systems with constant size ciphertext in cloud computing”. In: China Communications 13.2 (2016), pp. 146–162.
15. Hui Zhu et al. “SPEMR: A new secure personal electronic medical record scheme with privilege separation”. In: Communications Workshops (ICC), 2014 IEEE International Conference on. IEEE. 2014, pp. 700–705.

A Brief Study on Build Failures in Continuous Integration: Causation and Effect

Romit Jain, Saket Kumar Singh and Bharavi Mishra

Abstract Continuous Integration (CI) has successfully tackled the problem of bug fixing owing to which it has gained immense popularity among software developers. CI encourages to commit on the go so that each bug can be traced to its source and handled accordingly. However, CI remains a practice at its core, and only a part of it can be implemented. Anything which does not follow good CI practice would pave the way for a greater number of build fails. CI's continuous nature may cause a clutter in a big team, leading to one developer's build failing the other. Numerous consecutive build fails can put the project on a standstill till the build is made clean which may cause developers to lose interest eventually. We investigate, in this paper, causation and effect of build failure in CI. We first see whether a large team size contributes to more build failure and second, whether an increasing number of consecutive build failures have any impact on the productivity of developers. We have used data provided in TravisTorrent and analysed the 3,702,595 Travis builds which mostly contain Java and Ruby as the programming language used. For both the languages, we have made a comprehensive analysis of the problem we address.

Keywords Continuous integration · Build failures · Team size · Productivity

R. Jain (✉) · S. K. Singh (✉) · B. Mishra
The LNM Institute of Information Technology, Near Jamdoli, Rupa ki Nangal, Post Sumel, Jaipur, India
e-mail: y14uc236@lnmiit.ac.in

S. K. Singh
e-mail: y14uc238@lnmiit.ac.in

B. Mishra
e-mail: bharvi@lnmiit.ac.in

C. R. Panigrahi et al. (eds.), *Progress in Advanced Computing and Intelligent Engineering*, Advances in Intelligent Systems and Computing 714,
https://doi.org/10.1007/978-981-13-0224-4_2

1 Introduction

Continuous integration has seen its growth over the years along with CI development platforms like Travis CI. Earlier research on CI [1] summarizes the benefits of its introduction to a project. Pull requests by core team members are more easily merged. CI projects observe increased productivity, measured in many pull requests processed, in the project. The core members also observe increased bug reports without compromising external software quality.

However, there are reasons which make a CI environment less efficient. Typically when a build fails, developers have to leave the module they are working on and switch to making the build clean. As the team size increases, it would naturally be harder to synchronize the teams' commit frequency [2] as there can be teams with people distributed globally [3]. Research conducted by [4] suggests that having a build broken for too long would just make people switch to other projects, decreasing the external contribution and hence the overall productivity. Moreover, often developers ignore the build being broken and wait till someone else fixes the build [5]. We do a comprehensive analysis of Travis torrent data [6] and try to find whether productivity decreases after consecutive build failures in a particular project. We have taken data of unique 243,811 Java primarily, and 434,580 Ruby Travis builds (1,822 JavaScript builds were also present, but in comparison, the data was too less) for our analysis of the first research question [7]. In this research, we try to address two research questions:

- **RQ1**: What is the effect of team size (including external contributors) on build failures?
- **RQ2**: What is the effect of consecutive build fails on the productivity of the developers working in the branch?

We first give preliminaries about our analysis in Sect. 1, followed by addressing the two research questions in Sect. 2 and finally our conclusion in Sect. 3.

1.1 Preliminaries

In this section, we define each of the variables that we used from the data provided. Table 1 contains the variables and their corresponding description. The subsection ahead describes preprocessing done on the data used.

1.2 Preprocessing

The data contained many redundant rows corresponding to each Job Id; many parameters were just the same. First, we extracted all unique rows corresponding to

Table 1 Data description

Variable	Description
tr_ build_ id	The analysed build id, as reported from Travis CI
tr_ status	The build status (such as passed, failed, etc.) as returned from the Travis CI API
gh_ team_ size	Number of developers that committed directly or merged PRs from the moment the build was triggered and 3 months back
gh_ pushed_ at	Timestamp of the push that triggered the build (GitHub provided), in UTC
gh_ first_ commit_ created_ at	Timestamp of first commit in the push that triggered the build, in UTC
git_ prev_ built_ commit	The commit that triggered the previous build on a linearized history
tr_ prev_ build	The build triggered by git_ prev_ built_ commit
git_ all_ built_ commits	A list of all commits that were built for this build, up to but excluding the commit of the previous build, or up to and including a merge commit
gh_ sloc	Number of executable production source lines of code, in the entire repository
git_ diff_ src_ churn	Number of lines of production code changed in all git_ all_ built_ commits
git_ diff_ test_ churn	Number of lines of test code changed in all git_ all_ built_ commits
gh_ test_ lines_ per_ kloc	Test density. Number of lines in test cases per 1000 gh_ sloc

tr_ build_ id. After that, we segregated the builds into those of Java and Ruby. For the second problem, first we needed to process individual builds, and in the process, we found that some of the rows (data points) did not report *gh_ first_ commit_ created_ at* and *gh_ pushed_ at*. Since we needed these two variables to compute our reference productivity score in the second problem, and these data points would hence be an anomaly in our data, we did not consider them in our analysis. To have a better view of the data and ease of plotting in the end, we had sorted the data according to their chronological order of push that triggered the build. We used *gh_ pushed_ at* for this purpose which provides the timestamp of the final push that triggered the build.

2 Analysis

This section is divided into two subsections corresponding to the two research questions we addressed. Subsection A describes the procedure and results obtained of the first problem, i.e. build failures due to increasing team size. Subsection B describes the same for the second problem, i.e. effect of build failure on developers' productivity.

2.1 *Effect of Team Size on Build Failure*

We iterate through the entire data using the value of *gh_ team_ size* observed from 1 to 288. For each value, corresponding to the aforementioned range of team size, a total number of builds failed, success, errored and cancelled were obtained. Cancelled builds are not tested and only queued because they are cancelled by the developers before they can be triggered. Hence, we do not take it into account the total number of builds. For each team size, we decided to compare the ratio of build failures to total builds against each team size, because corresponding to a single team size there may be more builds than that for other team size, and hence, it would make the inference inaccurate in regard to the problem we are trying to address. Hence, the normalized build failures were calculated as build failures divided by the number of total builds, which contains passed, failed and errored. It is described in the following equation, for team size 't' as

$$normalized\,failure_t = \frac{(build\,fails)_t}{(total\,builds - cancelled\,builds)_t} \tag{1}$$

The resulting graphs for projects of Java and Ruby are shown in Figs. 1 and 2, respectively.

The data for team size 108–192 was unavailable in the data given for Ruby; hence, the graph has an empty space in between. However, the best fit line in Fig. 2 gives a similar figure to Fig. 1 with observations of Java. From both graphs, we can observe a depression in normalized failures at a point which is acting as minima as thereafter the graph has an increasingly positive slope. We may conclude that there is an optimal number of members a team can have. Any further increase in team size would cause an increase in build failure. The build failures maybe explained by the fact that large open source projects include people contributing from all over the world. Apart from pull requests, even the core team often does not work in the same office. This may sometimes lead to conflicts as the time zones are different and hence becomes difficult for the entire team to be in synch all the time.

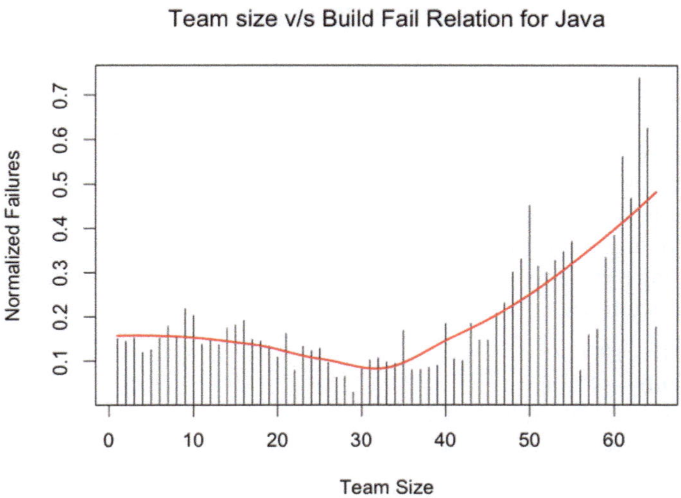

Fig. 1 Size-team ratio versus build failure for projects written in Java

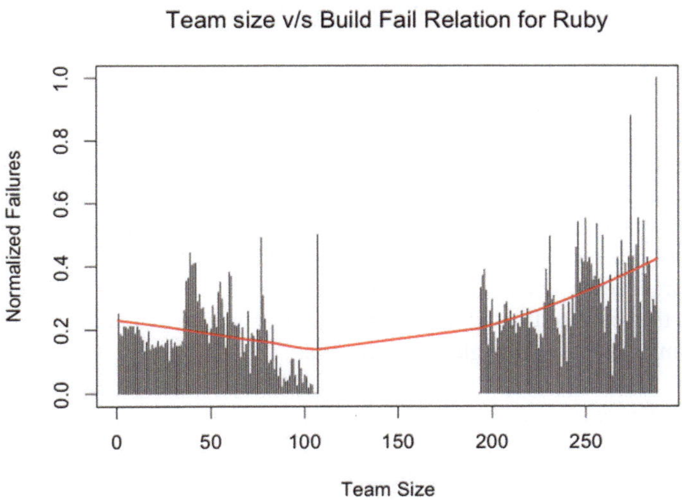

Fig. 2 Size-team ratio versus build failure for projects written in Ruby

2.2 Effect of Build Failures on Productivity

Measuring the productivity would require a comprehensive study of a single project as only there we can obtain build data serially. For observing a good amount of data, we took the most popular project from the languages, Java and Ruby which are apache/jackrabbit-oak, Cloudifysource/Cloudify, jruby/jruby and rails/rails, and mined the data corresponding to all branches. For our convenience, the data was sorted according to the chronological order of *gh_ pushed_ at* as described in the preprocessing part above. The total number of builds we have taken from the above projects is as follows:

- **rails/rails**—19,447
- **jruby/jruby**—12,085
- **apache/jackrabbit-oak**—8205
- **Cloudifysource/Cloudify**—5742

In order to aggregate all the above into a single score of productivity we first calculated the score according to the given formula for each build:

$$Score = \frac{\dfrac{(git\_diff\_src\_churn + git\_diff\_test\_churn)}{(gh\_sloc + (gh\_sloc/1000) * (gh\_test\_lines\_per\_kloc))}}{(gh\_team\_size)} \tag{2}$$

We have primarily observed the total change in test code and production code in that particular build and divided it by the total number of executable production source line and test density. This is done to take into account the size of the change in each build with respect to the size of the entire project. Each score is then divided by gh_ team_ size to negate differences that would come because of a different team size, i.e. larger team size contributing to a greater change than a smaller team size. After that, the score is divided by the time duration which we calculate by using gh_ first_ commit_ created_ and gh_ pushed_ at as

$$Duration = (gh\_pushed\_at - gh\_first\_commit\_created\_at) \tag{3}$$

for each build. We calculated the duration for which the team worked on a single build. We include this time to account for the fact that a team finishing its build in a lesser amount of time would naturally be more productive.

To calculate the total score, we divided the earlier score by duration. Hence,

$$Total\ Score = Score/Duration \tag{4}$$

To standardize the data for the whole project, we used the z-score corresponding to the total score. Z-score is defined by

$$Z = (x - \mu)/\sigma \tag{5}$$

where 'x' is the individual value of an instance, 'μ' is the mean and 'σ' is the standard deviation of the entire data. Here, x is the total score of any individual build and Z is the new standardized productivity score of that build. A greater z-score would imply more productivity in that build. To see the effect of consecutive build failures, we needed a method through which we could aggregate the impact of the past build fails and simultaneously measure the present productivity z-score in that reference, so we chose to calculate the cumulative moving average of the z-scores. In a cumulative moving average, the data is in an ordered stream, which in our case is ordered by gh_ pushed_ at, and the average of all of the data up until the current point is calculated. Hence for a sequence of n scores, Cumulative Moving Average (CMA) would be

$$CMA_n = \frac{(x_1 + x_2 + x_3 + \cdots x_n)}{n} \tag{6}$$

And for the element n + 1, we get the score as

$$CMA_{n+1} = \frac{(x_1 + x_2 + x_3 + \cdots x_n + x_{n+1})}{(n+1)} \tag{7}$$

We then plotted the graph for the cumulative moving average V/s build failure graph. As we mentioned earlier, we have taken into account the two most popular projects from the languages Java and Ruby. Figures 3 and 4 are the graph plotted for the two most popular project of Java and Figs. 5 and 6 are that of Ruby.

Refer to the legend in Table 2 for the significance of different lines in the plot. We used *lm* to obtain the generalized linear model to get a general inference from the data. We used *gam* method to plot the generalized additive model curve for the final inference. We used these methods since we only needed to observe the trend.

We observed a downward curve where there is consecutive build fails. In the beginning, with fewer build fails, the cumulative moving average of the normalized productivity score initially increased and quickly reached maxima, which suggests that initially the project is highly active but with time and increasing build failures, productivity, and activity on the project, both decrease.

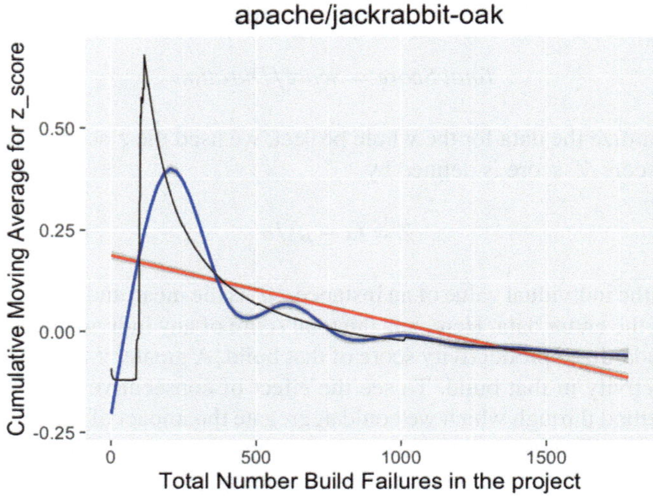

Fig. 3 Effects of consecutive build failures on productivity of developers working on apache

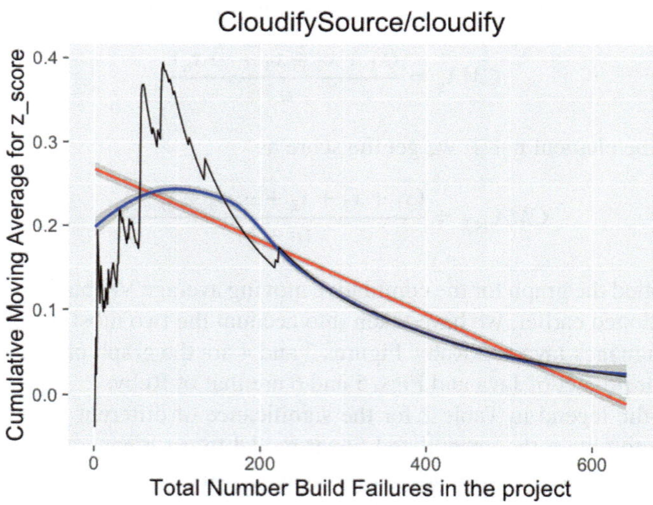

Fig. 4 Effects of consecutive build failures on productivity of developers working on Cloudify

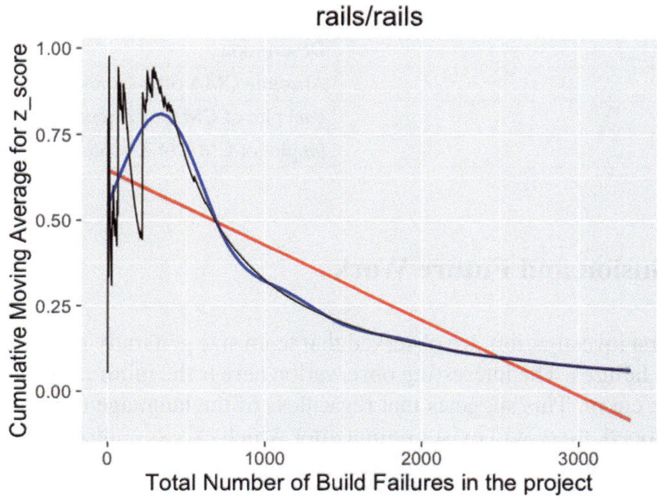

Fig. 5 Effects of consecutive build failures on productivity of developers working on rails

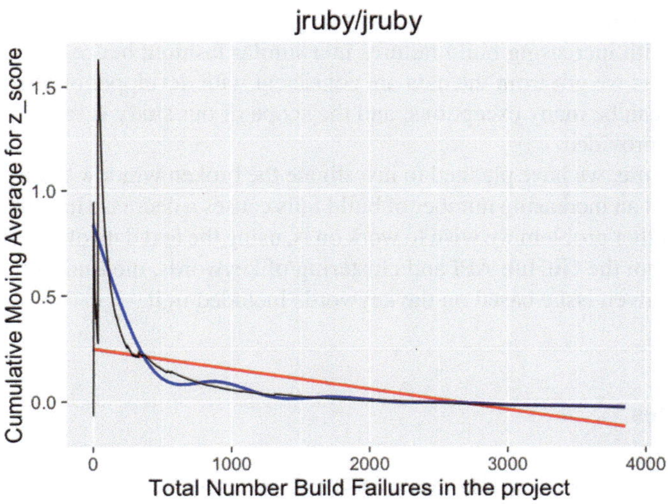

Fig. 6 Effects of consecutive build failures on productivity of developers working on jruby

Table 2 Plot legend

Line colour	Description
Black	Absolute CMA of Z-scores
Blue	*gam* plot of CMA of Z-scores
Red	*lm* plot of CMA of Z-scores

3 Conclusion and Future Work

From our first investigation, we observed that team size generally increases the number of build failures. The interesting observation here is the minima after which there is a positive curve. This suggests that regardless of the language used or the project being developed, there will exist minima after which excess contributors would also significantly contribute to build failures. A limitation which we must state here is that we have not taken into account the variety of sizes of a project which have the same team size. We plan to do more comprehensive work on this issue considering the various types and sizes of projects. In our second research problem, we see that productivity across four different types of projects written in two different languages decreases with increasing build failures in a similar fashion; hence, we can say that the inferences we got from the data are consistent with developer experience. However, there can be many exceptions, and the scope of our study is still limited to the data that is provided.

In the future, we have planned to investigate the broken window theory [8] which suggests that an increasing number of build fails causes a decrease in external contribution. Another problem we wish to work on is, using the text data of issue comments available from the GitHub API and clustering of keywords, measuring the response rate of any given issue based on the keywords included in it.

References

1. Vasilescu Bogdan et al., 'Quality and Productivity Outcomes Relating to Continuous Integration in GitHub', *Proceedings of the 2015 10th Joint Meeting on Foundations of Software Engineering Pages*, (2015) 805–816
2. 'Babysitting your Continuous Integration System', http://softwareengineering.stackexchange.com/questions/82632/babysitting-your-continuous-integration-system
3. Padhye Rohan et al., A study of external community contribution to open-source projects on GitHub, *Proceedings of the 11th Working Conference on Mining Software Repositories* (2014) 332–335
4. 'Please stop breaking the build', http://danluu.com/broken-builds/
5. 'Continuous Integration is Dead', http://www.yegor256.com/2014/10/08/continuous-integration-is-dead.html
6. 'travistorrent _11_1_2017.csv.gz', https://travistorrent.testroots.org/page_access/

7. Beller, Moritz and Gousios, Georgios and Zaidman, Andy, Travis Torrent: Synthesizing Travis CI and GitHub for Full-Stack Research on Continuous Integration, *Proceedings of the 14th working conference on mining software repositories* (2017)
8. 'Software Development and The Broken Window Theory' https://www.rtuin.nl/2012/08/software-development-and-the-broken-windows-theory/

Resource Monitoring Using Virtual Ring Navigation Through Mobile Agent for Heterogeneity Oriented Interconnected Nodes

Rahul Singh Chowhan and Rajesh Purohit

Abstract Expansion of Internet and increase in heterogeneity has right away brought in the need for more scalable, resourceful and available systems. With balancing and scheduling of various tasks happening within the network or being submitted by requests, the proper utilization of available resources and services at their full capacity can be advantageous. To achieve load balancing and scheduling a priori load information is required. Distributed applications are becoming mobile with mobile networks accommodating a broad range of services with different characteristics as users, services, databases and computer becoming increasingly mobile fading away the age of fixed networks. This transition in the technological trend has right away introduced the shift of paradigm from traditional client server to mobile agent paradigm. In this paper, the technique of resource monitoring is proposed to monitor the load on various interconnected nodes through mobile agents. To make navigation independent of topology on interconnected nodes, virtual ring is used. For this work, the experiments are carried out on virtual machine environment as well as on real machine environment.

Keywords Active objects · Distributed computing · Intelligent system
Autonomous computation agents

1 Introduction

In distributed systems, various autonomous nodes, resources, and interdependent machines are connected with each other to subject a uniform, logical and comprehensible view of a robust system. These connected devices can work in isolation, independently as well as cooperating with each other showing resource dependencies [1]. The shared execution of various tasks is kept transparent from the user isolating it from the internal functioning of system. To take complete advantage of

R. S. Chowhan (✉) · R. Purohit
MBM Engineering College, Jodhpur, India
e-mail: word2rahul@gmail.com

C. R. Panigrahi et al. (eds.), *Progress in Advanced Computing and Intelligent Engineering*, Advances in Intelligent Systems and Computing 714,
https://doi.org/10.1007/978-981-13-0224-4_3

the distributed system and their sharing functionalities, a good resource allocation scheme has always been a central need. Diverse load distribution algorithms work deliberately for an even distribution of the tasks/loads among all the partaking nodes in network so that the overall performance of the entire system is maximized [2].

Today, the central paradigm linking all distributed object technologies is a synchronous message-passing paradigm whereby all objects are stationary, but distributed and interact with each other through message-passing. These paradigms need to be enhanced in some fashion with much advanced paradigms like object mobility, asynchronous message-passing, and active objects [3].

Mobility, executability, and autonomy are the properties which also exist in computer viruses and worms, which has been a serious threat to computers in recent years. This is negative side of mobile agents being a malicious entity roaming on network if no protective countermeasures and security mechanisms are engaged with the underlying host or mobile agents [4]. Though it is an effective paradigm for distributed application that allows computing over partially connected devices such as smart phones, laptops, PDA, etc., as well as devices that are occasionally connected like home and business computers via dial-up connections [5].

K. Dhanalakshmi et al. have proposed a model called Matrix Hop Mobile Agent model (MHMA). In their approach the mobile agent visits nodes generating an offer. This offer is submitted to server every time before the mobile agent moves to next node in its itinerary [6]. Zhixin Tie et al. have proposed agent-based monitoring of server resources using Mobile-C library. Mobile-C is an IEEE FIPA complaint mobile agent system which they have used as base model. They have presented Mobile Agent Based Server Resource Monitoring System (MABSRMS) using monitoring agents to check status of various servers at periodic intervals for a specific amount of time [7]. A. Meera et al. have proposed a method of static non-automated agent based resource monitoring system for enhancement of monitoring services, resource and performance optimization in cloud computing environment. This system will be responsible of making its own decision on basis of resource status and reports. These monitoring services can further be extended to make watch of system vulnerabilities, network traffic analysis, and intrusion detection, etc., using mobile agent based strategies [8]. Maneet Kaur et al. have proposed scheme which provides a solution dealing with the lost of agent due to single server failure in information retrieval system. They achieved the fault tolerance by cloning the actual agent which follows the actual agent along in its itinerary. As long the actual agent is working, clone agent will be passive but it gets active if there is faulty server and change of state in actual agent [9].

The most of prevailing mobile agents are developed in Java, using its platform independence and security mechanism as integrated key features. Grasshopper, D'Agents and SMART are few Java based mobile agents studied in this report. Aglet is also one such mobile agent developed by IBM, for which the Java Aglet API (J-AAPI) is a proposed industry standard. J-AAPI was developed by a research team at the IBM Tokyo Research Laboratory in Japan in response to a call for a uniform platform for mobile agents in heterogeneous environments such as the Internet [10]. MASIF and FIPA are architectural standards which allow multiple agents to communicate, cooperate, and perform interoperations using agent communication languages, agent services and supporting managements of agent systems. They also help finding relevant architectural components that are present within publicly available mobile agent systems [11, 12].

In our work, mobile agents are used for purpose of monitoring of various resources like CPU utilization, memory usage, etc., in a heterogeneity oriented computing environment. This proposed work has implemented a virtual ring navigation mobile agent which goes on collecting and computing the resource information moving in its itinerary. It takes multiple hops in its fixed itinerary and on the fly stores the resultant information in sorted manner. The high priority thread counter is used with various delays that return the availability value for every machine in the network along with other system level parameters like CPU utilization, memory usage, etc.

2 Methodology and Working Principle

Mobile agents developed and used for experimentation purpose in this work is virtual ring navigation mobile agent. It moves in an itinerary to multiple hops computing various parameters and submitting their collected results to the server. To monitor the behavior of this mobile agent it is compared with existing client server based mobile agent for performance analysis. The methodology involves the various delays on different setups to figure out the relation between different matrices. To find out the relative availability, two setups were taken: Virtual machine setup and Real machine setup. Delays of 1, 2, 3, and 4 ms were provided for both the scenarios. A counter was also set to know the exact counting capability of a system. This capability of counting denoted the availability of machine, while performing the experiments a relation between CPU and memory utilization with availability is also noted for every machine for each of the setups. A free form equation deduced is as follows:

$$A = f(C, M, P), \tag{1}$$

where A is availability, C is CPU utilization, M is memory usage, and P is delay assigned for a machine in the network.

The proposed method works with a mobile agent moving in virtual ring topology. This mobile agent computes the utilization based on various parameters like CPU utilization, Memory Usage, etc. The comparative study is carried out with existing client server model of mobile agent. This structure of virtual ring navigation mobile agent allows freeing the server for some other tasks to initiate at its level like handling of more incoming client requests, managing network resources, scheduling and assignment of new jobs, etc. At same time, the mobile agent is moving in network from node to node without having any continuous open connection between server and its current state.

At the server node, a fixed time for mobile agent itinerary is taken care of, which waits for agent to arrive after a particular tick. This recovery mechanism uses accept and decline methodology—it simply accepts the agent if it returns within the specified time limit otherwise an agent is considered lost in network and a new agent is sent.

3 Proposed Algorithm for Load Monitoring and Event Flow Diagram

In this research work, the proposed algorithm may have its practicality in load-balancing mechanism depending on the domain of work. In this algorithm, Cn_i and Mn_i represents CPU usage and Memory Utilization, jointly called as the load information and Tn_i is a threshold value on the current machine n_i in the cluster, where i varies from 0 to 3 for four nodes.

To find out the load threshold two different matrices namely the CPU Load and Memory Load are combined. The weighted metric is used in 30:70 ratio for memory load and CPU load because memory load for any machine hardly varies while CPU load fickle frequently. The other ratio proportions are also tried to check the actual load but on increasing memory portion and decreasing the CPU load portion, the whole output moves to be a constant gradually.

$$\forall ni, Tn_i = \begin{cases} T_{H,} & \text{if } Tn_i \geq 80\% \\ T_{M,} & \text{if } 20\% < Tn_i < 80\% \\ T_{L,} & \text{if } Tn_i \leq 20\% \end{cases}$$

The other way round on decreasing the memory portion and increasing the CPU portion the resultant value gets more variable. These matrices can be used separately as and when needed.

A. **Load Monitoring Algorithm Using Virtual Ring Navigation Mobile Agent**

(a) Start

Mobile agents are created on network admin or server node to determine Cn_i, Mn_i, and Tn_i for each machine/node connected in the network.

$$\begin{aligned} Load\ Threshold\ (T) = 0.7 * (cpuLoad \\ = getSystemCpuLoad()) + 0.3 * (memLoad \\ = getSystemMemoryLoad()); \end{aligned}$$

(b) On arrival of MA on new machine
Switch(Tn_i)

Case HIGH: if (Tn_i = $T_{H)}$ then,
save its load information (i.e. Cn_i & Mn_i) & mark current node as "Heavily loaded";
insertionSort(array_of_integers[]){
Case LOW: if (Tn_i = T_M) then,
save its load information and mark current node as "Lightly loaded";
insertionSort(array_of_integers[]){
Case MED if (Tn_i < $T_{L)}$ then,
save its load information and mark current node as "Under-loaded";
insertionSort(array_of_integers[]){
Default: if (node failure|| no node) then
return to server; *

(c) MA keeps on moving to successive nodes as its only one MA moving in its itinerary. On the fly it sorts the results repeating step (b) using following algo for insertion sort:
   i. Begin with the first element as sorted element.
   ii. Compare the second element with first element.
   Tn_x < $Tn_{y,}$ i.e., element is already sorted. Where x is previous index and y is current index of element. This will sort the first element.
   iii. If anywhere Tn_x > $Tn_{y,}$ that means it is already sorted and on its correct index.
   iv. Similarly, on obtaining the new element (i.e., new Tn_i) repeat step ii. till the beginning of array to insert it at proper index.

(d) Lastly, the sorted results after completion of step (c) are submitted to the server;
(e) Now, server may choose to use these refined results for balancing the load further;
(f) End;

3.1 Event Flow Diagram

This event flow diagram for virtual ring navigation of mobile agents depicts that the mobile agent begins it execution from the server node/admin. The server node is called as home node and the mobile agent server installed on this node is called as home agent server/home agency. The itinerary for the mobile agent is defined on

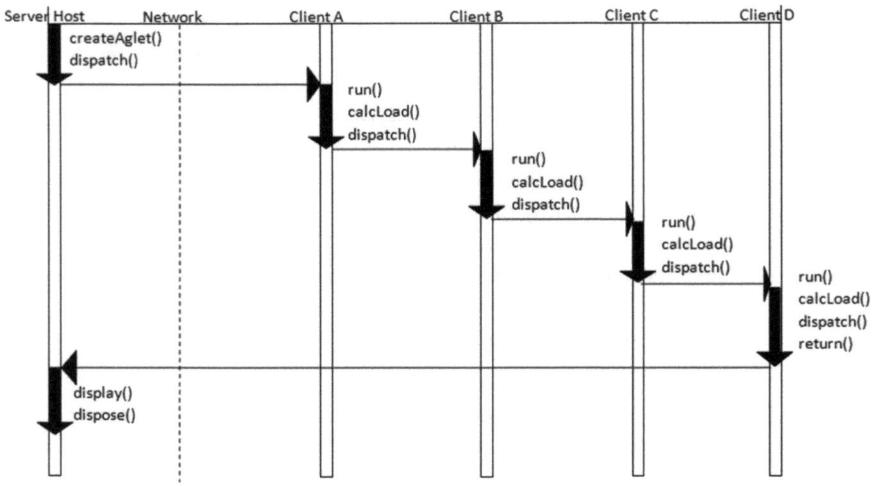

Fig. 1 Event flow diagram for virtual ring navigation based mobile agent

this node which mobile agent carries along with it. Using the dispatch() function, mobile agent moves to next hop, i.e., next node, in the network. When mobile agent arrives on the next node then next node becomes the current context of execution. On this node it calculates the load information for resources and dispatches to next connected hop/node in the network. Subsequently, covering all participating nodes in its itinerary and saving their load information, in the last it returns back to home agency and submits the sorted load information to the server (Fig. 1).

4 Experimental Setup and Execution Steps

In our experimental setup the various combinations of real machines and virtual machines for result analysis is carried out. Initially, we took two real machines and two virtual machines for moving mobile agent in its itinerary to know the utilization of various nodes in cluster. Then the experiment is repeated for three and four real machines and virtual machines separately. This we had carried out to move mobile agent in virtual navigation fashion.

4.1 For Virtual Machine Setup

The server setup installed on a server machine which is configured with Intel(R) Core(TM) i3-4005U CPU @1.70 GHz processer and 4 GB DDR-2L RAM, and installed with Windows 7 Ultimate 64-bit Operating System. The client setup is configured with Windows XP 32-bit Operating System, 512 MB RAM. To enable the usage of mobile agents for purpose of utilization, we have extended the mobile agent system by integrating it with the NetBeans, as the core framework of IBM Aglets only provides with the interface and overriding of methods.

4.2 For Real Machine Setup

The server and client machines are kept with same configuration with installation of Windows 7 Ultimate, 64-bit Operating System and Intel(R) Core(TM) i3-4005U CPU @1.70 GHz processer and 4 GB DDR-2L RAM. The programming set up was done using NetBeans to create mobile agents and interfaces of IBM Aglets are used.

The scope of resource monitoring includes fundamental parameters like CPU utilization, Memory utilization, Disk Space, I/O utilization, Bandwidth consumption and Network Utilization. We have considered two parameters CPU and Memory utilization for our mobile agent monitoring purpose which can be further used as either load balancing or for load scheduling and many more things.

4.3 Execution Steps for Virtual Ring Navigation MA

(1) Agency is started over all the client hosts which allow mobile agents to visit the node.
(2) Network Admin/Server creates a mobile agent (MA) to be sent to client hosts.
(3) MA is associated with an array of IPs that it has to visit in multiple hops.
(4) Each time MA visits the respective node, it computes the required parameters and store results in an array.
(5) The results stored in an array are inserted in sorted manner so no extra overhead of applying sorting algorithm is required.
(6) MA saves the computed result of every node and keeps on moving to next node in its itinerary.
(7) After finishing the visits of all nodes, it returns back to network admin with sorted results.
(8) Based on collected results, overloaded and idle node can be find for distribution of the load (Fig. 2).

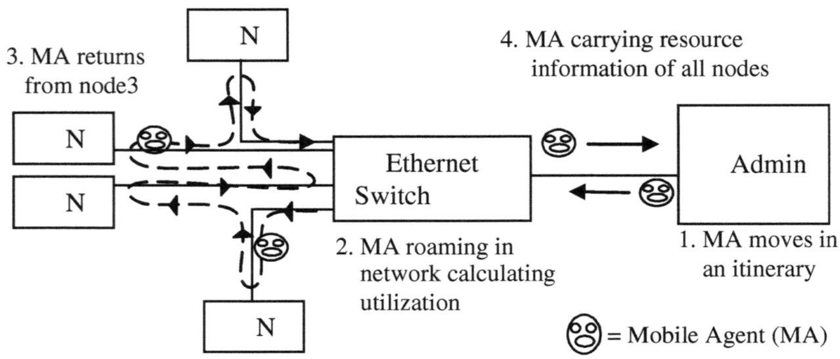

Fig. 2 Experimental setup showing virtual ring navigation mobile agent

5 Results Analysis and Observations

The primary objective of load monitoring is to ensure highest availability machine in network. Based on availability information, horizontal scaling on cluster of real machines can be achieved and vertical scaling in case of virtual machines can be attained for betterment of overall performance. For this various result scenarios were plotted in graphical forms that are as follows.

5.1 Scenario-1: Various Combinations of Real Machines Using Virtual Ring Navigation Mobile Agent

In this scenario, virtual ring navigation mobile agent moves in an itinerary to find the availability of real machines. There is not much deviation in the memory consumption of the system unless an application that consumes huge memory is started but CPU utilization factor shows much of variation every time as the processes and services running at OS level keeps on varying (Fig. 3).

The graphs (a)&(b), (c)&(d), and (e)&(f) shows utilization and availability for two real machines, three real machines, and four real machines respectively.

5.2 Scenario-2: Various Combinations of Virtual Machines Using Virtual Ring Navigation Mobile Agent

Virtual ring navigation mobile agent is an autonomous mobile agent that picks the IP address of next hop in its itinerary on-the-fly. This is computation agent that computes the utilization of machine locally and moves to next hop in the network. In this scenario utilization of virtual machines are calculated based on two, three

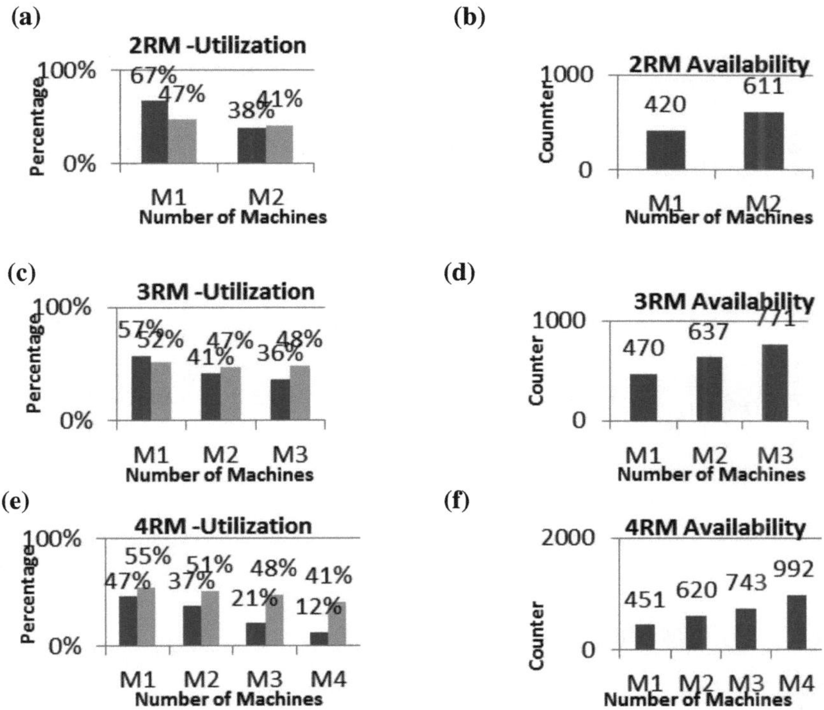

Fig. 3 The graphs (a)&(b), (c)&(d), and (e)&(f) shows utilization and availability for two real machines, three real machines, and four real machines respectively

and four virtual machines. The counter thread was run with delay difference to create the virtual load to find the availability (Fig. 4).

The graphs (a)&(b), (c)&(d), and (e)&(f) shows utilization and availability for two virtual machines, three virtual machines, and four virtual machines respectively.

6 Conclusion and Future Scope

In this work, a generalized autonomous mobile agent has been implemented that travels along its itinerary in Virtual Ring Navigation (VRN). For experimentation purpose, virtual machines as well as real machines were considered. For each of them the number of machine was varied from two to four. The experiment for availability concludes that memory usage and CPU utilization of a machine are inversely proportional to each other. That means the more CPU utilization and memory usage is, the lesser will be the availability of machine.

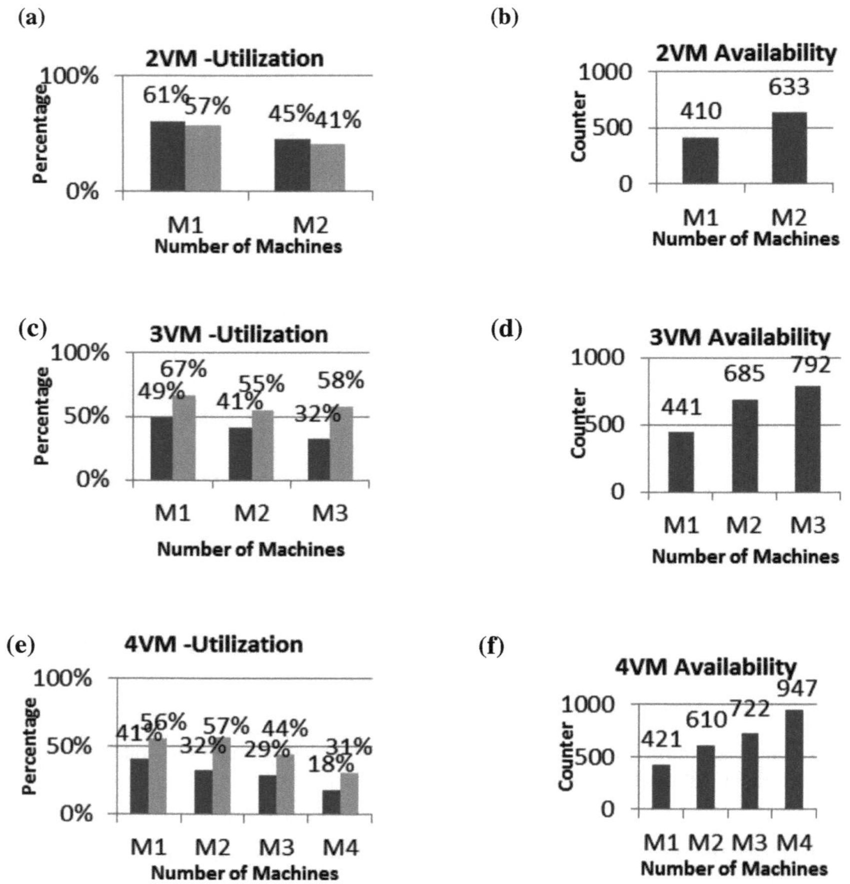

Fig. 4 The graphs (a)&(b), (c)&(d), and (e)&(f) shows utilization and availability for two virtual machines, three virtual machines, and four virtual machines respectively

As the virtual ring navigation mobile agent interacts locally with the client, it may leave the server free to handle other network tasks like handling of requests, load management, etc. This effort puts forth an initiation of solution to major problems like node failure, high bandwidth consumption, resource starvation, etc.

In a network of heterogeneous machines, it may happen that one node is overloaded with the work while some other node is waiting for jobs to execute. In this situation, an efficient monitoring technique is required that can provide load information of connected nodes in a network. This way the network admin can systematically organize the incoming request over the server. The monitoring is not only required to handle the request coming from outside the system but also to schedule the loads within the system.

The future scope of proposed work may include the recovery mechanism for mobile agent. This may also introduce a technique to send the monitored load information of systems in periodic fashion. Also, static agents can be implemented one each client nodes that may communicate to server for any malicious activity happening on that node. In this proposed work, security of mobile agent is dependent on security provided by the language in which the mobile agent is written, i.e., java. Security being biggest aspect on which the work has to be done, can play a vital role in future scope of the proposed work.

References

1. Chess, David, Colin Harrison, and Aaron Kershenbaum. "Mobile agents: Are they a good idea?" In Mobile Object Systems Towards the Programmable Internet, pp. 25–45. Springer Berlin Heidelberg, 1997.
2. Ahila, S. Sobitha, and K. L. Shunmuganathan. "Overview of mobile agent security issues—Solutions." In Information Communication and Embedded Systems (ICICES), 2014 International Conference on, pp. 1–6. IEEE, 2014.
3. Joanna Juziuk, "Design Patterns for Multi-Agent Systems", Linnaeus University, School of Computer Science, Physics and Mathematics, 2012.
4. Persson, Mats, "Mobile agent architectures." Division of Command and Control Warfare Technology, Defence Research Establishment [Avd. för ledningssystemteknik, Försvarets forskningsanstalt] (FOA), 2000.
5. Tatsiana Levina, "Mobile Agents", Rutgers University, Newark, December 2001.
6. Dhanalakshmi, K., and GM Kadhar Nawaz. "Matrix Hop Mobile Agent (MHMA) System for E-Service Applications." Procedia Engineering 30 (2012): 1171–1178.
7. Zhixin. Tie, "A Mobile Agent-Based System for Server Resource Monitoring." Cybernetics and Information Technologies, Volume 13, Issue 4, Pages 104–117, ISSN (Print) 1314–4081, DOI:https://doi.org/10.2478/cait-2013-0057, December 2013.
8. A. Meera and S. Swamynathan. "Agent based Resource Monitoring System in IaaS Cloud Environment." DOI:https://doi.org/10.1016/j.protcy.2013.12.353, Procedia Technology 10 (2013): 200–207.
9. Maneet Kaur, and Sandeep Sharma. "A dynamic clone approach for mobile agent to survive server failure." In Reliability, Infocom Technologies and Optimization (ICRITO) (Trends and Future Directions), 2015 4th International Conference on, pp. 1–5. DOI:https://doi.org/10.1109/icrito.2015.7359237, IEEE, 2015.
10. Schoeman, Marthie, and Elsabé Cloete. "Architectural components for the efficient design of mobile agent systems." Proceedings of the 2003 annual research conference of the South African institute of computer scientists and information technologists on Enablement through technology. South African Institute for Computer Scientists and Information Technologists, 2003.
11. Fragkakis, Michail, and Nikolaos Alexandris. "Comparing the trust and security models of mobile agents." Information Assurance and Security, 2007. IAS 2007. Third International Symposium on. IEEE, 2007.
12. Persson, Mats, "Mobile agent architectures." Division of Command and Control Warfare Technology, Defence Research Establishment [Avd. för ledningssystemteknik, Försvarets forskningsanstalt] (FOA), 2000.

Relating Vulnerability and Security Service Points for Web Application Through Penetration Testing

Rajendra Kachhwaha and Rajesh Purohit

Abstract In last decade, there have been enormous changes in the field of web applications. The phase has shifted from static to dynamic, and fixed layout has now taken the form of responsive layout, due to distribution of processing capabilities from server side to client side, mainly because of rich set of scripts for user interface and making request to server. This leads to reduction in network traffic. This is on the presumption of trustiness on client, eventually creating a web application more vulnerable. This paper will cover importance of each triad of web, mainly security with its service points. This will facilitate a developer to identify which service point is more important with respect to application requirements. It will also apply sufficient security checks at service point in each component of the application.

Keywords Web application security · Web application threats · Penetration testing · Security service point · Web vulnerability

1 Introduction

A web application is a task-oriented application which is deployed on a web server, accessed through a web browser, using an http/https connection for information retrieval/submission purpose. Web application has evolved from static-fixed layout application to dynamic-responsive layout application. For this, the response generated from the server is as per end user environment, i.e. customization of server response is achieved using AJAX/ NodeJS/ JQuery, etc. without overloading server [1–3].

R. Kachhwaha (✉) · R. Purohit
Department of CSE, MBM Engineering College, Jai Narain Vyas University, Jodhpur, India
e-mail: rajendra1983@gmail.com

R. Purohit
e-mail: rajeshpurohit@jnvu.edu.in

C. R. Panigrahi et al. (eds.), *Progress in Advanced Computing and Intelligent Engineering*, Advances in Intelligent Systems and Computing 714,
https://doi.org/10.1007/978-981-13-0224-4_4

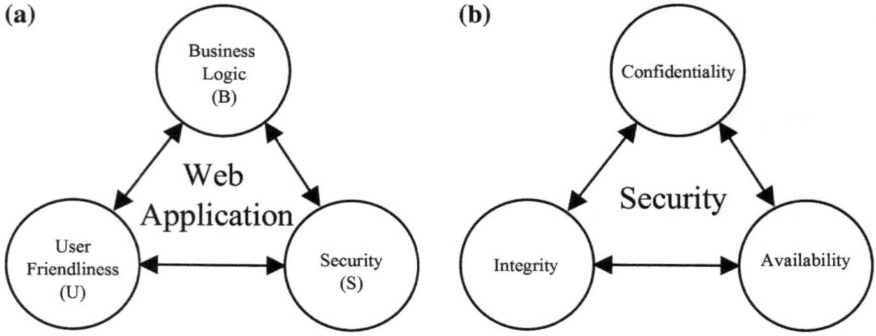

Fig. 1 **a** Web triad: BUS and **b** security triad: CIA

1.1 Web Application Triad and Security Triad

These days, web applications have acquired new dimensions due to easy availability of gadgets mainly smartphone, tablets and laptops with Internet facility. For any web application, it must satisfy web triad: BUS, that is, its basic functionality or business logic (B), user-friendliness (U) and security (S), as shown in Fig. 1a. Here, 'B' is functional attribute, and 'U' and 'S' both are non-functional attributes of web application.

The web application must do its basic functions as per its business logic for which it is developed while maintaining user-friendliness. User feels comfortable while using the web application with any Internet-enabled device. The web application must maintain its own data and its user data securely from unauthorized accesses and modifications, so that it will create a strong trust level of user in using the web application. Web application is developed in such a way that it maintains a balanced relationship between above three terms. For securing web and user data, it must follow the security triad: CIA. CIA stands for Confidentiality, Integrity and Availability, as shown in Fig. 1b [1, 2, 4].

2 Importance of Penetration Testing

Penetration testing or pen testing is a process of attempting to gain access to assets or resources without knowledge of user credentials and other normal means of access. The main thing that separates a penetration tester from an attacker is permission. The penetration tester will have permission from the owner of the computing resources or assets that are being tested. The goal of a penetration test is to increase the security of the computing resources being tested. Figure 2 shows a typical process of penetration testing [1, 2, 4, 5].

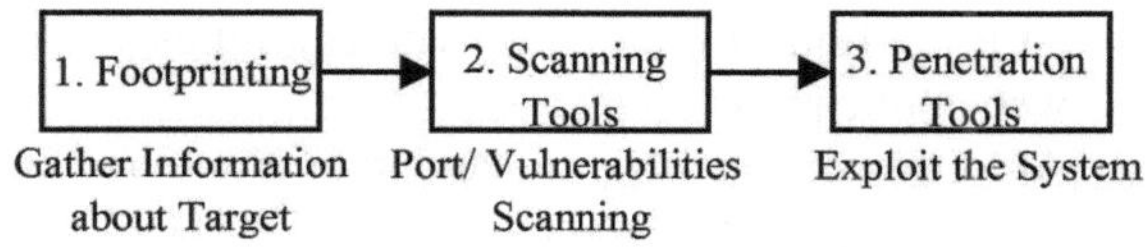

Fig. 2 Process of penetration testing

The very first step for performing penetration testing on any computer system or web application is footprinting. In this, we gather as much information about the target machine as we can. Few of the details are as follows: IP address, domain name server details, domain server start-end details, location of the server, registrar names, owner/ technical person contact details, etc.

At this point, we only have the IP address details, on which we perform scanning of using any scanning tool like NMAP [6]. We perform this scanning to identify any active port or service running or to find our any vulnerability on the server. If we have some knowledge of web application development, then we can test it manually also.

Now, we use the IP address details and other existing vulnerabilities of the web server and web application to penetrate into the either in web application or in web server. Following are some examples of vulnerabilities that may exist in a web server: a not required open port, a not required service is on and any vulnerability exists in the operating system or in dll/add-ons/plug-ins.

There are many tools exist which scan the actual web application for any user interface or business logic vulnerability, which can be exploited in the final step. Some of the tools are Acunetix, Vega, Havij and Sqlmap [6–8].

3 Threat Areas for Web Application

A threat is a possible danger that might exploit vulnerability in a system or in a network [1, 2, 4]. From web accessibility rule, there are mainly four component levels, which are vulnerable to any type of threat. These four levels are shown in Fig. 3.

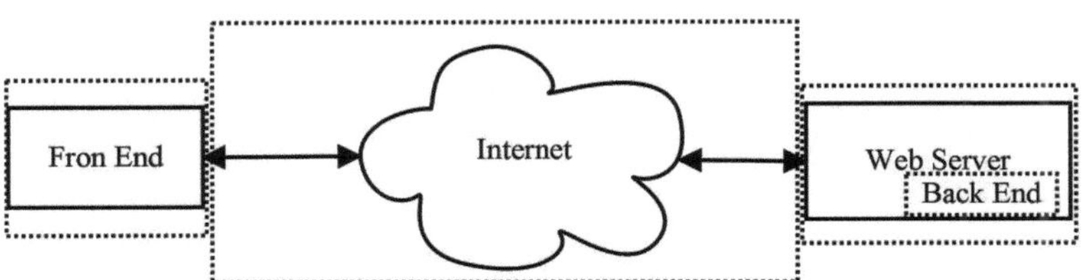

Fig. 3 Web accessibility with threat areas

1. At Front End: It involves all vulnerabilities possible in the user interface of the web application.
2. At Network: It involves all vulnerabilities that exist during transmission between client and the web server.
3. At Web Server: It involves all vulnerabilities with active services/ports running on the web server, with its running platform and its configuration.
4. At Back End: It involves all vulnerabilities left behind or not properly handled by the developer of the web application. It includes vulnerabilities of application's business logic layer, database layer or any other layer used.

With above four levels of threats, the following are seven security service points, which need to be handled by a developer [9]:

1. Validation (V): It is applied on that place of the application, which accepts inputs from the user. The web application must ensure each input supplied by the user is in correct format and length. It must have the capabilities to filter or reject each input, for its required format, before any additional processing.
2. Authentication and Authorization (AA): These are sub-parts of access control. Authentication defines who are you and authorization defines what can you do.
3. Session Management (SM): Session is a series of related interactions between a user and a web application. Session management means how a web application handles and protects user sessions.
4. Cryptography and Hashing (CR): It defines the functionality related with CIA (Confidentiality, Integrity and Availability). How a web application enforces CIA is defined under this.
5. Exception Handling (EH): In this, how the web application behaves, when a procedure call fails in web application, is defined.
6. Auditing and Log Management (AL): It is the sequence of security-related events, the web application and web server records. It defines who did what and when.
7. Configuration Management (CM): It defines mainly the configurations and operation issues of the web application and the web server.

Table 1 Relationship of security service points with component levels

Component levels	Security service points						
	V	AA	SM	CR	EH	AL	CM
Front end	Y	Y	–	Y	–	–	–
Network	–	–	Y	Y	–	–	–
Web server	–	Y	–	Y	Y	Y	Y
Back end	Y	Y	Y	Y	Y	Y	Y

Note *V* Validation, *AA* Authentication and Authorization, *SM* Session Management, *CR* Cryptography and Hashing, *EH* Exception Handling, *AL* Auditing and Log Management, *CM* Configuration Management

For a web application, we can derive a relationship for above-mentioned seven security service point with different component levels of an application. This relationship is illustrated in Table 1.

4 Experimental Setup

To demonstrate the effect of a vulnerability in web application, we select an online available web application named as http://demo.testfire.net/ [10]. This is a web application which is made to learn about web vulnerabilities. Here, we present the effect of vulnerabilities on web application and how an attacker can exploit those to gain sensitive information/data from the application. For this, we scan this web application manually, as the code files of this web application are not available to us. Once we find a vulnerability, we use available exploitation tool to penetrate into the web application and try to get as much sensitive information as we can get, without leaving any traces [5].

4.1 Test Performed and Observations

We applied both manual process and automated process (using Acunetix and Vega [6, 7]) to identify vulnerabilities. Table 2 presents a list of few manually observed vulnerabilities which includes webpages URLs with identified vulnerability in them. Those vulnerabilities can be exploited using another tool like Havij and Sqlmap [6, 8] to obtain sensitive information. In Table 2, we present webpage URL with observed vulnerability and small remark on that vulnerability [10–14].

4.2 Result Set

From above observed vulnerability, Table 2, we derive a notion that a particular vulnerability is due to improper handling of a one or more security service point at each component level of the application. This relationship of each security service point with observed vulnerabilities at each component level of the application is presented in Table 3.

4.3 Result Analysis

From Table 3, we can easily get an idea about which vulnerability is more severely affecting which service point at which component level of the application. This will

Table 2 Observed vulnerabilities with webpage URL

1.	Webpage	http://demo.testfire.net/bank
	Vulnerability	Directory service available
	Remark	We can get the directory structure of website
2.	Webpage	http://demo.testfire.net/bank/main.aspx
	Vulnerability	CSRF vulnerability
	Remark	<html><body><img src = “http://demo.testfire.net/bank/logout.aspx” height = “1” width = “1”/ ></body ></html>. This code will logout any user with active session on bank website
3.	Webpage	http://demo.testfire.net/bank/main.aspx
	Vulnerability	Plain text cookies
	Remark	Cookies contain data that is clearly plain text
4.	Webpage	http://demo.testfire.net/bank/main.aspx
	Vulnerability	Credentials in cookies
	Remark	Username and password, encoded by Base 64, are stored in cookies. They can be easily decoded
5.	Webpage	http://demo.testfire.net/bank/main.aspx
	Vulnerability	Network sniffing is possible
	Remark	Information sent over network is not secure. No use of HTTPS
6.	Webpage	http://demo.testfire.net/bank/login.aspx
	Vulnerability	Credential in plain text
	Remark	No encryption/hash for user information
7.	Webpage	http://demo.testfire.net/bank/login.aspx
	Vulnerability	Default admin username
	Remark	Username ‘admin’ exists
8.	Webpage	http://demo.testfire.net/bank/login.aspx
	Vulnerability	Common login page
	Remark	All users can login on same page
9.	Webpage:	http://demo.testfire.net/bank/login.aspx
	Vulnerability	Username and password are case insensitive
	Remark	Upper case converted username and password matched
10.	Webpage	http://demo.testfire.net/bank/login.aspx
	Vulnerability	No captcha
	Remark	Brute force can be performed for username/password combination
11.	Webpage	http://demo.testfire.net/bank/login.aspx
	Vulnerability	Authentication failed messages are invalid
	Remark	Different messages for wrong username and password
12.	Webpage	http://demo.testfire.net/bank/login.aspx
	Vulnerability	SQL injection
	Remark	Injection applied on both username and password

Table 2 (continued)

13.	Webpage	http://demo.testfire.net/bank/login.aspx
	Vulnerability	Query information dump
	Remark	A valid username and a single quote in password dumps lots of information
14.	Webpage	http://demo.testfire.net/bank/transaction.aspx
	Vulnerability	SQL injection
	Remark	A single quote in after filed will give SQL error. On using union clause in it, we got all username and password. Ex: 11/04/2017 union select username, password, user id, 4 from users
15.	Webpage	http://demo.testfire.net/search.aspx?txtSearch=something
	Vulnerability	XSS vulnerability on search page
	Remark	Enter this in search box: <script>alert ("Hello") </script >. Useful for stealing cookies of users
16.	Webpage	http://demo.testfire.net/cgi.exe
	Vulnerability	Executable download
	Remark	When clicked on location menu on left side, as executable is provided to download. It can be a malware
17.	Webpage	http://demo.testfire.net/pr/
	Vulnerability	Disclosure of sensitive information
	Remark	Sensitive information accessible directly to all
18.	Webpage	http://demo.testfire.net/default.aspx?content=../../../../boot.txt
	Vulnerability	Remote file inclusion vulnerability
	Remark	Any text or HTML file can be downloaded through this
19.	Webpage	http://demo.testfire.net/default.aspx?content=business_insurance.htm
	Vulnerability	Locally referenced file
	Remark	File that shows internal structure of server
20.	Webpage	http://demo.testfire.net/admin/application.aspx
	Vulnerability	Elevation of privilege. Access level is not checked
	Remark	We can access the admin pages through any account directly by typing above URL
21.	Webpage	http://demo.testfire.net/default.aspx?content=business.htm
	Vulnerability	Cross-site scripting
	Remark:	We can steal user sessions and cookies using <script>alert(document.cookie) </script>

Table 3 Relationship of security service points with vulnerabilities at each component level

Component level	Vulnerability																				
	1	2	3	4	5	6	7	8	9	10	11	12	13	14	15	16	17	18	19	20	21
Front end	–	V, AA	CR	CR	CR	CR	AA	AA	V, AA	V, AA	V, AA	V	V	V, AA	V, AA	–	AA, CR	AA	AA	AA	V, AA
Network	–	SM	CR	CR	CR	CR	–	–	–	–	–	–	–	–	–	–	CR, SM	–	–	SM	–
Web server	CM	AA	CR	CR	CR	CR	AA	AA	AA	AA	AA	–	EH	AA	AA	CM	AA, CR	AA, EH, AL, CM	AA, EH, AL, CM	AA	AA
Back end	CM	V, AA, SM	CR	CR	CR	CR	AA	AA	V, AA	V, AA	V, AA	V	V, EH	V, AA	V, AA	CM	AA, CR, SM	AA, EH, AL, CM	AA, EH, AL, CM	AA, SM	V, AA

Note V Validation, *AA* Authentication and Authorization, *SM* Session Management, *CR* Cryptography and Hashing, *EH* Exception Handling, *AL* Auditing and Log Management, *CM* Configuration Management facilitate a developer to think about security features for those service points at component-level development. We present this effectiveness of each service point for identified vulnerabilities at each component level in Fig. 4.

With the help of Table 3, we also derive Table 4, which shows the percentage (%) effect of identified vulnerabilities at each service point. This will help a developer to apply more security on that level, which has higher effect of percentage. This effect is shown in Fig. 5.

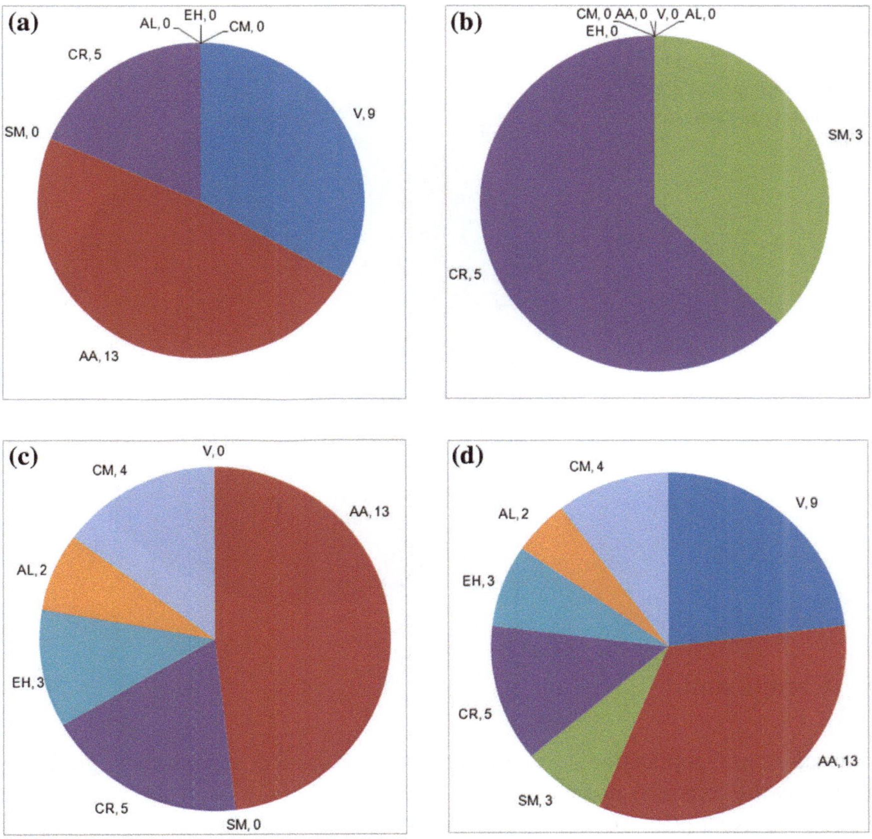

Fig. 4 Effectiveness of security service point for avoidance of vulnerability at **a** front end, **b** network, **c** web server and **d** back end

Table 4 Percentage effect of vulnerability at each security service point

Security service point	Vulnerability	Percentage effect (%)
V	18	17.82
AA	39	38.61
SM	6	5.94
CR	20	19.80
EH	6	5.94
AL	4	3.96
CM	8	7.92

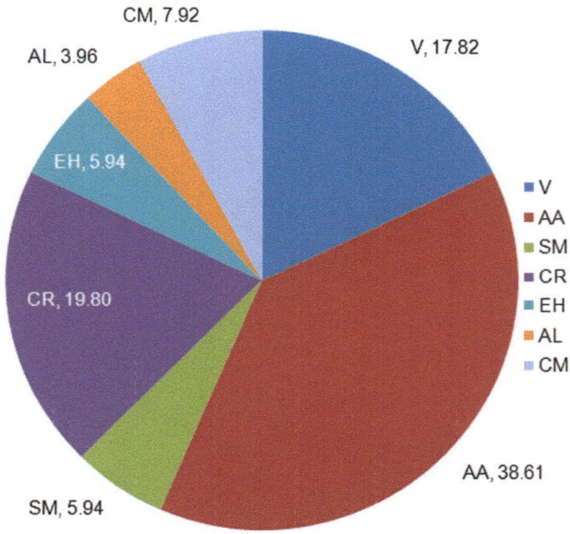

Fig. 5 Combined effect of vulnerability at each service point

5 Conclusion

Nowadays, web application development is very prominent area as many graduates are taking this as a start-up business area. For survival of new start-up companies as well as the existing companies, they have to strictly follow the web triad presented in this paper. We divide a web application into four components and seven service points, which are sub-parts of those components. The main intent of this paper is to present an idea of different service points, which effect the overall security feature of a web application. A developer must ensure that those seven service points are applied correctly at each component level and working as per application requirements. We use both manual and tool-based approach to find vulnerabilities in the selected web application: http://demo.testfire.net/. We identified some of the vulnerabilities with their impact. Once a vulnerability not removed at a level, it will affect more on subsequent levels. We present a tabular relationship of service points with various vulnerabilities at each component level. From that tabular relationship, we derive a relative efficiency of service point for vulnerability avoidance at each component level. This will help a developer to apply suitable protection at each component level for different types of vulnerabilities. We also present a combined impact of all identified vulnerability on each service point for a web application. This will help a developer for thinking which service point affects more in a web application. During our tests, Authentication and Authorization (AA) affects higher than any other service point. We cannot ignore remaining service points due to their lower values because they also affect the security feature of web application. As performing penetration testing is not an easy task of 1 day or 1 week. It requires practical knowledge of tools and identification of loopholes in coding methodology. There may be the possibility of existence of some more vulnerability in selected web application, which is not identified by us. We use both manual and open source tools for our experiment. There may be some other tools available in Kali Linux which exploit and penetrate more deeper.

References

1. Joel Scambray, Mike Shema: Hacking exposed: Web Application, McGraw-Hill (2002)
2. Dafydd Stuttard, Marcus Pinto: The Web Application Hacker's Handbook, Second Edition, Finding and Exploiting Security Flaws, John Wiley & Sons (2011)
3. R Kachhwaha, P Patni: Ajax enabled web application model with comet programming, International Journal of Engineering and Technology, Volume 2 No. 7, pp. 1155–1161 (2012)
4. Stuart McClure, Joel Scambray, George Kurtz: Hacking Exposed 7: Network Security Secrets & Solutions, McGraw-Hill (2012)
5. Stephen Northcutt, Jerry Shenk, Dave Shackleford, Tim Rosenberg, Raul Siles, Steve Mancini: Penetration Testing: Assessing your overall security before an attacker do, SANS Institute (2006)
6. Kali Linux Tools Listing https://tools.kali.org/tools-listing
7. Acunetix Web Vulnerability Scanner https://www.acunetix.com/
8. SQLMAP http://sqlmap.org/
9. John D. Meier, Web application security frame (Patents: US 7818788 B2), http://www.google.co.in/patents/US7818788
10. AltoroMutual, http://demo.testfire.net/
11. OWASP https://www.owasp.org
12. OWASP Top Ten Project https://www.owasp.org/index.php/Category:OWASP_Top_Ten_Project
13. OWASP Top Ten Vulnerabilities https://www.owasp.org/index.php/Top_10_2017-Risk
14. OWASP AltoroMutual https://www.owasp.org/index.php/AltoroMutual

Web Services Regression Testing Through Automated Approach

Divya Rohatgi and Gyanendra Dwivedi

Abstract Web services are a software technology which is based on Service-oriented architecture that is used to provide business functionalities on the web. Thus it is important to ensure proper quality and maintenance of Web Services. Maintenance activity is assumed to be the most expensive activity in software development. Regression testing is a part of maintenance which is done every time whenever a change is made to the software. Regression testing is challenging and time-consuming activity in web services because they are inherently distributed, heterogeneous and dynamic in nature. Thus it is important to reduce regression test effort thereby reducing software maintenance costs. In this paper we have given an efficient approach by which we can effectively carry out regression testing of a web-based application system whenever any changes is made to system.

Keywords Software maintenance · Regression testing · Automated testing
Web service

1 Introduction

A web service is a software technology which is based on service-oriented architecture and is used in various domains of applications. They use XML for tagging of information, WSDL for describing the service behavior, SOAP for communication between various parties, and UDDI to enumerate services present for use. The main advantage of using web service is that despite using different IT infrastructures, companies can communicate between them easily and efficiently with a

D. Rohatgi (✉)
Computer Science & IT Department, Sam Higginbottom Institute of Agriculture, Technology & Sciences, Allahabad, India
e-mail: divi.rohatgi@gmail.com

G. Dwivedi
Department of Quality Assurance, Ugam Solutions Pvt. Ltd., Mumbai, India
e-mail: gyanendra.mnnit@gmail.com

C. R. Panigrahi et al. (eds.), *Progress in Advanced Computing and Intelligent Engineering*, Advances in Intelligent Systems and Computing 714,
https://doi.org/10.1007/978-981-13-0224-4_5

web service [1]. They have revolutionized the approach in which applications are developed and with the help of which applications can have better productivity and efficiency across organizations. Prior to web services only static applications were built but now with this technology, organizations can develop business functionality which in the time of need can be made available and can also be combined with other available services in order to give a new functionality. Web services have an additional advantage that depending upon the changing business conditions or addition of new business constraints we can easily and effectively adapt to changed scenarios [2]. Thus according to changing business needs and conditions these services are also constantly evolving. So it is necessary to ensure quality and proper maintenance of web services as they represent essential business functionalities. This can be achieved through effective regression testing. In order to have better efficiency and productivity of business applications, it is required to remove the obstacles which are faced by software testers in the process of regression testing of such applications [3]. These problems are due to several reasons. Web services are distributed and heterogeneous in nature [4] operating upon different architectural platforms both in terms of hardware and software due to which testing particularly regression testing becomes difficult. Also due to dynamic nature of these services [4] we can always add, remove or collaborate with other services. This also makes regression testing a difficult and challenging process. Thus there is a need to have an efficient regression testing framework which can overcome the challenges posed by the inherent nature of web services.

In software engineering, software maintenance is considered as an expensive activity [5] and is performed every time we need to change the software either by fixing bugs or by adding or deleting preexisting functions or adapting the software to new platforms or architectures. Regression testing is a part of maintenance which aims at running the old test cases to ensure correctness of the program after changes applied. It is considered as a costly activity [6] as it encompasses considerable fraction of resources in terms of budget, schedule, and effort. Thus in order to cut short maintenance cost it is a need to reduce regression testing. One of the best approaches to reduce regression testing effort is selective retesting or regression test selection. There are different regression testing strategies proposed in literature but in case of web service it is a relatively less researched area.

This paper presents an automated approach for regression testing of web services based applications and to illustrate the approach we have used online survey reporting system.

2 Literature Review

The Regression testing is considered with running of all old test cases to ensure that new changes made to software had not introduced any new errors in software. There are two approaches or strategies to perform regression testing: retest all and selective retest strategy [7]. In retest all approach, we have to rerun entire earlier developed test suite on the modified program. This approach is safe as with this approach we can check entire modifications in the changed component. However it is not advisable to follow this practice for large software as it incurs large amount of time and budget. Another approach is selective retest in which first we select a subset from preexisting test suite and then run those test cases on only required part of modified software. Consequently we can minimize the time required for regression testing of modified program. Rothermel and Harrold [7] proposed two challenges in the selective retest techniques. First challenge is concerned with the strategy to select required test cases from previous test suite and secondly how to find when and where more test cases are needed to augment the existing one. In [8, 9] Rothermel and Harrold have given a set of metrics framework to measure the effectiveness of different regression test selection strategies. In the literature for different types of software paradigms there are different regression test selection strategies proposed. Regression test selection for web services is a relatively new research area. In [10] Chaturvedi and Gupta have proposed web service regression testing with three different categories for changes in WSDL, changes in code, and selective retesting of web service operations and for this authors have presented three types of WSDL: Difference WSDL to incorporate changes in WSDL, Unit WSDL for changes in code and Reduced WSDL for selective retest. In [11] Cladio et al. have shown an efficient regression testing by keyword analysis. In [12, 13] Ruth M. et al. have presented a framework which automates web service regression test selection by monitoring of service. Also authors have evaluated the framework for comparing the cost of using automated approach of regression test selection with the cost of without using regression test selection. In [14, 15] Masood et al. have proposed regression testing strategy which is fully automated and is safe based on original and changed WSDL. Li et al. [16] have presented an approach which chose test cases for regression testing of different versions of BPEL (business process execution language) composite service. In [17] Ruth M. has demonstrated empirical studies of Web Services Regression Testing for Privacy-Preserving. In [18] Izzat Alsmadi has analyzed the activities and challenges which arises in regression testing of web services. According to the author, it is very important to reduce test cases in case of web service whereas there is not so much urgency for reducing test suite in traditional programming paradigms. Most important is to optimize test execution in comparison to other processes as resources can be made available to them. For this, two strategies were proposed. First and foremost requirement is to produce a pretest execution component by which one can evaluate generated test cases and then minimize test case selection based on generated test cases. Secondly with the help of historical usage sessions we can minimize the process of test case selection. P. Bhuyan et al. [19] have proposed the usage of UML use case diagram for regression testing and UML activity diagram to generate test cases for service-oriented architecture. In [20] Tarhini et al. modeled a web application as two level abstract model provided through a Timed Labeled Transition Systems TLTS. They have also proposed an algorithm for regression testing of web applications which is proved to be safe. In [21] Mohanty et al. have proposed a control flow graph based approach with the help of which we can easily use safe regression test selection strategy to programs based on service-oriented architecture. A given approach of test selection is said to be safe if it selects all test cases which are capable of revealing modifications performed. To verify the strategy they have used a navigational subsystem comprising of three web services. In [22, 23] Ruth et al. and in [24] Khan et al. have proposed safe regression test selection strategies which utilizes analysis of control flow models. Ruth et al. [22, 23] have presented a gray box technique for regression test selection as it is sometimes problematic to follow white box approach as sometimes source code for components are not provided to the developer of the web service. The approach is based on the concept that each process in a web service is given by a CFG at the developer side After this CFG from all other components are summed up to have a global CFG for the entire system. Then identify the dangerous edges and select those test cases that pass through these dangerous edges. The techniques based on flow of control of program in [22, 23] have pros and cons which can be compared to traditional control flow based techniques for procedure-based programs. In [24] Khan et al. have presented a strategy based on model based approach where service interfaces are defined by visual contracts, i.e., pre- and post conditions illustrated as graph transformation rules. With the study of dependencies of these rules we can estimate the effect of a change and thereby decide the selection of test cases as a strategy for regression test selection.

3 Proposed Methodology

This paper presents an effective regression testing approach for web based applications. For demonstration we have used an online survey reporting application system as shown in Fig. 1.

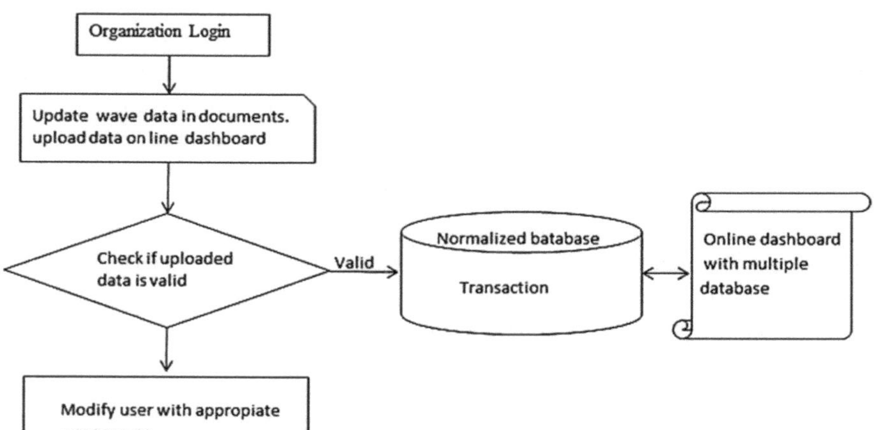

Fig. 1 Online survey reporting system

The online survey reporting system is used to have surveys of movie trailers. Earlier the historical data and key findings of the survey were kept as excel files. Online survey reporting system has replicated the current functionality of the excel files and has provided with GUI enabled dropdown selections, multi-select filters and comparison of data represented with the help of charts and graphs. Since data was kept in SPSS or other database format rather than SQL, so data is ported from other database to SQL which is in OLTP form. Then data is processed to OLAP form. On the basis of filters, OLAP data result set is shown to User Interface of web based application system through a web service. User Interface data can be exported to excel. Every time a new survey or existing survey tracker data is added, we need to test whole ETL process, OLTP to OLAP conversion and OLAP to UI conversion. This process is very time consuming and require large manual efforts as we have to run full test cases each and every time whenever we need to add data.

This paper presents an effective approach which is also automated by which we can reduce regression testing of web based application system. To do this, first match the frequency of responses for each question and variables from other database to OLTP database. Then on the basis of filters, the data is processed from OLTP to OLAP. On excel, we make database connectivity of analytical services using authentication credentials. Pivot service on excel is made. A macro service is made to pass the filters to web application as well as to pivot. In this approach, we have used iMacro testing tool for automation testing. Filter data is being kept in excel. Input from this excel sheet is passed to iMacro as well as to pivot. Since inputs to both are same, we should get the same output. Macro Service exports User Interface data to another excel and pivot data also to excel sheet. Finally it compares both excel sheet having UI data and pivot data. If everything is matched then it appends "True" to excel sheet name. If any cell is not matched then it appends "False" to excel sheet name.

4 Results

To test our approach, we have used an online web-based application system making use of a web service which is an online survey reporting system. The results of the implementation are shown. Figure 2 shows the snapshot of the pivot services made in excel.

A macro or custom service is made which is used to pass the filters to application system as well to pivot as shown in Fig. 3.

To automate the functional and regression testing, we have used iMacro testing tool which is used to read the filter data kept in excel, pass the same filter data to web application and export the result set in excel. Figure 4 shows the snapshot of iMacro tool.

Macro service matches data from the application excel with pivot excel data and appends "True" or "False" on comparison sheet which is another sheet in excel. Figure 5 shows the result after comparison.

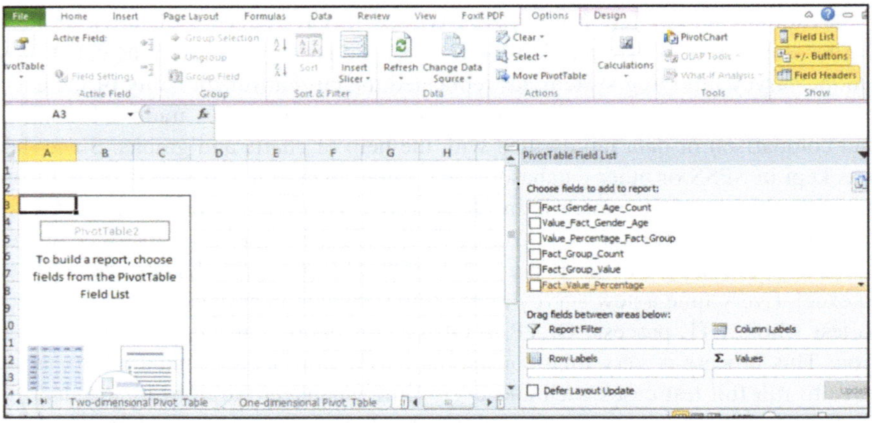

Fig. 2 Pivot of OLAP data in excel

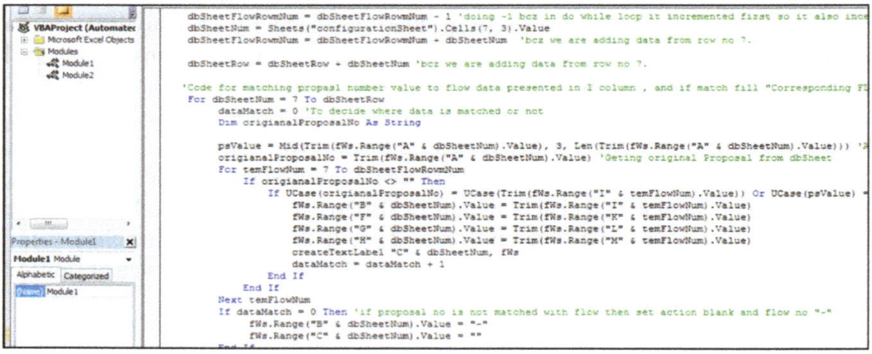

Fig. 3 Macro service or any customize application to pass the filters

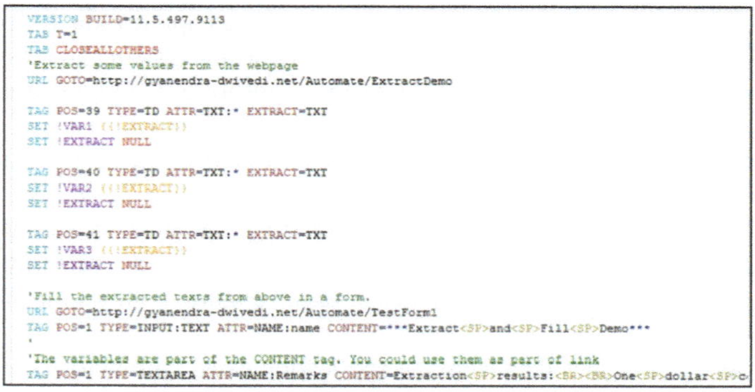

Fig. 4 iMacro or any custom testing tool with script

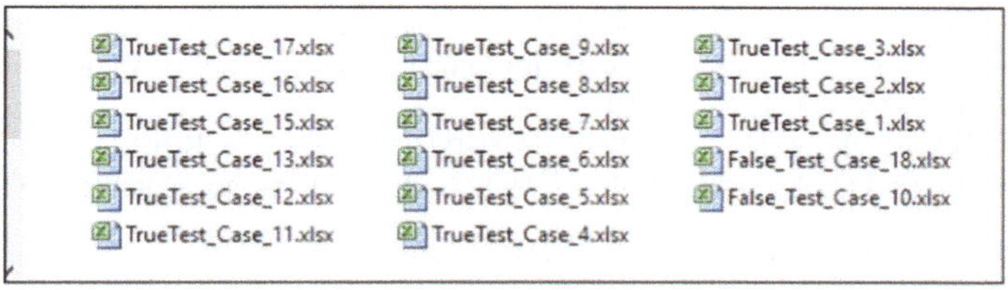

Fig. 5 Result after comparison of application excel and pivot excel

Thus with this approach, we attempted to reduce the regression testing of web-based application system containing web services as whenever a new job or new survey data is added to the application, we need not to test the full application manually. Instead, we automated the whole approach to save manual time and effort.

5 Conclusion

In this paper, we have analyzed the issues and challenges related to regression testing of web services which are a part of web-based application system. Also a novel approach of maintaining reliability and quality of such systems is proposed. The approach is evaluated on an online web based application system and results are reported thereby. Thus we can ensure to have quality and reduce maintenance costs of constant evolving web services. The approach presented is platform independent and provides fast processing of data for testing. The major disadvantage of the approach is that it is implemented for only those web based application containing structured databases. For unstructured databases, implementation is left as a future scope.

References

1. http://www.webopedia.com/TERM/Web_Services.
2. Chen, M., Chen, Andrew N. K., Shao, Benjamin B. M.: The Implications and Impacts of Web Services to Electronic Commerce Research and Practices. Journal of Electronic Commerce Research. VOL. 4 (2003).
3. Mohanty, R. K., Pattanayak, B K., Mohapatra, Durga Prasad: UML Based Web Service Regression Testing Using Test Cases: A Case Study. ARPN Journal of Engineering and Applied Sciences. Vol. 7, No. 11, ISSN 1819-6608 (2012).

4. Bassil, Youssef: Distributed, Cross-Platform, and Regression Testing Architecture for Service-Oriented Architecture. Advances in Computer Science and its Applications (ACSA). ISSN: 2166-2924, Vol. 1, No. 1(2012).
5. Seacord, R.C., Plakosh, D., Lewis G.A.: Modernizing Legacy Systems: Software Technologies, Engineering Process and Business Practices. Addison-Wesley Longman Publishing Co. Inc., Boston (2003).
6. Leung, H., White, L.: Insights into regression testing. Proceedings of the Conference on Software Maintenance. pages 60–69 (1989).
7. Rothermel, G., Harrold, M.: A safe, efficient regression test selection technique. ACM Transactions on Software Engineering Methodology (TOSEM), 6(2):173{210, 1997).
8. Rothermel, G., Harrold, M.: Analyzing regression test selection techniques. IEEE Transactions on Software Engineering 22(8):529–551 (1996).
9. Engström, E., Runeson, P., Skoglund, M.: A systematic review on regression test selection techniques. Information and Software Technology, 52(1):14–30 (2010).
10. Chaturvedi, A., Gupta, A.: A tool supported approach to perform efficient regression testing of web services. In Proceedings of IEEE 7th International Symposium on Maintenance and Evolution of Service Oriented and Cloud Based System. (2013).
11. Magalhaes, C., Barros, F., Mota, A., Maia, E.: Automatic Selection of Test Cases for Regression Testing. Proceedings of the 1st Brazilian Symposium on Systematic and Automated Software Testing. ACM (2016).
12. Ruth, Michael E.: Concurrency in a decentralized automatic regression test selection framework for web services. Proceedings of the 15th ACM Mardi Gras conference: From lightweight mash-ups to lambda grids. ACM (2008).
13. Ruth, Michael E., Tu, Shengru: Empirical studies of a decentralized regression test selection framework for web services. Proceedings of the workshop on Testing, analysis, and verification of web services and applications. Pages 8–14, ACM (2008).
14. Masood T., Nadeem A., Ali: An automated approach to regression testing of web services based on WSDL operation changes. In Proceedings of IEEE 9th International Conference on.
15. Masood T., Nadeem A., Lee: A Safe Regression Testing Technique for Web Services Based on WSDL Specification. Software Engineering, Business Continuity, and Education Communications in Computer and Information Science. Volume 257, pp 108–119, Springer (2011).
16. Li, B., Qiua, D., Leungb, H., Wanga Di: Automatic test case selection for regression testing of composite service based on extensible BPEL flow graph. The Journal of Systems and Software, 1300–1324 Science Direct, Elsevier Inc. (2012).
17. Ruth, M.: Empirical Studies of Privacy-Preserving Regression Test Selection Techniques for Web Services. Proceedings of the IEEE International Conference on Software Testing, Verification, and Validation Workshops. Pages 322–331 (2014).
18. Izzat Alsmadi, Sascha Alda: Test Cases Reduction and Selection Optimization in Testing Web Services. I.J. Information Engineering and Electronic Business. Vol 5, pp 1–8 (2012).
19. Bhuyan, P., Kumar, Abhishek: Model Based Regression Testing Approach of Service Oriented Architecture (SOA) Based Application: A Case Study. International Journal of Computer Science and Informatics. ISSN (PRINT): 2231-5292, Volume 3, Issue 2 (2013).
20. Tarhini, A., Fouchal, H., Mansour, N.: Regression Testing Web Services-based Applications. ACS/IEEE Int. Conf. on Computer Systems and Applications. pp. 163–170 (2006).
21. Mohanty, R.K., Pattanayak, B.K., Mohapatra, D.P.: A Regression Test Selection Technique for SOA Based Applications. International Journal of Software Engineering and Its Applications. Vol. 8, No. 3, pp. 65–72 (2014).
22. Ruth, M., Tu, S.: A safe regression test selection technique for web services. In Proceedings of the Second International Conference on Internet and Web Applications and Services. IEEE Computer Society (2007).

23. Ruth, M., Oh, S., Loup, A., Horton, B., Gallet, O., Mata, M., Tu S.: Towards automatic regression test selection for web services. In Proceedings of the 31st Annual International Computer Software and Applications Conference. Volume 02 pages 729–736. IEEE Computer Society (2007).
24. Khan, T. A., Heckel, Reiko: On Model-Based Regression Testing of Web-Services Using Dependency Analysis of Visual Contracts. FASE 2011, LNCS 6603, pp. 341–355, Springer Verlag Berlin Heidelberg (2011).

Securing Healthcare Information over Cloud Using Hybrid Approach

Kirit J. Modi and Nirali Kapadia

Abstract Cloud computing has increased the attention for accessing and storing information. To share and store healthcare information over Cloud is playing crucial role to provide cost-effective and flexible and reliable solution to the users. Despite advantages of Cloud-based Healthcare system, security of data is major factor, which restricts the acceptance of the Cloud-based model. As a solution to the security challenge, our work advocates the use of linear network coding and re-encryption based on ElGamal cryptography in the form of hybrid approach to secure healthcare information over cloud. To provide security and fault tolerance for cloud storage, we have considered linear network coding mechanism. To exchange the encoding key matrix securely with the receiver, ElGamal re-encryption scheme is used. As a proposed approach, we present how securely the data can be transferred between sender and the receiver over cloud.

Keywords Cloud computing · Linear network encoding · Proxy re-encryption ElGamal cryptography

1 Introduction

Cloud computing is gaining popularity as an emerging technology for sharing the resources over the Internet. Cloud computing provides flexibility, reliability, sustainability and cost effectiveness to the users. For existing healthcare systems, there

K. J. Modi (✉)
Department of Information Technology, U. V. Patel College of Engineering, Ganpat University, Gujarat, India
e-mail: kiritmodi@gmail.com

N. Kapadia
Department of Computer Engineering, U. V. Patel College of Engineering, Ganpat University, Gujarat, India
e-mail: niralijollykapadia@gmail.com

C. R. Panigrahi et al. (eds.), *Progress in Advanced Computing and Intelligent Engineering*, Advances in Intelligent Systems and Computing 714,
https://doi.org/10.1007/978-981-13-0224-4_6

is a key requirement to develop an approach that minimizes time-consuming work and expensive means to access a patient's medical record and integrating this changing set of medical information consistently to deliver it to the healthcare organization. Nowadays, healthcare providers have adopted the cloud platform that can perform their operations more efficiently. Cloud computing service enables a group of doctors to obtain an access to a patient's health record anytime, anywhere. Despite all the these advantages cloud computing provides to the healthcare systems, data security is among the major concerns, which make healthcare system move slowly towards the acceptance of Cloud-based healthcare technologies. Cloud computing benefits come at a cost of the emergence of various risks related to the information security that must be cautiously addressed. As a solution, we contribute our work as follows.

(i) To present the effective approach for securing the healthcare information over the cloud.
(ii) To perform the experimental work with the proposed approach and provide the results in the form of reliable solution.

The rest of this paper is organized as follows: in Sect. 2, we present the concepts of network encoding and proxy re-encryption using ElGamal cryptography. Section 3 discusses literature review related to secure healthcare system over the cloud. Section 4 proposes Cloud-based secure healthcare framework. Finally, Sect. 5 presents experimental work and results. Section 6 concludes this paper and discusses our future direction.

2 Background Concepts

In this section, we define network encoding [1] and Proxy re-encryption [2] using ElGamal cryptography concepts.

2.1 Network Coding

It is a technique in which coding is done at the nodes in a network. Network encoding is used to minimize the network delays and maximize the throughput of the network and make the network reliable and robust. Network encoding is used in the packet networks (where data is first fragmented into packets then transmitted to the destination). The network encoding is applied at the packets, so we can say that coding is done above the physical layer. Network coding improves the robustness, throughput, security and complexity of the network.

2.2 Proxy Re-encryption

It is technique which allows proxy to convert the cipher text generated by the sender's public key into such a form that can be decrypted by receiver's private key (without using sender's private key). There are many applications where we require proxy re-encryption. For example, Alice wants to send an encrypted email to Bob, without sharing her private key. In this case, Alice the sender uses a proxy re-encryption technique to re-encrypt the mail into a form that Bob the receiver can decrypt by using his own private key.

2.2.1 Proxy Re-encryption Using ElGamal Cryptography

The following steps show how we can perform Proxy re-encryption using ElGamal cryptography.

- Let us consider p be a prime number
- Let us consider g be a generator of $Zp = \{0, \ldots p - 1\}$
- Let y = (mod p), where x is a randomly selected private key
- Thus, the pubic key of ElGamal is a triplet {p, g, y}
- Private key = {x}

(a) **Encryption**
Generate a random value k and encrypt plaintext M as follows:

- a = (mod p)
- b = M*(mod p)
- Thus, encrypted text is (a,b)

(b) **Decryption**
The cipher text C = (a,b) is decrypted by using following modular operation:

- $M = b/a^{x}(\text{mod } p)$
- For using ElGamal in proxy re-encryption, the secrete key x is splitted into x1 and x2,
- such that x1 + x2 = x
- According to user's requirement x2 is splitted into x3 and x4 such that x3 + x4 = x2
- If we have cipher text C then using x1 we can have another text say M1 such that $M1 = b/a^{x1}$
- M1 can be converted into plaintext M2 such that $M2 = b/a^{x3}(\text{mod } p)$
- M2 can be converted into plaintext M such that $M = b/a^{x4}(\text{mod } p)$
- The correctness of proxy ElGamal encryption can be verified as follows:

$$
\begin{aligned}
M_2/a^{x4} \bmod p &= \left(M_1/a^{x3} \bmod p\right)/a^{x4} \bmod p \\
&= \left(b/a^{x1} \bmod p\right)/\,a^{x3+x4} \bmod p \\
&= \left(b/a^{x1} \bmod p\right)/\,a^{x2} \bmod p \\
&= \left(b/a^{x1} \bmod p\right)/\,a^{x2} \bmod p \\
&= b/\left(a^{x1+x2}\right) \bmod p \\
&= b/a^{x} \bmod p
\end{aligned}
$$

3 Literature Study

The following section presents the literature related to the security aspects for healthcare information over cloud.

Garg, Parul, and Vishal Sharma [3] have proposed an efficient mechanism to store data securely in cloud. Here the author uses RSA and Hashing cryptography tools to securely store data in cloud. A trusted third party is used where the data is present in unencrypted form. This is not suitable for healthcare data. All the computation and verification are offloaded to TPA so there is a need to make TPA more secure.

Rewadkar, D. N., and Suchita Y. Ghatage [4] have introduced a third-party auditor, who checks the integrity of data in cloud storage on the behalf of cloud customer. Before sharing the data with TPA, the data is encrypted by using homomorphic encryption method. During auditing process, TPA will know able to know anything about the data stored in cloud. The drawback here is that the data is stored over the cloud server in the form of blocks and these blocks along with their metadata are in unencrypted form. So, there is data integrity and confidentiality risk over that data as Cloud Service Provider (CSP) is considered trustworthy.

Khanezaei Nasrin and Zurina Mohd Hanapi [5], proposed a method in which they used the combination of RSA and AES encryption method to securely share data stored in cloud. Symmetric and asymmetric encryption respectively is used for both uploading and downloading file from cloud. The main drawback of system is that we have to do encryption and decryption twice for the same file stored in cloud which cause the overhead for the system.

Thiranant et al. [6] has designed a framework which provides security to e-healthcare system. This framework uses web services to provide security to the data stored in cloud. The application can be access through browser via the Internet. The data are stored in cloud is encrypted, but for security we have to trust on service providers. Since the system is accessed via internet stealing of data is one of the major challenges in such systems.

In [7], the author presents a hybrid approach by using RSA and AES encryption algorithm to securely store data in cloud server. In cloud system security is one of the biggest issues in this paper author focus on: (1) securely upload the data to cloud in such way that even administrator does not know about the contents. (2) Securely download the data from cloud in such a way that the data integrity is not affected. The drawback here is that the cloud service provider is partially trusted which is not acceptable for healthcare data.

In [7], the author has designed a "three-way mechanism" to increase the security in cloud by using AES and Diffie-Hellman key exchange algorithm and digital signature. In this mechanism author uses Diffe-Hellman key exchange thus if key is hacked while transmission it is useless to hacker because hacker doesn't have legitimate user's private key.

Gupta, Suneet K., Seema Rawat, and Pranaw Kumar [8] proposed a novel security architecture for access control in cloud computing. This scheme is advancement in CPASBE (cipher text-policy attribute-set-based encryption) scheme.

Louk, Maya, and Hyotaek [9] proposed a data security scheme for mobile multicloud computing (MMC) homomorphic encryption. This paper proves that homomorphic encryption is optimal for mobile multicloud computing. Improving security and performance is one of the future aspects for other researchers.

In [10], the author proposed a commutative encryption method based on the ElGamal encryption in which a plaintext is encrypted more than one time using different users' public keys. In this system, the computational result is not affected by the order of keys used in encryption and decryption.

In [11], the author proposed a novel Global Authentication Register System (GARS) to provide security in cloud system. They implemented the GARS algorithm in simulation environment and by analyzing the experimental result they show that their system provides effective security to cloud system.

In [12], author proposed an approach to provide security to cloud-based healthcare system. The patients and medical centers can store in cloud based centralized system. When data is stored in cloud security is one of the major issue thus to overcome that they use proxy re-encryption scheme in which allows proxy to convert the cipher text generated by the sender's public key into a such a form that can be decrypted by receiver's private key (without using sender's private key).

The above discussion concludes that there has been very less work done in the field which involves security and reliability of data at the same time. So, we focused on these two parameters for our work. We have concluded after literature study that the best strategy to provide security to data is to use symmetric and asymmetric algorithms on the data at same time. The reason behind it is that Symmetric algorithm takes less amount of time in cryptographic operations compared to asymmetric algorithm. Thus, we can encrypt our original data first by using symmetric algorithm and the key that we used to encrypt the data can be encrypted by asymmetric algorithm.

4 Proposed Work

In this section, we presented our proposed architecture and approach for Securing healthcare information over cloud.

4.1 Cloud-Based Secure Healthcare Information System

In Fig. 1, we have proposed framework for security to the healthcare information over Cloud which is divided into four main modules as follows. The functionality of each module is discussed here.

The framework is divided into four modules.

I. **Secure Data Storage**
   Secure data storage process using network coding technique is defined as follows which is presented in Figs. 2 and 3 as follows.

   - Network coding matrix EM1 and EM2 is generated.
   - File F is encoded using key EM1. Encode(F, EM1).
   - File F is encoded using key EM2. Encode(F, EM2).
   - Encoding Matrix EM = {EM1, EM2}.
   - ElGamal generates public key Pb and private key Pr.
   - Network coding matrix EM is encrypted using public key Pb of ElGamal. E (EM, Pb).
   - Private key is partitioned into two parts Pr1 + Pr2 = Pr.
   - EM is partially decrypted using Pr1. D(E (EM, Pb), Pr1).
   - Encoded Files and partially decrypted EM is sent to the cloud for storage.
   - Encoded files are P1, … P8.
   - The partially decrypted encoding matrix EM will be sent along with all this files as a metadata of the file.

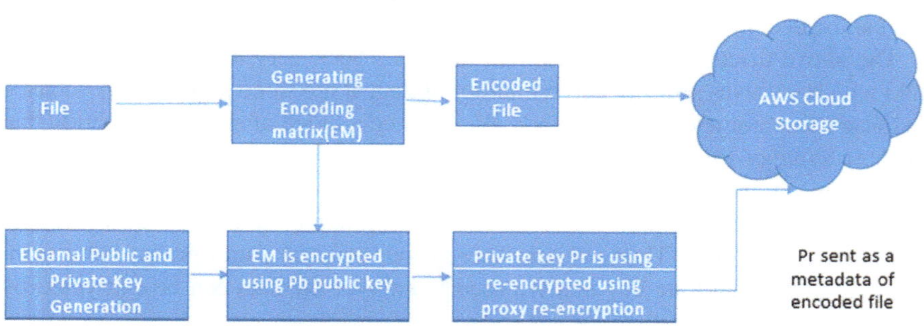

Fig. 1 Proposed framework of cloud-based secure healthcare system

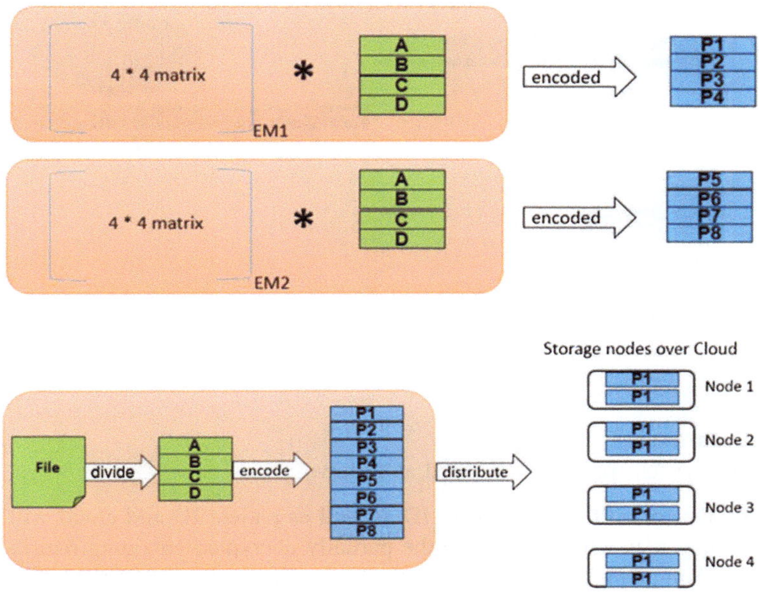

Fig. 2 Network coding

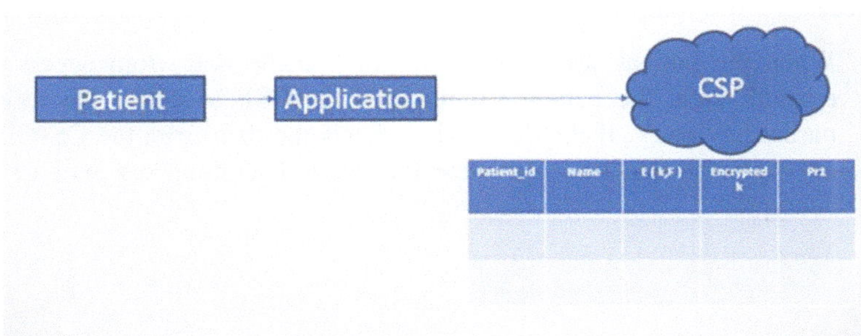

Fig. 3 Secure data storage

II. **Data Sharing**

Data sharing process is defined as follows which is presented in Fig. 4.

- When the doctor wants to download data, he makes request to the patient.
- Pr2 will be partitioned into two random parts. Such that Pr2 = Pr3 + Pr4.
- Pr3 will be sent to the storage node and will be stored as a metadata.
- The proxy will turn partially decrypted EM into another form using Pr3.
- Pr4 is send to the intended doctor.

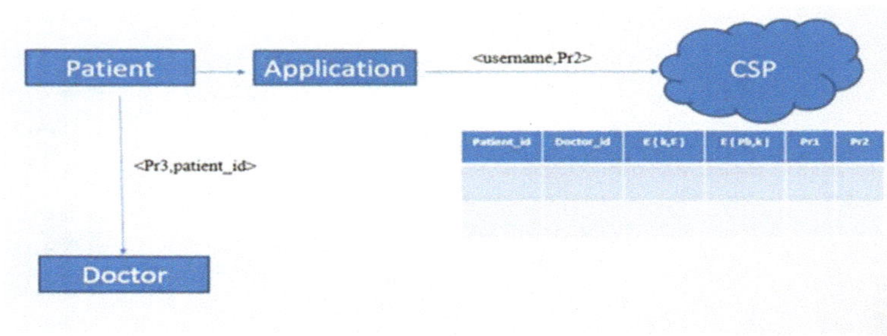

Fig. 4 Data sharing

III. **Data Access**

Data access process is defined as follows:

- Doctor will enter the user ID as well as patient ID and cloud will return any files which will have the partially decrypted encoding matrix EM.
- Using Pr4 symmetric key will be decrypted. D (D(D(E(EM, Pb), Pr1), Pr3), Pr4) = EM.
- Using inverse of EM, file F will be decrypted. Decode(F, EM).

IV. **Access Revocation**

Access revocation process is defined as follows which is presented in Fig. 5.

- When the patient wishes to withdraw specific data from access to his e-health data, the patient simply calls the CSP to delete the receiver's partial key entry. If the doctor downloads the data from the CSP, he will only get the encoded file since the network coding key will never be decrypted.

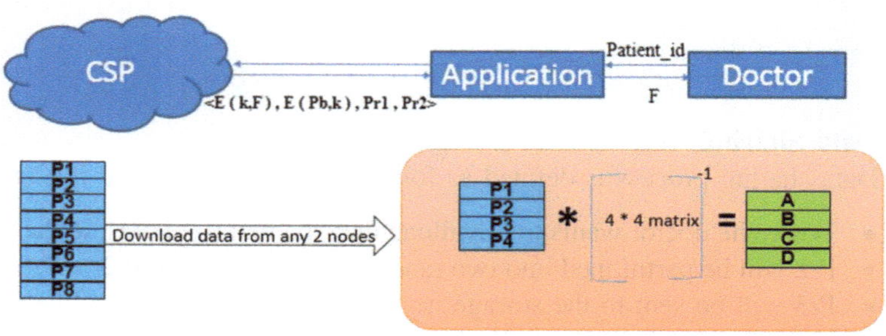

Fig. 5 Access revocation

- If the original file has n blocks of original data, then by downloading n blocks, instead of 2n blocks, we could get the original data blocks, by using inverse of the encoding matrix.

4.2 Reliability Proof Using Network Coding

Following example provides proof of reliability using Network coding.

- Suppose we have data [1], [2], then to do network coding over this data we need two 2 * 2 matrices as key matrix.

$$\begin{matrix} 1 & 2 \end{matrix} * \begin{matrix} 1 & 2 \\ 3 & 4 \end{matrix} = \begin{matrix} 7 & 10 \end{matrix}$$

$$\begin{matrix} 1 & 2 \end{matrix} * \begin{matrix} 3 & 4 \\ 9 & 6 \end{matrix} = \begin{matrix} 21 & 16 \end{matrix}$$

- Then we have encoded data as [7], [10], [21], [16] out of the original data [1], [2].
- If the lost data is [7], [21], then we could obtain the original data [1], [2] from the data [10], [16]

$$\begin{matrix} 10 & 16 \end{matrix} * \text{inverse of} \begin{matrix} 2 & 4 \\ 4 & 6 \end{matrix} = \begin{matrix} 1 & 2 \end{matrix}$$

5 Experimental Work and Results

The experiment is carried out on the machine having following configuration: Processor: Intel(R) Core(TM) i5-2467 M CPU @1.60 GHz, RAM: 4.00 GB, System Type: 32-bit OS Windows 8. The tools used for the implementation are: Eclipse kepler version 4.3, JDK 1.8, AWS SDK for Eclipse. We have implemented our work over AWS cloud services. Amazon Web Services (AWS), is one of the most popular cloud computing platform owned by Amazon. AWS offers different cloud computing solution that can be operated from 12 different geographical locations across the world. The well-known cloud services provided by Amazon are Amazon Simple Storage Service, also known as "S3" and Amazon Elastic Compute Cloud, also known as "EC2". AWS provides more than 70 cloud services such as storage, computing, networking, database, developer tools for Internet of Things mobile development tools, application services, etc. We have made the use of Amazon Simple Storage Service, also known as "S3" services of AWS.

Fig. 6 Comparison of reliability gained using LNC and replication

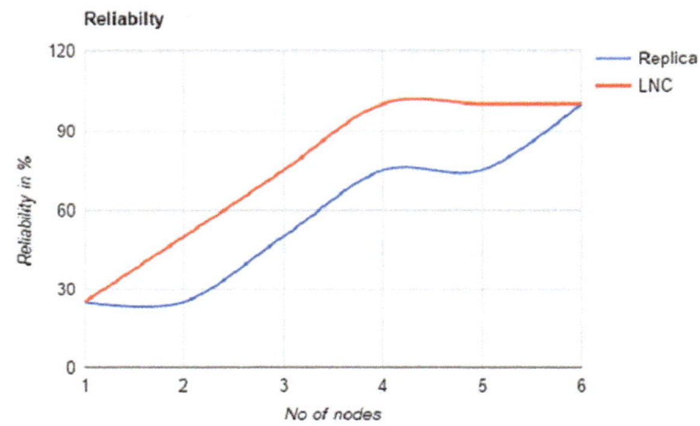

The steps are shown below how we can use S3 services

- Download AWS S3 SDK.
- Configure it in Eclipse EE. We have used Eclipse Kepler version 4.3.
- Downloading S3 API for Java.
- Creating Bucket across any region of the AWS Server.
- Applying the Proposed algorithm over the file.
- Adding the Metadata to the file contains the partially decrypted key.
- Upload the file along the Metadata over S3.

5.1 Comparison of Proposed LNC with Replication Approach

A comparative illustration has been depicted in Fig. 6 by considering number of nodes from 1 to 6. The level of reliability is increased with increased number of nodes.

The above results represent that by using Linear Network encoding (LNC) approach, we could always recover more amount of data compared to the nodes recovered using traditional replication approach.

6 Conclusion and Future Work

In this paper, we proposed hybrid approach using linear network coding and re-encryption based on ElGamal cryptography to secure healthcare information over the cloud. To provide security and fault tolerance for cloud storage, linear network coding is used. To exchange the encoding key matrix securely with the receiver, we have used ElGamal re-encryption scheme. We have presented how securely the data can be transferred between sender and the receiver. We also compared our coding scheme with the traditional replication scheme for achieving reliability. In this work, We have considered text data only.

As a future plan, this work could be extended for audio and video data over the cloud using the concept of P-Frame, B-Frame, and I-Frame. We could also work upon reducing the complexity of the operation carried out for achieving security and reliability of data.

References

1. Rathi G., Abinaya M., Deepika. M., Kavyasri. T.: Healthcare Data Security in Cloud Computing, IJIRCCE (2015).
2. Ahlswede, R., Cai, N., Li, S. Y., & Yeung, R. W.: Network information flow. Vol. 46 No. 4 IEEE Transactions on information theory (2000).
3. Garg, P., & Sharma, V.: An efficient and secure data storage in Mobile Cloud Computing through RSA and Hash function. In Issues and Challenges in Intelligent Computing Techniques (ICICT) International Conference on IEEE (2014).
4. Rewadkar, D. N., & Ghatage, S. Y.: Cloud storage system enabling secure privacy preserving third party audit. In Control, Instrumentation, Communication and Computational Technologies (ICCICCT), International Conference on IEEE (2014).
5. Khanezaei, N., & Hanapi, Z. M.: A framework based on RSA and AES encryption algorithms for cloud computing services. In Systems, Process and Control (ICSPC), 2014 IEEE Conference on IEEE (2014).
6. Thiranant, N., Sain, M., & Lee, H. J.: A design of security framework for data privacy in e-health system using web service. In Advanced Communication Technology (ICACT), 2014 16th International Conference on IEEE (2014).
7. Mahalle, V. S., & Shahade, A. K.: Enhancing the data security in Cloud by implementing hybrid (Rsa & Aes) encryption algorithm. In Power, Automation and Communication (INPAC), 2014 International Conference on IEEE (2014).
8. Gupta, S. K., Rawat, S., & Kumar, P.: A novel based security architecture of cloud computing. In Reliability, Infocom Technologies and Optimization (ICRITO) (Trends and Future Directions), 2014 3rd International Conference on IEEE (2014).
9. Louk, M., Lim, H.: Homomorphic encryption in mobile multi cloud computing. In Information Networking (ICOIN), 2015 International Conference on IEEE (2015).
10. Huang, K., & Tso, R.: A commutative encryption scheme based on ElGamal encryption. In Information Security and Intelligence Control (ISIC), 2012 International Conference on IEEE (2012).
11. Chen, C. Y., & Tu, J. F.: A novel cloud computing algorithm of security and privacy. Mathematical Problems in Engineering (2013).
12. Govinda, K.: Secure Framework for cloud environment in collaboration with customers (2015).
13. Zhang, Y., Qiu, M., Tsai, C. W., Hassan, M. M., & Alamri, A.: Health-CPS: Healthcare cyber-physical system assisted by cloud and big data (2015).
14. Sipos, M., Fitzek, F. H., Lucani, D. E., & Pedersen, M. V.: Distributed cloud storage using network coding. In Consumer Communications and Networking Conference (CCNC). IEEE (2014).
15. Heide, J., Pedersen, M. V., Fitzek, F. H., & Larsen, T.: Network coding for mobile devices-systematic binary random rateless codes. In Communications Workshops, 2009. ICC Workshops 2009. IEEE International Conference on IEEE (2009).

16. Ho, T., Médard, M., Koetter, R., Karger, D. R., Effros, M., Shi, J., & Leong, B.: A random linear network coding approach to multicast Vol. 52 No. 10. IEEE Transactions on Information Theory (2006).
17. Meier, A. V.: The elgamal cryptosystem (2005).
18. Fitzek, F. H., Toth, T., Szabados, A., Pedersen, M. V., Lucani, D. E., Sipos, M., Medard, M.: Implementation and performance evaluation of distributed cloud storage solutions using random linear network coding. In Communications Workshops (ICC), 2014 IEEE International Conference on IEEE (2014).
19. Hu, Y., Chen, H. C., Lee, P. P., & Tang, Y.: NCCloud: applying network coding for the storage repair in a cloud-of-clouds. In FAST (2012).
20. Fragouli, C., Le Boudec, J. Y., & Widmer, J.: Network coding: an instant primer Vol. 36. No. 1 ACM SIGCOMM Computer Communication Review (2006) 63–68.

Standardization of Intelligent Information of Specific Attack Trends

Ashima Rattan, Navroop Kaur and Shashi Bhushan

Abstract In recent days, cyber-attacks are rising rapidly by using various new techniques. These attacks have huge impact on organizational and an individual security. As many times an attack has been detected but it is too late to recover the damage perform by that attack. To study on previous attacks some organizations like Defense Advanced Research Project Agency (DARPA) provide offline dataset for researchers. KDD and DARPA dataset attributes was playing a good role in detection of many attacks and further useful in prevention of attacks also. But in recent days, dataset provided by them, become old one and not gives fruitful results. To keep in mind, the technique used in this research work is providing machine readable dataset attributes of specific attacks in a standard format which is CSV (Comma Separated Values) format. The attack data is captured by deploying various honeypot sensors. The achievement of this research work is "sharing of targeted attack data like Brute force Attack, Exploits etc., in machine readable form in standard format". This information is useful for security researchers, situational awareness programs and security communities. Security testing is another area, also needs some dataset attributes for security testing of the softwares or tools.

Keywords Attacks · Attack trends · Exploits · Brute force
Scans · CSV format and honeypots

A. Rattan (✉) · S. Bhushan
Department of IT, Chandigarh Engineering College Landran, Mohali, Punjab, India
e-mail: rattan_ashima@yahoo.com

S. Bhushan
e-mail: shashibhushan6@gmail.com

N. Kaur
CSTD Center for Development of Advanced Computing (CDAC), Mohali, Punjab, India
e-mail: navroop_kohli@yahoo.com

C. R. Panigrahi et al. (eds.), *Progress in Advanced Computing and Intelligent Engineering*, Advances in Intelligent Systems and Computing 714,
https://doi.org/10.1007/978-981-13-0224-4_7

1 Introduction

In recent days, cyber-attacks are rising rapidly by using various new techniques. The existing security works with the focus on finding the traditional protection and detection methods [1]. However attacker performs lot of attacks in very short time. These attacks have huge impact on organization and an individual security [2]. The first response to such campaigns is to detect them and collect sufficient information regarding tools, techniques used to exploit the vulnerability [3]. Hence effective capturing of the attack data and its timely dissemination to defenders is required for the mitigation and prevention of the large-scale attacks [4]. In this paper we have established the need for such an automated attack data capturing and sharing mechanism. The cyber threat information sharing is very important and very useful in the field of cyber security, where organization can take protective measures on time by watching the previous attack information. Such type of information is useful for security researchers, security agencies, etc., in a standard structured format which is readily usable\actionable by them [5–7]. We have also highlighted the fact that the format for sharing attack data is very crucial and the data sharing format should be machine digestible to reduce the human intervention and increase the response time [8]. As the threat landscape is ever changing, so as the techniques used for mitigation of those threats needs to be dynamic in nature [9]. In this research we tend to look for the feasibility of using proactive approaches for the mitigation of the dynamic threat to the security. The first level of defense in any deficient security set up is the firewall hosted [10, 11]. This device/software is responsible for catalog all the communication to and from the organization and allows and disallows the IP address based upon their reputation [12]. In this research work, attack data is collected through capturing and the event database is created. It helps to detect the malicious traffic.

2 Research Scope

The main focus of our research work involves in three steps: Situational Awareness, Attack Attribution and Cyber Security Researchers.

2.1 Situational Awareness

As the threat landscape is ever changing, so as the techniques used for mitigation of those attacks needs to be dynamic in nature. Cyber-attacks are increasing day by day and their impact is likely to be very much disturbing and harmful for the users. The term Cyber Situational Awareness refers to monitor all the unusual events and occurrence of bad activities which are specially performed by the attackers or the hackers [13]. The organization that works on Cyber Situational Awareness collects the current attack data, works on the collected attack data to find the refined and correct information about them as a result. Such information is helpful to aware the society about those new attacks and their possibilities by providing the refined data to harm over the cyber security network.

2.2 Attack Attribution

Attack attribution may be defined as in which it helps to provide the information of an attacker as well as attacker's channel [14]. The information includes the identity and the location of an attacker. Traditionally attack attribution is simply a process of trace back of an attacker. The identity of the attacker which is obtained by tracing includes the username, e-mail id, an account, an alias, password, an IP address, or geographic location [15]. The main principle behind the attack attribution technique depends upon the untrusted nature of IP protocol. If the source IP address is not authenticated then it is very easy to trace the location or address.

2.3 Security Researchers

As the threat landscape is ever-changing, so as the techniques used for mitigation needs to be dynamic in nature. Hence we are providing latest dynamic attacked data (i.e. IP reputation) and providing it in a standard structured format which is in a CSV format which is machine digestible. This format can be directly input for machine learning. Therefore, this kind of data provided in standard structured format and is hence useful for researchers in the domain of detection of malicious traffic. Standard sharing format can be used as CSV format and can be directly input into machine learning. Hence, this data is extremely useful for researchers. You will get the latest trends, reputation latest attack trends and IP reputation.

2.4 Security Testing

Security testing may be defined as the software testing which is useful to uncover the vulnerabilities (holes) of the system. Security testing is the technique to secure the data and information so that attackers could not able to attack or steal the important and personal data of the system. To protect and maintain data properly, it is the easiest way. It helps to find out the all means of escape and fault of the system which may results to the loss of information. It finds the way out to protect the mislay information.

3 Design Principles of Event Database

The principal behind event database is to provide readily available attack data in machine digestible form and provide broad prospective to researchers in the field of security. This basically covers specific types of attacks like Brute force attack, Exploits, etc., [16]. This is a big contribution to nation in cyber security by providing specific attack enrich data.

3.1 Network Architecture

In network architecture three machines are used, these are servers, each one is consisting of two ports, i.e., eth0 and eth1. Almost every server has two ports for various communications. The three servers are named as (Fig. 1):

- Broadband relational server.
- Elasticsearch server, i.e., Event DB is deployed on OS Ubuntu.
- Web server having OS Ubuntu—used for portal to access the data.

3.2 Repository Architecture

3.2.1 Honeypot

Honeypot is a system to trace the attacks by fooling the attackers and to get the information about how attackers exploit vulnerabilities in IT system [17–19].

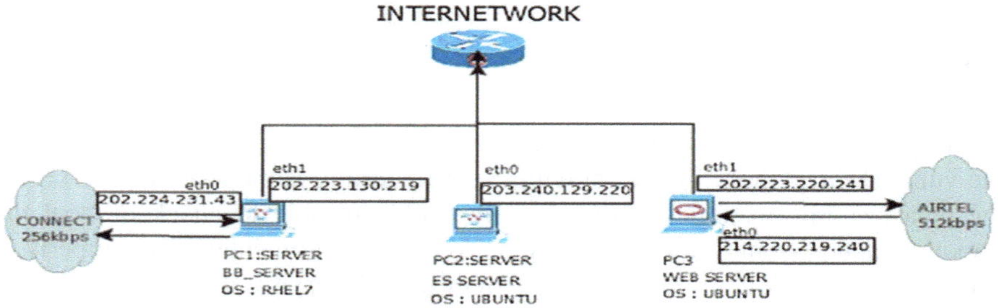

Fig. 1 Network architecture

Honeypot sensors do not let the attacker know about the system legitimacy. Attackers do not know that somebody has kept eye on them and are being monitored secretly.

- **Active Honeypot**—The term active (client) honeypot describes an advanced honeypot system. In contrary to traditional honeypots that undergo passively all attack attempts, active honeypot systems actively react to them.
- **Passive Honeypot**—Passive (server) honeypots offer services and wait for attacker to exploit the vulnerabilities.

3.2.2 Broadband Server

OS: RHEL7 having broadband connection to find the vulnerability in broadband network (Fig. 2).

3.2.3 Relational Database

In this research the relational database is used to store the data which is captured through various honeypots. Relational database is mandatory for establishing a relation between data captured from various honeypots and hence play a big role in specific types of data collection. The collected data further refined to provide the information about attacks such as Specific Attack Trend, example of collected data from various honeypot sensors, data capturing date and time, Attacker IP, Connection established, Services exploited, Malwares Downloads, Malware_virustotal_results, Generation_of_network_traffic, Events_detected.

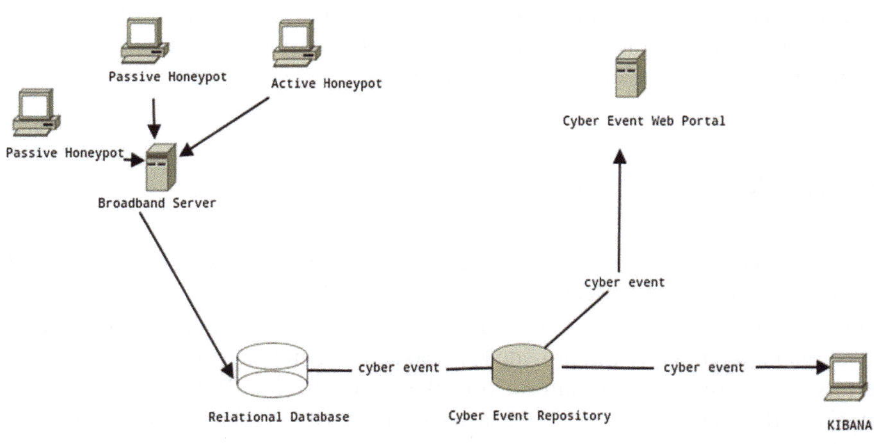

Fig. 2 Architecture of repository

Fig. 3 Kibana output

3.2.4 Cyber Event Repository

Cyber Event Repository is a repository that contains latest information about specific Attack Trends. This has been designed by keeping all the aspects of security where this type of information can be useful.

3.2.5 Elasticsearch, Logstash and Kibana (ELK)

In this research paper, the results are obtained by using the combination of Elasticsearch, Logstash and Kibana (ELK stack). Elasticsearch uses Apache Lucenewhich helps to generate and govern the inverted index. Elasticsearch is an approach to provide the fast responses according to user's search. It works on the real time platform; therefore it is the easiest and fastest approach to obtain the accurate results [20]. Logstash is an open-source tool which logs are collected, parsed, and stored for future use. Kibana is the web-based interface, the logs are indexed through Logstash which is helpful to display the results. Elasticsearch, Logstash and Kibana, when used together are known as an ELK stack. Therefore in this research ELK combination is used for better responses. Few of the kibana commands used to extract SMTP scans (Fig. 3).

4 Attack Data Results

Capturing and sharing specific attack trends is a big challenge which come up with many new problems like selection of standard format for sharing, to find out the most prominent features which attackers left with us, type of attack attacker prefer, what situational awareness we can provide [21]. So keeping all this in mind the attack data results gives various types of information like top 10 attacker IP along with the list of attacking IP's captured by our honeynet sensors, most top attacked port along with the details of various attacking port captures now a days, specific types of attacks captured using honeynet sensors, etc. [22]. Based on type of attack attacker is doing we have categorized and providing details of basically 3 types of attacks, i.e., Exploits, Scans and Brute force, etc. Details of which are mention below.

TOP 10 Attacker IP
See Fig. 4.

4.1 Specific Attack Trend Captured

As already discussed above about to find and segregation of the attack trends is a big challenge, when new attacks are coming day by day. Here shown graphical representation is shown of three specific attacks such as: Exploits, Brute force, and scans (Fig. 5).

4.2 Exploits

An exploit is an attack, when the attacker finds the vulnerability in the system then it takes the advantage to exploit the particular vulnerability. We keep check on

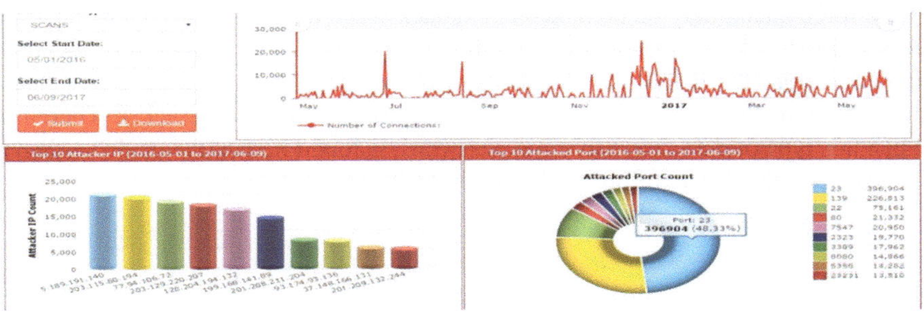

Fig. 4 Top Attacker IP

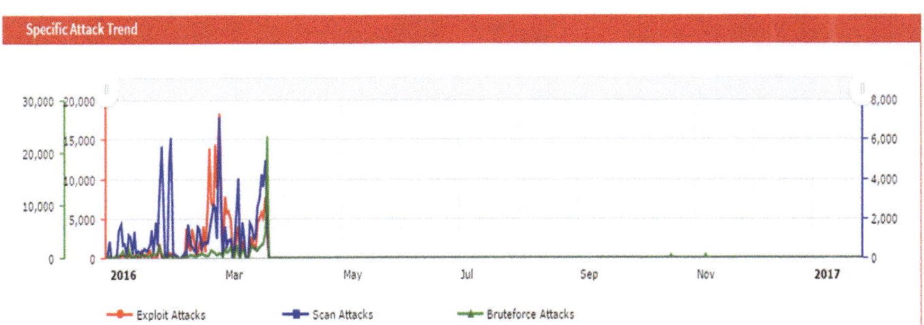

Fig. 5 Graph showing specific attacks

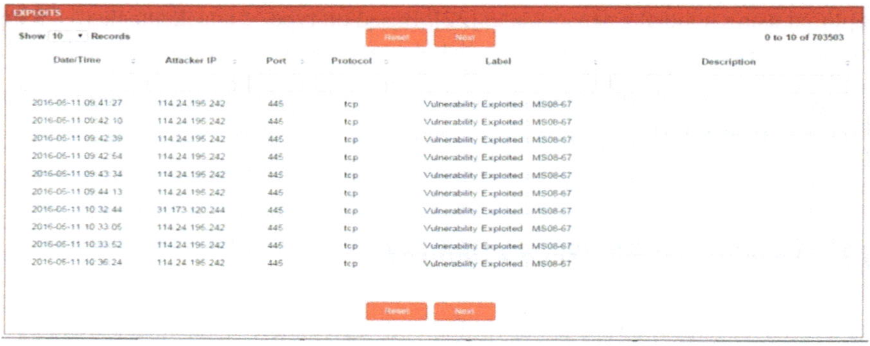

EXPLOITS

Show 10 Records Reset Next 0 to 10 of 703503

Date/Time	Attacker IP	Port	Protocol	Label	Description
2016-06-11 09:41:27	114.24.195.242	445	tcp	Vulnerability Exploited MS08-67	
2016-06-11 09:42:10	114.24.195.242	445	tcp	Vulnerability Exploited MS08-67	
2016-06-11 09:42:39	114.24.195.242	445	tcp	Vulnerability Exploited MS08-67	
2016-06-11 09:42:54	114.24.195.242	445	tcp	Vulnerability Exploited MS08-67	
2016-06-11 09:43:34	114.24.195.242	445	tcp	Vulnerability Exploited MS08-67	
2016-06-11 09:44:13	114.24.195.242	445	tcp	Vulnerability Exploited MS08-67	
2016-06-11 10:32:44	31.173.120.244	445	tcp	Vulnerability Exploited MS08-67	
2016-06-11 10:33:05	114.24.195.242	445	tcp	Vulnerability Exploited MS08-67	
2016-06-11 10:33:52	114.24.195.242	445	tcp	Vulnerability Exploited MS08-67	
2016-06-11 10:36:24	114.24.195.242	445	tcp	Vulnerability Exploited MS08-67	

Reset Next

Fig. 6 Exploit attacks

types of vulnerability exploited with the aim of finding the exploits. And at the end we are showing the count of exploits we have captured on daily and providing the record accordingly (Fig. 6).

4.3 *Brute Force Attacks*

Brute force attack may be defined as the attack when an attacker wants to steal the user's password or personal identification by attacking. When any type of authentication like if the user is asked for username or password on any website then user must be aware of it, that he is going to be a target of attacker. Therefore when this type of data (username and password) is found here in the payload, then it is clear that attacker is trying to do the brute force attack (Fig. 7).

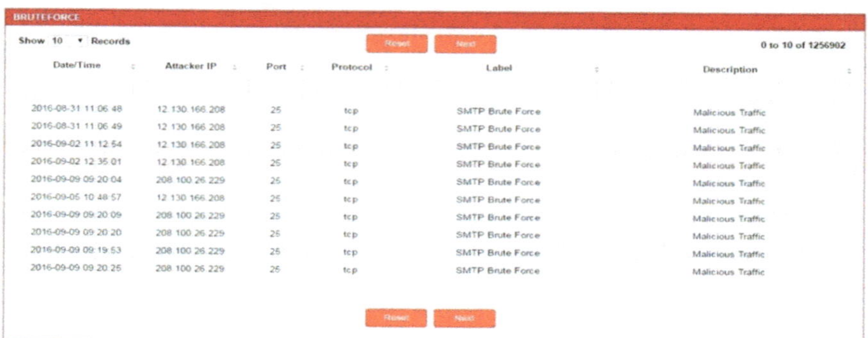

BRUTEFORCE

Show 10 Records Reset Next 0 to 10 of 1256902

Date/Time	Attacker IP	Port	Protocol	Label	Description
2016-08-31 11:06:48	12.130.166.208	25	tcp	SMTP Brute Force	Malicious Traffic
2016-08-31 11:06:49	12.130.166.208	25	tcp	SMTP Brute Force	Malicious Traffic
2016-09-02 11:12:54	12.130.166.208	25	tcp	SMTP Brute Force	Malicious Traffic
2016-09-02 12:35:01	12.130.166.208	25	tcp	SMTP Brute Force	Malicious Traffic
2016-09-09 09:20:04	208.100.26.229	25	tcp	SMTP Brute Force	Malicious Traffic
2016-09-05 10:48:57	12.130.166.208	25	tcp	SMTP Brute Force	Malicious Traffic
2016-09-09 09:20:09	208.100.26.229	25	tcp	SMTP Brute Force	Malicious Traffic
2016-09-09 09:20:20	208.100.26.229	25	tcp	SMTP Brute Force	Malicious Traffic
2016-09-09 09:19:53	208.100.26.229	25	tcp	SMTP Brute Force	Malicious Traffic
2016-09-09 09:20:25	208.100.26.229	25	tcp	SMTP Brute Force	Malicious Traffic

Reset Next

Fig. 7 Brute force attacks

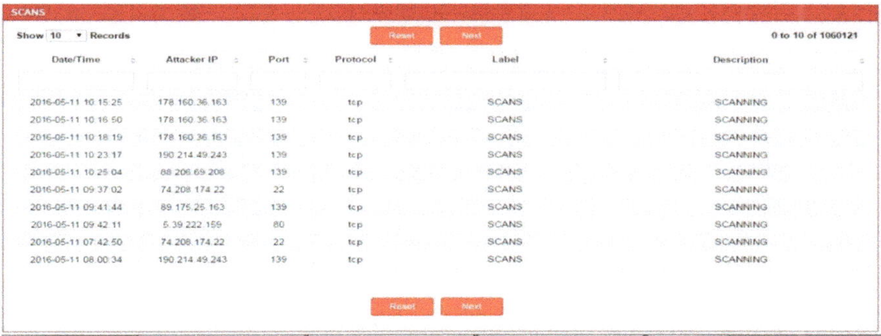
SCANS

Show 10 Records   Reset   Next   0 to 10 of 1060121

Date/Time	Attacker IP	Port	Protocol	Label	Description
2016-05-11 10:15:25	178.160.36.163	139	tcp	SCANS	SCANNING
2016-05-11 10:16:50	178.160.36.163	139	tcp	SCANS	SCANNING
2016-05-11 10:18:19	178.160.36.163	139	tcp	SCANS	SCANNING
2016-05-11 10:23:17	190.214.49.243	139	tcp	SCANS	SCANNING
2016-05-11 10:25:04	88.206.69.208	139	tcp	SCANS	SCANNING
2016-05-11 09:37:02	74.208.174.22	22	tcp	SCANS	SCANNING
2016-05-11 09:41:44	89.175.25.163	139	tcp	SCANS	SCANNING
2016-05-11 09:42:11	5.39.222.159	80	tcp	SCANS	SCANNING
2016-05-11 07:42:50	74.208.174.22	22	tcp	SCANS	SCANNING
2016-05-11 08:00:34	190.214.49.243	139	tcp	SCANS	SCANNING

Reset   Next

Fig. 8 Scan attacks

4.4 Scans

Scans identify the hosts who are active on the network for network security assessment. It is a way to recognize the running network services on the targeted hosts. The network services include User Datagram Protocol (UDP), Transmission Control Protocol (TCP), Operating System (OSs), and TCP sequence number predictability, etc. It is a method to tighten the system security and also an effective way to troubleshooting the system. It is a technique which is used to detect the known vulnerabilities computing system that are available on the network. So based upon this, we are trying to find the attacks which are executed through scanning (Fig. 8).

5 Standard Format for Researchers

Attack data is required to detect unauthorized activities to positively identify all true attacks and negatively identify all non-attacks, monitoring and analyzing user and system activities, to recognize known specific types of attacks and alerts, for statistical analysis of abnormal behavior model. In this research work we are providing the standard sharing format for researchers. They require this type of data for better response. The standard structured sharing format is provided in this research work which can be used as CSV format and can be directly input into the machine learning [23]. Hence the information provided in this format is extremely refined, qualitative, and useful in manners of latest attack trends for IP reputation.

One of the formats is CSV format (Comma Separated Values) is a file format for data storage which looks like a text file [24–26]. The information is organized with one record on each line and each field is separated by comma. CSV is human readable and easy to edit manually, simple to implement and parse, provides straight forward information schema (Figs. 9 and 10).

	A	B	C	D	E	F	G	H	I
1	date_time	attacker_ip	protocol	source_port	destination_port	label	others	description	payload
2	5/3/2016 14:28	222.186.21.57	tcp	6786	1433	Malicious MSSQL TRAFFIC		MSSQL Bruteforce	
3	5/3/2016 14:28	222.186.21.57	tcp	10254	1433	Malicious MSSQL TRAFFIC		MSSQL Bruteforce	
4	5/3/2016 14:41	222.186.3.52	tcp	5773	1433	Malicious MSSQL TRAFFIC		MSSQL Bruteforce	
5	5/3/2016 14:41	222.186.3.52	tcp	8520	1433	Malicious MSSQL TRAFFIC		MSSQL Bruteforce	
6	5/3/2016 15:45	124.193.177.29	tcp	6224	1433	Malicious MSSQL TRAFFIC		MSSQL Bruteforce	
7	5/3/2016 15:45	124.193.177.29	tcp	7358	1433	Malicious MSSQL TRAFFIC		MSSQL Bruteforce	
8	5/3/2016 15:54	222.186.56.21	tcp	16900	1433	Malicious MSSQL TRAFFIC		MSSQL Bruteforce	
9	5/3/2016 15:14	123.249.34.132	tcp	2194	3306	Malicious MySQL TRAFFIC		MySQL Bruteforce	
10	5/3/2016 15:33	123.249.45.166	tcp	3532	3306	Malicious MySQL TRAFFIC		MySQL Bruteforce	
11	5/3/2016 15:41	221.194.44.173	tcp	2302	3306	Malicious MySQL TRAFFIC		MySQL Bruteforce	
12	5/3/2016 15:59	173.254.236.104	tcp	1289	3306	Malicious MySQL TRAFFIC		MySQL Bruteforce	
13	5/3/2016 16:26	173.254.236.104	tcp	1567	3306	Malicious MySQL TRAFFIC		MySQL Bruteforce	
14	5/3/2016 16:58	120.24.177.101	tcp	50481	3306	Malicious MySQL TRAFFIC		MySQL Bruteforce	
15	5/3/2016 17:21	222.186.34.204	tcp	1353	3306	Malicious MySQL TRAFFIC		MySQL Bruteforce	
16	5/3/2016 17:54	23.88.177.135	tcp	4295	3306	Malicious MySQL TRAFFIC		MySQL Bruteforce	
17	5/3/2016 15:04	118.193.213.172	tcp	1921	3306	Malicious MySQL TRAFFIC		MySQL Bruteforce	
18	5/3/2016 15:05	173.254.236.104	tcp	4057	3306	Malicious MySQL TRAFFIC		MySQL Bruteforce	

Fig. 9 Machine digestible data in CSV format

811	5/3/2016 17:13	27.54.248.124	tcp	5081	1433	alicious MSSQL TRAFF		MSSQL Bruteforce	
812	5/3/2016 16:29	46.172.71.249	SSH	null	22	SSH BruteForce	user_name	Malicious SSH Traffic	cd ..
813	5/3/2016 15:52	183.3.202.88	SSH	null	22	SSH BruteForce	user_name	Malicious SSH Traffic	wget http://104.223.72.179:258/hlvip
814	5/3/2016 14:29	74.208.174.22	SSH	null	22	SSH BruteForce	user_name	Malicious SSH Traffic	wget http://104.223.72.179:258/hlvip
815	5/3/2016 14:29	74.208.174.22	SSH	null	22	SSH BruteForce	user_name	Malicious SSH Traffic	chmod 0777 hlvip
816	5/3/2016 15:16	183.3.202.88	SSH	null	22	SSH BruteForce	null	Malicious SSH Traffic	

Fig. 10 Result showing extracted payloads

6 Conclusion

Cyber security is a broad area and everything cannot be secured at the same time, when lots of attacks are propagating day by day with different intention of attacks [27]. In this research work the used technique is providing the specific attack trends in a machine digestible form, i.e., CSV (Comma Separated Value). For researchers, finding attack detection methodology and traditional prevention, "intelligent dataset attributes can be very useful". But nowadays, the dataset attributes provided by other standard organizations is not so fruitful as there is lack of research environment, privacy issues, and specific types of attack data available in industry or any other reason [28]. Taking this as a problem, our work starts from capturing the attacks from various vulnerable honeypot sensors deployed, refine the data using various technologies such as Snort, Wireshark, Sandbox, etc., along with our knowledge, further making the repository (in Elasticsearch, Logstash and Kibana) of attack data and at last result come up with dataset attributes of specific attacks in a standard format [29, 30]. In future work the researchers can expand the capturing strength to capture more of Ransomware and analysis strength to analyze more of Ransomware and also provide data to the researchers and academician's, with a challenge to clean the nation from Ransomware.

References

1. Masato Terada: Work on Cyber Security Measures for Collaboration between Organizations: Vol. 65, No. 1 in 2016.
2. Ashima Rattan, Navroop Kaur, Saurabh Chamotra and Shashi Bhusan: Attack Data Usability and Challenges in its Capturing and Sharing In the 3rd International Conference on Cyber Security (ICCS-2017) at Rajasthan Technical University Kota (Rajasthan), Published in "International Journal of Advanced Studies in Computer Science and Engineering" (IJASCSE): Vol-6-theme-based-issue-9.
3. http://www.icasi.org/cvrf/.
4. Vijay Varadharajan: On Malware Characterization and Attack Classification: Proceedings of the First Australasian Web Conference (AWC '13), Vol. 144, 43–47 in 2013.
5. Ashima Rattan and Shashi Bushan: IP Reputation Engine Based upon Malicious Events In the proceedings of the 11th INDIACom 2017 in the IEEE 4th International conference on "Computing for Sustainable Global Development", March 2017.
6. Sean Barnum: Standardizing Cyber Threat Intelligence Information with the Structured Threat Information expression (STIX™): Version-1.1, Revision-1 in Feb 20, 2014.
7. Panos Kampanakis: Security automation and threat information-sharing options: co-publish by the IEEE computer and reliability societies: Vol. 12, Issue-5, 42–51 in September/October 2014.
8. Kutub Thakur Meikang Qiu Keke Gai and Md Liakat Ali: An Investigation on Cyber Security Threats and Security Models in the IEEE 2nd International Conference on Cyber Security and Cloud Computing: 978-1-4673-9300-3/15, pp. 307–311, 2015.
9. Komal K. More and Prof. Pramod B. Gosavi: A Real Time System for Denial of service Attack Detection Based on Multivariate Correlation Analysis Approach in IEEE International Conference on Electrical, Electronics, and Optimization Techniques (ICEEOT): 978-1-4673-9939-5/16/, pp. 1125–1131, 2016.
10. Saoreen Rahman, Muhammad Ahmed and M. Shamim Kaiser: ANFIS Based Cyber Physical Attack Detection System in IEEE 5th International Conference on Informatics, Electronics and Vision (ICIEV): 978-1-5090-1269-5/16/, pp. 944–948, 2016.
11. D. Moore, V. Paxson, S. Savage, C. Shannon, S. Staniford, and N. Weaver: Inside the slammer worm: In Proceedings of IEEE Security and Privacy: Vol. 1, Issue: 4, 33–39 in June 2003.
12. Dhanashri Ashok Bhosale and Vanita Manikrao Mane: Comparative Study and Analysis of Network Intrusion Detection Tools: International Conference on Applied and Theoretical Computing and Communication Technology (ICATCCT), 312–315, in 2015.
13. Ulrik Franke, Joel Brynielsson: Cyber situational awareness A systematic review of the literature, Computers & Security, Volume 46, Pages 18–31 in October 2014.
14. Guodong Zhao, Ke Xu, Lei Xu, and Bo Wu; "Detecting APT Malware Infections Based on Malicious DNS and Traffic Analysis", IEEE 20 July 2015, pp. 1132–1142.
15. Jessica Steinberger, Anna Sperottoz, Mario Gollingy and Harald Baier: How to Exchange Security Events? Overview and Evaluation of Formats and Protocols in Biometrics and Internet Security: IEEE International Symposium on Integrated Network Management (IM2015), Darmstadt, Germany 2015.
16. M. Dacier, F. Pouget, and H. Debar: Attack processes found on the internet: NATO Research and technology symposium IST-041 "Adaptive Defence in Unclassified Networks", 19 April 2004, Toulouse, France.
17. Honeynet.org.
18. Dikshant Gupta, Suhani Singhal, Shamita Malik and Archana Singh: Network Intrusion Detection System Using various data mining techniques in IEEE International Conference on Research Advances in Integrated Navigation Systems (RAINS - 2016), April 06–07, 2016, R. L. Jalappa Institute of Technology, Doddaballapur, Bangalore, India: 978-1-4673-8819-8/16/, 2016.

19. Logrhythm Labs Embedded Expertise on Security Analysis Suite-Honeypot.
20. Sanjeev Kumar, Rakesh Sehgal and J.S. Bhatia: Hybrid Honeypot Framework for Malware Collection and Analysis in IEEE 7th International Conference on Industrial and Information Systems (ICIIS-2012), August 6–9, 2012, IIT Chennai, Published in IEEE Xplore.
21. Daniel Ramsbrock: Profiling Attacker Behavior Following SSH Compromises: Department of Computer Science University of Maryland, College Park in 2007.
22. Eric Ziegast, Paul Vixie: Domain Name Service Based block List in 1997.
23. CERT Polska and European Union Agency for Network and Information Security (ENISA) team: Standards and tools for exchange and processing of actionable information in November 2014.
24. Nazmul Shahadat, Imam Hossain, Anisur Rohman and Nawshi Matin: Experimental Analysis of Data Mining Application for Intrusion Detection with Feature reduction in International Conference on Electrical, Computer and Communication Engineering (ECCE), February 16–18, 2017, Cox's Bazar, Bangladesh, pp. 209–216.
25. Zhang, Xiaoming, and Guang Wang. "Hadoop-Based System Design for Website Intrusion Detection and Analysis." 2015 IEEE International Conference on Smart City/SocialCom/SustainCom (SmartCity), IEEE, 2015.
26. El Mostapha Chakir, Mohamed Moughit And Youness Idrissi Khamlichi: An Efficient Method for Evaluating Alerts of Intrusion Detection Systems in the conference of IEEE 978-1-5090-6681-0/17/ in 2017.
27. V. Yegneswaran, P. Barford, and D. Plonka: Design and use of internet sinks for network abuse monitoring: Lecture Notes in Computer Science book series (LNCS), Vol. 3224, Springer, Berlin, Heidelberg 2004.
28. Mahbod Tavallaee, Ebrahim Bagheri, Wei Lu, and Ali A. Ghorbani: A Detailed Analysis of the KDD CUP 99 Data Set in the conference of IEEE in 2009.
29. Mike Schiffman: Cisco Systems on The Common Vulnerability Reporting Framework An Internet Consortium for Advancement of Security on the Internet (ICASI) Whitepaper in 2011.
30. Abdul Razzaq, Ali Hur, H Farooq Ahmad, Muddassar Masood: Cyber Security: Threats, Reasons, Challenges, Methodologies and State of the Art Solutions for Industrial Applications 2013 in the conference of IEEE Eleventh International Symposium on Autonomous Decentralized Systems (ISADS), 1–6. 2013.

Challenges to Cloud PLM Adoption

Shikha Singh and Subhas Chandra Misra

Abstract The present global scenario calls for the collaboration of all stakeholders for the innovative and better product development and its production in manufacturing industry. Hence, product lifecycle management (PLM), is the urgent need of manufacturing organizations, which helps to manage the products throughout its lifecycle by collaborating all the product related data and stakeholders. Large manufacturing firms are facing several issues while implementing on-premise PLM systems such as high investment in IT infrastructure, technology upgradations, and interoperability with various other enterprise systems. Cloud PLM technology is offering many benefits to overcome these issues with low investment. But, still, the large manufacturing organizations are facing several challenges in adoption of cloud PLM. The present study aims at investigating the challenges of cloud PLM adoption in manufacturing firms. The work is supported by a case study in an aircraft manufacturing firm in India. The paper empirically investigates the causal challenges which large manufacturing firms face while adoption of cloud PLM.

Keywords On-premise PLM · Adoption · Cloud PLM · Manufacturing

1 Introduction

Product lifecycle management (PLM) is a business concept [1] which supports the better management of product-related knowledge in the organization. The product-related knowledge can be preserved in a shared repository which remains available to any stakeholder for the innovation and betterment of the product portfolio of the firm. The requirement of the PLM has been increased in last two decades

S. Singh · S. C. Misra (✉)
Department of Industrial and Management Engineering, Indian Institute of Technology (IIT) Kanpur, Kanpur 208016 Uttar Pradesh, India
e-mail: subhasm@iitk.ac.in

S. Singh
e-mail: shikhas@iitk.ac.in

C. R. Panigrahi et al. (eds.), *Progress in Advanced Computing and Intelligent Engineering*, Advances in Intelligent Systems and Computing 714,
https://doi.org/10.1007/978-981-13-0224-4_8

due to the multitudinous product-related data, the involvement of various information systems, and a large number of stakeholders. The institutionalization of PLM concept is supported by IT systems. IT support for PLM systems is of two types: on-premise PLM systems and cloud PLM solutions. On-premise PLM systems were coined first. They were adopted by most large manufacturing firms in the last decade, while the cloud PLM solutions are most recent. The development of cloud PLM solutions has been supported much in past few years because of the severe implementation issues of on-premise PLM systems such as high investment cost of IT infrastructure and its upgradations. Cloud computing is the fundamental technical concept of the cloud PLM development [2]. Cloud services are Internet-based services which utilize the cloud computing technology [2, 3]. Considering the sharing of the service pools, four types of clouds are presently known which are: public clouds, community cloud, private cloud, and hybrid cloud [2, 4–6]. Among various benefits offered by cloud computing solutions, the first is the lower investment costs which include IT infrastructure costs, operating costs, etc. [2, 7–9]. Second is scalability which helps in scaling up and down any time as per the requirement [2, 7, 8]. The third is the ease of access from anywhere at anytime [8, 9].

As the technology is evolving fast, manufacturing firms must also have to adopt the latest Internet-based cloud technology which helps to collaborate and design the innovative products. Hence, an attempt has been made in this paper for identification of the challenges which are alerting large manufacturing firms to adopt cloud PLM systems. Most of the challenges have been expressed by the cloud solution providers and consultants; scarce academic literature talks about the challenges to cloud PLM adoption. Hence, the present work focusses on the empirical investigation of the actual challenges to cloud PLM adoption in large manufacturing firms.

We have identified seven challenges from the available and relevant literature. Based on that, several discussions have been conducted with the industry experts to explore the practical aspects; later the four industry experts were asked to fill the questionnaire survey to assess the impact of the seven specified challenges on one another.

Further, the content of the present work is organized as follows: Section 2 reviews the literature on cloud PLM and its adoption. Section 3 discusses the DEMATEL method utilized to investigate the causal and impacted challenges empirically. Section 4 presents the data collection and analysis followed by the result and discussion in Sect. 5. At last, the findings are concluded in Sect. 6.

2 Literature Review

There are a limited number of academic papers available for cloud PLM adoption. Hence, the white papers and articles by consultant and service provider firms are also considered while identifying the challenges which affect the cloud PLM adoption in large manufacturing firms. Figure 1 expresses the identified challenges in a pictorial form.

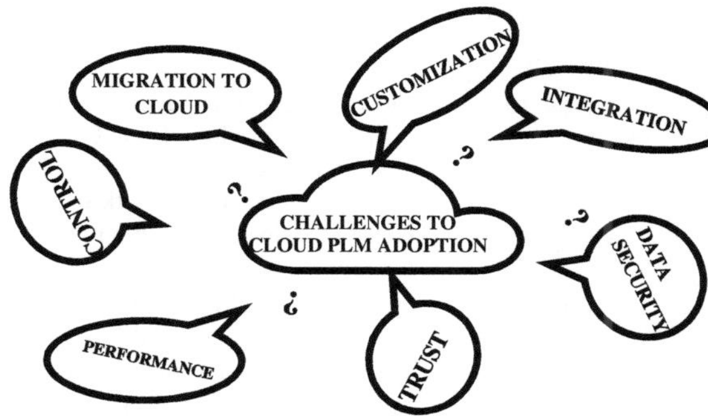

Fig. 1 Pictorial view of challenges to cloud PLM adoption

The literature for challenges to cloud computing has also been reviewed as the cloud computing is the underlying technology of cloud PLM. Oliveria et al. [10] have conducted empirical research to investigate the adoption issues of cloud computing and considered several aspects of adoption of this technology based on the firm's innovational, technological, environmental, and organizational level aspects. CIM data, a consulting firm has investigated the benefits and challenges of cloud PLM in manufacturing industry [11] and have concluded the integration of cloud PLM with existing enterprise systems, security issues, customization, and performance as the top four challenging issues to cloud PLM adoption in manufacturing firms. Noor et al. [12] have defined a set of trust characteristics which are expected to be fulfilled by the cloud service provider such as authorization control, data security, and privacy concerns, etc. The identified challenges have been reviewed from the literature and listed in Table 1 with their description.

3 Research Method: DEMATEL Technique

The decision-making trial and evaluation laboratory (DEMATEL) technique is a multicriteria decision-making technique (MCDM) which helps to investigate the interdependence among the criteria of a problem through a visual diagram, which reflects the prominence and relationship visually among all the criteria [15, 16]. DEMATEL technique has been adopted by various researchers to investigate a different kind of multicriteria problems in multiple areas. By using this method, key success factors to improve hospital services were identified by Shieh et al. [16], and critical factors to success of total quality management (TQM) were determined by Jamali et al. [17] in the Iranian context. By using the same method, an empirical study was performed by Lin et al. [9] to evaluate the core competencies and interrelationship of an IC design company. Wu and Tsai [15] have also utilized DEMATEL method to investigate the causal relations among the criteria in auto spare parts industry. In all these studies, the elements have been segregated into cause and impacted group, whereby it has been easy to take decisions on critical

Table 1 List of challenges to cloud PLM adoption in manufacturing firm

	Challenges	Description	References
A1	Migration to cloud	Mostly large firms have established on-premise PLM infrastructure, which leads to hybrid migration to the cloud which demands an overall change management in the organization to work on the hybrid model. Hence, the decision 'to migrate on the cloud or not' is a big challenge	[5, 7, 13]
A2	Integration and interoperability	Integration of cloud PLM to other enterprise systems may be a soft point, and integration of all applications are required for the success of PLM concept	[1, 4, 5, 10, 11, 13]
A3	Data security	The intellectual capital of the manufacturing firm, i.e., the product-related data must be protected from unauthorized sources	[1, 4–6, 11, 13, 14]
A4	Trust on solution provider	The selection of cloud PLM service provider is a challenge as the solution provider must be loyal to the data security and software upgradations timely	[5, 6, 11, 12]
A5	Authorization control	There is no control to make your data offline or limit the data access for certain periods as the service provider's policies may not be known and cannot be controlled	[4–6]
A6	Performance	The performance of all PLM systems' components such as CAD, CAM, ERP, etc. will be dependent on the internet speed. It must meet at least the on-premise PLM system performance	[4, 10, 11]
A7	Customization	The customization of vendor facilitated PLM systems as per the company's culture and the process will not be in control. The available solutions packages need to be adopted which will be a big challenge to institutionalize the PLM way in a manufacturing firm	[1, 11]

factors from the group of all identified criteria. Considering the capabilities of this technique in finding out the causal relationships among the criteria, the DEMATEL technique has been adopted here to investigate the causal relationship among the challenges to cloud PLM adoption in large manufacturing firms.

Following is the summary of the steps to DEMATEL method utilized by above researchers [9, 15–17]:

Step 1: Obtaining average relationship matrix

The impact of criterion is assessed on all the remaining criteria. Based on the initial relation matrix evaluated by an expert, the average relationship matrix is to be obtained by taking the average of all the inputs. The average relationship matrix is represented by Eq. (1).

$$\mathrm{A} = \left[a_{ij}\right]_{mXm}, \tag{1}$$

where, a_{ij} is the average element of all the corresponding elements in the respective direct relation matrix and 'm' represents the number of criteria.

Step 2: Establishing the normalized direct relation matrix

The normalized matrix is calculated by dividing each element of average matrix by the row sum or column sum whichever is maximum. Accordingly, the normalization factor is represented by Eq. (2)

$$\mathrm{u} = \min\left[\frac{1}{\max_{1\leq i\leq m}\sum_{j=1}^{m} a_{ij}}, \frac{1}{\max_{1\leq j\leq m}\sum_{i=1}^{m} a_{ij}}\right] \quad (2)$$

The normalized direct relation matrix

$$N_d = \mathrm{u} * \mathrm{A} \quad (3)$$

Step 3: Setting up total relation matrix

Total relation matrix is the summation of direct and indirect impacts of all the criteria. Considering all the relational impacts, the total relation matrix has been set up as in Eq. (4).

$$T = \lim_{r\to\infty}\left(N_d + N_d^2 + N_d^3 + \cdots + N_d^r\right) = N_d(I - N_d)^{-1}, \quad (4)$$

where I is the identity matrix.

Step 4: Calculating threshold value and Visual prominence-relation graph

The total relation matrix represents aggregate of all possible relations; those all may not be equally important. Hence, threshold value helps to filter out the most important relations which are more important. The threshold value is considered here as the average value of all the elements in the total relation matrix. This is represented in Eq. (5).

$$\alpha = \frac{\sum_{i=1}^{m}\sum_{j=1}^{m}\left[t_{ij}\right]}{m^2}, \quad (5)$$

where m^2 denotes the total number of elements in matrix T.

For visual representation of graphs, the graph is to be drawn between prominence $(S_r + S_c)$ and relation $(S_r - S_c)$ values [9]. Here, we calculate 'S_c' (the sum of the values of each column) and 'S_r' (the sum of each row in the total relation matrix). The criteria having positive value of relation, i.e., $(S_r - S_c)$ are the casual criteria, while the criteria which have negative values of relation are the impacted criteria.

4 Data Collection and Analysis

As described above, the investigations have been done in a large aircraft manufacturing firm, Transport Aircraft Division (TAD, Kanpur) of HAL as the case company. The firm has already implemented the on-premise PLM systems; we intend to investigate the adoption of the cloud PLM technology in future. On the basis of the literature survey, a list of challenges to cloud PLM adoption was first prepared, and then the practical existence of the same has been explored and discussed. A panel of four industry experts has given their inputs to the challenges in adoption of cloud PLM. Each of the four experts evaluated the impact of each challenge criterion on the remaining ones and rated on the scale of 0–4, i.e., no impact, very low impact, low impact, high impact, and very high impact respectively [9].

Based on the direct relation matrices obtained from the respondents (four industry experts), the average matrix has been calculated following Eq. (1) and represented in Table 2.

The normalization factor has been obtained following Eq. (2) and found as u = 0.0556. Hence, the normalization matrix is resulted as per the Eq. (3) and represented in Table 3.

Following the Eq. (4), all direct and indirect relations are aggregated to find the total relation matrix and shown in Table 4.

Table 2 Average direct relation matrix

	A1	A2	A3	A4	A5	A6	A7
A1	0	4	3.75	2.25	2.25	2.5	3.25
A2	2.75	0	2	1.75	2.5	3.25	3
A3	3	1.5	0	4	3	1.25	1
A4	3.5	1.75	3.75	0	0.25	0.5	0.75
A5	2.25	1.5	3.25	3	0	1	1.25
A6	2.75	1.5	0.25	2.25	0.25	0	2.75
A7	3.5	3	0.75	3	2.25	1.75	0

Table 3 Normalized relation matrix

	A1	A2	A3	A4	A5	A6	A7
A1	0.0000	0.2222	0.2083	0.1250	0.1250	0.1389	0.1806
A2	0.1528	0.0000	0.1111	0.0972	0.1389	0.1806	0.1667
A3	0.1667	0.0833	0.0000	0.2222	0.1667	0.0694	0.0556
A4	0.1944	0.0972	0.2083	0.0000	0.0139	0.0278	0.0417
A5	0.1250	0.0833	0.1806	0.1667	0.0000	0.0556	0.0694
A6	0.1528	0.0833	0.0139	0.1250	0.0139	0.0000	0.1528
A7	0.1944	0.1667	0.0417	0.1667	0.1250	0.0972	0.0000

Table 4 Total relation matrix

	A1	A2	A3	A4	A5	A6	A7
A1	**0.5686**	**0.6362**	**0.6454**	**0.6332**	**0.4771**	**0.4758**	**0.5573**
A2	**0.6173**	0.3862	**0.4978**	**0.5330**	0.4279	**0.4570**	**0.4919**
A3	**0.5843**	0.4254	0.3882	**0.5938**	0.4210	0.3290	0.3578
A4	**0.5332**	0.3839	**0.4966**	0.3407	0.2627	0.2590	0.3015
A5	**0.5117**	0.3914	**0.5026**	**0.5162**	0.2557	0.2937	0.3392
A6	**0.4693**	0.3514	0.2957	0.4104	0.2264	0.2098	0.3774
A7	**0.6307**	**0.5196**	**0.4406**	**0.5646**	0.4044	0.3799	0.3346

Table 5 Prominence-relation values

	Criteria	S_r	S_c	$S_r + S_c$	$S_r - S_c$
A1	Migration to cloud	3.9936	3.9151	7.9086	0.0785
A2	Integration and interoperability	3.4110	3.0942	6.5052	0.3169
A3	Data security	3.0994	3.2669	6.3664	−0.1675
A4	Trust on solution provider	2.5777	3.5920	6.1697	−1.0143
A5	Authorization control	2.8107	2.4752	5.2858	0.3355
A6	Performance	2.3406	2.4043	4.7449	−0.0637
A7	Customization	3.2743	2.7596	6.0340	0.5147

The threshold value is calculated following Eq. (5) and resulted as 0.4389, the relation values which are above the threshold value are marked in Table 4, and those relations are shown in the causal map. Further, the row sum (S_r) and column sum (S_c) of Table 4 have been calculated to find out the prominence and relation values and shown in Table 5.

Based on the $(S_r + S_c)$ and $(S_r - S_c)$ values, the prominence-relation map has been plotted and shown in Fig. 2.

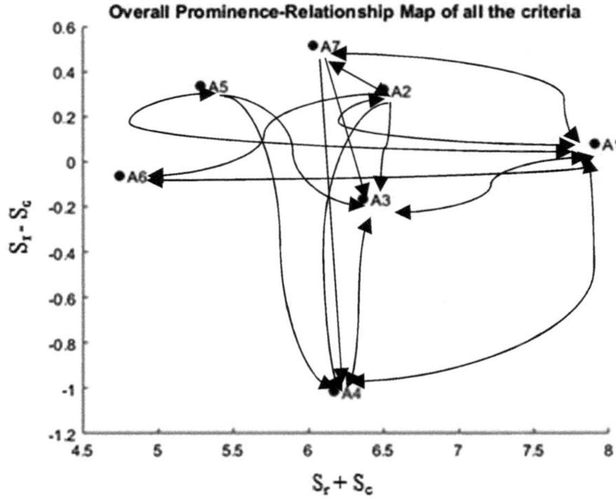

Fig. 2 Prominence-relation graph

5 Results and Discussion

The analysis resulted A1, A2, A5, and A7 as the causal challenges, viz. migration to cloud, integration, authorization control, and customization of cloud PLM services respectively. These causal challenges affect the occurrence of other challenges which are A3, A4, and A6, viz. data security, trust on solution provider, and performance respectively. The impacted group of the challenge criteria represents the affected challenges which does not impact others much. It is indicated that the data security depends on the type of integration among all the enterprise systems, cloud systems, and on the migration methodology to cloud. The data migration from on-premise PLM systems to cloud PLM systems is a challenging task which demands overall change management in the organization. The data migrations from one IT (Information Technology) platform to other IT platform may distort the data formats and may demand various revisions and reworks. Moreover, the uncontrolled authorization also impacts the data security.

Further, the performance (A6) is a challenge which depends on the Internet technology, its speed, and storage capacities. But considering the interdependence of the criteria, performance of cloud PLM systems may be affected by the improper migration of data to another platform, poor integration of systems, bad authorization controls, and improper customization issues viz. misalignment of the technology with process and company's culture. At last, the trust on solution provider (A4) is also affected by the various criteria such as improper migration, poor integration methodology, weak authorization and access control, and improper customization as per the organization's need. All these criteria A1, A2, A5, and A7, are the process and technology-based activities which the solution provider promises to provide while adopting the cloud PLM systems, according to the business contracts. Hence, not meeting the expected level of even one of the requirement is enough to affect the trust on the solution provider services.

The data analysis indicates that the very crucial challenges which are to be taken care most while making the decision to migrate to the cloud technology are as follows: the integration/collaboration of all previous and present information systems, interoperability, authorization control, and the customization of the cloud systems as per the organizations' culture and processes. These challenges are the causal ones and impact the remaining factors more, while the impacted challenges viz. data security, trust on solution provider, and performance have negligible impact on other challenges.

6 Conclusion and Future Scope

Cloud computing is the latest IT technology which offers the collaboration of scattered information systems and all stakeholders through the Internet. While cloud PLM provides enormous cost savings and efforts of the establishment of in-house IT infrastructure for on-premise PLM but still it is a challenge to adopt cloud PLM in case of large manufacturing firms. Rapidly, increasing rate of product-related data offers various challenges related to its required format, access control, etc. The present study is a guide to the cloud PLM solution providers who may work upon to improve the technology in such a manner that it may become accessible and adaptable to large manufacturing firms also. The results may help to prepare a checklist to managers in large manufacturing firms who are looking forward to adopt the latest technology of cloud PLM to be more competitive in the forthcoming digitized era.

Further investigations can be done with more number of organizations. The data analysis can be done using various other statistical decision-making techniques. The subjectivity of the respondents and their input can be considered by using the fuzzy or grey number theory. However, the present work explores and establishes the important step in academic literature in the area of cloud PLM adoption in the large aircraft manufacturing firm.

References

1. Saaksvuori, A., Immonen, A.: Product Lifecycle Management, Springer, Berlin, Germany, 2nd edition (2008)
2. Zhang, Q., Cheng, L., & Boutaba, R.: Cloud computing: state-of-the-art and research challenges. Journal of internet services and applications. 1(1) (2010) 7–18
3. NIST Definition of Cloud Computing, csrc.nist.gov/groups/SNS/cloud-computing/cloud-def-v15.doc
4. HP, white paper: PLM and cloud computing. (2010) retrieved from: www.plmworld.org/d/do/1576
5. Avram, M. G.: Advantages and challenges of adopting cloud computing from an enterprise perspective. Procedia Technology. 12 (2014) 529–534
6. Fernandes, D. A., Soares, L. F., Gomes, J. V., Freire, M. M., & Inácio, P. R.: Security issues in cloud environments: a survey. International Journal of Information Security. 13(2) (2014) 113–170
7. Brown, J.: Tech-Clarity e-book: Cloud Considerations for the PLM ISV. Retrieved from: https://www.nuodb.com/blog/product-lifecycle-management-and-cloud. (2017)
8. Ristova, E., Gecevska, V., Kuzinovski, M., & Mirakovski, D.: Cloud computing as business perspectives for product lifecycle management systems (2013)
9. Lin, Y. T., Yang, Y. H., Kang, J. S., & Yu, H. C.: Using DEMATEL method to explore the core competences and causal effect of the IC design service Company: An empirical case study. Expert Systems with Applications. 38(5) (2011) 6262–6268
10. Oliveira, T., Thomas, M., Espadanal, M.: Assessing the determinants of cloud computing adoption: An analysis of the manufacturing and services sectors. Information & Management. 51(5) (2014) 497–510
11. Making the Connection: The Path to Cloud PLM, CIM data e-book. PTC. (2017)
12. Noor, T. H., Sheng, Q. Z., Zeadally, S., & Yu, J.: Trust management of services in cloud environments: Obstacles and solutions. ACM Computing Surveys (CSUR). 46(1) (2013) 12
13. Cardoso, A., & Simões, P.: Cloud computing: concepts, technologies, and challenges. Virtual and Networked Organizations, Emergent Technologies and Tools (2012) 127–136
14. Lin, A., Chen, N. C.: Cloud computing as an innovation: Percepetion, attitude, and adoption. International Journal of Information Management. 32(6) (2012) 533–540

15. Wu, H. H., & Tsai, Y. N.: A DEMATEL method to evaluate the causal relations among the criteria in auto spare parts industry. Applied Mathematics and Computation. 218(5) (2011) 2334–2342
16. Shieh, J. I., Wu, H. H., & Huang, K. K.: A DEMATEL method in identifying key success factors of hospital service quality. Knowledge-Based Systems. 23(3) (2010) 277–282
17. Jamali, G., Ebrahimi, M., Abbaszadeh, M.A.: TQM implementation: an investigation of critical success factors. In International Conference on Education and Management Technology (ICEMT)-IEEE. (2010) 112–116

Millimeter Wave (MMW) Communications for Fifth Generation (5G) Mobile Networks

Umar Farooq and Ghulam Mohammad Rather

Abstract Millimeter wave (MMW) communication is envisioned to satisfy the need of high data rate wireless links for next generation 5G networks in addition of addressing the spectrum scarcity and capacity limitations of current wireless systems and enable a plethora of applications in near future. However the field of MMW communication is still in its infancy because of the various technical challenges faced in its practical implementation due to the fundamentally different propagation characteristics in this frequency band. The treatment of these challenges has initiated a lot of research activities. In this paper the MMW propagation characteristics have been studied in the context of free space loss, atmospheric loss and foliage loss and the effect of these losses on the performance of the MMW communication channel has been examined through simulation studies. Further in this paper the performance evaluation of the MMW communication link for different data rates for various Line of Sight (LOS) and Non-Line of Sight (NLOS) cases in a highly dense communication network scenario and the impact of various parameters on the performance of the communication channel are analyzed.

Keywords Millimeter wave communication · 5G · SNR · MIMO
LOS · NLOS

1 Introduction

The ever-increasing demand for multi-gigabit wireless applications has posed enormous challenges for next generation wireless technologies. In particular, the unbalanced temporal and geographical variations of spectrum usage along with the rapid proliferation of bandwidth-hungry mobile applications have inspired a new promising technology to alleviate the pressure of scarce spectrum resources for 5G

U. Farooq (✉) · G. M. Rather
Department of Electronics and Communication,
National Institute of Technology Srinagar, Srinagar 190006, India
e-mail: umarfarooq232@gmail.com

C. R. Panigrahi et al. (eds.), *Progress in Advanced Computing and Intelligent Engineering*, Advances in Intelligent Systems and Computing 714,
https://doi.org/10.1007/978-981-13-0224-4_9

mobile broadband applications. This technology is known as millimeter wave (MMW) communication technology [1]. It is envisioned as a key technology for fulfilling the high data rate requirements of the 5G wireless networks. Figure 1 shows the broad overview of next generation 5G system challenges that have initiated the different research activities by industries, academia, and research organizations [2]. The idea behind MMW communication is to exploit the huge unexploited spectrum from 30 to 300 GHz spectrum with the corresponding wavelength ranging between 10 and 1 mm [3] as it provides number of benefits like [4, 5]

- Availability of huge unexploited bandwidth that can provide high-speed links with throughput of ~10 Gbps.
- Higher degree of spectrum sharing as compared with lower frequencies.
- Smaller antenna size which can facilitate the fabrication of large antenna arrays over a small area like postage stamp.
- Better spatial resolution as the small wavelength allows modest size antennas to have a small beam width.
- Processors, memories along with other devices can be incorporated on a Monolithic Microwave Integrated Circuit (MMIC) chip and the received data can be quickly processed and stored.

However despite these attractive advantages, these high frequency bands have not been fully explored because at higher frequencies the signal generation, reception and propagation gets highly complex due to free space and atmospheric loss factors [6]. The distinguishing traits of MMW communications have a direct impact on the selection of antenna technologies, modulation, besides other technical issues like user dynamics and coexistence with other communication standards like Wi-Fi, LTE. Multiple approaches have been provided so far to overcome these technical difficulties so as to make MMW technology deployable in near future and enable a plethora of applications.

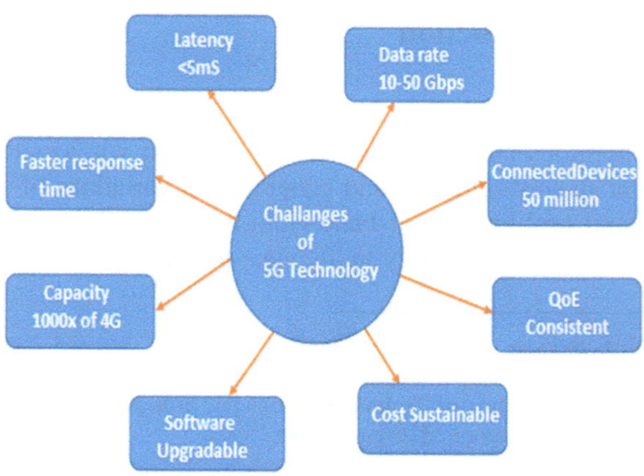

Fig. 1 General overview of next generation 5G system challanges

In the present study capacity of a MMW communication link is carried out for LOS and NLOS scenarios for different link parameters in a dense network scenario. The work is reported as technical paper and has been organized into number of sections. Section 1 gives the Introduction, Sect. 2 briefly summarizes the MMW propagation characteristics. Performance evaluation and mathematical analysis of the MMW communication link is carried out in Sect. 3. Section 4 concludes the paper.

2 MMW Propagation Characteristics

Attenuations and environmental losses in the MMW bands are much higher as compared to microwave frequencies, thus making 5G cellular network operations over these frequencies a much more challenging task. MMW link use LOS communication and these LOS communication links suffer from free space path loss which is given as [7]:

$$\mathrm{L_p}, dB = 20\, log_{10}\left(\frac{4\pi d}{\lambda}\right), \tag{1}$$

where 'λ' is the wavelength of signal and 'd' is the distance between transmitter and receiver.

This loss is quite high for longer distances. However these links can be utilized for short distance communication. The loss can be further mitigated by the use of highly directional antenna system, relay and multipath routing [8–10]

2.1 Atmospheric Loss Factors

In atmosphere millimeter waves are absorbed/scattered by molecules of oxygen, water vapor, and other atmospheric constituents including the rain, fog and clouds. These losses are greater at certain frequencies, coinciding with the mechanical resonant frequencies of the gas molecules. This atmospheric loss dB is given by [11]:

$$L_{atm} = -\alpha_{atm}(f) \cdot d, \tag{2}$$

where $\alpha_{atm}(f)$ is the measured value of atmospheric attenuation as the function of frequency operation 'f' and 'd' as the separation between transmitter and receiver.

Figure 2 shows the free space path loss and Fig. 3 shows the atmospheric loss for dry atmospheric conditions. Of all atmospheric conditions, rain causes the most significant signal loss depending on the intensity of the rainfall. Oxygen, humidity and fog also incur losses but they can be neglected to certain limits of the millimeter

Fig. 2 Free space path loss w.r.t. distance

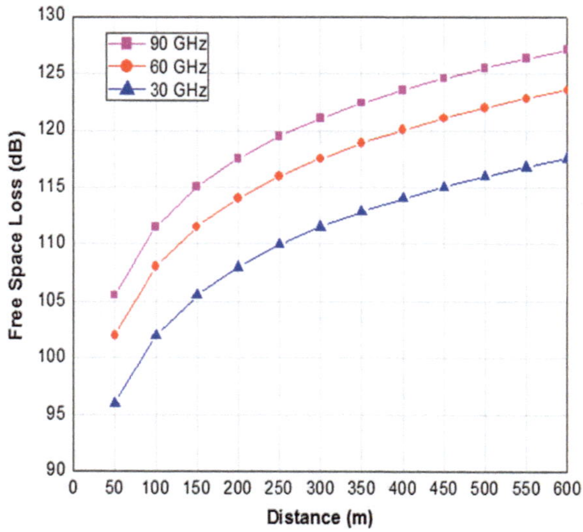

Table 1 Signal losses due to atmospheric conditions [12]

Factor	Description	Signal loss (dB/km)
Oxygen	Sea level	0.22
Humidity	100% @30°	1.8
Fog	1gm/m^3, 10 °C (50 m visibility)	3.2
Cloud burst	25 mm/hr rain	10.7
Light rain	1 mm/hr rain	0.9
Moderate rain	4 mm/hr rain	2.6
Heavy rain	25 mm/hr rain	10.7
Intense rain	50 mm/hr rain	18.4

wave link. The description of different atmospheric conditions and the losses incurred in dB/km are given in Table 1.

2.2 *Foliage Losses*

At millimeter wave frequencies these losses become a critical propagation impairment and are non negligible. If f is the frequency of signal in MHz, R is the foliage depth then Foliage losses are given as [13]

$$L_{foilage} = \frac{1}{5} f^{0.3} R^{0.6} \tag{3}$$

Equation 3 applies for the scenarios where R < 400 m. L $_{foliage}$ is measured in dB. The Foliage losses of the millimeter wave signals for different Foliage depths are plotted in Fig. 4.

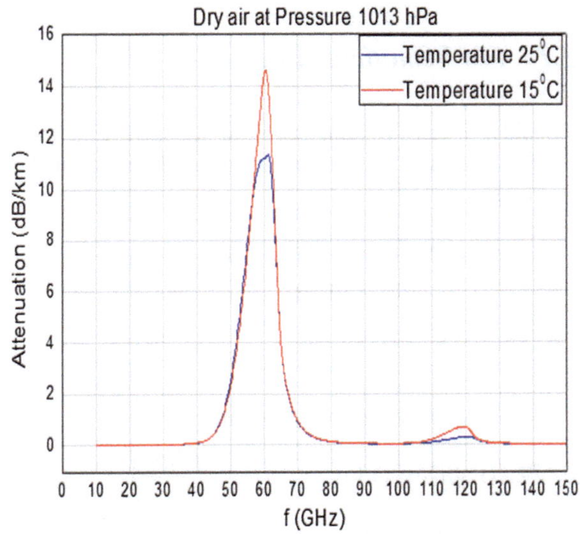

Fig. 3 Atmospheric losses for dry air

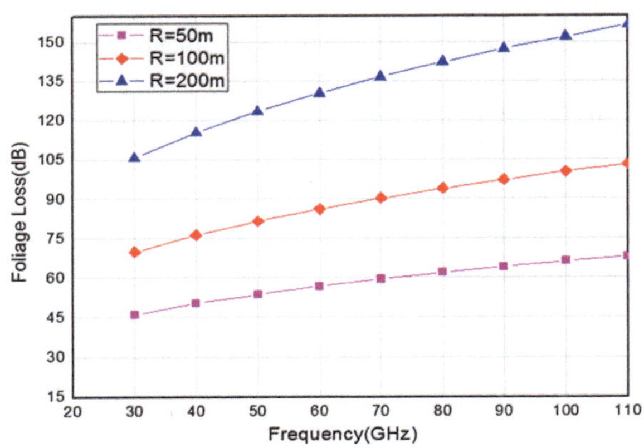

Fig. 4 Foilage loss as function of frequency

3 Performance Evaluation of MMW Link

The use of high gain directional antennas is an option to compensate the high path loss of millimeter waves however this method may be feasible only for clear LOS conditions. In scenarios where clear LOS is not guaranteed the antenna arrays, multipath routing, use of repeaters become highly desirable [14, 15].

In this section performance evaluation of the MMW communication channel for specific data rates is carried out for both LOS and NLOS case scenarios.

3.1 *Mathematical Analysis*

Consider a MMW link scenario as shown in Fig. 5. If ‘f’ is the frequency of operation and ‘d’ be the range of communication link. Then SNR at output of the receiver is

$$\begin{aligned} \mathrm{SNR} = {} & \mathrm{P_{Tx}} + \mathrm{G_{Tx}} + \mathrm{G_{Rx}} + \mathrm{G_{LNA}} + \mathrm{G_{BPF}} + \mathrm{G_{TFN}} + \mathrm{G_{RFN}} + \mathrm{G_{amp}} + \mathrm{G_{CSF}} \\ & - \mathrm{PL_0} - \mathrm{PL(d)} - \mathrm{N_{Rx}} - \mathrm{M_{SHAD}}, \end{aligned} \tag{5}$$

where

$$\mathrm{P_{Tx}} = \text{Transmitted Power (dBm)}$$
$$\mathrm{PL_0}\ (\text{Path loss at 1 m}) = 20\, log_{10}\left(\frac{4\pi d}{\lambda}\right)$$
$$\mathrm{G_{Tx}} = \text{Transmit antenna gain (dBi)}$$
$$\mathrm{M_{SHAD}} = \text{Shadowing link margin (dB)}$$
$$\mathrm{G_{Rx}} = \text{Receiver antenna gain (dBi)}$$
$$\mathrm{N_{Rx}} = \text{Input noise level at receiver.}$$

The gain due to the various components at the transmitter and the receiver of the link are mentioned in Table 2 [16].

In the study the other path loss parameters have also been considered which as per IEEE 802.15.3c standard [17] are given below

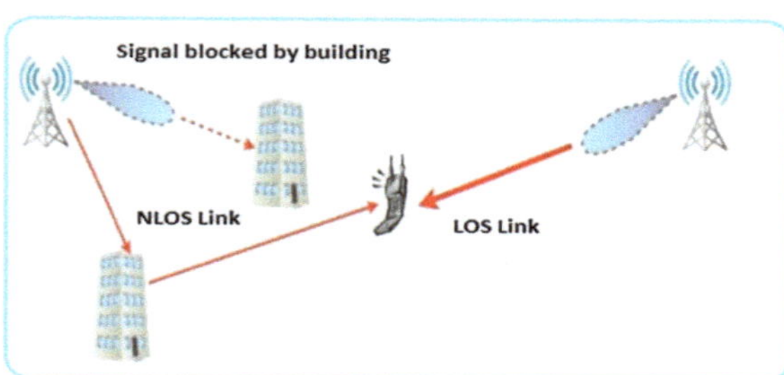

Fig. 5 MMW link in LOS and NLOS scenario

Table 2 Gain of various transmitter and receiver components

Component	Gain (dB)
Tx feeding network (G_{TFN})	−5
Rx feeding network (G_{RFN})	−5
Amplifier at receiver (G_{amp})	30
BPF at receiver (G_{BPF})	−1
LNA at receiver (G_{LNA})	20
Channel selector filter (G_{CSF})	−5

Path loss exponent: $n = \begin{cases} 2 & LOS\ Case. \\ 2.5 & NLOS\ Case. \end{cases}$

Shadowing Link Margin: $M_{SHAD} = \begin{cases} 1\,dB & LOS\ Case. \\ 5\,dB & NLOS\ Case. \end{cases}$

Assuming the noise to be AWGN and using Eq. 5 the data carrying capacity of the MMW link can be easily calculated by Shannon's Capacity formulae given by:

$C = B\ \log_2(1 + SNR),$ where B is the bandwidth of the communication link.

3.2 Simulation Results

The MMW link is simulated and analyzed for data carrying capacity. The parameters taken into account for the performance analysis of such system are: Range of the communication link (d), data rate, frequency of operation (f) and the bandwidth of the link (B). Figures 6 and 7 show the simulation results for LOS path scenario.

Figure 7 shows the data carrying capacity of the MMW communication link of 50 m as the function of bandwidth and it is observed from the figure that high gain

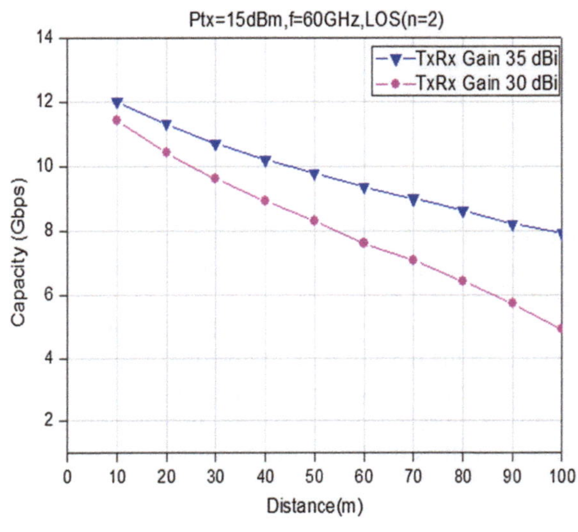

Fig. 6 Capacity as function of range of link

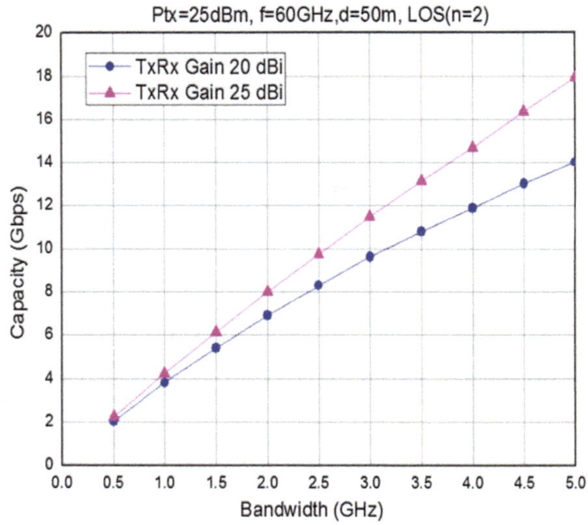

Fig. 7 Capacity as function of link band width

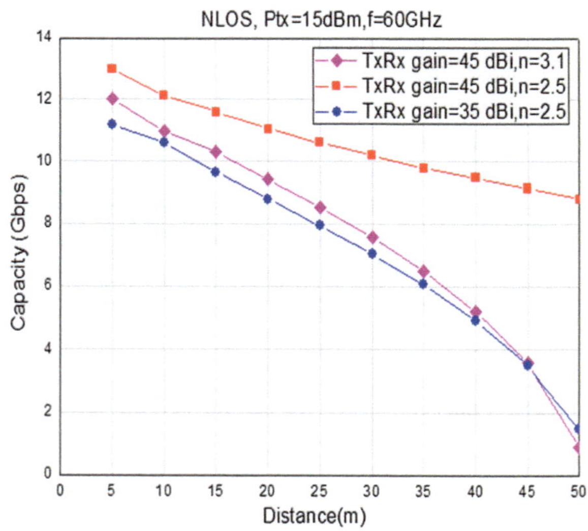

Fig. 8 Capacity for different antenna gain (NLOS Case)

antennas at the transmitter and the receiver are required to achieve the data rate of the order of Gbps for long distance communication applications even for LOS cases.

Similarly Figs. 8 and 9 show the simulation results in case NLOS Scenario. In either case the performance of the MMW channel is limited due to the system noise as well as the losses due to various atmospheric factors.

As can be seen from the figures, in both LOS and NLOS Scenarios, the impact of noise as well as the losses due to various atmospheric factors limit the performance

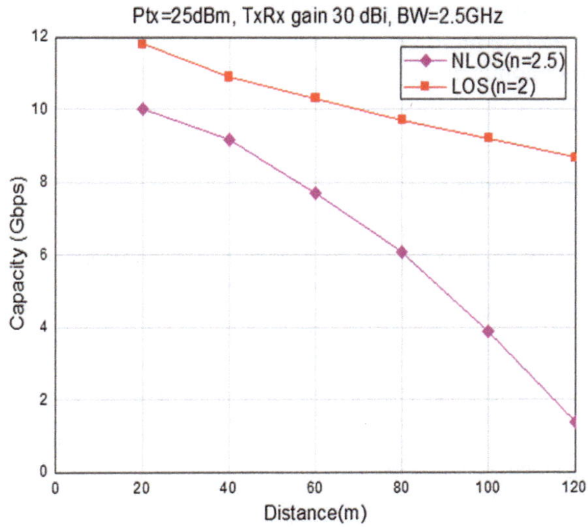

Fig. 9 Capacity for LOS and NLOS links

the MMW link however the antennas play a critical role in the enhancement of efficiency of such systems. It can also be observed that increase in the operating distance drastically decreases the capacity of the MMW link and this situation worsens in case of NLOS cases. Also it can be observed that Capacity increases as a function of the SNR or Bandwidth (B) or both; however as the operational distance of the link increases the SNR degrades due to the various atmospheric losses. Under such circumstances the capacity is improved with help of high gain directional antennas and hence highlighting the importance of such antenna configurations for very high data rate applications for next generation networks.

4 Conclusion

Millimeter wave (MMW) communication is a promising technology to cope up with the ever increasing demand for multi-gigabit wireless applications and in particular alleviate the problems of spectrum scarcity for 5G mobile communication applications. In this paper we broadly explore the MMW propagation characteristics and examine the effect of free space loss, atmospheric loss and foliage loss on the MMW communication channel. Performance evaluation of the MMW communication link is carried out for different data rates for both LOS and NLOS cases in a highly dense communication network scenario. In either case the performance of the MMW channel is limited due to the system noise as well as the losses due to various atmospheric factors. Since it is observed that the data carrying capacity is affected as a function of operational distance of the link, the use of high gain directional antennas becomes the first choice to overcome this issue. Hence the antennas play a critical role for improving the performance of such systems and therefore highlight the importance of such antenna configurations for very high data rate applications for next generation mobile networks.

References

1. Pi, Z., & Khan, F.: An introduction to millimeter-wave mobile broadband systems. *IEEE Communications Magazine*, *49*(6), (2011).
2. Gupta, A., & Jha, R. K.: A survey of 5G network: Architecture and emerging technologies. *IEEE access*, *3*, 1206–1232, (2015).
3. Pietraski, P., Britz, D., Roy, A., Pragada, R., & Charlton, G.: Millimeter wave and terahertz communications: Feasibility and challenges. *ZTE Communications*, *10*(4), 3–12, (2012).
4. Niu, Y., Li, Y., Jin, D., Su, L., & Vasilakos, A. V.: A survey of millimeter wave communications (mmWave) for 5G: opportunities and challenges. *Wireless Networks*, *21*(8), 2657–2676, (2015).
5. Saponara, S., & Neri, B.: mm-wave integrated wireless transceiver: enabling technology for high bandwidth short-range networking in cyber physical systems. *Microsystem Technologies*, *22*(7), 1893–1903, (2016).
6. Baldemair, R. et. al.: Evolving wireless communications: Addressing the challenges and expectations of the future. *IEEE Vehicular Technology Magazine*, *8*(1), 24–30, (2013).
7. Friis, H. T.: A note on a simple transmission formula. *Proceedings of the IRE*, *34*(5), (1946).
8. Singh, S., Ziliotto, F., Madhow, U., Belding, E., & Rodwell, M.: Blockage and directivity in 60 GHz wireless personal area networks: From cross-layer model to multihop MAC design. *IEEE Journal on Selected Areas in Communications*, *27*(8), (2009).
9. Wang, J., Prasad, R. V., and Niemegeers, I. G.: Exploring multipath capacity for indoor 60 GHz radio networks. In *Communications (ICC), 2010 IEEE International Conference*, IEEE, (2010).
10. Rusek, F., Persson, D., Lau, B. K., Larsson, E. G., Marzetta, T. L., Edfors, O., & Tufvesson, F. Scaling up MIMO: Opportunities and challenges with very large arrays. *IEEE Signal Processing Magazine*, *30*(1), 40–60, (2013).
11. am Atmospheric Model. Available Online https://www.cfa.harvard.edu/~spaine/am/.
12. Adhikari, P.: Understanding millimeter wave wireless communication. *Loea Corporation*, (2008).
13. Meng, Y. S., & Lee, Y. H.: Investigations of foliage effect on modern wireless communication systems: A review. *Progress In Electromagnetics Research*, *105*, 313–332, (2010).
14. Zhang, X., Zhou, S., Wang, X., Niu, Z., Lin, X., Zhu, D., & Lei, M.: Improving network throughput in 60 GHz WLANs via multi-AP diversity. In *Communications (ICC), IEEE International Conference*, pp. 4803–4807, (2012).
15. Lan, Z., Sum, C. S., Wang, J., Baykas, T., Gao, J., Nakase, H & Kato, S.: Deflect routing for throughput improvement in multi-hop millimeter-wave WPAN system. In *Wireless Communications and Networking Conference*, pp. 1–6, (2009).
16. Huang, K. C., & Edwards, D. J.: *Millimetre wave antennas for gigabit wireless communications: a practical guide to design and analysis in a system context*. John Wiley & Sons, (2008).
17. IEEE 802.15-05-0493-27-003c, "TG3c selection criteria." Jan. 2007.

MBA: Mobile Cloud Computing Approach for Handling Big Data Applications

Rajesh Kumar Verma, Chhabi Rani Panigrahi, V. Ramasamy, Bibudhendu Pati and P. E. S. N. Krishna Prasad

Abstract MBA is an efficient approach for handling big data applications using MCC, which also takes care of handling the offloading mechanism as may be required, based on the degree of computation and utilization of resources. Big data processing using Map-Reduce framework is very useful for several applications like pattern-based searching, sorting, log analysis, etc. A robust architecture which is cheap and viable at the same time has been proposed here using MCC, in which first offloading is done to the local cloud or cloudlet and subsequently, also offloads to the public cloud on a need basis for highly compute-intensive jobs.

Keywords Mobile cloud computing · Big data · Offloading

1 Introduction

Mobile Cloud Computing (MCC) mainly consists of three different types of components—mobile device, cloud and wireless network [1], and is fast catching up in today's world as the number of mobile devices being manufactured are increasing

---

R. K. Verma (✉)
Biju Patnaik University of Technology, Rourkela, Odisha, India
e-mail: rajeshverma_chicago2004@yahoo.com

C. R. Panigrahi · B. Pati
Department of Computer Science,
Rama Devi Women's University, Bhubaneswar, India
e-mail: panigrahichhabi@gmail.com

B. Pati
e-mail: patibibudhendu@gmail.com

V. Ramasamy
Park College of Engineering and Technology, Coimbatore, Tamil Nadu, India
e-mail: researchrams@gmail.com

P. E. S. N. Krishna Prasad
Prasad V. Potluri Siddhartha Institute of Technology, Vijayawada, Andhra Pradesh, India
e-mail: surya125@gmail.com

C. R. Panigrahi et al. (eds.), *Progress in Advanced Computing and Intelligent Engineering*, Advances in Intelligent Systems and Computing 714,
https://doi.org/10.1007/978-981-13-0224-4_10

at a fast rate and also their computing power is getting increased manifold. There are around 7.2 billion mobile devices, and also their number is growing at five times the present rate, which is far more than the human population of around 7.19 billion [2]. Also, big data is significantly changing this world today and is being used in almost every application area such as banking, agriculture, telecom, transportation, etc. Big data refers to mainly the 3 V's characteristics of data, mainly data Volume, Velocity, and Variety, and few more like Veracity, Validity, and Volatility [3]. Having these types of features, the data cannot be processed using traditional technologies of relational databases and hence the need to use big data technologies for tackling different problems in this fast and changing world.

In this work, authors have suggested an efficient architecture for processing of big data applications. We have refined the Hadoop Map-Reduce framework over MDFS and have refined it further to tackle the aspect of offloading as well [4–6]. As MCC is becoming very popular in the world today and can be used to process applications, we have used MCC for processing. Also, we will be using the power of the public cloud when required in order to take care of humongous tasks requiring huge computation power and vast degree of resources. In our architecture, the big data applications are processed very efficiency using Hadoop Map-Reduce framework. Also, we use offloading approach to use the power of the local cloud (cloudlet) or the public cloud, and increase the speed of computation and also facilitate storage of huge volumes of data (in order of petabytes and beyond) on the public cloud, and hence enabling the mobile devices to deliver results at a cheaper cost.

The rest of the paper is organized as follows: Sect. 2 presents the state of the art. Section 3 describes the proposed MBA architecture along with brief analysis. Finally, Sect. 4 concludes the paper.

2 Related Work

The authors [7] had proposed the cloudlet which is a combination of mobile devices group together and is used to run apps using large amounts of data and requiring huge computational power. It is also evident that both MCC and big data combined together can help to create a revolution in the computational field. The authors [8] had proposed the MDFS (Mobile Distributed File System) architecture which addresses issues of security and energy efficiency while performing big data computations on the mobile cloud. In [9], the authors talked about one of the most popular applications of cloudlet involving big data, which is the GigaSight, an IOT (Internet of Things) application that generates huge volume of data (big data) from sensor devices like camera, video cameras, etc. The videos can be useful to solve some of the hypothetical use cases like in the retail stores (which section in the outlet store attracts the people most?) and surveillance (studying people's behavior and tracking suspicious activity from the video clippings). Using cloudlet architecture, the huge data volumes can be analyzed and hence provide effective business solutions.

3 Proposed Architecture

The layers in the MBA architecture (as shown in Fig. 1) are detailed here. User Interface (UI) Layer consists of various types of users carrying different types of mobile devices through which the user mainly interacts with the system, and also it is mainly responsible for displaying the results back onto the mobile device after the computation has been completed. The Map-Reduce (MR) framework takes care of scheduling and execution of the MR tasks in an orderly manner. Also, memory and space management is taken into consideration before allocation of new tasks. It uses parallel computing which distributes the computational tasks to a large number of nodes which can be monitored through web interface. It is a fault-tolerant framework which can work even when few nodes fail while a particular job is running. It consists of a simple model of data processing in which the output is mainly the key-value pairs. Hadoop framework was originally written in Java and the MR application needs to be written in Java as well. The job of the programmer is to write the map and reduce function, which involves partitioning a large problem into many subproblems, process the subproblems in parallel, and then combine the solutions. MR provides abstraction that hides many system level details from the programmer. The storage, which is handled by HDFS, sits underneath MR. The internal Network layer (till Cloudlet only) is used such that the MR programs are able to converse with the Job Tracker (JT) and Task Tracker (TT) which does the task of actually running the MR jobs. The Name Node (NN) and Data Node (DN) are housed in this cloudlet layer and take care of running the MR jobs. The external Network layer (to the public Cloud) is used for connecting and transferring data from the cloudlet layer (local cloud) to the public cloud. The resource-intensive computations are offloaded to the public cloud such that the immense computational and storage power can be easily leveraged from the public cloud (like Amazon, Azure, etc.). The public cloud also contains the MR framework layer which houses the JT, TT, and both the NN and DN which handle the execution of the MR jobs. Subsequently, the end results are shown to the user in the UI layer.

The flowchart for MBA computation mechanism (as shown in Fig. 2) is as follows: At the beginning, it is necessary to find out if the mobile resources are sufficient enough to execute the big data application, that is, if the Map-Reduce application can run having sufficient resources using the huge data which is stored in large disk volume. In case it is possible to execute the application on the mobile devices itself (local cloud or cloudlet), we offload the computation to the local cloud and perform the job on the cloudlet. In case it is not sufficient, then we need to ensure that offloading to the public cloud is enabled via the architecture and the computation job is then offloaded to the public cloud and subsequently, the results are sent from the cloud to the mobile device.

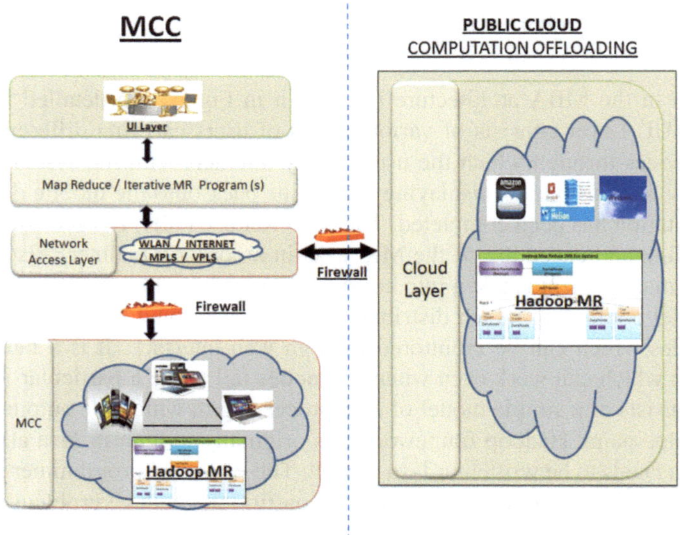

Fig. 1 Architecture of MBA computing

Fig. 2 Flowchart of MBA computing

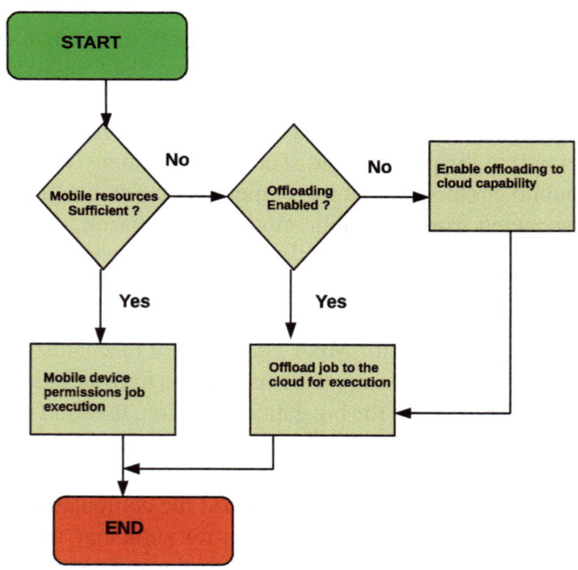

Definition 1 MBA technique considers efficient offloading approach by using the following conditions:

$$Exec(ME, CE, PC) = \begin{cases} ME, & if\ E_c(M) \le E_c(C) \\ CE, & if\ E_c(M), E_c(PC) > E_c(C) \\ PC, & if\ E_c(C) > E_c(PC) \end{cases}$$

where ME, CE, and PC denotes the mobile, local cloud, and public cloud execution, respectively, and $E_c(M), E_c(C)$, and $E_c(PC)$ are the energy consumption for processing the data over mobile device, local cloud, and public cloud execution, respectively.

Theorem 1 *MBA minimizes high latency by considering efficient offloading technique.*

Proof According to Definition 1, MBA considers the necessary big data applications offloading either to the local cloud (cloudlet) or to the public cloud. If the application is computed on the local cloud, then it minimizes the latency with respect to the public cloud [7, 10].

Theorem 2 *MBA minimizes energy consumption.*

Proof MBA considers an intelligent offloading decision based on Definition 1. If the big data application is computed only on the mobile devices, then it will consume large amount of energy. But MBA considers an intelligent decision about where to offload. Again, according to Definition 1, MBA considers big data application on the local cloud. The local cloud server takes an intelligent decision to either execute locally or on the public cloud.

4 Conclusion

Big data applications are very popular and find usage in various fields like medical, military agriculture, banking, telecom, etc. Processing using Hadoop MR provides valuable result outputs, which are used for analyzing several important aspects of that application. MBA approach helps to process big data applications using MR framework using MCC and CC technology using local cloud and offloading to public cloud. It is a cheaper option as we can use the local mobile devices and on few occasions use the public cloud if the computation is intensive in nature. Further research in this direction toward increasing the speed of MBA and minimizing the energy consumption during resource-intensive computations needs to be explored.

References

1. Abolfazli, S., Sanaei, Z., Gani, A., and Shiraz, M.: MOMCC: Market-Oriented Architecture for Mobile Cloud Computing Based on Service Oriented Architecture, 1st IEEE International Conference on Communications in China Workshops (ICCC), (2012).
2. http://www.independent.co.uk/: Last accessed: 18/02/2017.
3. Assuncao, M. D., Calheiros, R. N., Bianchi, S., Netto, M. A.S., Buyya, R.: Big Data computing and clouds: Trends and future directions. Journal of Parallel and Distributed Computing, 79–80, pp. 3–15 (2015).
4. Panigrahi, C. R., Pati, B., Tiwary, M., and Sarkar, J. L.: EEOA: Improving energy efficiency of mobile cloudlets using efficient Offloading Approach. IEEE International Conference on Advanced Networks and Telecommunications Systems, pp. 1–6 (2015).
5. Panigrahi, C. R., Sarkar, J. L., Pati, B. and Bakshi, S.: E^3M: An Energy Efficient Emergency Management System using mobile cloud computing. IEEE International Conference on Advanced Networks and Telecommunications Systems (ANTS) , Bangalore, India, pp. 1–6 (2016).
6. Pati, B., Sarkar, J. L., Panigrahi, C. R. and Debbarma, S.: eCloud: An Efficient Transmission Policy for Mobile Cloud Computing in Emergency Areas. Progress in Intelligent Computing Techniques: Theory, Practice, and Applications, pp. 43–49 (2018).
7. Satyanarayanan, M., Zhuo, C., Hat, K., Hut, W., Richtert, W., Pillai, P.: Cloudlets: at the Leading Edge of Mobile-Cloud Convergence. 6th International Conference on Mobile Computing, Applications and Services (MobiCASE) (2014).
8. Shu, P., Liu, F., Jin, H., Chen, M., Wen, F., Qu, y., and Li, b.: eTime: Energy-Efficient Transmission between Cloud and Mobile Devices. IEEE Infocom, pp. 195–199 (2013).
9. Satyanarayanan, M., Simoens, P., Pillai, Y. X., Chen, Z., Ha, K., Hu, W., Amos, B.: Edge Analytics in the Internet of Things. IEEE Pervasive Computing, 14(2), pp. 24–31 (2015).
10. Panigrahi, C.R., Sarkar, J. L., and Pati, B.: Transmission in mobile cloudlet systems with intermittent connectivity in emergency areas. Digital Communications and Networks (2017), https://doi.org/10.1016/j.dcan.2017.09.006.

Implementing Time-Bounded Automatic Test Data Generation Approach Based on Search-Based Mutation Testing

Shweta Rani, Hrithik Dhawan, Gagandeep Nagpal and Bharti Suri

Abstract Automatic test generation is a backbreaking task in software testing, and it is the need of the research community as well as for industry. Search-based mutation testing has been effectively applied for solving the testing problems. In this paper, an idea following the behavior of genetic algorithm with the benefits of mutation testing is proposed and implemented to generate the test cases automatically. For the sake of minimizing the cost incurred due to mutation testing, selective mutation technique is encouraged to generate the lesser number of mutants using delete mutation operators instead of all the traditional mutation operators. The process stops when it reaches the predefined time limit. In each iteration, it tries to optimize the size of the test suite by searching and eliminating the redundant less fit test inputs with the aim of mutation coverage. Results suggest that the generated test cases successfully detect more than 90% mutants.

Keywords Search-based mutation testing · Mutation testing · Genetic algorithm · Automatic test data generation

1 Introduction

Test data generation is a diligent task as it highly affects the efficiency of software testing and quality of the software product. Manual generation of test cases is generally the responsibility of the tester and it is extremely time-consuming and costly. Thus, automatic generation of the test data is essential and also a challenge for the research community as well as for the industry [1]. According to [2–4], it is devised as a search problem and can be elucidated with search-based techniques.

In the past few decades, search techniques have been employed to solve the test suite generation problem [5–8]. Search-based techniques were initially introduced

S. Rani (✉) · H. Dhawan · G. Nagpal · B. Suri
USICT, GGS Indraprastha University, New Delhi, Delhi, India
e-mail: shweta2610@gmail.com

B. Suri
e-mail: bhartisuri@gmail.com

C. R. Panigrahi et al. (eds.), *Progress in Advanced Computing and Intelligent Engineering*, Advances in Intelligent Systems and Computing 714,
https://doi.org/10.1007/978-981-13-0224-4_11

by Miller and Spooner [9]. These techniques search the complete search space for the solution and the process is enlightened by a fitness function. It is an objective function which indicates how close the solution is to the defined convergence criteria (indicates the stopping criteria of search). Search process stops when the solution is found, i.e., convergence criteria is met. Some of the search-based techniques are Hill Climbing (HC), Genetic Algorithm (GA), and Bee Colony Optimization (BCO) [7, 10]. In the current empirical research, authors are dealing with GA for test data generation problem.

Genetic Algorithm (GA) is initially adopted by J. H. Holland [11] in 1975 and follows the concept of Darwinian's theory of "survival of the fittest". The general process of GA includes generation of initial population P, fitness evaluation, selection, and recombination. It initializes with random initial population and then starts searching the solution in the search space. Each point in search space represents a possible solution. The search is driven by fitness function and recombination. Recombination acts as a key operator for natural evolution. It involves two operators: crossover and mutation. The crossover operator takes two chromosomes (individuals in initial population) and produces a new offspring (new individual in next population) via exchanging the properties of current individuals. Mutation operator takes a single chromosome and produces a new offspring by changing the value of chromosome at a particular location. The complete process is iteratively repeated until the optimal solution is found [12, 13].

With the aim to solve the test-related problems, fault-based testing is being used since 1970s as explained in surveys [14, 15]. Mutation testing is a structural software testing, which injects/seeds the faults in the program by making the use of mutation operators. This was suggested by DeMillo et al. [16] and Hamlet [17] to rank the test cases according to their quality in 1970s. Mutation operators [18] are nothing but the rules to create the mutants. They inject a single fault in the program, and this faulty form of the program is known to be a mutant. To detect the fault, test case is executed against the mutant; if it fails, then mutant remains live otherwise it is called as killed mutant. Mutation score or mutant killing capability is considered as the quality check parameter. Practitioners have given the evidence that mutants behave similar to realistic faults [19] and therefore, the faults/mutants coverage criteria can be utilized effectively to generate the test suite for testing purposes.

Search-Based Mutation testing (SBMT) has gained great attention and popularity to solve the test case problems like generation, prioritization, and optimization of test cases as stated in surveys [20–23]. The underlying principle of SBMT is to employ some search-based technique to generate the test cases with high mutation coverage. Mutation coverage is treated as an optimization problem and search technique is used to search the space for the optimal solution.

In this paper, an idea is proposed and implemented for automated test data generation following the principle of SBMT. Mutants are created using MuJava [24] mutation testing system. This approach provides the motivating results when evaluated with Trityp, a triangle classification Java program. It was found that the implemented automated approach leads to mutation coverage of 90% when examined for 50 times. The problem of equivalent mutants is not dealt in the approach and it may be the case that test suite could not achieve higher fault coverage. The results suggest to investigate and find the equivalent mutants for the improvement of the approach.

2 Related Work

The idea of search-based mutation testing is the realization of the search technique (GA) to automate and solve the test case problems. The problem is expressed as a search objective [4] which aims to produce the efficient test suite for a given problem. This section reviews the existing work presented in the field combining mutation testing and Genetic Algorithm (GA).

Jones et al. [25] initially utilized GA to detect the faults in the branch predicates. Boundary limits were used to generate the test inputs. The approach was empirically evaluated on quadratic equation solver and successfully attained 97% mutation score.

In 2001, Bottaci [26] proposed a new fitness function composed of three conditions based on mutant killing criteria. These conditions were reachability condition, necessity condition, and sufficiency condition. Later, Masud et al. [27] used a fitness function based on these conditions to generate the test data using GA and mutation testing.

Baudry et al. [28, 29] suggested the use of GA for generating the tests based on the integration testing concept. Mutation score was chosen as a fitness criterion. Later, they employed Bacteriological Algorithm (BA) for test data optimization and stated that BA is cost-effective than GA [30]. Mutation score-based fitness function was also implemented in [31–34].

The problem of generating the test data was also solved using GA in [35–40].

3 Proposed Time-Bounded Approach for Test Data Generation

This paper implements an approach mimicking the behavior of GA with the benefits of mutation testing. The steps of the approach are listed below:

Step 1: Source code is analyzed to mine the information like number of input variables, the range of these variables.

Step 2: Mutants are created only for some selective traditional level mutation operators (delete mutation operators) using MuJava mutation testing tool.

Step 3: Initial population is generated randomly and encoded in binary form.

Step 4: The initial population (from step 3) is then executed over the mutants (from step 2) for fitness evaluation that is the mutation score.

Setp 5: Based on the fitness value, obsolete (redundant) population is removed from the candidate solution. Highly fitted candidate solution is kept and forwarded for reproduction.

Table 1 GA parameters

Parameters	Value
Size of initial population	5 * num_var
Fitness function	Mutation adequacy Score
Parent selection	Fittest 50%
Crossover probability	1.0
Convergence criteria	Time limit 80 ms
Number of runs	50

Step 6: Crossover and mutation operations are performed, which leads to new population. In each iteration, crossover is performed, thus setting its probability to 1.0. On the other hand, mutation is performed if no new candidate solution is able to kill other live mutants. This new population (obtained from crossover and mutation) is then executed over the mutants (generated in step 2) for fitness evaluation.

Step 7: Goto step 5. Candidate solution with higher fitness score is kept in each iteration. Each iteration tries to improve the mutation score/fitness value of the test set. Best fit individuals are also kept from the previous iteration. In this way, properties of GA are used with memorization function.

Step 8: Repeat steps 5 to 7 until execution time reaches its predefined limit.

MuJava mutation testing tool [24] is employed to automatically generate the selective traditional mutants. Only delete mutation operators [41, 42] are preferred to use over all traditional mutation operators; it leads to low cost as less number of redundant mutants are generated. To set the execution time limit, the complete process was executed for different time limits 40, 60, 80, and 100 ms. It was noticed that when the time limit is set to 80 ms, the process is able to generate the test data that successfully detected up to 95% delete operator-based mutants. Therefore, the time limit of 80ms is set to get the optimum results for the complete process.

Figure 1 depicts the complete procedure of the proposed time-bounded GA and mutation testing. The parameters of GA are set as given in Table 1. Initially, delete mutation operators-based mutants are generated for the subject program. Initial population, T[i] is generated randomly and is evaluated for fitness against all the generated mutants. On the basis of fitness score of each individual in initial population, redundant tests are deleted. The process stops when execution time reaches the time limit of 80 ms; otherwise, crossover and mutation are operated on the top 50% population of test cases. Repeat the complete process until termination condition is satisfied.

GA provides the solution based on some parameters for guiding the search process. The size of the initial population, crossover rate, mutation rate, parent selection criteria, and number of iterations are the main parameters. Each test case is encoded using an array of 24 bits. In each iteration, test cases are executed against the mutants and are evaluated for fitness. This fitness evaluation guides the search process. As the process completes, a final, nonredundant test suite is obtained. Due to the spec-

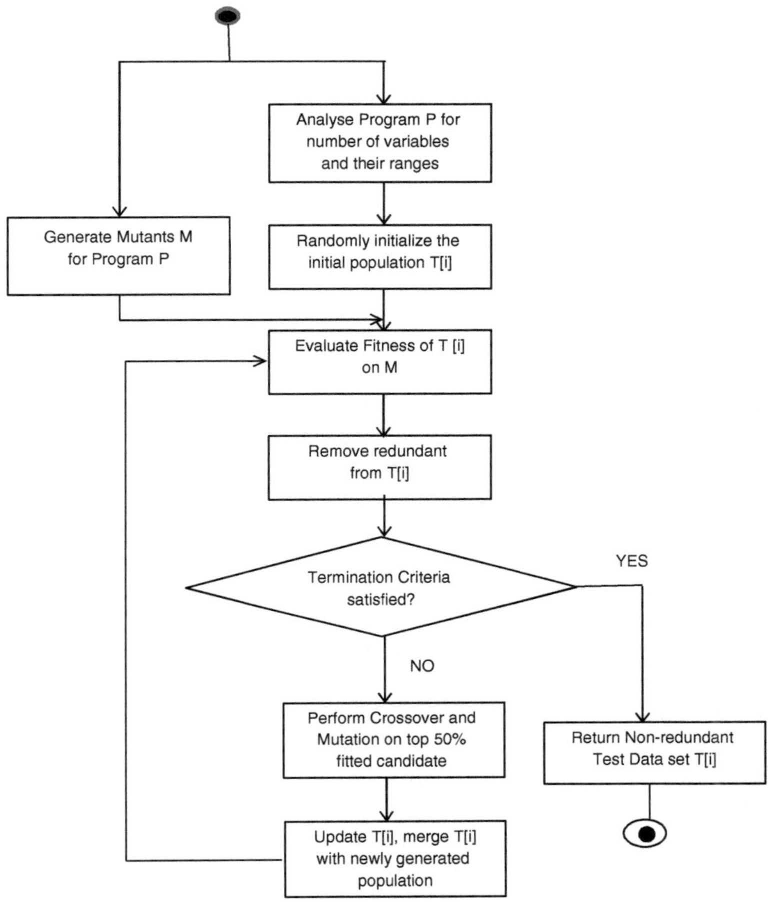

Fig. 1 Adopted process based on GA and mutation testing for test data generation

ulative nature of GA, the complete procedure is repeated several times in order to alleviate the effects of random variation. In this paper, experiments are repeated 50 times.

4 Results

The most popular Java program is selected and used for evaluating the proposed approach for automatic test case generation. Trityp program is utilized in the empirical study [15]. Trityp is taken from SIR (Software-artifact Infrastructure Repository) http://sir.unl.edu/portal/index.php. This triangle program takes the three inputs with

Table 2 Mutants information for Trityp

Parameter	Count
LOC	73
M_{SDL}	33
M_{VDL}	18
M_{CDL}	3
M_{ODL}	35
M_{sel}	89
M_{all}	474

reference to each side of the triangle and generates an outcome that indicates the triangle type.

Delete operators-based mutants are generated using MuJava [24]. Only four mutation operators (SDL, VDL, CDL, and ODL) are selected to generate mutants M_{sel} for the purpose of test generation while all traditional mutants M_{all} are generated to assess the quality of generated test data. Information about mutants is given in Table 2.

The complete approach is executed till it reaches the time limit of 80 ms. This is repeated 50 times. For each iteration, the test data are evaluated in terms of the mutation score for quality assessment. Initially, mutants for Trityp class are created and saved for further use. For each generation of the test data, mutation score is recorded for its further improvement in the next generation. In order to show that mutation coverage metric can replace other code coverage metrics, we find the correlation between these metrics.

Results obtained from the approach test data are listed in Table 3. To analyze the performance of the proposed approach, test data is executed against all the traditional mutants M_{all}; it leads to MS_{Mall} (mutation score when test data executed over all traditional mutants). In order to prove that the approach successfully improves the initial random test data set, it is executed over both types of mutants: traditional set of

Table 3 Quality of test data obtained from the approach (averaged over 50 runs). I_MS_{Msel} represents the mutation score for initial test suite over M_{sel}, and I_MS_{Mall} represents the initially obtained mutation score over M_{all}

Parameter	Value
Test data size	18
Number of delete mutants killed	80
MS_{MDel}	90
MS_{Mall}	76
I_MS_{Msel}	35
I_MS_{Mall}	33

Table 4 Mutation coverage and code coverage obtained by final test suite (averaged over 50 runs). MS_{Mall} is the mutation score when test data executed over all traditional mutants

Metric	Percentage
MS_{Mall}	76
Branch coverage	95
Statement coverage	96

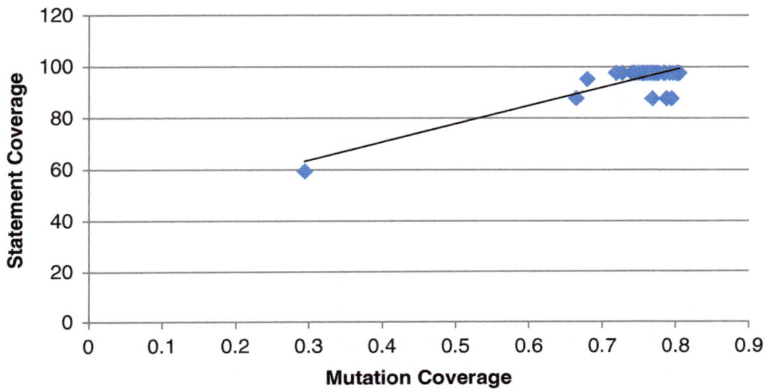

Fig. 2 Correlation between statement coverage and mutation coverage

mutants M_{all} and delete operators-based mutants M_{sel}. As can be seen from Table 3, initial test data could only achieve 35% mutation score that is iteratively improved by the approach. The significant improvement of 90% mutation score is recorded by the final test suite. However, the problem of equivalent mutants is not dealt in this work, and it might be the reason that the approach could not achieve 100% mutation coverage.

To measure the correlation between mutation coverage and code coverage metrics, we also measured branch coverage and statement coverage for 50 runs using Eclemma, eclipse-based plug-in for coverage analysis [43] as listed in Table 4. Figures 2 and 3 depict that the code coverage metrics are highly correlated with mutation coverage. Therefore, mutation coverage can be used instead of code coverage metrics.

5 Conclusion

The test generation problem has been inspected since 1976. Search techniques have been considered effective to solve the test-related problems. In this paper, the authors propose an approach, adapting genetic algorithm with the objective of maximizing

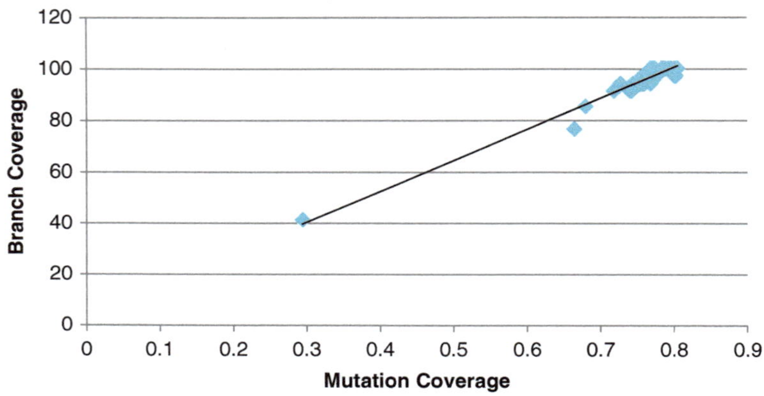

Fig. 3 Correlation between branch coverage and mutation coverage

the mutation score for test data generation. The initial population is generated randomly and evolved over a number of iterations till it attains 80 ms time limit. The mutants are created using delete mutation operators; thus minimizing the cost of mutation testing. The approach is completely automated in Java.

In order to evaluate the approach, a triangle classification program is used. As per the findings, the approach successfully produces the evolved test data with reference to the initial population. Test cases from our approach are found to reveal 90% traditional mutants in lesser number of iterations. Comparing mutation coverage with branch coverage and statement coverage, the research reveals that the mutation coverage criteria can be used as an adequacy criterion for measuring the code coverage also.

Acknowledgment The authors would like to acknowledge Ministry of Electronics and Information Technology, Govt. of India for supporting this research under Visvesvaraya Ph.D. Scheme for Electronics and IT.

References

1. D. C. Ince, "The automatic generation of test data," *The Computer Journal*, vol. 30, no. 1, p. 63, 1987.
2. M. Harman and P. McMinn, "A theoretical & empirical analysis of evolutionary testing and hill climbing for structural test data generation," in *Proceedings of the 2007 International Symposium on Software Testing and Analysis*, ser. ISSTA '07. ACM, 2007, pp. 73–83.
3. J. Wegener, A. Baresel, and H. Sthamer, "Evolutionary test environment for automatic structural testing," *Information and Software Technology*, vol. 43, no. 14, pp. 841–854, 2001.
4. J. Clarke, J. J. Dolado, M. Harman, R. Hierons, B. Jones, M. Lumkin, B. Mitchell, S. Mancoridis, K. Rees, M. Roper, and M. Shepperd, "Reformulating software engineering as a search problem," *IEEE Proceedings - Software*, vol. 150, no. 3, pp. 161–175, 2003.

5. M. Dave and R. Agrawal, "Search based techniques and mutation analysis in automatic test case generation: A survey," in *2015 IEEE International Advance Computing Conference (IACC)*, 2015, pp. 795–799.
6. P. McMinn, "Search-based software test data generation: A survey: Research articles," *Software Testing, Verification and Reliability*, vol. 14, no. 2, pp. 105–156, 2004.
7. P. McMinn, "Search-based software testing: Past, present and future," in *Proceedings of the 2011 IEEE Fourth International Conference on Software Testing, Verification and Validation Workshops*, ser. ICSTW '11. IEEE Computer Society, 2011, pp. 153–163.
8. O. Sahin and B. Akay, "Comparisons of metaheuristic algorithms and fitness functions on software test data generation," *Applied Soft Computing*, vol. 49, pp. 1202–1214, 2016.
9. W. Miller and D. L. Spooner, "Automatic generation of floating-point test data," *IEEE Transactions on Software Engineering*, vol. SE-2, no. 3, pp. 223–226, 1976.
10. Z. Li, M. Harman, and R. M. Hierons, "Search algorithms for regression test case prioritization," *IEEE Transaction on Software Engineering*, vol. 33, no. 4, pp. 225–237, 2007.
11. J. H. Holland, *Adaptation in Natural and Artificial Systems: An Introductory Analysis with Applications to Biology, Control and Artificial Intelligence*. Cambridge, MA, USA: MIT Press, 1992.
12. S. N. Sivanandan and S. N. Deepa, *Intorduction to Genetic Algorithms*. Springer-Verleg, 2008.
13. S. Luke, *Essentials of Metaheuristics*. Lulu, 2009, available for free at http://cs.gmu.edu/~sean/book/metaheuristics/d.
14. M. Delamaro, M. L. Chaim, A. M. R. Vincenzi, M. Jino, and J. C. Maldonado, "Twenty-five years of research in structural and mutation testing," in *2011 25th Brazilian Symposium on Software Engineering*, Sept 2011, pp. 40–49.
15. Y. Jia and M. Harman, "An analysis and survey of the development of mutation testing," *IEEE Transaction on Software Engineering*, vol. 37, no. 5, pp. 649–678, 2011.
16. R. A. DeMillo, R. J. Lipton, and F. G. Sayward, "Hints on test data selection: Help for the practicing programmer," *Computer*, vol. 11, no. 4, pp. 34–41, 1978.
17. R. G. Hamlet, "Testing programs with the aid of a compiler," *IEEE Transactions on Software Engineering*, vol. SE-3, pp. 279–290, 1977.
18. R. H. Untch, "On reduced neighborhood mutation analysis using a single mutagenic operator," in *Proceedings of the 47th Annual Southeast Regional Conference*. ACM, 2009, pp. 71:1–71:4.
19. J. H. Andrews, L. C. Briand, and Y. Labiche, "Is mutation an appropriate tool for testing experiments?" in *Proceedings of the 27th International Conference on Software Engineering*, ser. ICSE '05. ACM, 2005, pp. 402–411.
20. S. Ali, L. C. Briand, H. Hemmati, and R. K. Panesar-Walawege, "A systematic review of the application and empirical investigation of search-based test case generation," *IEEE Transactions on Software Engineering*, vol. 36, no. 6, pp. 742–762, 2010.
21. R. A. Silva, S. d. R. S. d. Souza, and P. S. L. d. Souza, "A systematic review on search based mutation testing," *Information and Software Technology*, vol. 81, pp. 19 – 35, 2017.
22. N. Jatana, B. Suri, and S. Rani, "Systematic literature review on search based mutation testing," *e-Informatica Software Engineering Journal*, vol. 11, no. 1, pp. 61–78, 2017.
23. N. Jatana, S. Rani, and B. Suri, "State of art in the field of search-based mutation testing," in *2015 4th International Conference on Reliability, Infocom Technologies and Optimization (ICRITO) (Trends and Future Directions)*, 2015, pp. 1–6.
24. Y.-S. Ma, J. Offutt, and Y. R. Kwon, "Mujava: An automated class mutation system," *Software Testing, Verification and Reliability*, vol. 15, no. 2, pp. 97–133, 2005.
25. B. F. Jones, D. E. Eyres, and H.-H. Sthamer, "A strategy for using genetic algorithms to automate branch and fault-based testing," *The Computer Journal*, vol. 41, no. 2, p. 98, 1998.
26. L. Bottaci, "A genetic algorithm fitness function for mutation testing," in *Proceedings of the SEMINALL-workshop at the 23rd international conference on software engineering, Toronto, Canada*, 2001.
27. M. Masud, A. Nayak, M. Zaman, and N. Bansal, "Strategy for mutation testing using genetic algorithms," in *Canadian Conference on Electrical and Computer Engineering, 2005*. IEEE, 2005, pp. 1049–1052.

28. B. Baudry, V. L. Hanh, and Y. L. Traon, "Testing-for-trust: The genetic selection model applied to component qualification," in *Proceedings of the Technology of Object-Oriented Languages and Systems (TOOLS 33)*, ser. TOOLS '00. IEEE Computer Society, 2000, pp. 108.
29. B. Baudry, V. Le Hanh, J.-M. Jézéquel, and Y. Le Traon, "Trustable components: Yet another mutation-based approach," in *Mutation testing for the new century*. Springer, 2001, pp. 47–54.
30. B. Baudry, F. Fleurey, J.-M. Jézéquel, and Y. Le Traon, "From genetic to bacteriological algorithms for mutation-based testing," *Software Testing, Verification and Reliability*, vol. 15, no. 2, pp. 73–96, 2005.
31. Y. M. Ben Ali and F. Benmaiza, "Generating test case for object-oriented software using genetic algorithm and mutation testing method," *International Journal of Applied Metaheuristic Computing*, vol. 3, no. 1, pp. 15–23, 2012.
32. H. Haga and A. Suehiro, "Automatic test case generation based on genetic algorithm and mutation analysis," in *2012 IEEE International Conference on Control System, Computing and Engineering*. IEEE, 2012, pp. 119–123.
33. L. Louzada, C. G. Camilo-Junior, A. Vincenzi, and C. Rodrigues, "An elitist evolutionary algorithm for automatically generating test data," in *2012 IEEE Congress on Evolutionary Computation*. IEEE, 2012, pp. 1–8.
34. S. Subramanian and N. Natarajan, "A tool for generation and minimization of test suite by mutant gene algorithm," *Journal of Computer Science*, vol. 7, no. 10, pp. 1581–1589, 2011.
35. G. Fraser and A. Zeller, "Mutation-driven generation of unit tests and oracles," *IEEE Transactions on Software Engineering*, vol. 38, no. 2, pp. 278–292, 2012.
36. M. B. Bashir and A. Nadeem, "A fitness function for evolutionary mutation testing of object-oriented programs," in *2013 IEEE 9th International Conference on Emerging Technologies (ICET)*, 2013, pp. 1–6.
37. K. K. Mishra, S. Tiwari, A. Kumar, and A. K. Misra, "An approach for mutation testing using elitist genetic algorithm," in *2010 3rd International Conference on Computer Science and Information Technology*, vol. 5, 2010, pp. 426–429.
38. C. Molinero, M. Nunez, and C. Andres, "Combining genetic algorithms and mutation testing to generate test sequences," in *Proceedings of the 10th In-ternational Work-Conference on Artificial Neural Networks: Part I: Bio- Inspired Systems: Computational and Ambient Intelligence*. Springer, 2009, pp. 343–350.
39. R. Nilsson, J. Offutt, and J. Mellin, "Test case generation for mutation-based testing of timeliness," *Electronic Notes in Theoretical Computer Science*, vol. 164, no. 4, pp. 97–114, 2006.
40. C. P. Rao and P. Govindarajulu, "Genetic algorithm for automatic generation of representative test suite for mutation testing," *International Journal of Computer Science and Network Security*, vol. 15, no. 2, pp. 11–17, 2015.
41. L. Deng, J. Offutt, and N. Li, "Empirical evaluation of the statement deletion mutation operator," in *Proceedings of the 2013 IEEE Sixth International Conference on Software Testing, Verification and Validation*. IEEE Computer Society, 2013, pp. 84–93.
42. M. E. Delamaro, J. Offutt, and P. Ammann, "Designing deletion mutation operators," in *Proceedings of the 2014 IEEE International Conference on Software Testing, Verification, and Validation*. IEEE Computer Society, 2014, pp. 11–20.
43. *Java Code Coverage for Eclipse*. Available for free at http://www.eclemma.org/index.html.

A Novel Approach to Minimize Energy Consumption in Cloud Using Task Consolidation Mechanism

Sanjay Kumar Giri, Chhabi Rani Panigrahi, Bibudhendu Pati and Joy Lal Sarkar

Abstract Task consolidation is a process to increase usage of cloud computing resources. Maximizing the utilization of resources provides numerous advantages like the customization of IT services, quality of service, and candid services. However, increasing the utilization of resources does not mean optimal energy usage. Most of the researches indicate that the consumption of energy and the utilization of resources in clouds are exceptionally conjugated. The idea of performing the consolidation of tasks is to decrease the usage of resources in order to save energy, while another effort is to maintain a balance between the usage of energy and utilization of resources. In this work, we propose an architecture for minimizing energy consumption in cloud. We used an algorithm for task consolidation in the proposed architecture to minimize energy consumption.

Keywords Cloud computing · Energy consumption · Task consolidation

S. K. Giri (✉)
Department of Computer Science and Engineering, RITE, Bhubaneswar, India
e-mail: girisarkar@gmail.com

C. R. Panigrahi · B. Pati
Department of Computer Science, Rama Devi Women's University, Bhubaneswar, India
e-mail: panigrahichhabi@gmail.com

B. Pati
e-mail: patibibudhendu@gmail.com

J. L. Sarkar
Central University of Rajasthan, Ajmer, India
e-mail: joylalsarkar@gmail.com

C. R. Panigrahi et al. (eds.), *Progress in Advanced Computing and Intelligent Engineering*, Advances in Intelligent Systems and Computing 714,
https://doi.org/10.1007/978-981-13-0224-4_12

1 Introduction

Cloud computing is becoming the transformative change in the business landscape and gaining competency in the technology. It serves businesses as well and positions them for the perspective with respect to information technology. It is gaining popularity because of the advancement of technologies like network agents, software applications, and hardware capacities. All the resources in a cloud are present in distributed fashion. Cloud computing uses several methods to manage all the resources [1]. In recent years, researchers focus on how to efficiently use the resources of the cloud and to minimize energy usage.

Virtualization is one of the main technologies used in cloud computing. Virtual Machines (VMs) are created dynamically which is done on demand. It provides a mechanism to manage all the resources efficiently in cloud [2]. To maximize resource utilization, many methods have been proposed such as discriminating requests, compression of memory, defining the threshold for any resource utility, and distributing the tasks among various VMs [3–5]. There exist a number of researches which show the relationship between usage of various resources and energy consumption [2–4, 6–9]. Some work focused on how to increase the utilization of resources and at the same time, others focus on how to decrease the consumption of energy. The main purpose of both the approaches is to reduce the cost. Every organization wants to reduce the cost. The usage of energy changes with respect to CPU utilization [8]. This encourages researchers to save energy by limiting the CPU utilization. In this work, authors proposed an architecture which aims at reducing the energy consumption in the cloud by using task consolidation in VMs.

The rest of the paper is structured as follows. In Sect. 2, we present the related work proposed in the literature. The proposed architecture is explained in Sect. 3. Lastly, the conclusion and the future work are given in Sect. 4.

2 Related Work

Every consumer and industries use energy for their production. If a system is large and complex, then the system consumes more energy. This results in more financial burden on the organization. Like networks having more systems, cloud having more number of clusters takes the burden of higher cost due to the higher energy usage. There exist a number of researches on how to decrease energy consumption in any network, cloud, or in any industry as ultimately it helps in huge financial benefit to the organizations. So, usage of power is considered as an important issue in any organization.

Gunaratne et al. [1] described a technique to decrease the misuse of energy in cloud. The authors argued that whenever personal computers and network links are futile, they are full of power and energy is wasted unnecessarily. There is no mechanism to manage the personal computers, switches, and network links whenever they are unproductive. This costs in millions to an industry or a country. According to authors view, all the links should be disabled if the devices are unproductive, and power management techniques should be added to the personal computers and network authorities should be more alert. Vasić and Kostić [5] described a mechanism to minimize power on web. According to the authors, on web a large number of links are ideal. So, they should be sent to sleep sate. They proposed a mechanism named as Energy Traffic engineering (EATe) to promote working condition which overlooked energy usage. EATe is a technique which easily sends the link to sleep state and handles the modification of network load without altering the traffic rate. According to the researchers, some services such as streaming, videos on demand, large files on web, etc. enhance the energy usage and the acceptance of cloud computing services among all entities help to consume more energy. EATe and ETC [10] are both energy-aware approaches that help to minimize energy usage on web. EATe technique works with network traffic and hardware components while ETC works in cloud computing.

Various components of hardware and software consume energy to work. Alizai et al. [11] identified items according to their energy usage and found that CPU consumes more energy. Lien et al. [7] gathered data regarding energy usage and CPU utilization and tried to establish a relationship between this two. They developed virtual instrumentation software to find out the energy usage of any streaming components on web in real time. The relationship between CPU and usage of energy shows a nonlinear graph.

Nathuji et al. [9] suggested a mechanism for energy distribution to the VMs and to all the virtualized components of VMs. The authors named their approach as virtual power approach. Their research suggested nearly 34% betterment in consumption of energy. Torres et al. [4] advocated a unification approach for the cloud data center by integrating two components. These are memory compression and request discrimination method. Their evaluation plan was based on two factors, i.e., a real workload and a representative workload scenario. Srikantaiah et al. [3] analyzed the correlation among performance decline, power usage, CPU performance, and disk usage. They represented the problem as a Bin-Packing problem and identified the performance of each server. Finally, they found out which server has better performance and minimum energy usage. Song et al. [2] suggested a utility analytical method for servers which are related to Internet. They found the upper bound of physical server depend on Quality of Service (QoS) and calculated the energy and its utility. Lee and Zomaya [6] recommended two methods to minimize power usage named as ECTC and MaxUtil. These two methods conserve energy without degrading the performance in a cloud atmosphere with uniform resources in terms of capacity and computing power. Both the methods consolidate cloudlets to save energy. MaxUtil helps in decreasing the energy usage by increasing the CPU utilization. It assigns as many cloudlets to the VMs. ETC is opposite to MaxUtil. In ETC, the rate of energy usage increases with the increase in CPU utilization. Hsu et al. [10] proposed an Energy-aware Task Consolidation (ETC) approach that decreases the energy consumption. According to the authors, if there will be a threshold limit for the utilization of CPU, then the usage of energy will be minimized. ETC binds cloudlets among the virtual clusters.

3 Proposed Architecture

In this section, the proposed architecture is presented. Figure 1 shows the proposed architecture having a cloud structure with more than one cluster. Millions of users may be present in a cloud and submit tasks to the cloud manager. A cloudlet is a small-scale data center in a cloud environment used to indulge services quickly.

A cloud manager acts as a mediator which mediates between users and clusters. The cloud manager provides services such as managing, monitoring, and backing up of data. Monitoring means real-time reporting, anticipation, and should be vigilant to all the components. Cloud manager has a job queue where all the submitted tasks by the users are stored. It keeps all the relevant information about a task such as ID of the task (T_j), CPU handling time of task T_j (T_{pj}), arrival time of task T_j (T_{aj}), data size of task T_j (T_{dsj}), and the utilization of CPU. Cloud manager sorts the entire set of tasks according to the size. Normally, the execution time of a task depends on its data size. Higher is the data size, higher is the execution time. The task having least data size is present in the rear part of the job queue and the task having highest data size is present in the front end of the job queue. It uses Shortest Job First (SJF) scheduling algorithm to send the tasks to the cluster. It manages the entire task dynamically. Tasks may come at any time on web. The job of cloud manager is to place the task in the appropriate place in the job queue. Cloud manager checks three constraints before sending a task having least data size to cluster, and they are as follows: First, it checks which cluster is available. In the next step, it checks whether resources are available in the cluster or not. Finally, it

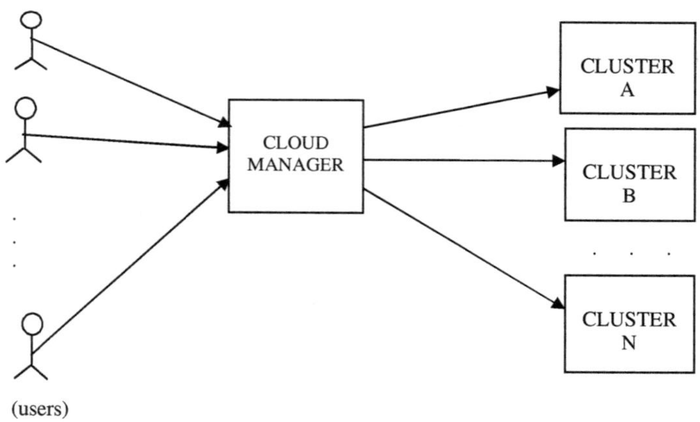

Fig. 1 Proposed architecture

verifies whether the cluster is within the average threshold limit of its CPU utilization. The average threshold of CPU utilization is decided by the researchers. If all the three conditions are satisfied, then a task is sent to the identified cluster.

Cluster is a collection of computers which are integrated loosely or tightly. They work together as if a single computer. All the parts of a cluster are linked through fast LAN. The primary focus of a cluster is to upgrade the performance. VMs are the basic unit of a cluster. A single VM can execute more than one task. Each VM has a CPU utilization limit. If one VM is engaged, then the task is assigned to another VM. In a cluster, tasks are shared among VMs, i.e., workload is shared among CPUs. This encourages keeping the CPU utilization below a threshold limit.

A single cluster is shown in Fig. 2 where more than one VMs are integrated. A number of tasks are assigned to a cluster and similarly, more than one task may be assigned to single VM. Every VM has a threshold limit. Suppose, VM_1 has a threshold limit a_1, i.e., $0 \leq a_1 \leq 100$, VM_2 has a threshold limit b_1, i.e., $0 \leq b_1 \leq 100$, and VM_n has a threshold limit n_1, i.e., $0 \leq n_1 \leq 100$. We can find the average threshold limit of the cluster by using the formula, i.e., $(a_1 + b_1 + \cdots + n_1)/n$, where a_1, b_1, …, n_1 are the CPU utilization threshold limit of VM_1, VM_2, …, VM_n, respectively, and n is total number of VMs present in the cluster A. As the average CPU utilization depends on the users, his/her proper decision leads to minimization of energy consumption.

The steps for execution of task consolidation algorithm for the proposed architecture are given as follows.

Step 1: Cloud_manager keeps the list of tasks [T1,T2, …, Tn] and list of clusters [CLUSTER A, CLUSTER B, …, CLUSTER N].

Step 2: Cloud_manager checks the size of each task.

Step 3: Cloud_manager applies the shortest job first to select the task that needs to be sent to the VM for execution that means task [T_i] having min size is identified.

Step 4: Cloud_manager checks the properties of the selected task and identifies the cluster that provides similar platforms.

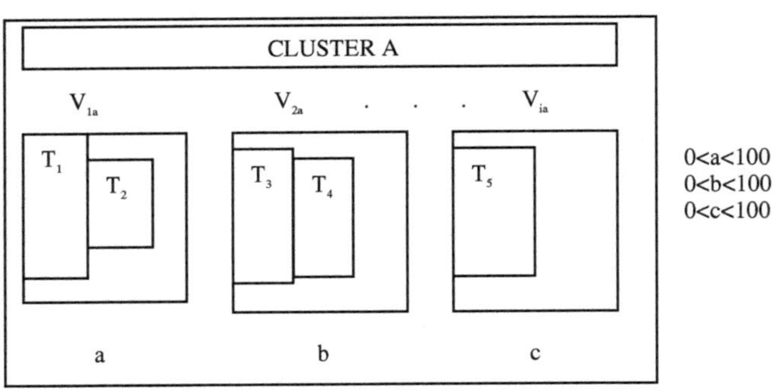

Fig. 2 Average energy consumption (V_{1a},V_{2a} … V_{ia}) < T

Step 5: Cloud_manager checks whether the resources are available with the cluster and average CPU utilization threshold.

Step 6: If steps 4 and 5 are satisfied, then task T_i is sent to the cluster let VC_j.

(i) If cluster VC_j has more than one VMs available to execute the task T_i, then the task T_i will be executed by one of the VMs, which consumes least energy to execute.

(ii) Otherwise, task T_i is sent to another cluster where step (i) is performed.

4 Conclusion and Future Work

Cloud is a collection of various heterogeneous resources. It is the technology in which a network of remote servers is hosted on the Internet to store, manage, and process data rather than using a local server or a personal computer. Resources in a cloud may be virtualized and diversified. The main goal of any cloud computing supported organization is to maximize profits. This can be achieved in various ways. In this work, an architecture is proposed to maximize profit by conserving energy usage in cloud. As a future work, we will check the performance of the proposed algorithm and will compare it with the existing algorithms.

References

1. Gunaratne, C., Christensen, K., and Nordman, B.: Managing energy consumption costs in desktop pcs and lan switches with proxying split TCP connections and scaling of link speed. International Journal of Network Management 15 (5), (2005).
2. Song, Y., Zhang, Y., Sun, Y., and Shi, W.: Utility analysis for internet-oriented server consolidation in VM-based data centers. In Proceedings of IEEE International Conference on Cluster Computing, pp. 1–10, (2009).
3. Srikantaiah, S., Kansal, A., and Zhao, F.: Energy Aware consolidation for cloud computing. In Proceedings of the 2008 Conference on Power Aware Computing and Systems, (2008).
4. Torres, J., Carrera, D., Hogan, K., Gavaldà, R., Beltran, V., and Poggi, N.: Reducing wasted resources to help achieve green data centers. In Proceedings of IEEE International Symposium on Parallel and Distributed Processing, pp. 1–8, (2008).
5. Vasic´, N., and Kostic´, D.: Energy-aware traffic engineering. In Proceedings of the 1st International Conference on Energy-Efficient Computing and Networking, pp. 169–178, (2010).
6. Lee, Y. C., and Zomaya, A. Y.: Energy efficient utilization of resources in cloud computing systems. The Journal of Supercomputing, pp. 1–13, (2010).
7. Lien, C.-H., Bai, Y.-W., Lin, M.-B., Chang, C.-Y., and Tsai, M.-Y.: Web server power estimation, modeling and management. In Proceedings of 14th IEEE International Conference on Networks, vol. 2, pp. 1–6, (2006).

8. Lien, C.-H., Liu, M. F., Bai, Y.-W., Lin, C. H., and Lin, M.-B.: Measurement by the software design for the power consumption of streaming media servers. In Proceedings of the IEEE Instrumentation and Measurement Technology Conference, pp. 1597–1602, (2006).
9. Nathuji, R., and Schwan, K.: VirtualPower: coordinated power management in virtualized enterprise systems. In Proceedings of Twenty-First ACM SIGOPS Symposium on Operating Systems Principles, pp. 265–278, (2007).
10. Hsu, C.-H., Slagter, K. D., Chen S.-C., Chung Y.-C.: Optimizing energy consumption with task consolidation in clouds. The information Sciences 258, pp. 452–462, (2014).
11. Alizai, M. H., Kunz, G., Landsiedel, O., and Wehrle, K.: Power to a first-class metric in network simulations. In Proceedings of the Workshop on Energy Aware Systems and Methods, (2010).

Part II
Machine Learning and Data Mining

Classification of Spam Email Using Intelligent Water Drops Algorithm with Naïve Bayes Classifier

Maneet Singh

Abstract The paper proposes an emerging evolutionary and swarm-based intelligent water drops algorithm for email spam classification. The proposed algorithm is used along with the machine learning classification technique known as naïve Bayes classifier. The intelligent water drops algorithm is used for feature subset construction, and naïve Bayes classifier is applied over the subset to classify the email as spam or not spam. The result of the hybrid method is compared with other evolutionary algorithm used with machine learning classifiers. The proposed algorithm outperforms the other hybrid algorithms.

Keywords Intelligent water drops · Naïve Bayes classifier · Email spam classification

1 Introduction

Email is one of the most commonly used means of communication. Email is a kind of data transmission done over the Internet, which involves transfer of messages in digitized form. This transmission generally leads to two kinds of problems. One is the messages are highly vulnerable to various kinds of direct attacks by the hackers or intruders. Another problem that is faced by the user is the receiving of unwanted emails. This is a sort of indirect attack. These unwanted emails are usually sent to the user by providing one-click link to some phishing websites, thus making the user as a victim of the online security attack. The unwanted emails are generally referred as spams.

M. Singh (✉)
Department of Computer Engineering and Applications,
GLA University, Mathura, India
e-mail: maneetsingh88@gmail.com

C. R. Panigrahi et al. (eds.), *Progress in Advanced Computing and Intelligent Engineering*, Advances in Intelligent Systems and Computing 714,
https://doi.org/10.1007/978-981-13-0224-4_13

1.1 Spam Classification

The different kinds of email are categorized into different classes by the several mail engines. The spam folder is specifically used for containing the spam messages. There are a number of cases when the message which is not unwanted is sent to the spam folder which may lead to any consequences. This task of sending message to the usual inbox or to the spam folder requires the classification of email as spam or not spam. The problem of spam classification has been under consideration since decades. There are certain words or characters which are common in spam messages which can be used for classification. Thus, spam classification is a two-class problem, i.e., binary classification.

1.2 Intelligent Water Drops Algorithm

A single agent may not be able to do a certain task effectively whereas when multiple agents work together to perform certain task, they give efficient results. This type of intelligence is termed as swarm intelligence. There are several techniques which come under swarm intelligence such as ant colony optimization, particle swarm optimization, bee colony optimization, etc. Intelligent water drops is also a swarm-based optimization algorithm. The natural phenomenon of water drops of finding path to flow is simulated by intelligent water drops algorithm. The IWD algorithm could be used to find shortest path problem as well as for selecting a subset of features and eliminating redundant features.

1.3 Naïve Bayes Classifier

The machine learning provides a variety of techniques for classification problem. Some of them are decision trees, k-nearest neighbor, naïve Bayes classifier, etc. The performance of naïve Bayes classifier as compared to others has always been encouraging such that it is widely being used for various classification problems. The naïve Bayes classifier is a probability-based algorithm which follows the Bayes theorem.

2 Related Work

The email spam classification problem has been solved by hybridization of machine learning algorithms and evolutionary algorithms. The task of evolutionary algorithms is to form a subset of features out of the given entire feature set. The machine learning algorithm is then applied for classification on reduced set of features.

Ant colony optimization has been used with naïve Bayes classifier by Renuka et al. [1]. The ACO algorithm is used for feature selection and naïve Bayes classifier for further classification process. The results are compared with the hybrid genetic algorithm with Naïve Bayes classifier.

Wang et al. [2] used the genetic algorithm along with SVM for spam filtering. The feature selection is done by GA, and the results were also compared with the use of SVM alone.

Alijila et al. [3] used intelligent water drops algorithm for rough set feature selection. The use of IWD with RS gave a very good result as compared to other competing methods.

3 Proposed Methodology

The intelligent water drops algorithm as proposed by Hoseini [4] has been used to eliminate the redundant features from the given set of features. The two main properties of water drops have been used, i.e., soil and velocity. The water drops prefer path which contains less soil and always carry some soil of the path which leads to reduction in the velocity of water drops. The intelligent water drops algorithm always requires the representation of any given problem in the form of a graph. Here, nodes represent features and edge between the nodes represents the selection of subsequent features. The water drops are required to flow from one node to another node until the performance of the naïve Bayes classifier is under tolerance level. This stopping criterion is based on the performance of the classifier as well as a number of features used for classification.

The steps of the hybrid algorithm are as follows:

- Random Distribution
  - Randomly distributing all the IWDs over the distinct features.
  - The number of IWDs could be taken as same as the number of features.

- Next Feature Selection
  - An IWD move from one node to the other, i.e., select another feature.
  - The selection is based on the probability of selecting the next feature.
- Local Soil Updation and Velocity Updation
  - The velocity of each IWD is updated based on the soil on the edge connecting the current and the next feature.
  - Soil of the selected edge and soil of each IWD is updated based on the performance of classifier and the number of feature selected.
- Global Soil Updation
  - The current iteration best solution is found and its feature subset is retained.
  - The soil of the path traversed by the best performing IWD is updated.
- Total Best Solution Updation
  - The total best solution is compared with the iteration best solution and updated if the quality of the iteration best solution is better than total best solution.
- The above steps are repeated till they meet the stopping criteria.

The proposed method is also elaborated with the help of a flowchart depicted in Fig. 1.

4 Results

The proposed method was tested on a standard database provided by the UCI Repository. The spambase dataset was used. The dataset comprises 4601 observations and 57 features, out of which 1813 are spam and 2788 are not spam. The algorithm when implemented on the above dataset gives very good results. The performance of the method is evaluated on the famous parameters used in classifiers, i.e., precision, recall, accuracy, and f1 score. All these parameters are computed with the help of confusion matrix obtained from the classifier results (Table 1).

The result of the proposed algorithm is also compared with GA and ACO. The proposed algorithm outperforms the above methods in terms of all evaluating parameters. The comparative chart is shown in Table 2.

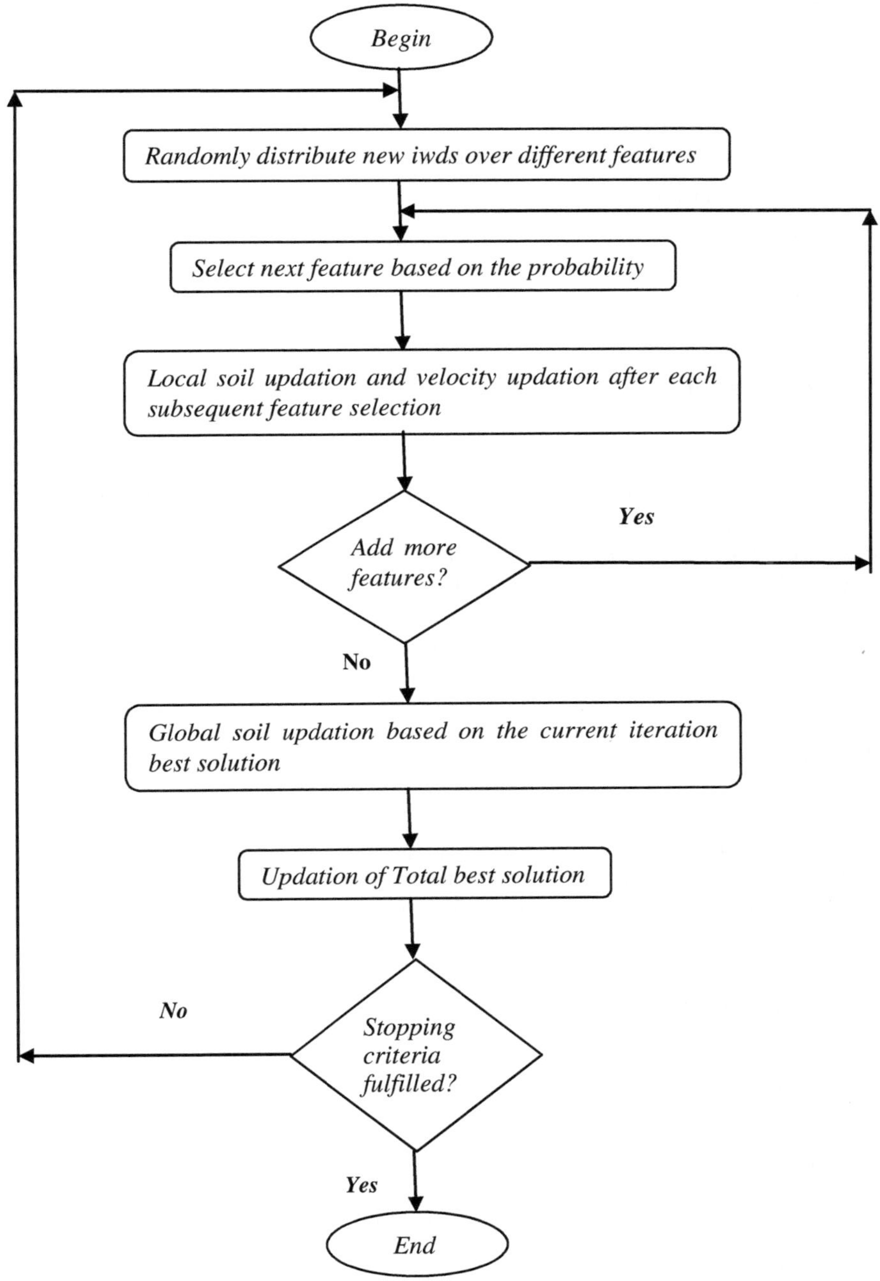

Fig. 1 Flowchart of the proposed methodology

Table 1 Confusion matrix obtained after applying the proposed method

Desired/Actual->	Positive	Negative
Positive	1723	90
Negative	192	2596

Table 2 Comparison of IWD with other algorithms

Parameters	GA-naïve Bayes	ACO-naïve Bayes	IWD-naïve Bayes (proposed)
Precision	85	89	90
Recall	71	78	95
Accuracy	77	84	94
F1 Score	75	87	92

5 Conclusion

The IWD algorithm is an emerging algorithm that has been applied successfully on various problems [5] which may be interdisciplinary in nature. The IWD has shown always the capability to replace the much famous another swarm-based method known as ACO. The results of applying IWD with naïve Bayes classifier have been very much encouraging and better than GA and ACO when used with naïve Bayes classifier. These results motivate to apply the proposed method on various other datasets as well as if required enhanced the IWD algorithm [6] to get better results.

References

1. Renuka D.K., Visalakshi P., Sankar T.: Improving E-mail Spam Classification using Ant Colony Optimization. International Conference on Innovations in Computing Techniques (2015).
2. Wang H., Yu Y., Liu Z. (2005) SVM Classifier Incorporating Feature Selection Using GA for Spam Detection. In: Yang L.T., Amamiya M., Liu Z., Guo M., Rammig F.J. (eds) Embedded and Ubiquitous Computing – EUC 2005. EUC 2005. Lecture Notes in Computer Science, vol 3824. Springer, Berlin, Heidelberg.
3. Alijla B.O., Peng L.C., Khader A.T., Al-Betar M.A. (2013) Intelligent Water Drops Algorithm for Rough Set Feature Selection. In: Selamat A., Nguyen N.T., Haron H. (eds) Intelligent Information and Database Systems. ACIIDS 2013. Lecture Notes in Computer Science, vol 7803. Springer, Berlin, Heidelberg.
4. Hosseini S.H.: The intelligent water drops: a nature - inspired swarm-based optimization algorithm. Int J. Bio-Inspired Computation, Vol 1, Nos. ½, 2009.
5. Singh M., Saini S. (2014) Optimization of Complex Mathematical Functions Using a Novel Implementation of Intelligent Water Drops Algorithm. In: Babu B. et al. (eds) Proceedings of the Second International Conference on Soft Computing for Problem Solving (SocProS 2012), December 28–30, 2012. Advances in Intelligent Systems and Computing, vol 236. Springer, New Delhi.
6. Kumar M., Jayaraman S., Bhat S., Ghosh S., Jayaraman V. (2014) Variable Selection and Fault Detection Using a Hybrid Intelligent Water Drop Algorithm. In: Babu B. et al. (eds) Proceedings of the Second International Conference on Soft Computing for Problem Solving (SocProS 2012), December 28–30, 2012. Advances in Intelligent Systems and Computing, vol 236. Springer, New Delhi.

Evaluation of Neuropsychological Tests in Classification of Alzheimer's Disease

N. Vinutha, R. Jayasudha, K. S. Inchara, Hajira Khan, Sonu Sharma, P. Deepa Shenoy and K. R. Venugopal

Abstract Many neuropsychological tests are available to measure cognitive declinement in a person affected by Alzheimer's disease. To evaluate his/her current stage in dementia and also to find the disease progression, it is necessary to perform a serial assessment of tests. As a result, the huge amount of data gets collected which depends on the number of neuropsychological tests performed to examine the patient and also with the number of visits to the clinic. From the previous correlation studies, it is observed that high computational time is required to process many neuropsychological tests. Therefore, the scores obtained from these tests are subjected to attribute selection algorithms. The six different attribute selection algorithms are used to rank the attributes, but the top four ranked attributes are consistent with InfoGain and OneR attribute evaluators. So, we subject the ordered attributes from these two attribute evaluators to different classifiers with 10-fold cross-validation. The random forest classifier performed better with InfoGain and OneR attribute evaluators. Therefore, an accuracy of 99.1% and ROC area of 0.999 is obtained from the set of top four attributes, and similar results are obtained from the set of top six and seven attributes with the combination of Infogain and OneR with BayesNet classifier.

Keywords Alzheimer's disease · Attribute selection · Cognitive domains Classification · Neuropsychological test

---

N. Vinutha (✉) · R. Jayasudha · K. S. Inchara · H. Khan · S. Sharma
P. D. Shenoy · K. R. Venugopal
Department of Computer Science and Engineering, University Visvesvaraya College of Engineering, Bangalore University, Bangalore, India
e-mail: vinutha1v@gmail.com

C. R. Panigrahi et al. (eds.), *Progress in Advanced Computing and Intelligent Engineering*, Advances in Intelligent Systems and Computing 714,
https://doi.org/10.1007/978-981-13-0224-4_14

1 Introduction

Alzheimer's Disease (AD) [1] is an irreversible form of dementia that occurs among adults who fall in the age group of 40–90, but most commonly seen after 65 years. It is caused by the deposition of amyloid plaques and neurofibrillary tangles in different regions of the brain. As a result of deposition, it shows the impact on the brain size and further, the functional ability of neurons is reduced and gradually gets destroyed. These changes are noticeable and can be measured before the development of symptoms. The symptoms begin with short-term memory loss which continues further to long-term memory loss and change in behaviors and language (aphasia) that become severe day by day as the disease progresses. There are several factors like age, family history, smoking, obesity, diabetes, and high blood pressure that also increase the risk to AD. Hence, a wide range of techniques such as medical imaging, neuropsychological tests, medical history, physical, and neurological examination are performed to assess the clinical diagnosis of the patients.

Neuropsychological testing [2] is a measure of cognitive declinement in a person. It is utilized to identify the ability of an individual to perform day-to-day activities in the diseased state. It is necessary to follow the serial assessments of the tests to evaluate the performance of a person after subjecting to medication. From the previous studies, it is inferred that a specific pattern is developed in an affected patient of the similar age group that can be differentiated from the normal aging. Further, better discrimination can also be done by the fusion of neuroimaging data with neuropsychological scores or by the combination of genetic risk factors with neuropsychological scores. Table 1 shows the list of neuropsychological tests with their associated domains.

Each test follows a standardized protocol and is conducted with the help of pencil, paper, visual aids, and computer. Due to multiple cognitive deficits, it is

Table 1 List of neuropsychological test

Cognitive domains	Neuropsychological test
Episodic memory	Logical memory (WMS-III)
Semantic memory	Wechsler Adult Intelligence Scale (WAIS)-II
Attention/working memory	WAIS-III, Number span
Executive function	Trail B, Digit symbol, Trail A
Depression	Geriatric Depression Scale (GDS)
Dementia severity	Mini Mental State Examination (MMSE)
Language (verbal fluency)	Animal list, Vegetable list
Language (naming)	Boston Naming Test (BNT)

always better to utilize the combination of tests from various domains that are helpful to characterize the pattern developed due to cognitive impairments and also to make a better decision in clinical diagnosis.

But there is a quick rise in the cost of medical and healthcare system. This is due to accumulation of a large amount of data and lot of time requirement by an expert to process the collected data and to make a decision in diagnosis and treatment of patients. All the problems mentioned above can be handled by machine learning approach [3, 4] as it plays a significant role in feature reduction and also retains only those features that lead to high performance.

1.1 Motivation

Neuropsychological scores have tremendous scope to integrate and validate under various domains. However, consideration of all the clinical scores requires more computational time. Thus, identifying a small subset of scores is very crucial for the correlation studies with either neuroimaging data or genetic risk factors.

1.2 Contribution

- To identify the visit with more number of demented cases.
- Identify suitable attribute selection algorithms based on ranking method.
- Evaluate different machine learning algorithms for classification of AD.
- To identify a minimum set of attributes with better performance.

1.3 Organization

The paper is organized as follows: Literature survey is presented in Sect. 2. Proposed system architecture for Alzheimer's disease classification is explained in Sect. 3. Experiments and results are presented in Sect. 4. The paper concludes in Sect. 5.

2 Literature Survey

Enormous techniques are developed by researchers to focus toward the prediction of AD. The papers [5–9] provide state-of-the-art survey on clinical scores to measure the progression of disease using longitudinal data, correlation studies, replacement of the existing neuropsychological tests with equivalent new tests, and the contribution of single and multiple predictors toward prediction of AD.

McCutcheon et al. [5] have evaluated whether AD pathology and depression are related to each other in MCI and Mild Dementia. The study requires clinical and neuropathological data. The GDS is obtained as a result, by subjecting the covariates Neuritic Plaque (NP) score and Braak stages of Neurofibrillary (NF) to the regression model. The outcome showed that GDS is not related to NP score or cognitive decline or their combination. Hence, it can be said that depression in early AD is evident to be independent of NP and NF pathology.

Authors in [6] have suggested four new non-proprietary tests in NACC's UDS neuropsychological battery. The suggested tests can be used as a replacement for the existing tests by measuring the correlation factor between them. To assess the correlation between each of the previous and new tests, a crosswalk study is conducted. Tests having good correlation are said to have high prediction accuracy. These equivalent scores can be considered for the longitudinal analysis.

The authors of the paper [7] have proposed the development of a multi-domain model to predict the progression of dementia in Alzheimer's disease. The data obtained from NACC are used in the evaluation of transition probabilities between the health states based on the behavioral, functional abilities, cognitive function, and also to analyze the status of symptoms. From the above analysis, it is inferred that there is a transition in the stages of AD within a time span of 12 months and the model helped in the assessment of AD.

Lee Gavett et al. [8] considered the longitudinal data of healthy older adults from NACC dataset. The longitudinal data between two and three annual visits were considered for each subject. The followup scores and baseline test scores of eleven neuropsychological tests are used in linear mixed effect regression to obtain Reliable Change Intervals (RCI) and also to calculate the cumulative frequency of the raw scores. It is inferred that age, education, and baseline test scores are good predictors. Tests related to attention and executive functioning are significant to healthy aging, and tests related to episodic and semantic memory are effective with relevance to practice effects.

According to John et al. [9], the cognitive performance of neuropsychological tests from UDS dataset has been interpreted in two approaches, namely shared variance and unique variance. In the first approach, the latent factor is used as a single predictor for measures of severity, whereas the second approach utilizes 12 raw scores from the neuropsychological tests as the predictors of dementia diagnosis. A logistic regression analysis is performed on single and multiple predictors to obtain a log-odd ratio, model fit statistics, and classification accuracy. The results thus obtained from logistic regression revealed the significance of each test in the diagnosis of dementia.

3 Proposed Work

Figure 1 illustrates the architecture of the proposed work. It consists of four modules:

(1) Data collection,
(2) Preprocessing,
(3) Attribute selection and
(4) Classification.

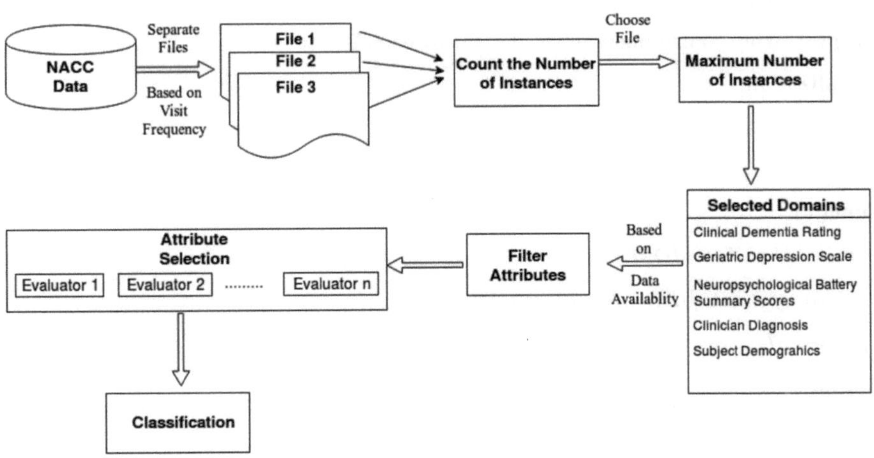

Fig. 1 System architecture for classification of Alzheimer's disease

3.1 Data Collection

Data for the research work are collected from National Alzheimer's Coordinating Center (NACC), as it comprises the data of various Alzheimer's Disease Centers (ADCs). The collected NACC data constitute of subject demographics, health history, global staging, clinical dementia rating, neuropsychiatric inventory questionnaire, geriatric depression scale, functional activities questionnaire, clinician judgment of symptoms, clinical diagnosis, and neuropsychological battery summary scores for 11,735 unique instances.

3.2 Preprocessing

Patients' visits are available from 1 to 12. The preprocessing step begins by mapping unique IDs of the instances to the set of consecutive integers and to identify the number of visits available for each patient as shown in Algorithm I.

Algorithm I: ***To count the number of visits by each ID***

Input: NACC Data

Output: Array containing number of visits by each ID

Initialization:

Map Unique ID's To 0, 1, 2.....N

No_of_Visits [0 to N] $\leftarrow$*0*

Function

for i$\leftarrow$ *[1 to Total ID's]*

    No_of_Visits [ID] $\leftarrow$*No_Visits [ID] +1*

    i$\leftarrow$*i+1*

end for

The identification of number of visits for each instance is followed with the determination of demented cases from each visit time. After the determination, it is observed that the number of demented patients is increased with the higher visit times. Therefore, we group the instances with visits three to twelve into separate files as shown in Algorithm II.

Algorithm II: ***To group, the ID's with 3 to 12 visits***

Input: *No_of_Visits*

Output: *2D Array Group, Size of Group*

Initialize:

Group [3 to 12][0 to N]←0

Size_of_Group [0 to 12] ←0

Function:

```
for i←[3 to 12]
    for j←[0 to N]
        if No_of_Visits[j]==i
            push j into Group[i]
            Size_of_Group[i]++
        end if
        j←j+1
    end for
    i←i+1
end for
```

After the separation of files based on visit times, we count the number of patients for each selected visit and chose the file with the largest number of instances as shown in Algorithm III. From analysis of all the visits, we infer that the three times visited data are the largest with 1345 unique instances as shown in Fig. 2. Therefore, we consider three times visit data in our study.

In the next step, following domains such as subject demographics, global staging, clinical dementia rating, geriatric depression scale, clinician diagnosis, and neuropsychological battery scores are selected for the third visit data. From the above domains, we select the attributes that have data availability greater than 50%. So attributes such as *CDRSUM (Clinical Dementia Rating Sum Of Boxes)* (100%), *CDRGLOB (Global Clinical Dementia Rating)* (100%), *MEMORY* (100%), *COMPORT* (98.88%), *CDRLANG (Language)* (98.88%), *NACCGDS (Geriatric Depression Scale)* (91.59%), *NACCMMSE (Mini Mental State Examination)* (58.73%), *LOGIMEM (Logical Memory)* (58.43%), *MEMUNIT (Logical Memory IIA-Delayed)* (58.36%), *DIGIF (Digit Span Forward)* (58.28%), *DIGIB (Digit Span Backward)* (58.21%), *ANIMALS (Animals List)* (91.07%), *VEG (Vegetables List)*

Algorithm III: ***To find the group with maximum number of ID's***

```
Input: Group, Size_of_Group
Output: Group having maximum no of ID's
Initialize:
Max←0, Group_No←0
Function:
for i←[3 to 12]
    if Size_Of_Group[i]>=Max
        Max←Size_Of_Group[i]
        Group_No←i
    end if
    i←i+1
end for
```

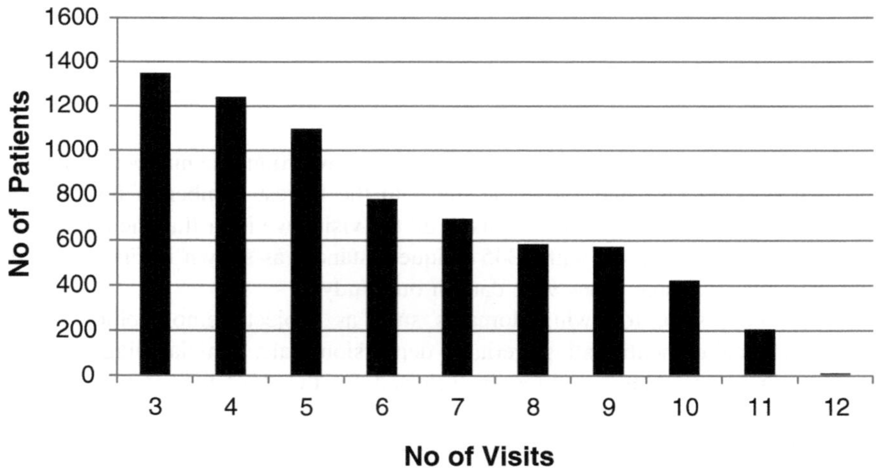

Fig. 2 Total number of instances based on number of visits

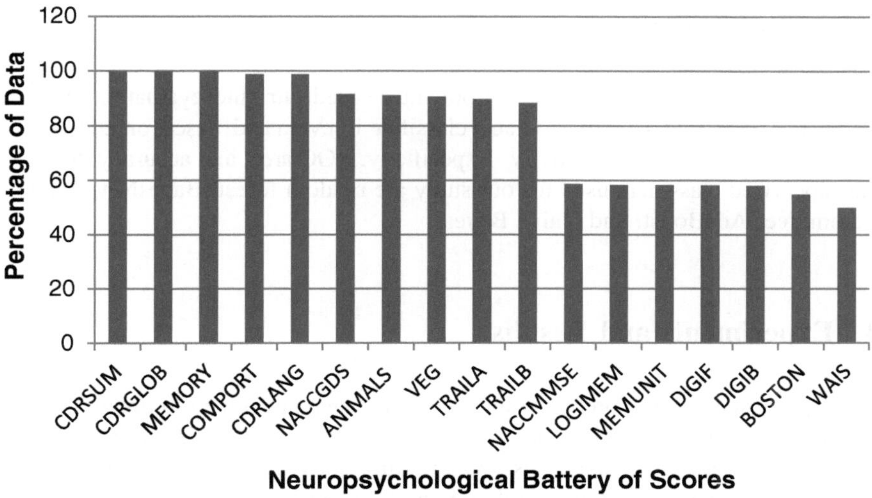

Fig. 3 The neuropsychological scores with data availability greater than 50%

(90.70%), *TRAILA (Trail Making Test Part A)* (89.73%), *TRAILB (Trail Making Test Part B)* (88.40%), *WAIS (Wechsler Adult Intelligence Scale)* (50.96%), and *BOSTON (Boston Naming Test)* (57.91%) are considered due to sufficient data availability as shown in Fig. 3.

3.3 Attribute Selection

Attribute selection is a process of searching the best subset of attributes from a given dataset. Various measures considered for attribute selection are correlation, distance, information, dependence, and consistency. The two different approaches to attribute selection are wrapper and filter method. In wrapper method, the subset selection is based on the learning algorithm, so the computational time increases for every subset that is evaluated in the context of the learning model, whereas, in filter method, the relevance of attribute is measured by using their correlation with the dependent variable and it is computationally faster since it does not involve training of the model. In our study, filter-based attribute evaluators are used to order the attributes based on the obtained rank.

3.4 Classification

All the ordered attributes obtained from filter-based attribute evaluators are subjected to supervised classifiers. Each classifier is evaluated based on the performance measures such as sensitivity, 1-specificity, ROC area, and accuracy. Some of the supervised classifiers used for our study are random forest, BayesNet, Random Committee, AdaBoost, and Naive Bayes.

4 Experiments and Results

In the proposed system, preprocessing is the first step performed to obtain the data required for our study. The preprocessed data are then subjected to attribute selection algorithms such as OneRAttributeEval, InfoGainAttributeEval, GainRatioAttributeEval, ReliefAttributeEval, SymmetricalUncertAttributeEval, and CorrelationAttributeEval. The ranked attributes obtained from these algorithms are further subjected to classifiers with 10-fold cross-validation. The classifiers, random forest, and BayesNet performed better with an accuracy of 99.4 and 99.1% for all the 22 attributes, that were ordered based on ranks obtained for Infogain and oneR attribute evaluators. However, our aim is to predict AD with a minimum number of attributes. Hence, the least-ranked attribute is removed each time and subjected to above classifiers until the minimal subset with the highest accuracy and ROC area is obtained.

The top-ranked attributes from InfoGainAttributeEval and OneRAttributeEval are *{CDRSUM, MEMORY, CDRGLOB, NACCMMSE, ANIMALS, MEMUNITS, VEG, LOGIMEM…} and {CDRSUM, MEMORY, CDRGLOB, NACCMMSE, LOGIMEM, CDRLANG, TRAILB…}*, respectively. It is observed that top four ranked attributes are common in these two attribute evaluators, so the performance is measured by the classifier with a minimal set of attributes.

An accuracy of 99.1% and ROC area of 0.999 is obtained from the top six attributes for the combination of InfoGainAttributeEval with BayesNet classifier, and same results are obtained from top seven attributes for the combination of OneRAttributeEval with BayesNet classifier. Figures 4 and 5 show the plot of ROC area versus number of attributes for BayesNet classifier.

However, the combination of InfoGainAttributeEval and OneRAttributeEval with random forest classifier results with an accuracy of 99.1% and ROC area of 0.999 from the top four attributes. Figures 6 and 7 show the plot of ROC area versus number of attributes for random forest classifier. The comparison of performance measures is shown in Table 2.

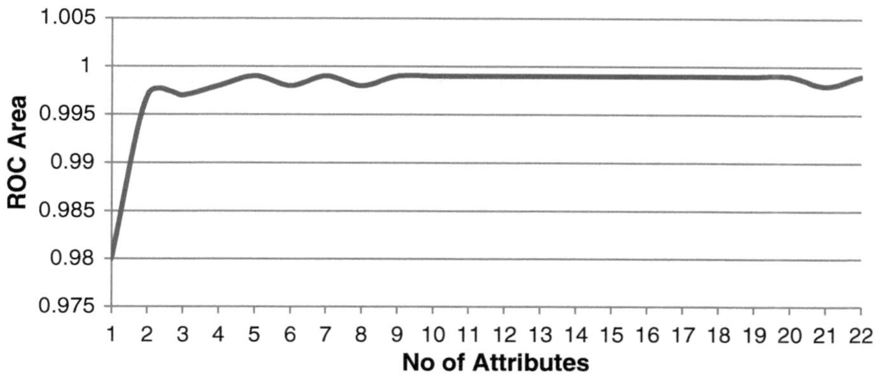

Fig. 4 The plot of ROC area versus number of attributes for OneR with BayesNet

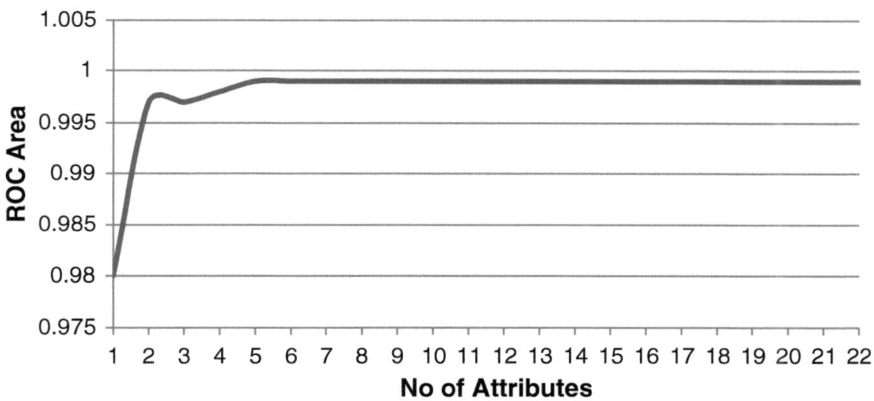

Fig. 5 The plot of ROC area versus number of attributes for Infogain with BayesNet

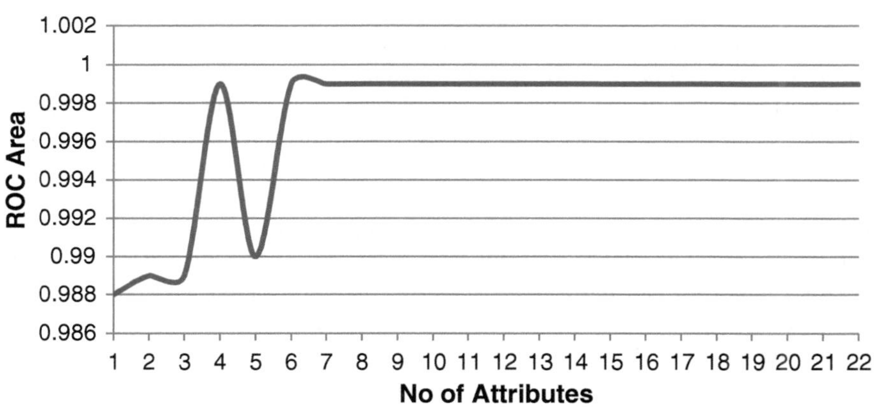

Fig. 6 The plot of ROC area versus number of attributes for OneR with Random forest

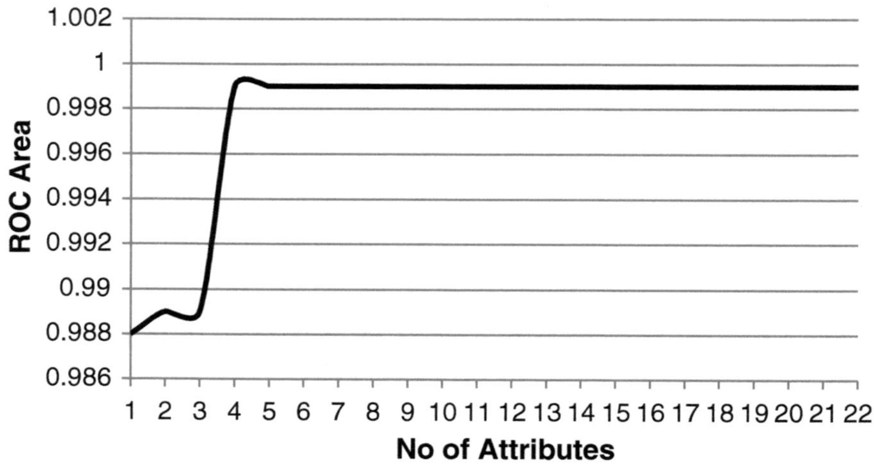

Fig. 7 The plot of ROC area versus number of attributes for Infogain with random forest

Table 2 Comparison of classification accuracy

Attribute evaluator	Classifier	No of attributes	Accuracy	ROC area	Sensitivity	1-Specificity
InfoGain	Random Forest	4	99.1	0.999	0.991	0.101
	BayesNet	6	99.1	0.999	0.99	0.068
OneR	Random Forest	4	99.1	0.999	0.991	0.084
	BayesNet	7	99.1	0.999	0.99	0.068

5 Conclusions and Future Work

The neuropsychological data for the instances with same visit number are significant to classify AD patients. Hence, we consider various domains for the selected visit as a measure of the cognitive declinement in a person. For each domain considered, the attributes only with sufficient data availability are selected to further analysis. After the refinement of data, it is subjected to six attribute selection methods. The performance of the ranked attributes is evaluated based on the metrics: sensitivity, 1-specificity, ROC area, and accuracy. An accuracy of 99.1% and ROC area of 0.999 is obtained from top four attributes by using OneRAttributeEval and InfoGainAttributeEval in combination with random forest classifier. Thus, it is inferred that these top four attributes have a significant role in the classification of AD.

The neuropsychological scores with data availability less than 50% are excluded in our study. Therefore, our focus is to handle a missing data and to study their significance in the classification of AD and in future, we extend our study on fourth visit data.

Acknowledgements The authors would like to thank curators of the NACC database for providing the data to conduct the research.

References

1. Emilien, G., Durlach, C., Minaker, K.L., Winblad, B., Gauthier, S., Maloteaux, J.M.: Alzheimer Disease: Neuropsychology and Pharmacology. Birkhauser (2012).
2. Harvey, P.D.: Clinical Applications of Neuropsychological Assessment, Vol. 14. Dialogues in Clinical Neuroscience (2012).
3. O'Kelly, N.: Use of Machine Learning Technology in the Diagnosis of Alzheimer's Disease (2016).
4. Joshi, S., Shenoy, P.D., Venugopal, K.R., Patnaik, L.M.: Evaluation of Different Stages of Dementia Employing Neuropsychological and Machine Learning Techniques. In First IEEE International Conference on Advanced Computing (2009) 154–160.
5. McCutcheon, S.T., Han, D., Troncoso, J., Koliatsos, V.E., Albert, M., Lyketsos, C.G., Leoutsakos, J.M.S.: Clinicopathological Correlates of Depression in Early Alzheimer's Disease in the NACC. International Journal of Geriatric Psychiatry (2016).
6. Monsell, S.E., Dodge, H.H., Zhou, X.H., Bu, Y., Besser, L.M., Mock, C., Hawes, S.E., Kukull, W.A., Weintraub S.: Results from the NACC Uniform Data Set Neuropsychological Battery Crosswalk Study, Vol. 30. Alzheimer Disease & Associated Disorders (2016) 134–139.
7. Green, C., Zhang, S.: Predicting the Progression of Alzheimer's disease Dementia: A Multidomain Health Policy Model, Vol. 12. Alzheimer's and Dementia (2016) 776–785.
8. Gavett, B.E., Ashendorf, L., Gurnani, A.S.: Reliable Change on Neuropsychological Tests in the Uniform Data Set, Vol. 21. Journal of the International Neuropsychological Society (2015) 558–567.
9. John, S.E., Gurnani, A.S., Bussell, C., Saurman, J.L., Griffin, J.W., Gavett, B.E.: The Effectiveness and Unique Contribution of Neuropsychological Tests and the Latent Phenotype in the Differential Diagnosis of Dementia in the Uniform Data Set, Vol. 30. Neuropsychology (2016) 946.

Brain Visual State Classification of fMRI Data Using Fuzzy Support Vector Machine

S. Kavitha, B. Bharathi, S. Pravish and S. S. Purushothaman

Abstract The fMRI (Functional Magnetic Resonance Imaging) technology is a revolutionary tool that has lit up the studies of human cognitive processing with the help of efficient methods of image and data analysis. Machine learning classifiers are widely employed to extract all sorts of information from neuroimaging data. This study aims to identify tangible patterns in the fMRI data for visual activity and perform multivariate pattern analysis. It is done by selecting relevant features to indicate the response to visual stimulus of a set of objects belonging to eight different categories. The task intends to identify the nature of the response to the stimuli and classify them according to the brain's neural activation to the visual stimuli. An SVM (Support Vector Machine) classifier and an FSVM (Fuzzy Support Vector Machine) classifier are implemented to perform the classification based on the features. The training of the classifiers involved 72 test samples per category. The 24 test samples of each category were tested with each of the classifiers. Conclusively, for this dataset, the FSVM classifier performs better than SVM classifier with an increased accuracy of 4% and classifying certain categories with improvement.

Keywords Neuroimaging · fMRI · Support vector machine · Fuzzy support vector machine · Classification

---

S. Kavitha (✉) · B. Bharathi · S. Pravish · S. S. Purushothaman
Department of Computer Science & Engineering, SSN College of Engineering, Kalavakkam, Tamil Nadu, India
e-mail: kavithas@ssn.edu.in
URL: http://www.ssn.edu.in

B. Bharathi
e-mail: bharathib@ssn.edu.in

S. Pravish
e-mail: pravishsainath@gmail.com

S. S. Purushothaman
e-mail: purushoth1109@gmail.com

C. R. Panigrahi et al. (eds.), *Progress in Advanced Computing and Intelligent Engineering*, Advances in Intelligent Systems and Computing 714,
https://doi.org/10.1007/978-981-13-0224-4_15

1 Introduction

Given the multitude of activities made possible by the brain, the visual object recognition is an extremely intriguing task. For so many years, neuroscientists have been trying to further their understanding of the various cognitive processes. It is increasingly possible by solving the problem of brain-mapping wherein a relationship is established between the perceptual state and the specific patterns in the brain.

Functional Magnetic Resonance Imaging (fMRI) is an imaging technology which is primarily used to record the brain activation during any activity by measuring neural activity in the brain. Its non-invasive, safe and easy-to-use nature powered with its promising spatial and good temporal resolution have contributed immensely to its popularity in medicine, research and industry. It has been instrumental in empowering studies that have thrown light on the functional aspects of the brain with respect to memory, language, pain, learning and emotion as elaborately discussed in [1]. Multiple methods of data analysis when applied on the fMRI data can give deeper insights into the patterns represented by these images of the brain. Researchers have now employed fMRI to conduct hundreds of studies that identify which regions of the brain are activated on average when a human performs a particular cognitive task. Research publications have enumerated the summary statistics of brain activity in various locations.

As elucidated in [2–4], a number of machine learning techniques can be effectively employed to draw certain scientific results. These depict how computing is used as tool to delve deeper into the patterns that are generated in the brain. These patterns have to be used by computing algorithms to draw inferences about many useful things. Pattern analysis is the key to solve the brain-mapping problem.

1.1 Related Studies

Many approaches have been developed for pattern analysis. Multivariate Pattern Analysis (MVPA) is described and used in [5] and [6]. It involves analysing the pattern considering the fMRI data as a whole. It has proven to be more sensitive and more informative about the functional organization of cortex than in univariate analysis with the General Linear Model (GLM). The multivariate pattern analysis allows us to study how specific stimuli are encoded in detailed activity patterns in specific parts of the brain.

Classifying the stimuli for a particular activity is a fundamental task in dealing with the brain. Machine learning makes this possible with data classification algorithms. A slew of classifiers have been used across various works. LDA is implemented by the work in [7]. A technique of using a collection of machine learning algorithms to train classifiers of specific stimuli is adopted in [8]. Here, GNB and kNN classifiers are combined to achieve more than 95% accuracy. It points out that the high dimensionality and intrinsically low signal-to-noise ratio of fMRI data raises a need for using alternate and collective methods of classification. The approach seems to improve multiple subject experiments by reducing the high inter-subject variability in brain function. The work in [9] uses LDA and SVM classifiers with the SVM classifiers achieving 53% accuracy for restricted voxels. It explains the classification with specific reference to the visual cortex. The work [10] also uses an SVM classifier to predict the orientation of the stimulus. Jeiran Choupan [11] compares SVM, NN and CRF classifiers under various conditions for the same dataset used in our work. Song [12] is a comprehensive study of SVM classification for fMRI data with different voxel selection schemes. Weili Zheng [13] points out the need for these classifiers to optimally select brain regions. It is evident that, owing to the high-dimensional nature and volume of the fMRI data, the performance of SVM classifier seems to have outsmarted all the other classifiers as in [14]. Hence, the motivation of this work is to incorporate an SVM classifier. In order to further enhance the performance of an SVM classifier, a different version of the same can be employed.

1.2 Fuzzy SVM

Fuzzy SVM (FSVM) is a classification methodology that can be incorporated as an extension of the SVM classifier with additional conditions for classification. A clear explanation about the underpinnings of the SVM and FSVM classifiers formulation were detailed in [15]. It explains about the handling decisions of classification based on certain rules whenever the distribution of the test data in the feature space does not yield a decisive classification.

The interest is to train classifiers to automatically decode the subjects' visual cognitive state over an interval in time. When such classifiers are trained reliably, they can be made as virtual sensors of cognitive states to use them for further analysis or usage. This study investigates the utility of methods in improving the prediction accuracy of classifiers trained on functional neuroimaging data taken from [16].

1.3 Scope of Our Work

This work explores the use of a classification method—FSVM in the context of an event-related functional neuroimaging experiment where participants viewed images of objects in intervals. It requires to train support vector machines on functional data to predict with a greater accuracy the objects viewed by the participants. It shows that the classifier achieves better than random predictions and the average accuracy is close to that of the actual stimuli. Here, the classification method consists of feature extraction, feature selection and classification parts, and it also employs a feature extraction method based on the mean change in the intensity from baseline condition to the sample.

To process the fMRI data corresponding to the task of visualizing objects belonging to finite categories one after the other, classifiers are built to classify input fMRI image volumes into their corresponding categories. It involves performing statistical corrections and analysis on the data, selecting and extracting the characteristic features as voxels and training the data for classification by an SVM classifier, followed by the constructed FSVM classifier. N-fold cross-validation mechanism is used with the training of these classifiers. The performance of the two classifiers is compared with respect to their relative accuracies in predicting the different categories corresponding to the data.

2 System Design

The system implemented in this work involves the fMRI data of the visual one-back task dataset downloaded from [16] and the acquired fMRI data of two additional subjects that are employed as test data in the classification. The image volumes are preprocessed applying many techniques and the category representing features are extracted. The feature set is given as input to the SVM and FSVM classifiers for categorizing the data into the corresponding categories of objects that were viewed by the subject during data acquisition.

2.1 Dataset

During the task of recording the fMRI images for the dataset, the subjects see the eight objects presented as greyscale photographs for 24-s, followed by 12-s of rest. Each of the stimuli is held for 500 ms with an inter-stimulus interval of 1500 ms. Twelve time series volumes are extracted for each of the eight subjects.

Additional real test data was acquired by us by carrying out the same task (only for two categories—shoe and bottle images) with two healthy volunteers under the same experimental conditions [repetition time (TR) = 2500 ms, 40 3.5-mm-thick sagittal images, field of view (FOV) = 24 cm, echo time (TE) = 30 ms, flip angle = 90°].

Currently, the dataset consists of visual identification of eight different categories of objects: House, Scrambled, Cat, Shoe, Bottle, Scissors, Chair and Face as greyscale images by eight different subjects and additional test data.

2.2 Preprocessing

A series of operations is applied to correct and normalize the data to make it compatible for extracting features and further processing. This helps in preparing our data for classification. They are summarized in Table 1.

Table 1 A summary of the preprocessing steps applied

	Preprocessing step	Reason
1	Brain extraction	Elimination of non-brain tissues with highly variable contrast
2	Motion correction	Adjustment of the variation of intensity due to head movements
3	Spatial filtering	Increase in signal-to-noise ratio and smoothness
4	Temporal filtering	Discarding noise due to very high and very low frequencies
5	Detrending	Ensuring that there is significant intensity change over time
6	Intensity normalization	Transformation of data into a normal distribution

2.3 *Feature Extraction*

Extracting the features with respect to the baseline condition using feature space reduction and a searchlight technique to construct data that can be used for training.

The major steps in feature selection and extraction are explained as follows:

Examples Creation The brain images corresponding to each category are distributed across time in independent blocks. This step combines the images across time points as an example. It is done by block averaging, i.e. averaging the images within each block of time in a run.

Spherical Searchlight The image volume examples in a trial is analysed by applying a searchlight to compare each voxel with the neighbouring voxels. In this process, it is inferred if the voxel is representative of the features of the category. Hence, a set of voxels which represent the features are selected and the pattern is generated by formulating it as a feature vector labelled by the category it represents.

To reduce feature input dimensionality feature representing voxels need to be selected using a similar approach used in [3].

i. A fixed sphere is moved over the brain image volume, voxel-by-voxel.
ii. The mean intensity of all the voxels within the sphere is computed.
iii. Fixing the mean value within the sphere as a threshold, all the voxels with higher intensity are assigned a score based on ranking.
iv. This scoring information is corrected for multiple comparisons as each data point is used multiple times.
v. Finally, all those voxels with the maximum score are selected.

Voxel Reduction The set of voxels returned by the searchlight are huge in number and contains voxels that are trivial. The features that represent the visual activity are localized around the visual cortex. A brain atlas that provides a spatial mask of the visual cortex is used as an anatomical mask to select the voxels that are confined around the visual cortex and reject the remaining voxels. This process yields a reduced list of voxels based on the Region of Interest (ROI).

Generation of Training and Test Data The reduced set of voxels is converted into a form of data which can be used to train a classifier. The x, y and z coordinates of the voxels are indicated along with the category label whose features the voxels represent. The unequal number of voxels of each category is adjusted by padding with out-of-bound values. The data is then optimally split into test and training data.

2.4 Classification

Support Vector Machine (SVM) and Fuzzy Support Vector Machine (FSVM) classifiers with linear kernels have been used for the classification. A cross-validation mechanism is used to determine the best possible subset for training.

SVM leads to good generalization performance [17] even in case of high-dimensional data and a small set of training patterns. It reduces the problems due to dimensionality by reducing the risk of overfitting the training data when the number of voxels is reduced.

FSVM follows the same principle of SVM, but certain additional computations are performed to add more decision rules to classify data that are either unclassified or classified in an overlapping fashion.

In FVSM, for an m-dimensional input $\mathbf{x_i}(i = 1, \ldots, M)$ belonging to a class y_i, and assuming the data to be separable linearly, the decision function is given by

$$D_i(\mathbf{x}) = \mathbf{w}^t\mathbf{x} + b \tag{1}$$

where $\mathbf{w}$ is an m-dimensional vector and b is a scalar with the separating hyper-plane satisfying:

$$y_i(\mathbf{w}^t\mathbf{x_i} + b) \geq 1 \tag{2}$$

As stated in [15], the procedure of classification is as follows:

i. If $D_i(\mathbf{x}) > 0$ for just one class, the input is classified into the class.
ii. If $D_i(\mathbf{x}) > 0$ for more than one class $i\epsilon(i = i_1, \ldots, i_l, l > 1)$, classify the datum into the class with the maximum $D_i(\mathbf{x})(i \in i_1, \ldots, i_l)$.
iii. If $D_i(\mathbf{x}) \leq 0$ for all the classes, the datum is assigned to the class with the minimum absolute value of $D_i(\mathbf{x})$.

The corresponding category is determined by the decision function is output. This classification result for the test data belonging to all of the categories is output by constructing a confusion matrix by the classifiers.

3 Results and Discussion

The features are extracted from the fMRI data in the dataset to construct the training set and tested with test data for classification. The results obtained from the classifiers are analysed to measure their performance.

The neuroimaging data exists as anatomical image volume and functional image volumes. The functional image volumes are the acquired data that reflect the intensity change as the stimulus events take place. The 4D time series for each subject consists of 1452 volumes with $40 \times 64 \times 64$ voxels, corresponding to a voxel size of $3.5 \times 3.75 \times 3.75$ mm and a volume repetition time of 2.5 s.

A sequence of preprocessing steps were applied using FSL [18], to the four-dimensional images to refine them and highlight the features. The brain portion is extracted from the image volumes and the corresponding masks are generated. Motion correction and filtering are done on them to correct recording errors and remove noise.

The resulting image volumes were further preprocessed to normalize the intensities across the voxel space. Detrending was performed on consecutive image volumes in the time series. After applying the other preprocessing steps, it appears like Fig. 1.

The final preprocessed fMRI data is used for creating examples of the average image volumes for the corresponding stimuli conditions. Further, the spherical searchlight technique is applied on them using PyMVPA [19] to extract the voxels which represent the features.

The voxels in the features of the corresponding runs are reduced in number using ROI representing the visual cortex and corresponding feature data of the 577 voxels per object category is generated. The feature data is represented as training and test data.

The input data is split into training data consisting of nine runs and test data consisting of three runs per subject.

An SVM classifier with a linear kernel is invoked using PyMVPA with the generated training and testing data as input. It performs N-fold cross-validation by selecting various combinations of the training and test data to come up with the best possible classification.

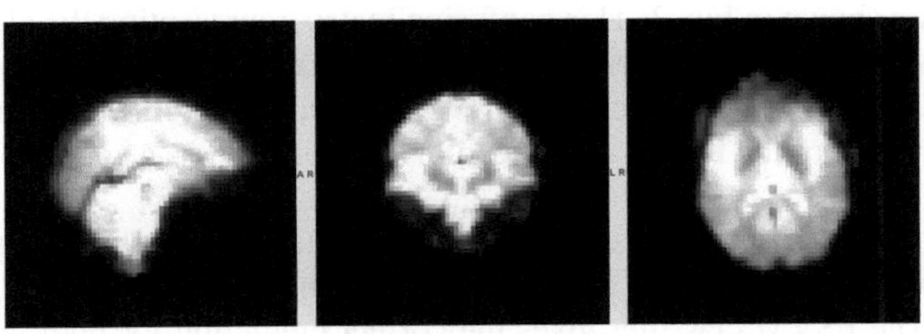

Fig. 1 A slice of the final preprocessed fMRI image volume

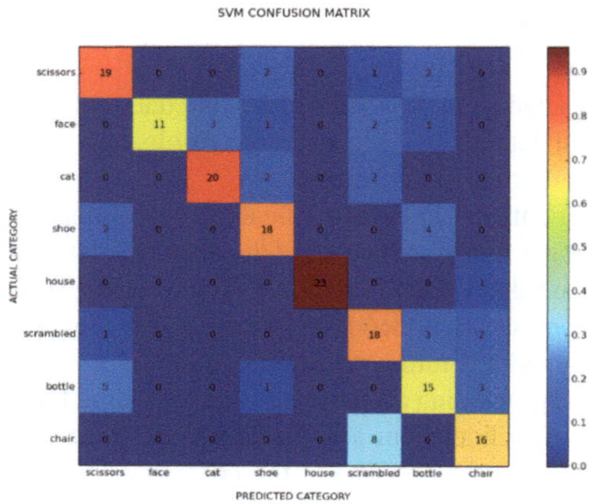

Fig. 2 SVM confusion matrix

Table 2 Results of testing SVM and FSVM classifiers with acquired dataset

Test subject	Test subject 1		Test subject 2	
Actual category	Shoe	Bottle	Shoe	Bottle
SVM classification result	Shoe	Scissors	Bottle	Scissors
FSVM classification result	Shoe	Bottle	Shoe	Scissors

The 12 runs of 8 subjects are split into 9 runs for training data and 3 runs of test data. There are eight categories of visual objects. The classifier outputs the classified label in each case. The category labels predicted by the classifier for test samples are compared with the actual categories they belong to.

The test samples of each category were tested with the SVM classifier. Out of 192 total samples, 140 were correctly classified. The results of the SVM classifier are summarized as the number of test samples predicted per categories versus the actual categories are shown in Fig. 2. This confusion matrix representing the classification results of the SVM classifier for each of the 24 test samples for the eight categories.

In the case of FSVM classifier, out of 192 total samples, 146 were correctly classified. The results of the FSVM classification are presented as a confusion matrix is shown in Fig. 3.

The real test data acquired for the categories: shoe and bottle were tested with the SVM and FSVM classifiers and the results of the classification are summarized in Table 2. The FSVM classifier gives the correct prediction for both the subjects, indicating a better generalization over SVM.

Table 3 compares the number of test samples that were classified correctly in each category by the SVM and FSVM classifiers.

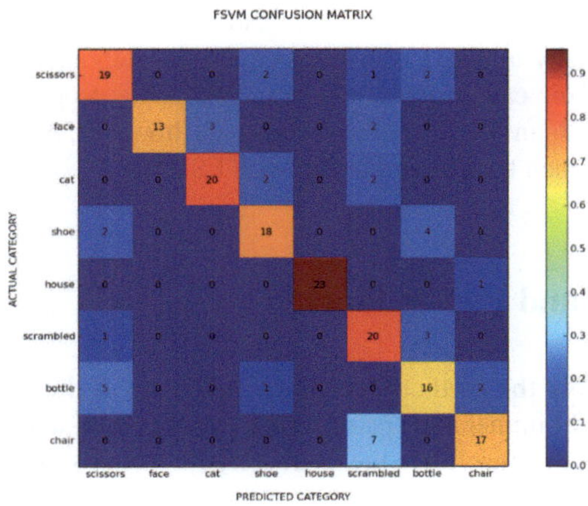

Fig. 3 FSVM confusion matrix

Table 3 Classification results of SVM and FSVM classifiers for various categories

Category of visual object	No. of test samples correctly classified by SVM	No. of test samples correctly classified by FSVM
Scissors	19	19
Face	11	13
Cat	20	20
Shoe	18	18
House	23	23
Scrambled	18	20
Bottle	15	16
Chair	16	17

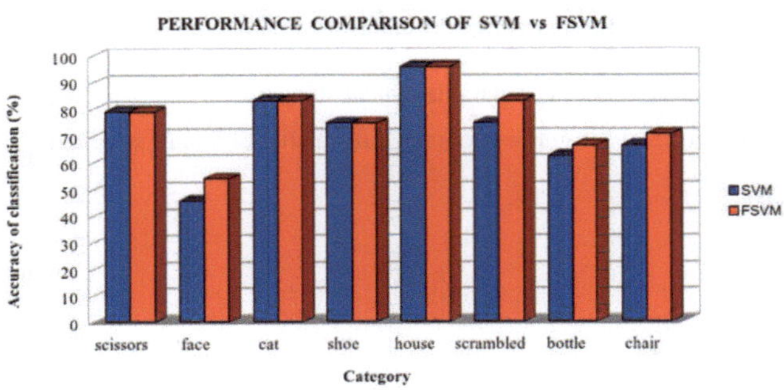

Fig. 4 Performance comparison of SVM and FSVM classifiers

The overall accuracy percentage of SVM was 72.92% and that of FSVM was 76.04%. Figure 4 shows that FSVM has considerably enhanced the overall percentage of accuracy along with the accuracy of certain specific categories. The category—*face*, whose accuracy was 45.83% with SVM had improved crossing the halfway mark to 54.16%. *Scrambled*, which had 75% accuracy in previous SVM has increased the accuracy to 83.33%. It is especially a category that is hard to generalize. *Bottle* and *chair* categories also saw considerable progress in accuracy with an increase of more than 4%. The other categories, however, perform with the same accuracy as SVM when trained and tested with FSVM.

4 Conclusion and Future Work

This work carried out the prediction of the visual state of the subject according to the object viewed by him/her and classified the visual stimuli into various categories. The major task was to consolidate the characteristic features of each of the stimulus object into a number of voxels to use for multivariate pattern analysis. The extracted features were used to train an SVM classifier and was tested to understand which categories were predicted accurately and which categories were mistaken for other categories by the classifier. The accuracy of the classifier was noted down. To minimize the effect of wrongly classified or unclassified data, Fuzzy SVM classifier was built by modifying the existing classifier and performing training and testing for the same data. This work demonstrates the improvement in the classification accuracy of the presently existing SVM algorithm when a Fuzzy SVM (FSVM) is used. This work is aimed at highlighting the possibility of applying computational methods to further the current medical diagnosis practices. It can be extended to building human–computer interfaces and understanding brain visual information encoding.

Acknowledgements We would like to thank the management of our institution, SSN College of Engineering, for funding our work and providing all necessary support. We would also wish to extend our gratitude to Bharat Scans, Chennai for facilitating in the capture of fMRI dataset for additional subjects.

Ethical approval All procedures performed in studies involving human participants were in accordance with the ethical standards of the SSN research committee and with the 1964 Helsinki declaration and its later amendments or comparable ethical standards.

Informed consent Informed consent was obtained from all individual participants included in the study.

References

1. P. Jezzard, P.M. Matthews, S.M. Smith.: Functional Magnetic Resonance Imaging: An Introduction to Methods, Oxford Medical Publications (2006)
2. T.M. Mitchell, R. Hutchinson, R.S. Niculescu, F. Pereira, X. Wang, M. Just, S. Newman.: Learning to decode cognitive states from brain images, Machine Learning, vol. 57, issue. 1, pp. 145–175. Springer (2004)
3. Francisco Pereira, Tom Mitchell, and Matthew Botvinick.: Machine learning classifiers and fMRI: a tutorial overview, Neuroimage, vol. 45, issue. 1, pp. 199–209 (2009)

4. Lemm S., Blankertz B., Dickhaus T., Muller K.R.: Introduction to machine learning for brain imaging. Neuroimage, vol. 56, issue. 2, pp. 387–399 (2001)
5. Y. Fan, D. Shen, C. Davatzikos.: Detecting cognitive states from fMRI images by machine learning and multivariate classification. In: Conference on Computer Vision and Pattern Recognition Workshop, New York (2006)
6. Norman K, Polyn SM, Detre G, Haxby JV.: Beyond mind-reading:multi-voxel pattern analysis of fmri data, Trends in Cognitive Sciences, vol. 10, issue. 9, pp. 424–430 (2006)
7. Haynes J.D., Rees G. Decoding mental states from brain activity in humans. Nat. Rev. Neurosci vol. 7, issue. 7, pp. 523–534 (2006)
8. Carlos Cabral, Margarida Silveira, Patricia Figueiredo.: Decoding visual brain states from fMRI using an ensemble of classifiers. Pattern Recognition, vol. 45, issue. 6, pp. 2064–2074 (2004)
9. David D. Cox., Robert L. Savoya.: Functional magnetic resonance imaging (fMRI) "brain reading": detecting and classifying distributed patterns of fMRI activity in human visual cortex. NeuroImage, vol. 19, pp. 261–270 (2003)
10. Kamitani, Y., Tong, F.: Decoding the visual and subjective contents of the human brain. Nature Neuroscience, vol. 8, issue. 5, pp. 679–685 (2005)
11. Jeiran Choupan, Julia Hocking, Kori Johnson, David Reutens, Zhengyi Yang.: Brain Decoding Based on Functional Magnetic Resonance Imaging Using Machine Learning: A Comparative Study. International Journal of Machine Learning and Computing, vol. 3, issue. 1, pp. 132–136 (2013)
12. Song S, Zhan Z, Long Z, Zhang J, Yao L.: Comparative Study of SVM Methods Combined with Voxel Selection for Object Category Classification on fMRI Data. PLoS ONE, vol. 6, issue. 2, e17191 (2011)
13. Weili Zheng, Elena S. Ackley, Manel Martinez-Ramon, Stefan Posse.: Spatially Aggregated Multi-Class Pattern Classification in Functional MRI using Optimally Selected Functional Brain Areas, Magn Reson Imaging, vol. 31, issue. 2, pp. 247–261 (2013)
14. J.M. Miranda, L.A.W. Bokde, C. Born, H. Hampel, M. Stetter.: Classifying brain states and determining the discriminating activation patterns:support vector machine on functional MRI data, Neuroimage, vol. 28, issue. 4, pp. 980–995 (2005)
15. Abe Shigeo, Inoue Takuya.: Fuzzy support vector machines for pattern classification. In: International Joint Conference on Neural Networks, pp. 1449–1454 (2001)
16. OpenfMRI data repository, http://openfmri.org/dataset/ds000105
17. Bilwaj G., Christos D.: Analytic estimation of statistical significance maps for support vector machine based multivariate image analysis and classification. Neuroimage, vol. 78, pp. 270–283 (2013)
18. FMRIB Software Library, http://www.fmrib.ox.ac.uk/fsl
19. Python Multivariate Pattern Analysis toolbox (PyMVPA), http://www.pymvpa.org/

Brain Tumor Classification for MR Imaging Using Support Vector Machine

Monika, Rajneesh Rani and Aman Kamboj

Abstract Nowadays, brain tumor segmentation is most challenging task in the field of medical image processing. Manual segmentation of these images by the domain experts is a time-consuming process. There is numerous automatic algorithm for MRI image segmentation and classification but still, they need to develop an efficient and fast algorithm. Accurate segmentation of tumor helps in early diagnosis of the tumor. This paper presents an efficient approach for brain tumor for MRI image using support vector machine (SVM) and Otsu thresholding. We tested the performance of fuzzy c-means clustering, k-means, and KIFCM (integration of k-means and fuzzy c-means). Our proposed method outperforms the existed algorithm in terms of accuracy and execution time.

Keywords Brain tumor · MR Images · PCA · Segmentation
SVM · Thresholding

1 Introduction

Image segmentation plays a vital role in medical image analysis because of its usefulness in detecting and treating the brain tumor. Image segmentation subdivides the medical image into the meaningful portion for extracting the relevant information for treatment of the patient [1]. Brain tumor nowadays is a leading cause of death of many people. There is a significant research to prove that if the tumor is detected in the early stage it can be cured. A brain tumor occurs when cells grow

Monika (✉) · R. Rani · A. Kamboj
Department of Computer Science and Engineering, Dr. B. R. Ambedkar National Institute of Technology, Jalandhar, Punjab, India
e-mail: monika.pu30@gmail.com

R. Rani
e-mail: ranir@nitj.ac.in

A. Kamboj
e-mail: amankambojj@gmail.com

C. R. Panigrahi et al. (eds.), *Progress in Advanced Computing and Intelligent Engineering*, Advances in Intelligent Systems and Computing 714,
https://doi.org/10.1007/978-981-13-0224-4_16

disorderly in the brain. The complete examination of the human brain can be scanned by computed tomography (CT) scan or magnetic resonance imaging (MRI) scan. MRI scan is better than CT scan because of its greater range and it does not use any radiation. On other hand, CT scan uses X-rays. So, MRI scan is the best option for detecting a tumor in the clinical diagnosis [2]. There are two types of tumors, i.e., benign tumor (moderate developing) and malignant tumor (rapidly developing) [3]. The most common type of the tumor is a benign tumor which is low glioma; this type of tumor looks moderately typical, develops gradually, and does not spread (metastasize) to other solid tissues in the body or attack cerebrum tissue. It does not cause serious injury [4]. Therefore, many segmentation techniques are used for classification of MRI images such as thresholding methods, histogram-based methods, edge-based methods, clustering methods (Mean shift, k-means, and fuzzy c-means), and hybrid methods (merging of two or more techniques) [5–7].

The rest of this paper is organized as follows: Sect. 2 includes related work in tumor detection and classification techniques. Section 3 includes our proposed work for classification of a brain tumor. Section 4 includes experimental setup used in this research. Finally, the conclusion is conferred in Sect. 5.

2 Related Work

There are numbers of studies in the field of medical image segmentation, which are used for brain tumor detection. Due to the effect of noise, distortion, and bad intensity values, there is significance aftermath on the performance of these methods. Some of them are described in our literature survey below.

Deepa et al. [8] proposed a hybrid genetic-fuzzy method for detection of the tumor. The proposed method is achieved by two algorithms: FCM clustering algorithm and genetic algorithm. They first input the brain MRI images. Then to remove the noise and denoising, they apply preprocessing and after that perform clustering based on hybrid techniques (FCM and Genetic algorithm). At last, tumor is detected using above steps. They proved that the accuracy and computation time on the given datasets images are very impressive.

Maksoud et al. [9] proposed an efficient image segmentation approach using integration of k-means with fuzzy c-means algorithm to provide an accurate brain tumor detection. K-means gives the best output for large datasets and it is fast but if the tumor is malignant then it undergoes incomplete detection. Also, the strategy can get the perfect condition of k-means in the parts of computation time. However, fuzzy c-means can get the perfect condition in the part of accuracy for detecting malignant tumor cells. The proposed algorithm was performed on three datasets and is based on four steps: preprocessing (denoising using a median filter and skull removal using BSE), clustering (integration of k-means and fuzzy c-means), extraction, and contouring stage (thresholding segmentation and active contour by level set) and validation.

Wu et al. [10] proposed a method based on color-based k-means clustering-based brain tumor segmentation. In this technique, k-means is used to convert a grayscale MRI image to color space image using pseudocolor transformation. In this step, all the pixels map one by one to the predefined color map. After that, this RGB color space is converted to CIE lab color model. In the last step, this model is used to separate tumor region using k-means clustering and histogram clustering. Luminosity feature is used to obtain the final segmented image.

Majumder and Kshirsagar [11] proposed a brain tumor segmentation approach which is a combination of k-means, adaptive mean shift, and expectation maximization. In this proposed method, first, the median filter is applied for preprocessing of the image. After that, k-means, adaptive mean shift, and expectation maximization are used to segment the image that features are extracted from the segmented image and load into gray level co-occurrence matrices. In the last step, SVM classifier is used to check whether the segmented tissue is normal or abnormal.

3 Proposed Methodology

The main idea of this proposed work is to utilize the advantages of both thresholding and SVM for efficiently detect the tumor region in MRI input image. There are eight stages of our algorithm as shown in Fig. 1.

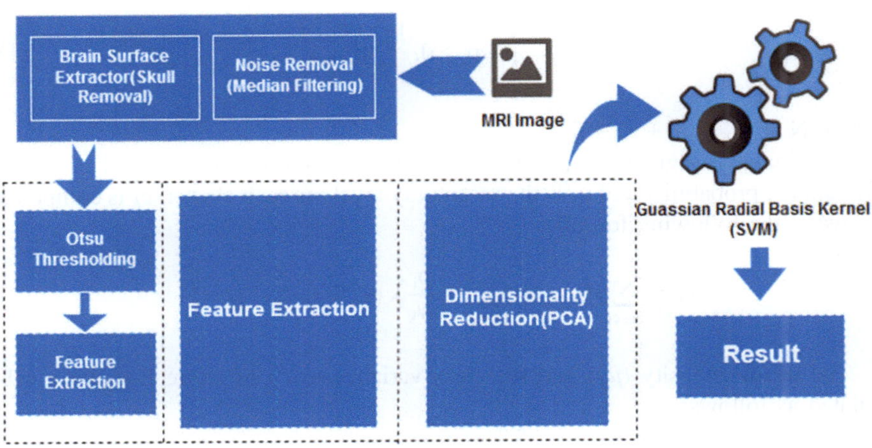

Fig. 1 Proposed algorithm for brain segmentation

3.1 Preprocessing

The first stage of brain tumor extraction is to preprocess the MRI image to enhance the quality of Input image. Since MRI image is more sensitive than any other images in the medical field, they should have maximum quality and minimum redundancy. These images are primarily degraded due to Gaussian and Poisson noise. Median filter is used to remove the noise from MRI images. It works by moving pixel by pixel through the image, supplanting each an incentive with the median of neighboring pixels. Most of the recent research proves that median filter is a better method as compared to linear filter for removing noise from the image. The final image of this step is a noise-free image. In the second step of preprocessing, brain skull is extracted from the image using brain surface extractor (BSE) algorithm. BSE used dilation and erosion for removing the irregularities from the image. Skull, eyes, scalp, background, and all the unwanted structure are filtered out in this step.

3.2 Otsu Thresholding

In this step, we use intensity-based thresholding. It is the most important technique in image computer vision and image processing. It converts the grayscale image into the binary image that makes it easier to extract the important features from the MRI image. It is used to highlight the tumor region from the background. The resultant image of this step provides high processing speed and smaller storage space.

For the gray level, L from 0 to L − 1. f_i stands for the number of pixels with gray level of i. The probability is calculated by the following formula:

$$p_i = \frac{f_i}{N},\ p_i \geq 0,\ \sum_{i=0}^{l-1} p_i = 1 \tag{1}$$

where $N = (f_0 + f_1 + \cdots + f_{L-1})$

If an image is segmented in k clusters, then selected threshold be k − 1. w_k is the cumulative probability and μ_k is the mean gray level for each cluster. c_k is computed by using the following formula:

$$w_k = \sum_{i \in c_k} p_i,\ \mu_k = \sum_{i \in c_k} i \cdot \frac{p_i}{w_k} k \in \{0, 1, \ldots, K-1\} \tag{2}$$

The mean intensity (μ_T) and the class variance ($\sigma_B{}^2$) of entire image are calculated as follows:

$$\mu_T = \sum_{i=0}^{L-1} i \cdot p_i = \sum_{k=0}^{K-1} \mu_k . w_k \tag{3}$$

$$\sigma_B^2 = \sum_{k=0}^{K-1} w_k . (\mu_k - \mu_T)^2 \tag{4}$$

3.3 Feature Extraction

After preprocessing is done, the features are extracted from the image. We are considering both intensity and texture features for segmentation because it helps us to differentiate between benign and malignant tumors. In this step, we are using combination of Gabor, GMM, and GLCM features for segmenting and classifying the tumor region. The following features are used for this purpose:

1. Contrast: It is used to differentiate between darker and the lighter areas of the image.

$$\text{Contrast} = \sum_{i,j=0}^{n-1} P_{i,j}(i-j)^2 \tag{5}$$

2. Correlation: Correlation of the pixels is computed using the coefficient between ranges −1 and +1.

$$\text{Correlation} = \sum_{i,j=0}^{n-1} P_{i,j} \frac{(i-\mu)(j-\mu)}{\sigma^2} \tag{6}$$

3. Homogeneity: It is defined as homogeneous state of pixels.

$$\text{Homogeneity} = \sum_{i,j=0}^{n-1} \frac{P_{ij}}{1+(i-j)^2} \tag{7}$$

4. Entropy: It is a measurement to check the random variable's uncertainty.

$$\text{Entropy} = \sum_{i,j=0}^{n-1} -\ln(P_{ij})P_{ij} \tag{8}$$

5. Energy: It is defined as sum squared on GLCM elements. It is known as uniformity of elements.

$$\text{Energy} = \sum_{i,j=0}^{n-1} \left(P_{ij}\right)^2 \tag{9}$$

6. GMM features: GMM features are previously used for the face identification [12] which provides better results in terms of complexity, robustness, and classification. These features also provide better accuracy of classifier to differentiate between the clutter parts [13]. It is a probability density function to represent weighted sum of K Gaussian component densities according to

$$p(\chi \backslash \lambda) = \sum_{i=1}^{K} \omega_i g(\chi \backslash \mu_i, \Sigma_i) \tag{10}$$

where χ is the N-dimensional vector, ω_i is the mixture weight, and g is the component of Gaussian densities.

3.4 Feature Reduction (Principal Component Analysis)

In feature extraction phase, there are numerous features which take longer computational time and memory storage to perform the various operations. So, dimensionality reduction techniques imply to reduce the feature set. In this research, we are using principal component analysis (PCA) to reduce the feature set.

PCA projects the original features into the eigenvectors to compute the eigenspace of features from where features with higher variance are selected which can linearly distinguishable. First, linear transformation I_n^T is created by converting feature vector space $X \in R^m$ into lower dimensional space $Y \in R^m$, $m > n$, where m is the dimension of input data and n is the number of eigenvectors. The covariance matrix is used to construct the feature space. Top n eigenvectors are used to construct a transformation matrix A_n of size $m \times n$. The output feature vector contains the features with the higher variance among them. The number of feature used in the proposed method is kept using hit-and-trial method.

3.5 Extraction of Tumor Region

For extraction of tumor, we are using a support vector machine. It is the best classifier for brain tumor detection as suggested by many research studies. It is independent of any dimension and feature set. It transforms the higher dimensional data into nonlinear map function by constructing the new hyperplane with maximum margin from the training dataset. In linear SVM, it tries to find out all hyperplanes that can minimize the error in the training data and separate it to maximum distance from the closest points.

$$W \cdot X + b = 0 \tag{11}$$

In above equation, W is the weight parameter and b is the bias parameter.
The maximum margin hyperplane is defined as

$$\text{Minimize}\frac{1}{2}\|\boldsymbol{w}\|^2 \boldsymbol{with}\, \boldsymbol{y}_i(\boldsymbol{w}\cdot\boldsymbol{x}_i+\boldsymbol{b})\geq 1 \tag{12}$$

In our proposed work, we are using nonlinear SVM because it can easily separate nonlinear data from hyperplane. We are using Gaussian radial basis kernel (GRBK) SVM for transforming nonlinear function into higher dimensional space by fitting it to maximum margin hyperplane. The kernel function of GRBK is defined as

$$K(x_i, x_2) = \exp\left(-\frac{(\|x_i - x_j\|)^2}{2\sigma^2}\right) \tag{13}$$

4 Experimental Results

4.1 Dataset

To evaluate our proposed algorithm accuracy, we conduct the experiment on five datasets. The first one is **Brain Web** (Simulated Brain database) dataset [14] which consists of 152 images that contain brain tumors with ".MNC" extension. The second dataset is **Digital Imaging and Communications in Medicine (DICOM)** [15] which consists of 22 brain tumor images. All these DICOM records are compacted in JPEG2000 transfer syntax with ".DCM" extension. The third dataset is **BRATS** dataset [16] which consists of 81 images. The fourth dataset is **Medinfo** [17]. The last dataset is **Harvard** dataset [18] that contain 17 images which are collected from the Harvard Medical School website. All these files were opened by Medical Image Processing, Analysis, and Visualization (MIPAV) [19] and changed over to ".JPG" extension. MIPAV is used to analyze and visualize the medical images. The MIPAV application is platform independent since it is composed in java.

4.2 Training Phase

In this step, training dataset selected manually. From DS1 (BrainWeb) out of 157 images we choose 70 images, for DS2 (DICOM) out of 22 images we select 11 images, for DS3 (Harward) out of 17 images 9 are selected, from DS4 (BRATS) out of 81 images 40 is selected, and finally for DS5 (Medinfo) out of 31 images 15 images are selected. These images contain tumor region of different classes. We choose training images from all the datasets so that problem of overfitting can be avoided. If we use only one dataset for training the classification model, accuracy of the method will be quite low for another dataset. Therefore, we combine all the dataset training images and create a ".mat" file for training the SVM model. Once the training ".mat" file containing all classes of tumor is generated, SVM is trained by using the features of MRI images. Finally, classification is done on the basis of Gaussian Kernel SVM function. The pixels are classified into Benign, Malignant according to label assigned during the training phase.

4.3 Result and Discussion

In this section, we demonstrate the consequences of our proposed image segmentation technique. The experiment was executed utilizing MATLAB 7.12.0 R2013a. We used i7 core computer with 4 GB RAM and an NVIDIA/(1 GB VRAM) VGA card for result evaluation.

Table 1 shows the result of four proposed methods step by step. First, a median filter is applied and after that BSE applied. In the second step, thresholding is applied and after that feature is selected and reduced. In the final step, SVM classifier is applied to segment the tumor and detect the certain type of tumor.

Table 2 shows the performance metrics of our proposed method on five datasets. Proposed method shows highest accuracy among all the methods described earlier in this section. The computation time is also very low in case of the proposed method. We also find that in some cases our proposed method can accurately find the tumor region and its type.

Tables 3, 4, and 5 show the performance metrics of k-means, fuzzy c-means, and KIFCM [10]. From the results, we can observe that k-means and fuzzy c-means have equal accuracy for DS2. However, the accuracy of DS1 and DS3 is higher in the case of fuzzy c-means but k-means takes less time for computation. We additionally found that without skull removal, time of computation is increased.

For the performance measure, we use four techniques as follows:

$$\text{True positive (TP)}: \frac{\text{Number of images having brain tumor}}{\text{Total number of images}} \quad (14)$$

Table 1 Stages of the main framework applied to five datasets

Data sets	Input image	BSE	Filtered image	Otsu threshold	Segmented image
DS1 (Brain web)	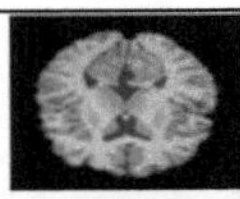	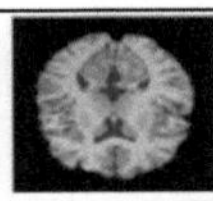	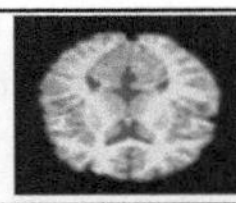	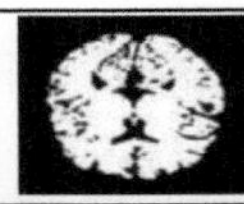	
DS2 (DICOM)	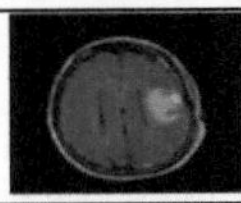	NO Skull Removal	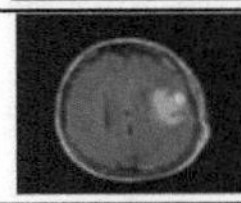		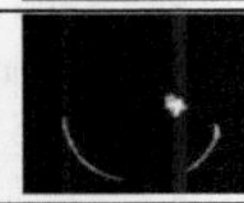
DS3 (Brats)	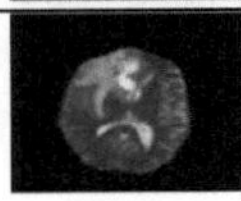	Already Skull	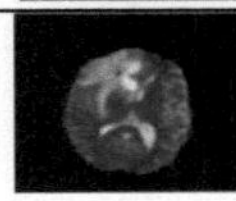		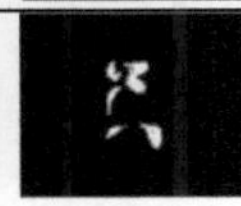
DS4 (Medinfo)	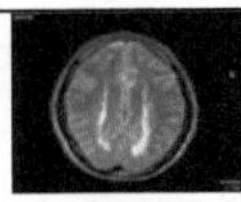	NO Skull Removal	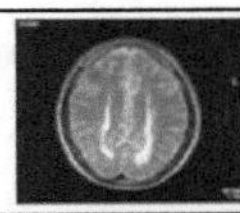		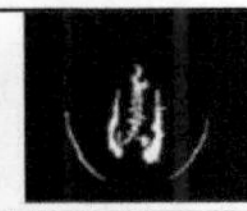
DS5 (Harvard)	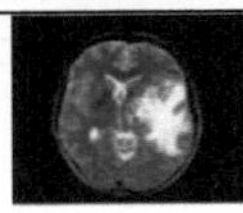	NO Skull Removal	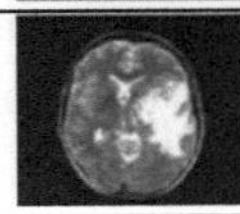		

Table 2 The performance matrices of the proposed method

Dataset	TP	TN	FP	FN	Accuracy	Precision	Recall
DS1 (Brain web)	100	0	0	0	100	100	100
DS2 (DICOM)	95.45	0	0	4.55	95.45	100	95.45
DS3 (Brats)	100	0	0	0	100	100	100
DS4 (Medinfo)	100	0	0	0	100	100	100
DS5 (Harvard)	92.85	0	0	7.15	92.85	100	92.85

Table 3 The performance matrices of k-means

Dataset	TP	TN	FP	FN	Accuracy	Precision	Recall
DS1 (Brain web)	96.7	0	0	3.3	96.7	100	96.7
DS2 (DICOM)	85.7	0	0	14.3	85.7	100	85.7
DS3 (Brats)	95.06	0	0	4.94	95.06	100	95.06

Table 4 The performance matrices of fuzzy c-means

Dataset	TP	TN	FP	FN	Accuracy	Precision	Recall
DS1 (Brain web)	100	0	0	0	100	100	100
DS2 (DICOM)	85.7	0	0	14.3	85.7	100	85.7
DS3 (Brats)	100	0	0	0	100	100	100

Table 5 The performance matrices of KIFCM

Dataset	TP	TN	FP	FN	Accuracy	Precision	Recall
DS1 (Brain web)	100	0	0	0	100	100	100
DS2 (DICOM)	90.5	0	0	9.5	90.5	100	90.5
DS3 (Brats)	100	0	0	0	100	100	100

$$\text{True negative (TN)}: \frac{\text{Number of images that dont have tumor}}{\text{Total number of images}} \tag{15}$$

$$\text{False positive (FP)}: \frac{\text{Number of images that don't have tumor and detected positive}}{\text{Total number of images}} \tag{16}$$

$$\text{False negative (FN)}: \frac{\text{Number of images having tumor and not detected}}{\text{Total number of images}} \tag{17}$$

Precision: It is also referred as exactness. It is percentage of the tuple that is actually classified as positive and actually positive. It is defined as

$$\text{Precision} = \frac{\text{TP}}{\text{TP} + \text{FP}} \tag{18}$$

Recall: It is also referred as completeness. It is percentage in which classifier is labeled as positive. It is measured using the following equation:

$$\text{Recall} = \frac{\text{TP}}{\text{TP} + \text{FN}} \tag{19}$$

Accuracy: It is the measure that defines how well the classification test correctly identifies. It is calculated as

$$\text{Accuracy} = \frac{\text{TP} + \text{TN}}{\text{TP} + \text{TN} + \text{FP} + \text{FN}} \tag{20}$$

Figure 2 shows the level-wise accuracy of five different techniques. Figure 3 shows the execution time of different techniques on DS2 (DICOM). It is clearly shown in Fig. 3 that our proposed method has the lowest execution time as compared to other techniques for DS2.

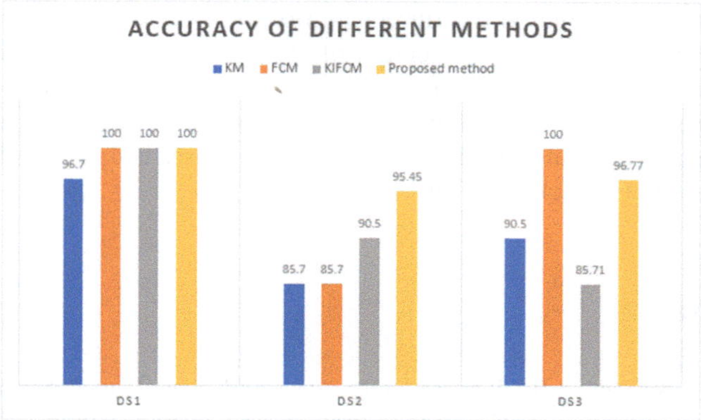

Fig. 2 The clustering techniques accuracies for the five datasets

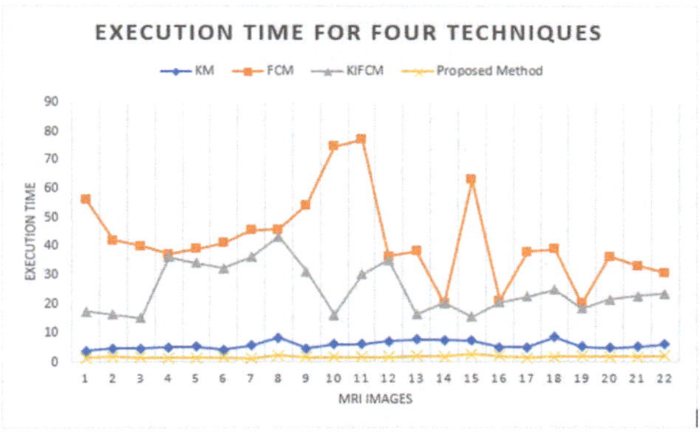

Fig. 3 The execution time for the tested four techniques for DS2

From the above results, we can conclude that our proposed method is less time-consuming than any other existed methods and it outperforms another algorithm in terms of accuracy of segmentation.

5 Conclusion

In this research work, a new approach is developed for image segmentation based on intensity and textual features than processing them with PCA and SVM. Our method efficiently detects the tumor region with the actual type of tumor. We use five datasets for the evaluation of our proposed algorithm. The computation time is relatively low than any other techniques.

We compare our proposed algorithm with fuzzy c-means, k-means, and KIFCM. As shown in Table 2, we get the accuracy of 95.45 in DS2 and computation time is very low for detecting the brain tumor. In future work, we implement our method on 3D images and different techniques for dimensionality reduction are used for benchmarking. We can also extend our work for diagnoses and identify brain stroke that happens due to blockage of blood vessels.

References

1. Viajy, V., Kavitha, A.R., Rebecca, S.R.: Automated Brain Tumor Segmentation and Detection in MRI using Enhanced Darwinian Particle Swarm Optimization (EDPSO). In: International Conference on Intelligent Computing, Communication & Convergence. 92 (2016) 475–480
2. Patel, J., Doshi, K.: A study of segmentation methods for detection of a tumor in brain MRI. Advance in Electronic and Electric Engineering. 4 (2014) 279–284
3. Pareek, P.: A Survey on Cerebrum Tumour Detection in MRI using Medical Imaging Techniques. Int. J. of Advanced Research in Computer and Communication Engineering. 5 (2016) 634–638
4. Kavithal, A.R.: Chitra, L., Kanaga, R.: An approach for brain tumor segmentation and classification using genetic algorithm with SVM classifier. International Conference on Emerging Engineering Trends and Science. (2016) 1468–1471
5. Naik, D., Shah, P.: A review on image segmentation clustering algorithms. Int. J. of Computer Science and Information Technologies. 5 (2014) 3289–3293
6. Christe, S.A., Malathy, K.: Improved hybrid segmentation of brain MRI tissue and tumor using statistical features. ICTACT J. on Image and Video Processing. 1 (2010) 43–49
7. Seerha, G.K., Kaur, R.: Review on recent image segmentation techniques. Int. J. of Computer Science and Engineering (IJCSE). 5 (2013) 109–112
8. Deepa., Singh, A., Singh, K.K.: Brain tumour detection from MRI images using Hybrid Genetic FCM. Int. J. of Engineering Applied Sciences and Technology. 1 (2016) 179–184
9. Maksoud, E.A., Elmogy, M., Awadi, R.Al.: Brain tumor segmentation based on a hybrid clustering technique. Egyptian Informatics J. 16 (2015) 71–81
10. Wu, M.N., Lin, C.C., Chang, C.C., Brain tumor detection using color based K-means clustering. Third International Conference on Intelligent Information Hiding and Multimedia Signal Processing. 2 (2007) 245–250
11. Majumder, P., Kshirsagar, V.: Brain Tumor Segmentation and Stage Detection in Brain MR Images using K AMS EM Algorithm. Int. J. of Computer Applications. 95 (2014) 31–38
12. Cardinaux, F., Sanderson, C., Bengio.: User authentication via adapted statistical models of face images. IEEE Transaction on Signal Processing. 54 (2006) 361–373
13. Kilaru, V., et al.: Gaussian mixture modeling approach for stationary human identification in through-the wall radar imagery. In: Journal of Electronic Imaging. 24 (2015)
14. Brain Web: Simulated Brain Database. http://brainweb.bic.mni.mcgill.ca/brainweb/
15. DICOM Samples Image Sets. http://www.osirix-viewer.com/datasets/
16. MICCA Nice 2015. http://www2.imm.dtu.dk/projects/BRATS2015/data.html
17. Medinfo dataset http://www.medinfo.cs.ucy.ac.cy/
18. Harvard dataset.http://www.harvad.edu/aanlib/home.html
19. NIH Center for Information Technology. http://mipav.cit.nih.gov/

Intelligent Mobile Agent Framework for Searching Popular e-Advertisements

G. M. Roopa and C. R. Nirmala

Abstract e-Advertising is the rapidly growing e-commerce application which acts a significant entity for user browsing behavior/patterns and is employed by majority of ad platforms to evaluate the ad selection process for choosing the right product. Existing search engines adopt other link structure/content-oriented and do not consider the browsing patterns. In link structure, rank scores are applied evenly irrespective of the links as target web pages are self-descriptive and use links for navigating. In a content-oriented, the web page lacks in having the rich content description to match with the search query. Thus, generating the top priority list for the user query is a difficult task. Here, mobile agent architecture is proposed to perform the search engine process by tracking the ad relevance and to estimate the probability of views/clicks to distribute the rank scores based on the ad popularity. Mobile agents extend to apply classification technique to classify the ad list into three classes and display the most relevant ads which improve the result list.

Keywords e-Advertising · Classification · Information retrieval
Mobile agents · Search engine · User relevance

1 Introduction

In this highly competitive world, online advertising supports a huge segment of the current Internet eco-system, where there is a broad usage of the World Wide Web for searching the user needs and retrieving the relevant information. It also acts a prominent economic force and main source of income for websites and services [1]. Search engine based concept of displaying the advertisements on web pages has

G. M. Roopa (✉) · C. R. Nirmala
Department of Computer Science & Engineering, Bapuji Institute of Engineering & Technology, Davangere, Karnataka, India
e-mail: roopa.rgm@bietdvg.edu

C. R. Nirmala
e-mail: crn@bietdvg.edu

C. R. Panigrahi et al. (eds.), *Progress in Advanced Computing and Intelligent Engineering*, Advances in Intelligent Systems and Computing 714,
https://doi.org/10.1007/978-981-13-0224-4_17

become the driving force behind the large-scale monetization process of web services through e-marketing [16].

Recently, WWW supports one of the world's largest information repositories for knowledge reference and such information is constantly changing over time. With regard to the literature review, we have explored that most of the traditional search engines methods adopt client/server paradigm for information retrieval, which requires a good bandwidth to return back the user-relevant information based on the search query; network traffic increases with the increase in the client request which loads the server; and the search engine needs to wait between the requests, which increases the time required for requesting, searching, and retrieving the required information [2].

In this research work, we have made an attempt to add artificial intelligence for user-relevant advertisement search process to improve the quality of the result list. The mobile agent technique has been proposed for generating the most popular advertisements for the given user query, and the static search agents will perform the task of searching the information on behalf of the user, according to the user preference. A combination of search engine process with the knowledge of mobile agent technology increases the efficiency of required information retrieval, which improves the network bandwidth resource occupancy. These search agents do not require a continuous connectivity, as there is a re-establishment of the network connection only while returning the user required information from the remote machines, which reduces the network traffic. As mobile agents operate autonomously and asynchronously, the users need not keep track of the dispatched agents, which saves the user time and reduces the network communication costs.

The analysis, design, and implementation of the proposed work concentrate on creating the mobile agent/static agent to perform the task of advertisement search process to estimate the probability of views/clicks on every advertisement. At the first stage, we retrieve relevant advertisements for the given user query and later we rank the advertisements based on the user relevance. Next, mobile agents apply the classification technique to classify the generated list into three classes and display the most relevant result list for the given user query. The use of mobile agent theme applied for e-advertising for displaying the best matching advertisements for the given user query involves the combination of both the probability of views/clicks and user relevance which effectively reduces the network load and traffic and efficiently increases the usage of the bandwidth allocated and improves the extraction of the user-relevant advertisements.

2 Related Work

Bhanu C. Vattikonda et al. [3] showed that by using the feature to capture, the advertiser's intent can significantly improve the performance of the relevance ranking. The support for the search engine interprets the ad keyword by submitting the advertisement keyword as an independent query and then incorporated the results as features when determining the relevance of the advertiser's sponsored result to the user's original query.

K. Dave and V. Varma [4] focused on the problems and solutions pertaining to the information retrieval, machine learning, and statistics domain of contextual advertising. They have addressed the core problem of finding the best matching advertisements in the given context based on targeting scheme by combining the content, user profile, demographics, and other contextual aspects.

Weinan Zhang et al. [1] proposed a novel algorithm for recommending the advertising keywords for short-text web pages by supporting the content of Wikipedia, which contains a huge collection of entities that are related to each other for the given topic. They have adopted content-based PageRank on the Wikipedia graph to rank the related entities. Their proposed approach produces a substantial improvement in the precision of the top 20 recommended keywords on short-text web pages over the existing approaches.

Jin-Yong Jung et al. [5] suggested a novel ads classification method which handles the lack term features to classify ads with short text. They have utilized a vocabulary expansion technique using semantic associations between terms learned from large-scale 4.0–9.7% improvements in terms of hierarchical f-measure over the baseline classifiers without vocabulary expansion.

Michael Bendersky et al. [6] investigated various strategies for compact, hierarchical-aware indexing for sponsored search advertisements through adaptation of standard IR indexing techniques. Here, the advertisement corpus is transformed into a collection of hierarchically structured text documents and then adapts standard IR indexing techniques to construct a compact yet efficient ad index.

3 Problem Definition

The core problem to be addressed is that as the amount of e-advertisement data is tremendously increased over time, it is very difficult for most of the generic search engines to provide the users with the updated and timely information for the user search query on a target page/target website; rather than retrieving the generic advertisements, it is preferable to retrieve the ads related to provide an improved user experience and to increase the probability of views/clicks. The problem defined is, for a given target page p, with set of advertisements $A = (a_1, a_2, \ldots, a_n)$ and the user search query Q, we need to estimate the probability views/clicks as $P(click|p_i, a_i)$ to record the quality of the match between the given page by tracking the user relevance by framing the <query, ad_i> pairs in the set of advertisements, apply the rank score, and estimate the probability of views/clicks to retrieve the top relevant advertisements for the display.

4 System Description

The architecture consists of advertiser module, aggregator module, ad publisher module, and user module as shown in Fig. 1. The static/mobile agent is used to upload the advertisements to ad publisher sites in parallel and also to retrieve relevant advertisements to the user-defined keyword search query.

4.1 Advertiser Module

Advertiser module acts as an ad provider and participates in the ad campaigns with a specific goal of promoting the products and services online. In such ad campaigns, the advertiser provides the ad description with multiple keywords as a parameter for deciding the popularity of advertisements based on user views/clicks as shown in Fig. 2a.

In order to participate in the ad campaigns, the advertiser must provide the necessary information:

- Bid phrase: This is decided by the advertiser and is used to indicate the advertisement content which is invisible to the user.
- Bid amount: For the bid phrase, the advertiser needs to decide the bid amount which is masked from the user.

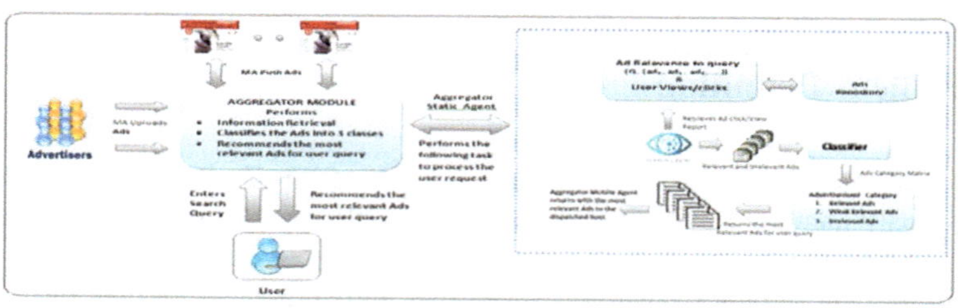

Fig. 1 System architecture for retrieving the user-relevant advertisements

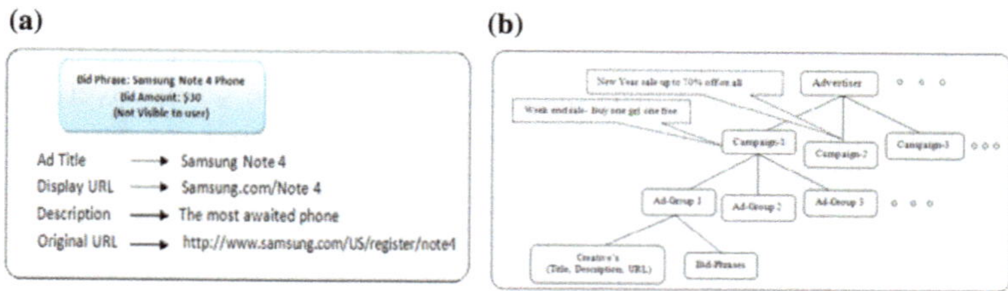

Fig. 2 **a** Structure of the textual advertisement. **b** Hierarchical scheme to store the textual information

- Title: It refers to the ad title.
- Description/Creative: It includes short ad description which must be creative and innovative to attract the users.
- Display/Original URL: Display URL refers to the URL where the advertisement is displayed.

4.2 Aggregator Module

The aggregator static/mobile agents in this module act as a mediator between registered advertisers, ad publishers, and users and the tasks carried out by the aggregator static agent/mobile agents are listed below:

- Collects, aggregates, and stores the advertisements with a description.
- Creates, loads, and dispatch the mobile agents for publishing.
- Maintains and updates the ad repository over time.
- Retrieves the information about the user browsing behavior of the mobile agents to track the user relevance to the advertisements and record the user clicks/views.
- Accepts and processes the user search query for retrieving the desired advertisements.
- Classifies the generated list into three classes and finally loads the mobile agent to push the desired relevant advertisements which are closer to the user given query.

This module consists of the following major components:

- *Ad Repository*: Stores the textual information of the advertisements which are defined and organized in a hierarchically structured manner with various entities, as shown in Fig. 2b. Here, each advertiser holds an account and can participate in the ad campaigns which contains an ad group that includes the set of creatives (visible text) and the bid phrase. Such, hierarchically stored product information is used for the expansion of advertisement and is applied in user relevance stage to record the <query, advertisements> pairs. The static agent is responsible for maintaining and updating the ad repository over time.
- *Search Engine process*: The three-staged approaches adopted by the search engine process are (i) to find advertisements relevant for the user query, (ii) to estimate click-through rate for retrieving advertisements and apply rank scores, and (iii) to display the advertisements on the search page.

  1. First, the static agent computes the score by measuring the relevance of advertisements to query <query, ad> pairs. The stages involved in tracking the user relevance to the advertisements are shown in Fig. 3.

Fig. 3 Advertisement pipeline for selecting the most relevant advertisements

- Previous to the relevance stage, the user query (Q) is expanded and matched with the product description provided by the advertiser. The query expansion is used to enhance the retrieval performance by expanding the user query with the additional relevant terms and re-weighting the terms.
- On the relevance stage, each advertisement (ad_1, ad_2, …) identified in the previous stage is evaluated to estimate the closer relevance of each advertisement to the user query.

An existing ranking function is based on the probabilistic retrieval of the relevant advertisements.

Here,

$$score(ad, q) = \sum_{T \in Q} idf(q_i) \cdot \frac{(k_1 + 1) \cdot dtf}{K + dtf} \cdot \frac{(k_3 + 1) \cdot qtf}{k_3 + qtf} \tag{1}$$

$$idf(q_i) = log \frac{N - n(q_i) + 0.5}{n(q_i) + 0.5} \tag{2}$$

where

N	refers to the entire set of advertisements in the collection.
$n(q_i)$	refers to the number of advertisements relevant to the given query.
Q	refers to the user query and contains a set of advertisements T.
K	is given by $k_1 \cdot \left((1-b) + b \cdot \frac{adl}{avadl}\right)$.
K_1, b, k_3	The set of parameters which depends on the nature of the given query.

The default values for K_1, b, and k_3 are 1.2, 0.75, and 7.

dtf	refers to the frequency of occurrences of the advertisements for a given the target web page.

qtf refers to the frequency of terms associated with the given query Q.
adl and avadl refer to the ad list and the average ad list length.

The advertisement relevance probability can be estimated by $P_q(rel|y)$ for the given advertisement(y) and a query q. Assume that, if the $P_q(rel|y) = 0$, then the relevance probability given the advertisement is 0 and hence is not retrieved.

Next, we can refine by considering the non-relevance probability for the given advertisement(y) and a query q as $P_q(\overline{rel}|y)$. If $P_q(rel|y) > P_q(\overline{rel}|y)$, then it is assured that the relevance probability is greater than the non-relevance probability and hence the advertisement must be retrieved. This can be extended by using the thresholds for relevance probability and non-relevance probability, $((P_q(rel|y) - P_q(\overline{rel}|y)) > \text{threshold})$, to restrict the further retrieval process.

The higher the relevance and non-relevance probability ratio in (Eq. 3) for given advertisements, then it is likely that the advertisement y is relevant to the user interest and must be retrieved.

Bayes theorem is used to compute $P_q(rel|y)$ and $P_q(\overline{rel}|y)$ and Eq. 4 demonstrates the case for relevance.

$$\frac{P_q(rel|y)}{P_q(\overline{rel}|y)} \tag{3}$$

$$P_q(rel|y) = \frac{P_q(y|rel) \cdot P_q(rel)}{P(y)} \tag{4}$$

where

$P_q(rel)$ refers to the prior probability that any given advertisement on the list is relevant to the given query q.
$P_q(y|rel)$ refers to the probability of the observed advertisement y given the relevance information.
P(y) refers to the probability of the observed advertisement y irrespective of the relevance.

But the relevance $P_q(rel)$ and non-relevance $P_q(\overline{rel})$ probability are assumed to be same for all the advertisements in the given list. In such case, the ranking depends on values of $P_q(y|rel)$ and $P_q(y|\overline{rel})$ and not on the $P_q(rel)$ and $P_q(\overline{rel})$. Hence, these terms can be ignored and final odds observed for relevance and non-relevance can be given by Eq. 6.

$$\frac{P_q(y|rel) \cdot P_q(rel)}{P_q(y|\overline{rel}) \cdot P_q(\overline{rel})} \tag{5}$$

$$\frac{P_q(y|rel)}{P_q(y|\overline{rel})} \tag{6}$$

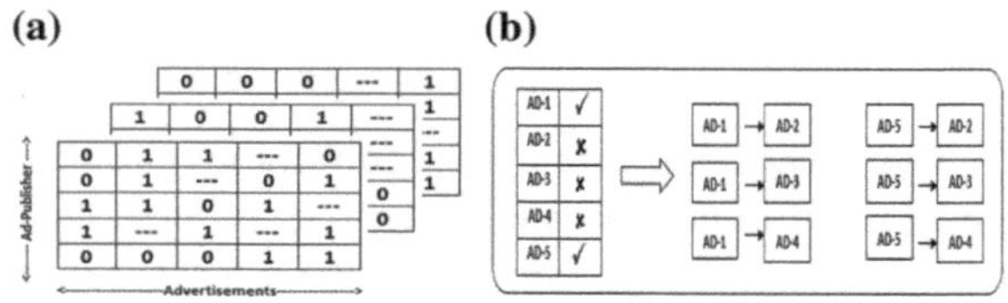

Fig. 4 **a** Aggregated view of ad publisher, user, and advertisements. **b** Pair-wise user preferences to the advertisements

2. Second, the user clicks/views logs are recorded by generating the aggregated view for the advertisements, ad publishers, and the users as shown in Fig. 4a and the pair-wise user preferences for the advertisements are shown in Fig. 4b.

The probability that an advertisement is clicked/viewed by the user mainly depends upon two main factors: (a) probability that the advertisement is viewed and (b) probability that the advertisement is clicked on, by considering it is viewed based on the position.

$$\mathrm{p(clicks|ads, pos) = p(clicks|ad, pos, view)\ p(view|ad, pos)}$$

But here we have made an assumption that the probability that an advertisement is clicked is independent of its position. Thus, we estimate that the probability of an advertisement would be clicked or viewed at any given display position.

$$\mathrm{P(clicks|ad, pos) = p(clicks|ad, view)\ p(view|pos)}$$

Finally, by considering the probability of clicks/views and the ad relevance to the user interest, the set of advertisement list generated by the search engine still includes the advertisements that are both relevant and irrelevant to the user query.

- Classifier

The classifier provides an opportunity for exact match and passes only the most desired advertisements for the given user query which reduces the result list by a huge margin. The filter-in process of non-relevant advertisements is important, where the display of unsuitable advertisements can lead to the decline of click-through rate on the intact list [6]. This classifier utilizes the J48 decision tree algorithm to classify the generated list into three class labels like most relevant advertisements (R-Ad), weak relevant advertisements (Weak-Ad), and irrelevant advertisements (IRR-Ad). The set of fields and descriptors associated with the advertisements is given in Table 1.

The retrieved advertisements from the search engine process are split into training and test sets which comprise <query, ad> pairs. Next, using the set of pairs with the more similarity features is used to retrieve the advertisements closer to the user query. Finally, we pick the tree with the best performance compared to all the constructed trees and display the advertisements which belong to the class with label R-Ad that provides exact match for the user search query.

Table 1 Description of advertisement field

Field	Descriptions
Ad campaign	The campaign to which the advertisement belongs to
Ad number	Identifier for ads
Date	Advertisement posted date
Ad position	The display position of the advertisements on the given target web page
Key phrase	The key phrase used to trigger the advertisements
Ad impressions	The total number of ad impressions based on the keyword
Clicks	The total number of ad clicks
Cost	Advertisement cost
Sales	The revenue generated based on the number of sales
Orders	The number of orders
URL	The landing page URL

```
Algorithm: build_decision_tree
create node 'n';
If tuples in the ad_list belong to the same class (C), then return 'n' as the leaf
node with labelled as class C ;
        if ad_attribute_list is empty, then return 'n' as the leaf node with
        majority class label in ad-list; // majority of voters
        apply the attribute_selection_criteria ( ad_list , ad_attribute_list )
        find best splitting_criteria;
        label the node 'n' with splitting_criteria;
        if value of splitting_attribute is discrete and multilevel splits is allowed
        then
        ad_attribute_list = split attribute-list; //remove the splitting-attribute
        for each j outcome of splitting_criteria //extend the sub-tree
        Let AD_j be the set of ad-data tuples in AD, which satisfies the j
        outcome;
            if AD_j is empty then,
            attach the leaf-node with majority class in AD to node 'n';
            else
            attach the node 'n' returned to generate the decision tree
            (AD_j , ad_attribute_list) to node 'n';
           end for
        return 'n';
```

5 Results and Discussions

The initial experiment was conducted for 4000 total advertisement records to show the percentage of retrieval. Here, we have carried out two-stage ad filtration process: First, during the search engine stage followed by the classification technique.

By using static/mobile agents for retrieving the desired relevant advertisements for the user keyword search query reduces the search result list by a huge margin.

1. The total number of advertisement list retrieved from the search engine process with regard to the user relevance was 1712 instances. This list includes both relevant and irrelevant advertisements, and the users will not examine all the retrieved instances; this affects the CTR on the displayed advertisements and reduces the publisher revenue.
2. To improve the retrieved result list, we apply the classification technique for further filtration of advertisements. At this stage, the previously generated advertisement list is given to the classifier which classifies the list into three defined classes. To measure the effectiveness of retrieving the desired relevant advertisements, we consider precision and recall values.

Precision (P) is given by the fraction of retrieved advertisement which is relevant.

$$(P) = \frac{\textbf{Number of relevant advertisements retrieved}}{\textbf{retrieved advertisements list}} = \mathrm{P(relevant|retrieved)}$$

Recall (R) is given by the fraction of relevant advertisements which are retrieved.

$$(R) = \frac{\textbf{Number of relevant advertisements retrieved}}{\textbf{relevant advertisements list}} = \mathrm{R(retrieved|relevant)}$$

$$\text{Precision } = (t_p)/(t_p + f_p) \quad \text{Recall} = (t_p)/(t_p + f_n)$$

For such set, we plot the precision/recall curve as shown in Fig. 5a, b that shows the detailed accuracy for three defined classes. From Fig. 5b, if the next [K + 1]th advertisement from the retrieved list is non-relevant, then the recall value remains the same, whereas there is a drop in the precision value. If the current advertisement is relevant to the user query, then there is an increase in both precision and recall values.

(a)

TP Rate	FP Rate	Precision	Recall	Class
0.97	0.014	0.813	0.797	R-Ad
0.305	0.066	0.351	0.805	Weak-Ad
0.95	0.137	0.502	0.957	IRR-Ad

(b)

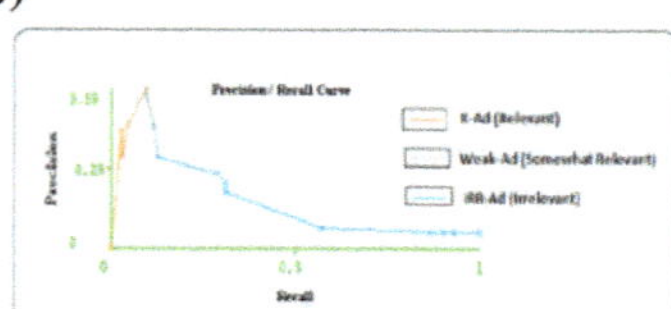

Fig. 5 **a** Detailed accuracy by class. **b** Precision/Recall curve for exact match

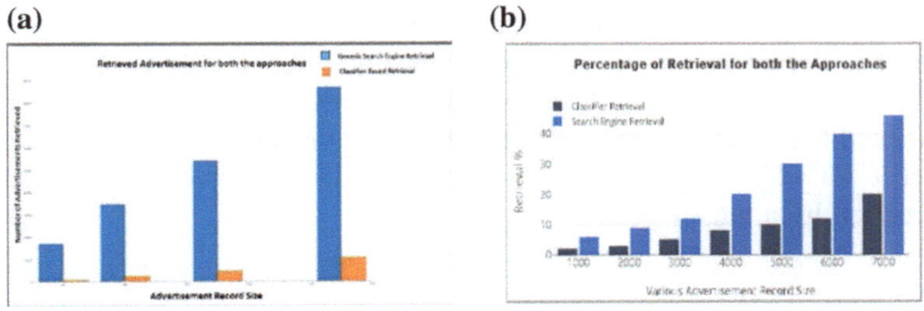

Fig. 6 **a** The final retrieved and displayed advertisement list. **b** The percentage of retrieval for both the approaches

5.1 Observations

Figure 5b plots the precision–recall curve for the values recorded in Fig. 5a. Higher precision–recall value indicates that the larger fraction of advertisements selected is most relevant to the user query and enhances the user browsing experience and in-turn increases the CTR value, whereas higher recall values with the drop in the precision values indicate that the fraction of advertisements selected is weak and irrelevant advertisements.

- Figure 6a shows the ad list displayed to the user.

Larger fraction of the ad list displayed to the user leads to the reduction in CTR and ad publisher revenue as the user will not tend to examine all the ad list. A smaller fraction of ad list displayed to the user leads to the increases CTR and ad publisher revenue.

The final list displayed to the user reduces the search space by a huge margin. Figure 6b shows the percentage of retrieval for both the cases.

6 Conclusion

The click-through rate prediction plays a major role in the field of e-advertising because ad ranking, ad filtration, ad placement, and ad pricing relay on it. The three major contributions are as follows:

- First, adopt the mobile agent paradigm in the field of e-advertising for retrieving the most relevant advertisements. This approach overcomes the major drawbacks of the existing model in terms of network traffic, network load, and efficient bandwidth utilization.
- Second, the search engine process with a broader matches to rank the advertisements based on the user relevance and retrieves the relevant advertisements list which includes relevant and irrelevant advertisements to the user interest.

Huge retrieved list leads to the reduction in CTR and directly affects the ad publisher's revenue and the advertiser's intent.

- Finally, we apply the classification technique with exact match to classify the generated list into three class labels and display the advertisements with a class label R-Ad which are desired advertisement to the user interest and reduce the search list by a huge margin.

References

1. X. Wang, W. Li, Y. Cui, R. Zhang, and J. Mao, "Click-through rate estimation for rare events in online advertising," in Online Multimedia Advertising: Techniques and Technologies, pp. 1–12, 2011.
2. Cheng. H. and Cantu-Paz E. 2010: Personalized click prediction in sponsored search. In the proceedings of 3rd ACM International Conference on WEB search and Data Mining 351–360 (18).
3. Georgios Theocharous, Philip S. Thomas, Mohammad Ghavamzadeh: Ad Recommendation Systems for Life-Time Value Optimization WWW 2015 Companion, May 18–22, 2015, Florence, Italy. ACM 978-1-4503-3473-0/15/05.3.
4. Long-Sheng Chen, Tai-Cheng Kuo: Using Decision Trees to identify key Factors of Keyword Advertisements. Proceedings of the International Multi-Conference of Engineers and Computer Scientists 2014 Vol-1 IMECS 2014, March 12–14, 2014, Hong Kong.
5. Nirmala C R, Dr V Rama Swamy and Jyoth id M N: Mobile Agents for Audio Search and retrieval. 2010 International Journal of Computer Applications (0975–8887) Vol. 1. No. 23 (17).
6. Michael Bendersky, Evgeniy Gabrilovich, Vanja Josifovski and Donald Metzlre: The Anatomy of an Ad: Structures Indexing and Retrieval for Sponsored Search. WWW 2010, April 26–30 Raleigh, North Carolina, USA (16).
7. Zhipeng Fang, Kun Yue, Jixian Zhang, Dehai Zhang, and Weiyi Liu: Predicting Click-Through Rates of New Advertisements Based on the Bayesian Network Hindawi Publishing Corporation Mathematical Problems in Engineering Volume 2014, Article ID 818203, 9 pages http://dx.doi.org/10.1155/2014/818203.
8. Kushal Dave, Vasudeva Verma: Computational Advertising- Techniques for Targeting Relevant Ads. Foundations and Trends informational Retrieval Vol. 8, No. 4–5 (2014) 263–418.
9. Bernard J. Jansen and Zhe Liu: The Effect of Ad Rank on the performance of Keyword Advertising Campaigns. Journal of the American Society for Information Science and Technology 2013.
10. Weinan Zhang, Dingquan Wang, Gui-Rong Xue, Hongyuan Zha: Advertising Keywords Recommendation for Short-Text Web Pages Using Wikipedia, ACM Transactions on Intelligent Systems and Technology, Vol. 3, No. 2, Article 36, Publication date: February 2012.
11. Jin-Yong Jung, Jung-Hyun Lee, Jong Woo Ha and Sang Keun Lee: Vocabulary Expansion Technique for Advertisement Classification, KSII Transactions on Internet and Information Systems Vol 6 No. 5 May-2012.
12. Andrei Broder, Marcus Fontoura, Vanja Josifovski and Lance Riedel: A semantic Approach to Contextual Advertising SIGIR'07 Copyright 2007 ACM. (23).
13. Cristo M., Ribeiro Neto B., Golgher P B., and De Moura E. 2006: Search Advertising in the proceedings of the StudFuzz Conference 197, 259–285 (26).

14. Chuan-Feng Chiu, Timothy K. Chi and Ying-Hong Wang "An Integrated Analysis, Strategy and Mobile Agent Framework for Recommendation System in EC Over Internet" Tamkang Journal of Science and Engineering, Vol. 5 No. 3, pp 159–174 (2002).
15. Bhanu c, Vattikonda, Santhosh Kodipaka, Hongya Zhou, Vacha Dave, Saikat Guha and Alex C Snoeren: Interpreting Advertiser Intent in Sponsored Search Publication rights licensed to ACM 978-1-4503-3664-2/15/98.

A Bayesian Approach for Flight Fare Prediction Based on Kalman Filter

Abhijit Boruah, Kamal Baruah, Biman Das, Manash Jyoti Das and Niranjan Borpatra Gohain

Abstract Decision-making under uncertainty is one of the major issues faced by recent computer-aided solutions and applications. Bayesian prediction techniques come handy in such areas of research. In this paper, we have tried to predict flight fares using Kalman filter which is a famous Bayesian estimation technique. This approach presents an algorithm based on the linear model of the Kalman Filter. This model predicts the fare of a flight based on the input provided from an observation of previous fares. The observed data is given as input in the form of a matrix as required to the linear model, and an estimated fare for a specific upcoming flight is calculated.

Keywords Flight fare · Observation · Prediction · Kalman filter
Linear model

1 Introduction

The prices of air tickets change frequently and randomly. There are various factors for change in prices of flight tickets such as days, timing of the flight, time of the year in which flight tickets are booked, stocks of that airways, etc. [1]. It will be a

A. Boruah (✉) · K. Baruah · B. Das · M. J. Das · N. B. Gohain
Department of Computer Science and Engineering, DUIET, Dibrugarh University,
Dibrugarh, Assam, India
e-mail: abhijit.btcs06@gmail.com

K. Baruah
e-mail: kamalb83990@gmail.com

B. Das
e-mail: bimaaaan100@gmail.com

M. J. Das
e-mail: mjdasmanash94@gmail.com

N. B. Gohain
e-mail: niranjangohain9435@gmail.com

C. R. Panigrahi et al. (eds.), *Progress in Advanced Computing and Intelligent Engineering*, Advances in Intelligent Systems and Computing 714,
https://doi.org/10.1007/978-981-13-0224-4_18

big relief to get an estimation of dates when fares will be low, so that traveling can be planned in advance. Of course, in case of flight fare, we can easily search online and book our tickets in advance as they are already provided in the airline websites. But those flight fares change within short period, i.e., they are dynamic. Flight fares are often seen to be tripled in an hour or decrease to half in a minute. This uncertainty in flight fare pricing usually doesn't affect the population falling in the higher income categories. Hence, the prime objective in this work would be to derive a method by which we can easily predict the upcoming fares of the desired flight. We tried to derive a prediction method for flight fare using Kalman filter which is a Bayesian estimation technique.

Different airlines vary prices based on numerous factors whose information is not available on the Internet. An airline may reduce the flight fare for a period of time if the number of unsold seats, on a particular date, is high relative to usual situations. A consumer is always unaware of the factors responsible for the change in flight fare or to the number of available seats on the flights. So price changes could appear random or unpredictable to a consumer tracking prices over the web. Our model for flight fare prediction is based upon the principle of Kalman filter, which allows our method to take set of observation for the prediction. Since state-space model is time-dependent, therefore parameters for our model are linear. These change in parameters can be represented by linear equations. A Kalman filter has two basic models—(a) Linear model and (b) state-space model.

In our work, we have used the linear model over the state-space model. Here, a particular state is dependent on only the previous state. Linear state model can take limited inputs due to which we can predict fare for only 1 day. We have taken the observations of a particular flight daily at different times for a certain period and input those observations in the Kalman filter implementation for the desired output.

The idea of flight fare prediction revolves around the fact that a flight fare changes many times in a single day and also over duration of time. Although the airlines provide the price chart to their customer through various websites, that chart only provides the fare during that specific time of the day. Let us take an example to understand this better (Figs. 1 and 2).

Suppose on day 1, the fare of a flight from location A to location B which is to be deported on day 2 is Rs. 1450 (Let us assume this fare as per shown in a private

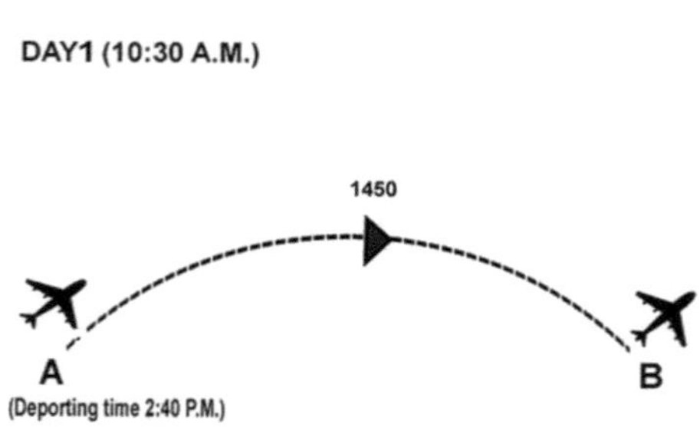

Fig. 1 Fare on day 1 from destination A to B

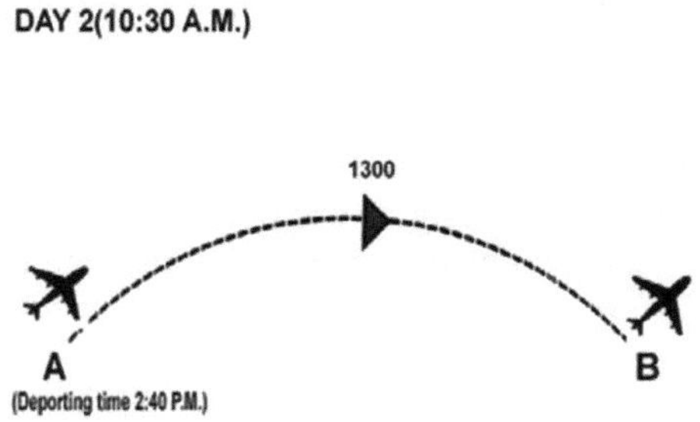

Fig. 2 Fare on day 2 from destination A to B

airline website). On the next day, i.e., on day 2, the fare of the same flight changes to Rs. 1300. Even an airline website cannot assure us about a fixed price for a particular day. Our model predicts the flight fare with a close approximation. It will help the customer to get the near to exact fare for the day.

For our work, we have taken observations of a flight of Indigo airways as Indigo is one of the economic and comfortable airlines for domestic air travel in India. We have taken observations of a domestic flight from Dibrugarh to Kolkata.

2 Related Work

Airlines apply different policies on a flight fare based on the yield management. Some related works are being done in the field of probabilistic reasoning. All the related works are a bit different from each other. Some works are done from customer's point of view, while some are from the airline's point of view. Several efforts are made previously in the game theory community to model aspects of the airline ticket domain. Different methods have taken different factors and issues into account. There has been work done on minimizing ticket purchase price using data mining techniques. The model basically allowed customers to choose a suitable air ticket by analyzing a variation pattern in available data. Some work has been done on building prediction models for airfare prices using machine learning techniques [2–4]. A dynamic programming model was developed to determine optimal fare class on a flight [5]. There was a small disadvantage in that model which was extending to more flights resulted in increase in complexity. We have developed a method to predict flight fare based on previous data which is then analyzed through Kalman filter.

3 Pre-requisites

The Kalman filter is a probabilistic estimation technique for the linear quadratic problem, which focuses on the estimation of immediate "state" of a linear dynamic system consisting of white noise [6].

The two models of Kalman filter are linear model and state-space model [7].

3.1 Linear Model

Linear models are widely used to describe dynamical behavior throughout the natural sciences and also in other fields such as economics [8]. A model represents the equation of a given environment. This equation helps us to understand and to predict the behavior of a complex system. The variables present in the equation represent the entity or the function of the system. Linear model is a unique state model. Here, a particular state is dependent on only the previous state, which explains why we can predict the next day's flight fare only from the fare of the previous day. Following are the equations which are involved in a linear model [9].

$$X_t = A_t X_{t-1} + B_t U_t + \epsilon_t \tag{1}$$

$$Z_t = X_t + \delta_t \tag{2}$$

Components

$\mathbf{A_t}$	n × n matrix describing how the state evolves from t − 1 to t without controls or noise.
$\mathbf{B_t}$	n × n matrix describing how the control U_t changes the state from t − 1 to t.
$\mathbf{U_t}$	Control matrix.
$\boldsymbol{\epsilon_t}$ **and** $\boldsymbol{\delta_t}$	denote the random variable representing the process and measurement noise that is assumed to be independent and normally distributed with covariance.
$\mathbf{X_{t-1}}$ **and** $\mathbf{X_t}$	is the previous and current state.

Equation 1 is the transition equation, and Eq. 2 is the observation equation [10]. The transition equation gives a schematic result of the state $t-1$ to state t, and the observation equation will give us an approximate value of the state t from the value of the state $t-1$; here, $t-1$ is the previous and t is the current state. Matrix A_t and B_t can change at every point of time t and they allow us to map from the previous state to the next state using the control command. The matrix involved in these equations has $n \times l$ dimensions, where n is the dimension of the state and l is the dimension of the command.

3.2 State-Space Model

State-space model has some advantage over the linear state model. A particular state in a state-space model is dependent on previous states; hence, this model is time-dependent [11]. State-space model is a *linear* combination of the prior state at time $t - 1$ as well as *system noise* (random variation) [12]. Equations 3 and 4 involved in this model are quite similar with the linear model.

$$X_t = A_t X_{t-1} + B_t U_t + \epsilon_t \tag{3}$$

$$Z_t = C_t X_t + \delta_t \tag{4}$$

The extra component $\mathbf{C}_t$ is the variance–covariance matrix for the multivariate normal distribution from which the system noise is drawn. All the other components are same as that of the linear model.

4 Survey and Observations

For our observations, we choose Indigo's regular flight 6E206 from Dibrugarh to Kolkata. The reason to choose this specific flight is based on the economic motivations discussed in the introduction section. Observations were taken on daily basis of the selected flight from November 2, 2016 to January 15, 2017 at the time 12:40 p.m., 5:14 p.m., and 11:40 p.m., respectively. Table 1 is a sample of observations taken so far.

From the complete observations, it has been seen that the fare has changed from Rs. 3180 to Rs. 12480. Table 2 shows some fares within this range.

After a thorough analysis of all the readings, it is seen that there exists a relationship between the flight fare and the weekdays. This relationship between the flight fare and the weekdays will assist in deducing the control matrix, i.e., U_t matrix.

After taking all the observations and analysis of the data, the pattern of variation of fare in between the weekdays is prominent. The variation bar graph in Fig. 3

Table 1 Sample observations

Observations for →	11/11/2016	12/11/2016	…	26/11/2016
Observations on ↓	Fare (Rs.)	Fare (Rs.)	…	Fare (Rs.)
10/11/2016 (12:40 p.m.)	6789	10370	…	3389
10/11/2016 (5:15 p.m.)	6889	9389	…	3889

Table 2 Range of flight fares (in rupees)

3180	3480	3880	4380	4780	8780	…	9480	10370	11480	12480

shows this change of pattern. From the bar graph, it is seen that the fares on Monday, Tuesday, Wednesday, and Thursday are same on all the 4 weeks of a month, whereas on Friday and Saturday, the fare varies on first and third week from second and fourth week. The variation in the graph for Friday shows that on first and third week, the flight fare is high as compared to the second and fourth week. But on Saturdays, pattern is a bit different. On Saturdays, the fare of second and fourth week is higher than the fare of first and third week.

The deviation of the fare on a daily basis is shown in Fig. 4, where it is seen that the deviation can be positive or negative. Positive deviation means that the fare has been increased on that day, whereas negative deviation means that the fare has been decreased on that particular day in that month in comparison to the base fare. The space between two bars for a specific day signifies zero deviation. For example, in Sunday, there is no negative deviation (neither for first and third week, nor for second and fourth week). So there are blank spaces on the graph for that day.

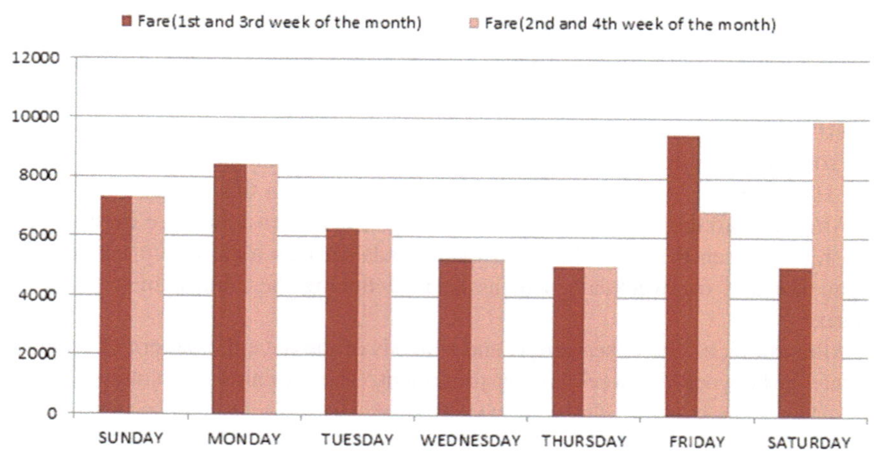

Fig. 3 Weekly fare chart

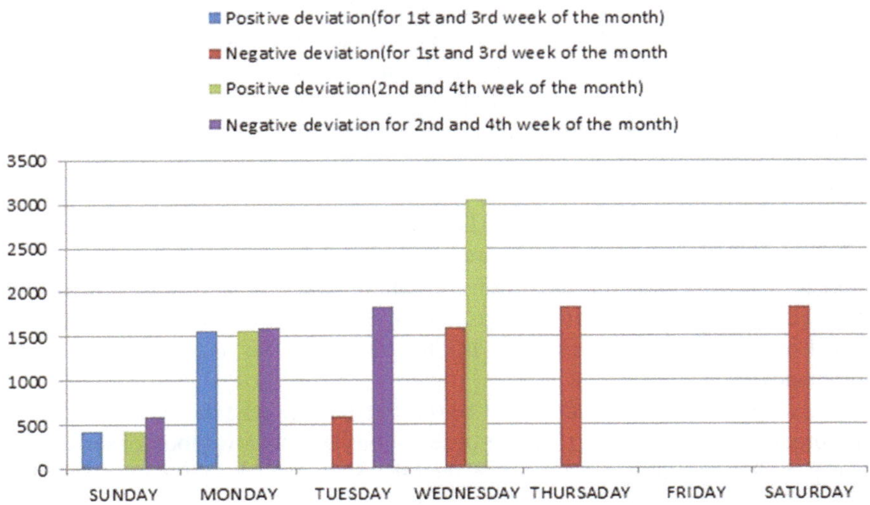

Fig. 4 Weekly deviation graph

5 Algorithm and Implementation

This section discusses some basic matrices based upon the previous observations and defines a methodology for fare prediction based on Kalman filter. For predicting flight fares based on previous observations, some stepwise procedures have been proposed. The following example provides a scenario for better understanding.

At first, it is required to find a base fare from which it can be determined by how much the flight fare will vary at each and every moment. Base fare means the average flight fare which can be calculated from Table 2. To predict the flight fare of a particular day of the week, it is required to analyze the previous data of that day in different weeks. Only after analyzing the data, a pattern can be viewed in which the fare changes for that particular day. Suppose it is required to find out the fare of Monday. So we have to check the fare of Monday on the previous Sunday. The observations with same procedure for different Mondays have to be taken. After getting some decent set of observations, an average price from the observed data can be calculated.

Let us assume M1, M2, and M3 for first, second, and third Monday similarly S1, S2, and S3 for Sunday. Fare of M1 checked on S1 is 8300, fare of M2 checked on S2 is 8400, and fare of M3 checked on S3 is 8500. Therefore, the average fare value for Monday is 8400. Now, as the average price of Monday is present, the deviation of the average price of Monday from the calculated base fare (base fare 6870 and average fare of Monday is 8400, so the deviation is 1530) can be deduced.

Now, the control matrix (U_t) and the change matrix (B_t) are to be calculated. U_t will depend upon three factors for flight fare prediction: Time, day, and deviation ("+1" for increase from the base fare and "−1" for decrease from the base fare).

As U_t is control matrix, consider a constant value for time since the time factor (here time = 1) cannot be controlled. Day value will also be 1 as in the linear model only the fare of next day can be predicted. Third factor is the deviation of the average fare from the base fare. Take deviation value +1 if average fare is more than the base fare and −1 if average fare is less than the base fare. U_t will be a 3 × 1 matrix (3 is the dimension of the state and 1 is the dimension of the command).

In change matrix B_t, the time will be 0 as we have to take the same time for the fare prediction at which we took the observations (suppose we took observation at 5:30 p.m. for three different Mondays, then the predicted fare will be for a Monday at time 5:30 p.m. only). Value of day will be 1 in B_t, as in linear model we have to predict fare of the next day only. The third factor for change matrix is the deviation value. It is the value of the difference between the average fare and the base fare. The deviation value will be a particular constant value for a particular day. Change matrix will be a 1 × 3 matrix where 1 is the dimension of state and 3 is the dimension of the command.

Based on this example, an algorithm is proposed for predicting similar environment with the help of linear Kalman filter model. Next subsection describes the steps for prediction of the flight fare.

5.1 Procedure for the Implementation

The various steps involved in the prediction are as follows:

- **Step 1**: Take the average value of all the 17 fares; in our case the average fare is 6870. For the sake of convenience this value is named as the base fare.
- **Step 2**: For each day's data, deduce the variance or the deviation of the fare (of that day) from the calculated base fare.
- **Step 3**: Calculate A_t matrix, which will be a unitary matrix for linear model (in this example it is a 3 × 3 matrix).

$$A_t = \begin{bmatrix} 1 & 0 & 0 \\ 0 & 1 & 0 \\ 0 & 0 & 1 \end{bmatrix}$$

- **Step 4**: Calculate the control matrix (U_t) and the change matrix (B_t).

$$U_t = \begin{bmatrix} Time \\ Day \\ +1 \vee -1 \end{bmatrix} \qquad B_t = [Time \quad Day \quad Deviation]$$

- **Step 5**: Calculate the prediction using Eq. 1.
- **Step 6**: ϵ_t and δ_t are constant values for different weekdays. After predicting the flight fare, the error percentage can be calculated; by comparing with the original fare, this will give a relative error which will be applied for any changes or to get more accurate results.

5.2 Algorithm

The prediction algorithm is as follows:

```
Start
Read present day fare
Read prediction data and day;
For a specific day i;
  Calculate Xti = input * At + Bt * Ut+ ∈
  If Xti >  MaxVali
    Crti= Xti - cei
End
```

where $MaxVal_i$ is the maximum price for day i and ce_i is the correction constant on day i. Both ce_i and $MaxVal_i$ can be deduced from observations. Cr_{ti} is the corrected value for day i if prediction increases above maximum price.

5.3 Matrices

After detailed analysis of the observations, the following matrices are defined to be used in the implementation. All the matrices are n * m matrix, where n(row) represents the days on which we will predict and m(column) represents the days for which prediction is done.

- **Observation matrix**: Values are from initial survey.
  Ovr[i][j]—value observed on day i for day j.
- **Prediction matrix**: This matrix is generated after prediction by Kalman filter on observation matrix.
  Pr[i][j]—value predicted on day i for day j.
- **Correction matrix**: This matrix is generated after correction on prediction matrix by respective correction constants ce_i's.
  Cr[i][j]—corrected value for prediction on day i for day j.

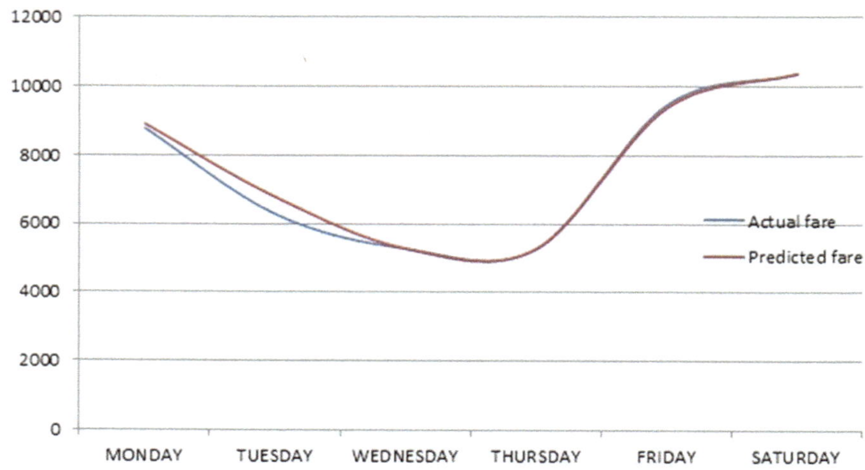

Fig. 5 Comparative deviation with actual price

6 Results and Analysis

A comparative graphical analysis of the implementation is shown in Fig. 5. The graph shows the deviation of actual fare from the predicted fare. There are only 6 days in this graph as Sundays are excluded from the domain of this work. Sunday requires much more data as compared to the other weekdays. The fare variation is very quick and random on Sundays which will require a nonlinear model for prediction.

Since there is a variation in actual fare and fare predicted in this approach, therefore, there will be an error for each day, which has been taken as the correction constant. Table 3 shows the relative error percentage of days from Monday to Saturday. The relative error percentage can be calculated by dividing actual error by actual fare and then multiplying the result with 100 which finally gives the actual error percentage.

The variation in the flight fare generally follows the normal distribution rule, which means that the variation is within a Gaussian range. This can be proved by

Table 3 Table showing relative error percentage

	Actual fare	Predicted fare	Actual error	Relative error (%)
Monday	8780	8911	211	2.4
Tuesday	6280	6781	501	7.9
Wednesday	5280	5281	1	0.01
Thursday	5280	5281	1	0.01
Friday	9480	9391	89	0.93
Saturday	10370	10381	11	0.1

Fig. 6 Observation matrix

	SUN 20	MON 21	TUE 22
MON 14	4889	4389	4389
TUE 15	5389	4389	3889
WED 16	5889	4889	3889
THU 17	8789	8089	3889
FRI 18	8089	9489	6789

Fig. 7 Prediction matrix

	SUN 21	MON 22	TUE 23
MON 14	6520	5010	3390
TUE 15	7020	5010	2890
WED 16	7520	5510	2890
THU 17	9830	8710	2890
FRI 18	9130	7010	5790

Fig. 8 Correction matrix

	SUN 21	MON 22	TUE 23
MON 14	6100	3970	3180
TUE 15	4980	3180	3180
WED 16	7520	4890	3180
THU 17	8410	7290-	3180
FRI 18	8590	6980	5760

the following method, where three matrices (observation matrix, prediction matrix, and correction matrix), and their bell curves have been shown in Figs. 6, 7, 8, and 9. All the matrices will have a unique standard deviation value as shown below.

This method works on the observed data, which have been converted into a 5 × 3 matrix form. Rows represent the day of the week and the columns represent the day to be predicted. The first matrix is the observation matrix (Fig. 6).

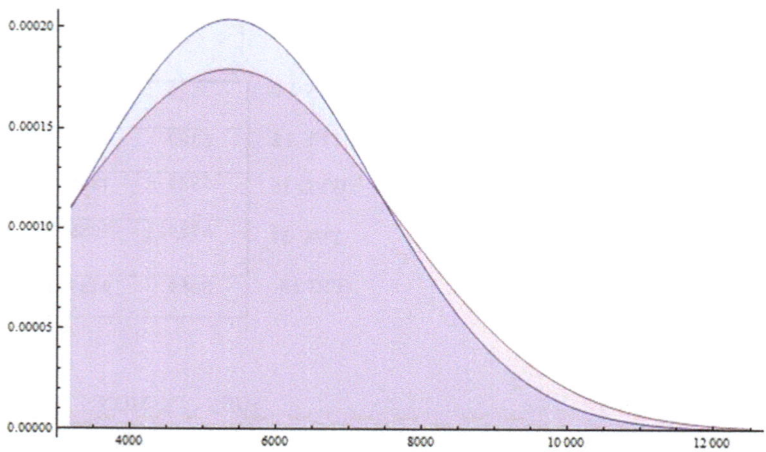

Fig. 9 Normal distribution curve for prediction (purple region) and correction (blue region)

In this observation matrix, MON 14, TUE 15, WED 16, THU 17, and FRI 18 are the rows, which represent the days of a particular month in which the observations on SUN 20, MON 21, and TUE 22 have to be taken. The values in this matrix are taken as inputs in order to create the prediction matrix (Fig. 7) with the help of the algorithm discussed in earlier section.

In prediction matrix, Monday (21), Tuesday (22), and Wednesday (23) are the predicted days, and this new matrix is formed with the help of the observation matrix.

After prediction matrix, the next matrix formed is the correction matrix as shown in Fig. 8. The correction matrix is formed by eliminating errors from the prediction matrix using the correction constant. The normal distribution curve for prediction and correction is shown in Fig. 9. The curves are generated by using mean (μ) value as 5359.32 and deviation values of 1958.18 (σ for correction) and 2228.1 (σ for prediction).

Table 4 provides the information for mean, standard deviation, and probability density for the abovementioned matrices.

Table 4 Mean standard deviation and probability density table

Matrix	Mean (μ)	Standard deviation (σ)	Probability density ($f(x\|\mu, \sigma^2)$)
Prediction matrix	5941.3	2228.1	0.8741
Corrected matrix	5359.3	1958.18	0.8176

7 Conclusion

In this work, we have tried to predict the fight fares in the short periods by implementing observations in a proposed algorithm based on Kalman Filter. The implementation is done in a linear model of Kalman Filter.

In the algorithm presented in this paper, observed data and the technical data (time, day, etc.) were given as input to the prediction filter. After providing input to the algorithm based on detailed observations, the user gets an estimated flight fare within a range of dates to select the best combination of date and price. The fare for Sunday cannot be calculated in this proposed model, as Sundays have the most random fare variation in all the days in a week and will require more factors and a nonlinear model for efficient prediction, which will be the future scope of work to be done for this proposed method.

References

1. Feng, Youyi, and Baichun Xiao. "A dynamic airline seat inventory control model and its optimal policy." Operations Research 49.6 (2001): 938–949.
2. Etzioni, Oren, et al. "To buy or not to buy: mining airfare data to minimize ticket purchase price." Proceedings of the ninth ACM SIGKDD international conference on Knowledge discovery and data mining. ACM, 2003.
3. Rama-Murthy, Krishna. Modeling of United States Airline Fares–Using the Official Airline Guide (OAG) and Airline Origin and Destination Survey (DB1B). Diss. Virginia Tech, 2006.
4. Groves, William, and Maria Gini. "A regression model for predicting optimal purchase timing for airline tickets." Technical report (2011).
5. Diebold, Francis X. Elements of forecasting. South-Western College Publ., 2006.
6. Maybeck, Peter S. "The Kalman filter: An introduction to concepts." Autonomous robot vehicles. Springer New York, 1990. 194–204.
7. "QuantStart." State Space Models and the Kalman Filter - QuantStart. N.p., n.d. Web. 10 May 2017.
8. "An Explanation of the Kalman Filter." Mathematics Stack Exchange. N.p., n.d. Web. 10 May 2017.
9. Harrison, Jeff, and Mike West. Bayesian forecasting & dynamic models. New York: Springer, 1999.
10. "State Estimation with a Kalman Filter." https://courses.cs.washington.edu/courses/cse466/11au/calendar/. Web. 10 May 2017.
11. "Linear Time-Invariant Systems." Stanford University. N.p., n.d. Web. 10 May 2017.
12. Kosanam, Srikiran, and Daniel J. Simon. "Kalman filtering with uncertain noise covariances." (2004): 375.)

Crop Suitability and Fertilizers Recommendation Using Data Mining Techniques

Archana Chougule, Vijay Kumar Jha and Debajyoti Mukhopadhyay

Abstract Economy of India highly depends on agriculture. Still traditional ways of recommendations are used for agriculture. Currently, agriculture is done based on various approximations of fertilizers quantity and the type of crop to be grown or planted. Agriculture highly depends on the nature of soil and climate. Therefore, it becomes important to make advancement in this field. The paper proposes development of an ontology-based recommendation system for crop suitability and fertilizers recommendation. It bridges the gap between farmers and technology. The system predicts suitable crop for the field under consideration based on region in Maharashtra state of India and type of soil. It provides proper recommendation of fertilizers to the farmers. Fertilizer recommendation is done based on nitrogen, phosphorus, and potassium (NPK) contents of soil and using past years research data that is stored in ontology. Along with fertilizer recommendation system also provides suggestions about crop suitability in particular region. Recommendation system uses random forest algorithm and k-means clustering algorithm.

Keywords NPK · K-means clustering · Fertilizer recommendation
Random forest algorithm · Ontology

A. Chougule (✉)
Walchand College of Engineering, Sangli, India
e-mail: chouguleab@gmail.com

A. Chougule · V. K. Jha
Birla Institute of Technology, Mesra, Ranchi, India
e-mail: vkjha@bitmesra.ac.in

D. Mukhopadhyay
Adamas University, Kolkata, India
e-mail: debajyoti.mukhopadhyay@gmail.com

C. R. Panigrahi et al. (eds.), *Progress in Advanced Computing and Intelligent Engineering*, Advances in Intelligent Systems and Computing 714,
https://doi.org/10.1007/978-981-13-0224-4_19

1 Introduction

Agriculture is the main source of income and survival in India for majority population. Agriculture is done from ages. Hence, a rich collection of agricultural past data is available. Information technology can be used to process such a large amount of data and then for recommendation. Various data mining techniques can be used for finding recommendations about crops and fertilizers. Outputs of these techniques can be communicated to the smartphones. This paper focuses on the implementation of data mining algorithms which can help in building an effective recommendation system using available observation data.

The paper describes a system which recommends the crops suitable for particular region based on crop yield history of last 3 years in that region and the fertilizers suitable for specific crop based on soil measurements to farmers. It can help farmers for increasing their crop production. The paper shows how information available with government about yearly production in various areas can be used for crop recommendations to farmers. As information represented in the form of ontology can be easily shared and reused, the knowledge base of recommendation system is maintained in the form of ontology. The system uses random forest algorithm for crop recommendation as it works efficiently on huge dataset and can handle missing values. Paper describes how k-means clustering can be used for predicting best suitable fertilizer for the crop based on given available NPK content in the soil.

2 Related Work

Limin Chuana and Ping Hea proposed a fertilizer recommendation system for wheat in China [1]. Two parameters namely yield response and agronomic efficiency are used by the recommendation system. Limin Chuana and Ping Hea also consider the nitrogen, phosphorous, and potassium (NPK) contents for fertilizer recommendation of wheat. It helps to prevent the inappropriate application of fertilizers in wheat production systems in China. Yield response and agronomic efficiency were incorporated as part of the nutrient expert for wheat fertilizer recommendation decision support system.

Department of agriculture, government of West Bengal has developed soil test-based fertilizer recommendation system (STFRS) for farmers in West Bengal [2]. Information about soil testing laboratories, availability of nutrients, and recommendations by experts are provided to farmers through SMS service on mobile phones. Smart soil health card provides access to cloud-based data on mobile phones. Display of digital soil maps is also an important feature of the system.

QUEFTS model was used for calculating soil fertility required for fertilizer recommendation [3]. Web-based decision support system for fertilization application on wheat, maize, and peanut is provided by Hao Zhang et al. [4]. It is specifically developed for villages in China. Maps of villages are taken and location-specific recommendations are provided using ArcView in ArcGIS. Three parameters are considered for recommendations namely soil measurements, farm production level, and target yield for the crop. Three types of databases as system database, spatial database, and attribute database are maintained here. Along with other attributes, meteorological information is also considered for decision-making.

Recommendations for purchasing fertilizers from online portal based on history of past purchases are proposed by Mansi Shinde et al. [5]. Apriori algorithm is used for this purpose. Fertilizers analysis services are provided by Spectrum Analytics Inc. Washington [6]. It provides recommendations for 250 types of crops. It provides fertilizers schedule along with its quantity at each growth stage of crop under consideration.

Precision Fertilization Management Information System (PFMIS) [7] is fertilization recommendation system based on GIS and GPS. ArcGIS is used for maps on soil resources. Recommendations are done by applying data mining techniques on information collected by GIS and GPS.

Mansi Shinde et al. [8] have proposed a crop recommendation and fertilizer purchase system which uses a priori algorithm for recommendation. Based on previous history of fertilizers, purchase recommendations are provided. For crop recommendations, they have used market trends data and applied random forest algorithm on that. Kiran Shinde et al. [9] have developed web-based recommendation system for crop and fertilizers recommendation by considering past data about market price.

3 Design and Implementation

The recommendation system is developed as android-based application connected to server and has ontology knowledge base as shown in Fig. 1. The farmer has to create an account and log into the android application for accessing the system. After that, he can either get two services as crop recommendation or fertilizer predictions.

Crop prediction is done using algorithm called random forest; fertilizer recommendation is done through k-means algorithm. For getting these outputs, farmer needs to enter some input values which act as test data. The trained data is stored in the ontology which helps in creating a machine learning model for crop prediction. All these processes take place at the server end and output is displayed back to the farmer on his android device.

3.1 Crop Recommendation

Specific regions in Maharashtra state of India have specific soil characteristics, weather conditions, and crop production history. The information about types of

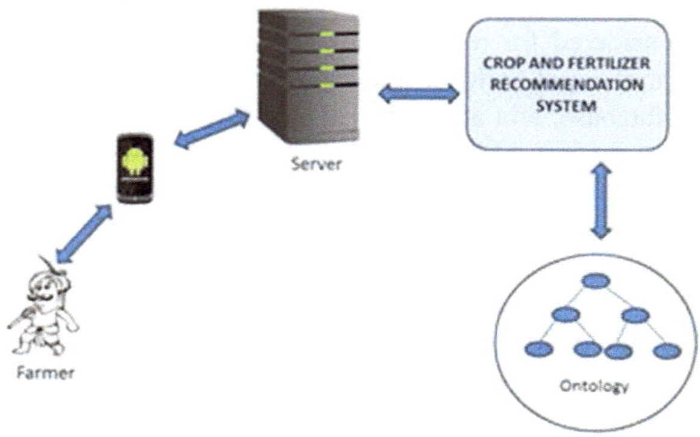

Fig. 1 System architecture

crops harvested and yield of those crops in those areas can be used for making decision on which crops give maximum yield in specific region. For recommending the most suitable crop for the field, knowledge base contains past 3 years data about yields of crops in Maharashtra. It is taken as training set. This knowledge is collected from department of agriculture, Government of Maharashtra. This knowledge is stored in the form of ontology. Web Ontology Language (OWL) is used for ontology representation.

OWL [10] is a semantic web language. It is designed to represent rich and complex concepts, groups of concepts, and relationships between concepts. In OWL file, data properties about the crops are stored, which represents the relationship between the crop and their attributes. As shown in Fig. 2, cropid, cropname, district, taluka, season, etc. are the data properties of the OWL class named as crop. All records received from government in excel sheets are converted

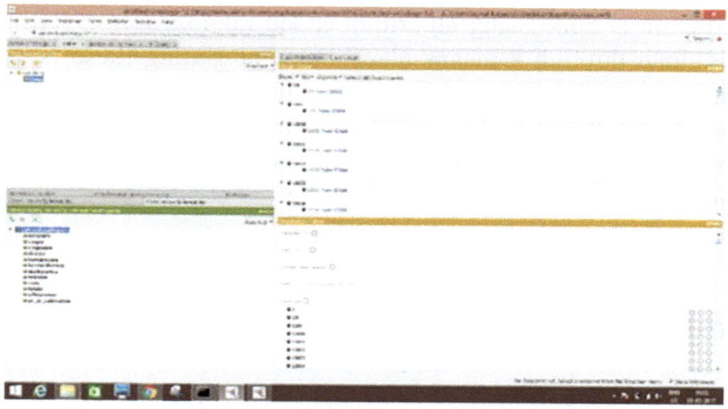

Fig. 2 Crop yield history stored in ontology

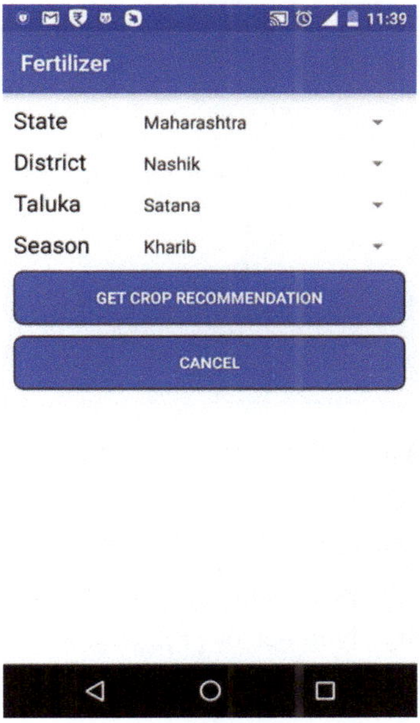

Fig. 3 Input form for crop recommendation

into ontology data property values using APIs available for data conversion in ontology form.

System applies random forest algorithm using knowledge base in ontology as training set for crop recommendation. Random forest algorithm is used as the accuracy of it is found to be higher than ID3 algorithm for a given dataset. This is because ID3 algorithm constructs only a single tree. If one node/crop is not included in the tree accurately, the entire prediction may be wrong. A random number of trees are constructed by random forest algorithm and output of random trees is calculated. Final output of random forest algorithm is aggregation of output random trees. Decision criteria for crop recommendation are based on production quantity of the crop and market price of the crop in the specific area (Fig. 3).

District, state, and season are the input parameters for random forest. Random forest generates many number of decision trees by extracting training data stored in ontology, and each tree predicts a crop for given test data. Final output is calculated as the probability of a particular crop predicted by random trees. Figure 4 shows example of predicted probabilities for crops suitability.

Probability of a particular crop is calculated as

$$P_a = n_a / n_t$$

Fig. 4 Output of k-means clustering for suitable crop recommendation

where

P_a Probability of a particular crop (a)
n_a Number of predicted trees of crop (a)
n_t Total number of trees.

3.2 *Fertilizer Recommendation*

Recommendation of fertilizers is based on nitrogen, phosphorous, and potassium measurements from soil. Nitrogen in the soil is responsible for color of leaves. If low quantity of nitrogen is found in the soil, then plants will have slight yellowish leaves and if quantity is moderate or high, it will have greener leaves. The phosphorous content in the soil is responsible for the reproductive system of the plant. Its value will predict the growth of fruits and flowers of the plants. The potassium content of soil is responsible for its overall growth. Its value will predict how stronger the plant roots will be and will also determine the overall growth process of the plant.

For recommending fertilizer to the farmer, K-means clustering algorithm is used here. It is an unsupervised learning algorithm used to find out fertilizers with NPK contents nearest to requirements for specified crop. Crop name and soil contents nitrogen (N), phosphorous (P), and potassium (K) are given as an input to the clustering algorithm (Fig. 5).

There are two main steps in the algorithm implementation. First, algorithm calculates the required amount of fertilizer as follows:

$$R_a = S_a - M_{ta} \tag{1}$$

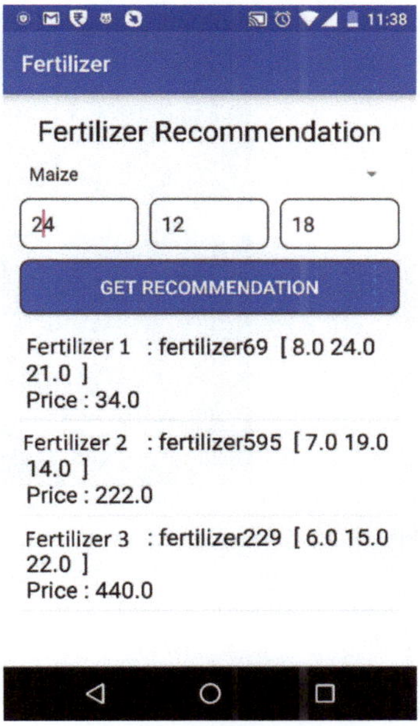

Fig. 5 Fertilizers recommendation for selected crop

where

R_a Required NPK for crop "a"
S_a Standard NPK for crop "a"
M_{ta} Measured NPK for crop "a"

In second step, algorithm forms clusters of nearby fertilizers with the help of Euclidean distance. It is the difference between NPK values. Fertilizers in clusters with minimum distance are recommended to farmer.

4 Performance Evaluation

Implemented algorithms are checked for performance and accuracy with the help of farmers. Standard precision measure is used for calculating accuracy. Precision is a fraction of the retrieved information that is relevant. It is marked for evaluation of accuracy and exactness. Here, accuracy of predicted crops and fertilizers is compared against actual values given by experts for those fields and set of crops for each farmer. If the crops recommended by the system belong to the expert's recommendation sets, then those crops are relevant crops. Considering these relevant crops, precision of the recommendation system is calculated. The graphical representation of precision versus number of users is shown in Fig. 6.

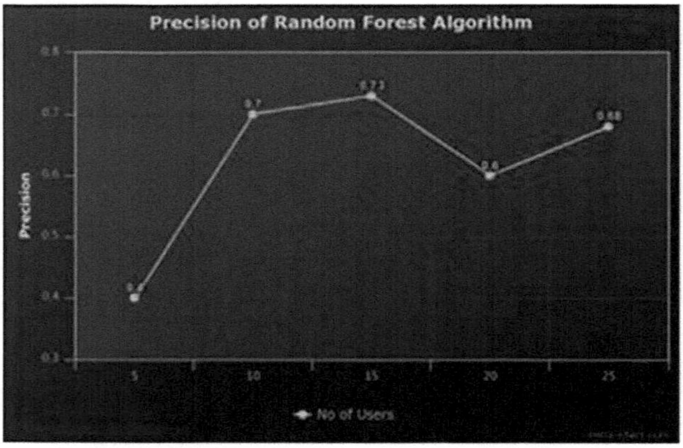

Fig. 6 Performance evaluation of recommendation system

$$\text{Precision} = \text{A}/\text{T} \tag{2}$$

where

A Number of users who got relevant predictions
T Number of users.

5 Conclusions

As recommendation of fertilizers and crops is important for farmers in farming decision-making, the paper proposes the use of two data mining techniques to provide recommendations. Recommendations of suitable crop in the field and fertilizers for crops to farmers are provided with the help of data stored in ontology. Proposed system provides crop recommendation based on region, type of crop, and fertilizer recommendation based on NPK content of soil, available on their mobile phones. Thus, the aim of this system is to increase the production of crops by recommending correct crop and fertilizer. The performance evaluation shows that the accuracy of developed system is reasonably high. In future, android application will be developed in regional language.

References

1. Limin Chuana, Ping Hea; Establishing a scientific basis for fertilizer recommendations for wheat in China: Yield response and agronomic efficiency; Field Crops Research; Volume 140; January 2013; pp. 1–8

2. Jitendra Roy; Soil Test based Fertilizer Recommendation System (STFRS), Department of Agriculture, Government of West Bengal, 2015
3. Models Library (http://models.pps.wur.nl)
4. Hao Zhang, Li Zhang, Yanna Ren, Juan Zhang, Xin Xu, Xinming Ma, Zhongmin Lu; Design and Implementation of Crop Recommendation Fertilization Decision System Based on WEBGIS at Village Scale; In: Li D., Liu Y., Chen Y. (eds) Computer and Computing Technologies in Agriculture IV. CCTA 2010. IFIP Advances in Information and Communication Technology, vol. 345. Springer, Berlin, Heidelberg
5. Mansi Shinde1, Kimaya Ekbote, Sonali Ghorpade, Sanket Pawar, Shubhada Mone; Crop Recommendation and Fertilizer Purchase System; International Journal of Computer Science and Information Technologies, Vol. 7 (2), 2016; pp. 665–667
6. http://www.smart-fertilizer.com/
7. Zhimin Liu, Weidong Xiong, Xuewei Cao; Design of Precision Fertilization Management Information System on GPS and GIS Technologies; CCTA 2011, Part I, IFIP AICT 368, SpringerLink; pp. 268–277; 2012
8. Mansi Shinde, Kimaya Ekbote, Sonali Ghorpade, Sanket Pawar, Shubhada Mone; Crop Recommendation and Fertilizer Purchase System; International Journal of Computer Science and Information Technologies, Vol. 7 (2), 2016; pp. 665–667
9. Kiran Shinde, Jerrin Andrei, Amey Oke; Web Based Recommendation System for Farmers; International Journal on Recent and Innovation Trends in Computing and Communication; ISSN: 2321–8169 Volume: 3 Issue: 3; pp. 1444–1448
10. Pascal Hitzler, Markus Krötzsch, Bijan Parsia, Peter F. Patel-Schneider, Sebastian Rudolph, eds.; OWL 2 Web Ontology Language: Primer (Second Edition); W3C Recommendation, 2012, http://www.w3.org/TR/2012/REC-owl2-primer-20121211/

Medicinal Plant Information Extraction System—A Text Mining-Based Approach

Niyati Kumari Behera and G. S. Mahalakshmi

Abstract In this paper, we have discussed on applying text mining techniques to extract information on health benefits of medicinal plant from text article. The presence of multi-term phrases and complex sentences, i.e., the sentences that include clauses, implicit semantic relations in a text article, has always raised the complexity of information mining process. We have proposed a simple pattern-based semi-supervised approach powered by NLP techniques to deal with these issues. We have evaluated our methodology for a set of web documents on medicinal plants. Performance (in terms of recall) of our method was observed to considerably higher than existing relation extraction methodologies.

Keywords Medicinal plants · Plant part · Disease · NLP · Text mining

1 Introduction

Though we have evolved through ages in terms of culture, tradition, and technology, alternative medicines have always been people's first choice because of their easy accessibility and less or no side effect. These alternative therapies use various medicinal plants directly or indirectly as a base of their medicine. Technological advancement has made it possible to locate details about numerous such plants online in different web formats.

Text mining has played a key role to extract and analyze the biomedical literature to bring out useful information about medicinal plants, i.e., how they are helpful for curing different diseases or details about plant pertaining to a particular geographical region, etc. For instance, authors in [1] have used text mining to analyze multiple electronic databases like PubMed/MEDLINE, Scopus, Science

N. K. Behera (✉) · G. S. Mahalakshmi
Department of Computer Science and Engineering, Anna University, Chennai, India
e-mail: niyatibehera@yahoo.co.in

G. S. Mahalakshmi
e-mail: gsmaha@annauniv.edu

C. R. Panigrahi et al. (eds.), *Progress in Advanced Computing and Intelligent Engineering*, Advances in Intelligent Systems and Computing 714,
https://doi.org/10.1007/978-981-13-0224-4_20

Direct, and Wiley to identify Indian medicinal plants for diabetes. Similarly, research was done on plants that can be applied for malaria [2–4], cough [5], cancer [6], skin infection [7], and eye treatment [8], which are just few to name.

Given a disease name, existing systems have used tf-idf method, co-occurrence analysis, or pattern-based approach to mine information from biomedical literature. Though statistical methods have been used, in a sentence, the phrase between the considered plant or herb name and disease name can be crucial to analyze the health benefits of the former.

For example: "**Powdered ginger can cause heartburn, bloating, gas, belching and nausea**."

So if we can identify the correct and precise declarative phrases, this will surely improve the system performance.

2 Objective

Main objective of this paper is to employ text mining to identify common health benefits of the plants like diseases they cure or human body parts which are benefited by the particular plant and also the parts of the plant used in the healing process. We are dealing with multi-term phrase extraction, i.e., for the disease term "joint pain" rather than extracting the noun head "pain" as a disease name, our approach can identify "joint pain" as a whole term, thereby increasing the efficiency of the system. Our study includes Ayurvedic text articles for plants like turmeric, aloe vera, amla, ginger, curry leaves, neem, basil or tulsi. The following sections of the paper discuss the methodology in detail including the outcome and shortcomings of the framework.

3 Herbal Information Extractor

The entire knowledge acquisition process consists of the following steps:

(i) Creation of herb and disease lexicons,
(ii) Extraction of Treatment Specific Phrase Patterns (TSPP), and
(iii) Analyzing new text document to extract relevant information.

3.1 Create Lexicon

A highly comprehensive and accurate lexicon is considered as prerequisite for any kind effective information extraction tasks. We have used traditional knowledge digital library (TKDL),[1] "AYUSH" of National Health Portal (NHP),[2] and BioPortal[3] to build the lexicons for both herb and diseases.

The herb lexicon includes the terms listed under "Parts used" section of the plant description. The final list for herb lexicon consists of English name of 95 medicinal plants along with around 25 plant parts that are commonly used in traditional medicine.

The human body lexicon has been created by referring Wikipedia[4] and includes the glossary terms from the referred webpage.

3.2 Extract Treatment-Specific Phrase Pattern

The main motive behind this task is to identify the English phrases that commonly used to describe relation between herb name and disease. The phrase connecting these two entities is often very complicated. Sometimes, the phrase can be very general such as "*Herb* **cures** *Disease*" or very specific like "*Herb* **is used as an antiseptic to** Disease", and "*Herb* **are widely used as a local application for** *Disease*". Though there can be different ways to express the semantic association between the drug name and the disease names, due to the flexibility of expression in natural language, our system is based on the assumption that those semantic relations are not very commonly used with every other entity pair. Here, in this paper, we define those phrases as **Treatment Specific Phrase Pattern (TSPP)** which takes the form:

(Herb/medicinal plant) **TSPP** (Disease/Body Part)

In recent studies, authors in [9] have proposed to automatically learn treatment-specific textual patterns using known drug–disease pair from MEDLINE corpus. In order to collect such phrase patterns, we have used an e-book on "Medicinal Plants"[5] listed in the reference section of National Health Portal and also a website on medicinal plants.[6] Since our work focuses on extracting information like diseases healed by the herb and herb part, used to cure, we have referred the important section of each plant description in the e-book. Similarly, in the referred web documents (Footnote 6), we have focused only on the "uses and benefits" section for our purpose. We have used general description of 90 medicinal plants listed in Fig. 1 and manually tagged around 475 sentences from 1200

[1]http://www.tkdl.res.in/tkdl/langdefault/common/Home.asp?GL=Eng.

[2]https://www.nhp.gov.in/ayush_ms.

[3]https://bioportal.bioontology.org/ontologies/MESH/.

[4]https://en.wikipedia.org/wiki/Outline_of_human_anatomy.

[5]http://joyppkau.tripod.com/PDFs/Bk%20Medicinal%20Plants.PDF.

[6]http://www.iloveindia.com/indian-herbs/basil-herb.html.

Alstonia	chicory	Garlic	khus khus	Pomegranate
Ambrette	Cinamon	Glory lily	Kudzu	Pruriens
Amla	Cinchona	Greater ammi	Lemongrass	Purging croton
Ashoka	Common Henbane	greater galangal	Liquorice	Quinine
Ashwagandha	Common indigo	Green chiretta	Long pepper	Reetha
Asparagus	Coomb teak	Guggul	Malabar nut	Rosy flowered leadwort
Brahmi	Costus	Gymnema	Manjishta	Sandalwood
Bael	Curcuma	Hibiscus	marigold	Sarsaparilla
Baheda	Curry leaf	Holostemma	Mesua	Senna
Bamboo	Custard Apple	Indian bdellium	Moringa	Serpentwood
Bauhinia	Datura	Indian beech	Mulberry	Shankapushpi
Bhringaraj	Desmodium	Indian crocus	Nagadanti	Soapnut
Buckwheat	Eclipta	Indian ginseng	Neem	Strobilanthes
Calamus	Ephedra	Indian sarasapa-rilla	Oblonga	Strychnine tree
castor	Fenugreek	Indian senna	oleander	Tinospora
Catechu	Flax Seeds	Ipecac	Papaya	Walnut
Cedarwood	Garden rue	Jackfruit	Periwinkle	White flowered Leadwort
Chamomile	Ginger Lily	Kantakari	Podina	Worm killer

Fig. 1 List of medicinal plants used for collecting seed TSPPs

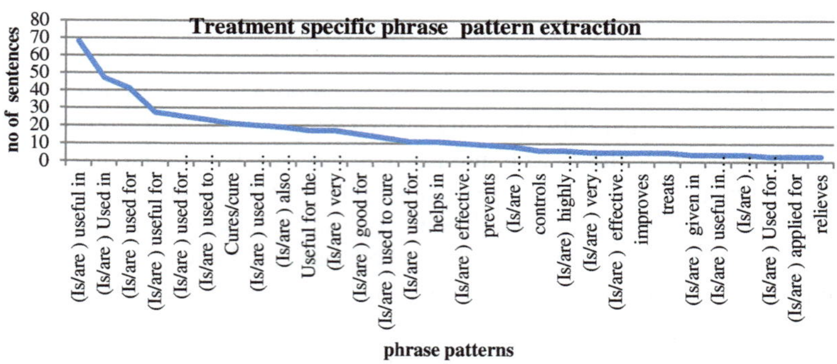

Fig. 2 Treatment-specific phrase pattern extraction

sentences that discuss their medicinal benefits. Once such treatment-specific phrase selection is done, we ranked the phrases based on their frequency of occurrence and considered top 30 phrases which have occurred at least 2 times in the set of selected sentences as shown in Fig. 2. Then, these patterns were used to extract information from any new document on a specific herb for further analysis.

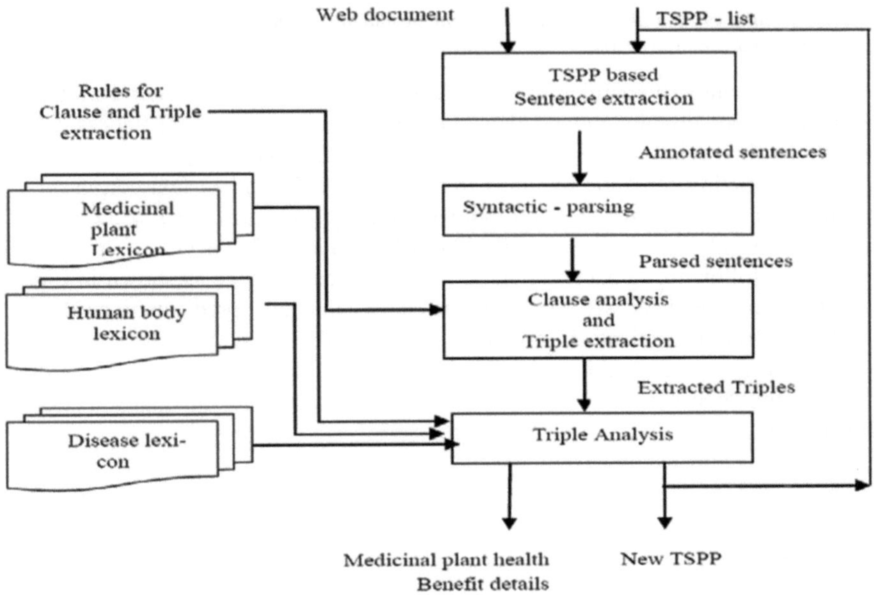

Fig. 3 Overall process details of herbal information extractor

3.3 *Obtain Medicinal Plant Health Benefit Details from Text*

In this section, we have discussed a text mining-based approach to automatically identify the information about an herbal plant from the text article. After the pre-processing phase, for any new medicinal plant, the process begins with querying the web with the "common English name" or "botanical name" of plant.

As shown in the block diagram Fig. 3, the framework considers a set of seed TSPP = {*P*1, *P*2, …, Pm} and a text article on a specific medicinal plant as input. The framework learns set of diseases D= {d1, *d*2 … d*n*} cured by the plant, a set of plant parts used for healing the diseases, and new treatment-specific patterns from the text (if exists).

3.4 *Identify Clause and Triples*

Clause present in a complex sentence can be identified by analyzing the sentence structure. A complex sentence can be divided into clauses based on conjunctions such as "and", "that", "which", and "it". The sentence structure is analyzed thoroughly from the parse information of a parser.

For example, consider the sentence regarding herb "**basil**" (Footnote 6).

The essential oil has also been used to fight headaches, reduce hay fever, allergies, or asthma, and it can even relieve the symptoms of hiccoughs.

A set of handwritten rules used to generate clauses is summarized in Fig. 5. These rules are domain independent and can handle complex sentences which involve implicit relations. Similar rule-based approach has been proved efficient in [10] for extracting semantic relations from domain-independent text corpuses

The semantic relation extraction tool would assign "***has also been used to fight***" as a relation only between ***essential oil*** and ***headaches***. However, our framework derives the following semantic relations for the given sentence as shown in Fig. 4.

After clause analysis, we select only the clauses that have predicate in predefined TSPP list. But if predicate is not present in the TSPP list, then first the VP is checked for any preceding negative sentiment words like "not", "rarely", "hardly", "never", and "occasionally". Then, they are further analyzed for final selection as shown in Fig. 6.

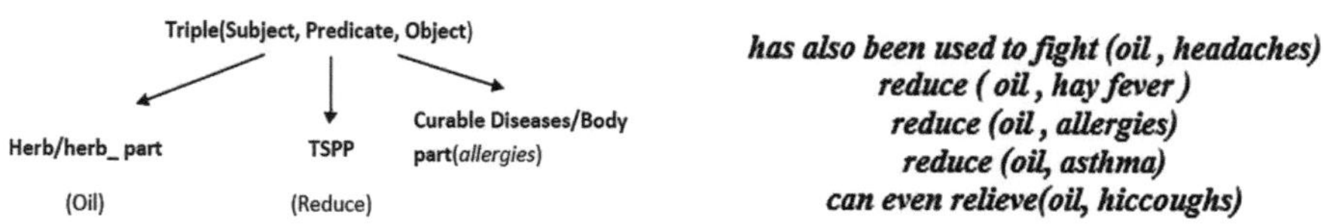

Fig. 4 Triple analysis details

Fig. 5 Rules for clause analysis

Sl No	Rule	Subject	Predicate	object
1	If (NP1 VP1 NP2)	NP1	VP1	NP2
2	If ((NP1, NP2..... and .NPn) VP1 NP0)	NP1	VP1	NP0
		NP2	VP1	NP0
		NPn.	VP1	NP0
3	If (NP0 VP1 (NP1, NP2, ... and NPn))	NP0	VP1	NP1
		NP0	VP1	NP2
		NP0	VP1	NPn
4	If (NP0 VP1 NP1,VP2 NP2, ...(and|,) VPnNPn)	NP0	VP1	NP1
		NP0	VP2	NP2
		NP0	VPn	NPn
5	If (NP0 VP1 NP1 and it VP2 NP2)	NP0	VP1	NP1
		NP0	VP2	NP2
6	If (NP0 VP1 NP1 'which' VP2 NP2)	NP0	VP1	NP1
		NP0	VP2	NP2
7	If (NP0 VP1 NP1 'that' VP2 NP2)	NP0	VP1	NP1
		NP1	VP2	NP2

Fig. 6 Psudocode for clause analysis

```
Input: a set of candidate sentences S, set of rules R  for triple analysis,
TSPP list,   lexicons
Output: set of triple recognized as having health
Benefits, updated TSPP list

BEGIN
  For each Si ∈ S,
    Apply rules Ri ∈ R, resolve clauses
    Generate new Candidate sentence set Snew
  End loop

  For each Sinew ∈ Snew,
      Apply rules Ri∈ R and extract triples                    T={t1,t2,…,tn}
where ti(si, vi, oi)
        [si: subject    vi: verb phrase  oi: object]
  End loop

  For each ti∈ T
       Check vi in TSPP list
     If vi ∈TSPP list
            store triple
     Else
      check if vi is preceded by any negation word
          if 'yes'
              discard the triple
          else
              check subject (si) in herb lexicon and                 object
(oi) in anatomy  lexicon as well as                 in disease lexicon
            if both conditions are satisfied
                  add vi to TSPP list , store ti
            else
                  discard ti
   End loop
 END
```

4 Implementation and Experiment

We have evaluated our system for medicinal plants like "neem", "aloe vera", "amla", "ginger", "tulasi", "curry leave", and "turmeric" by querying in web with their botanical name and also common English name. The statistical details of collected text documents are given in Table 1. The corresponding web documents are preprocessed and POS-tagged using pattern POS-tagger[7]. Then, sentences which include the pattern "#medicinal plant/plant part name TSPP #disease/body part name" are selected. Relaxing n-grams on the left side of the phrase patterns increases the overall recall from 45 to 71% in terms of number of relevant sentences extracted w. r. t the actual number of relevant sentences present in all the considered web documents. By relaxing n-gram, we mean that the phrase pattern is changed from "is used in" to "used in" or "is also used for" to "used for", and so on. Here, we have considered only up to n = 1, 2, i.e., unigram and bigram relaxation. For illustration of the process, consider a text on **Indian Gooseberry**[8] which contains the following sentence:

> ***it may be used as a rasayana (rejuvenative) to promote longevity, and traditionally to enhance digestion (dipanapachana), treat constipation (anuloma), reduce fever (jvar-aghna), purify the blood raktaprasadana), reduce cough (kasahara), alleviate asthma (svasahara), strengthen the heart (hrdaya), benefit the eyes (chakshushya), stimulate hair growth (romasanjana), enliven the body (jivaniya), and enhance intellect (medhya).***

The parsed output of the sentence in Fig. 7 shows that these tools have been proved effective for analyzing simple sentences; it fails to identify semantic

[7]http://www.clips.ua.ac.be/pattern.

[8]https://en.wikipedia.org/wiki/Phyllanthus_emblica.

Table 1 Details of collected text dataset

Name of medicinal plant	Neem	Aloe Vera	Ginger	Turmeric	Basil	Amla	Curry leaves
Sentences in text document	15	74	96	118	28	33	18

'SBJ': {1: Chunk ('amla/NP-SBJ-1')},

'VP': {1: Chunk('may be used/VP-1'), 2: Chunk('to promote/VP-2'), 3: Chunk('to enhance/VP-3'), 4: Chunk('reduce/VP-4'), 5: Chunk('purify/VP-5'), 6: Chunk('reduce/VP-6'), 7: Chunk('alleviate/VP-7'), 8: Chunk('strengthen/VP-8'), 9: Chunk('benefit/VP-9'), 10: Chunk('stimulate/VP-10'), 11: Chunk('enliven/VP-11'), 12: Chunk('enhance/VP-12')}

'OBJ': {2: Chunk('longevity/NP-OBJ-2'), 3: Chunk('digestion/NP-OBJ-3'), 4: Chunk('fever/NP-OBJ-4'), 5: Chunk('the blood/NP-OBJ-5'), 6: Chunk('cough/NP-OBJ-6'), 7: Chunk('asthma/NP-OBJ-7'), 8: Chunk('the heart/NP-OBJ-8'), 9: Chunk('the eyes/NP-OBJ-9'), 10: Chunk('hair growth/NP-OBJ-10'), 11: Chunk('the body/NP-OBJ-11'), 12: Chunk('intellect/NP-OBJ-12')}}

Fig. 7 Parsed output of the sentence structure

relations present in complex sentences that involve clauses and hidden relations. However, our methodology resolves the clauses using the rules listed in Fig. 5 and extract triples as shown in Table 2. The final selection of triples after referring the lexicons is shown in Table 3. The seed TSPP list contained 30 treatment-specific phrases. As our algorithm is based on the assumption that the "phrase patterns" are

Table 2 Triples extracted from sentence after applying clause handling rules

Subject	Predicate	Object
["amla"]	["may", "be", "used", "as"]	["rasayana", "rejuvenative"]
["amla"]	["to", "promote"]	["longevity"]
["amla"]	["to", "enhance"]	["digestion", "dipanapachana"]
["amla"]	["reduce"]	["fever", "jvaraghna"]
["amla"]	["purify"]	["blood", "raktaprasadana"]
["amla"]	["reduce"]	["cough", "kasahara"]
["amla"]	["alleviate"]	["asthma", "svasahara"]
["amla"]	["strengthen"]	["heart", "hrdaya"]
["amla"]	["benefit"]	["eyes", "chakshushya"]
["amla"]	["stimulate"]	["growth", "romasanjana"]
["amla"]	["enliven"]	["body", "jivaniya"]
["amla"]	["enhance"]	["intellect"]

Table 3 Final selected triples

Subject	Predicate	Object
["amla"]	["reduce"]	["fever", "jvaraghna"]
["amla"]	["purify"]	["blood", "raktaprasadana"]
["amla"]	["reduce"]	["cough", "kasahara"]
["amla"]	["alleviate"]	["asthma", "svasahara"]
["amla"]	["strengthen"]	["heart", "hrdaya"]
["amla"]	["benefit"]	["eyes", "chakshushya"]
["amla"]	["enliven"]	["body", "jivaniya"]

very situation specific, the predicate is found in seed list and the corresponding triple is selected as relevant.

It was observed that few predicates like "is/are used in", "is/are used for", "is/are helpful in", "is/are good for", and "is/are useful in" though are obtained higher ranking in seed pattern list and generated few triples which did not describe any health benefits. Then, during the second phase of selection process, the subject is looked for in medicinal plant lexicon and the object is searched in disease and human body lexicon. If both the search results return "Success" and the predicate does not contain any negative sentiment words, then the triple is selected as relevant and the TSPP list is updated with the new predicate. We have chosen the basic evaluation parameters used in IR, i.e., precision and recall to evaluate our system performance. Here, precision measures the fraction of triples that are retrieved correctly and recall counts the number of correctly extracted triples The overall relevant triple retrieval efficiency of the proposed methodology is around 42% with original TSPP and 58% with relaxed TSPP(Rel_TSPP). Since we are using existing POS tagger and chunker, errors in these will have sure impact on the performance of our approach. With every new document, the proposed algorithm enables us to learn new treatment-specific phrases (if present in the document). It was found that our methodology works very effective in terms of learning new patterns from documents involving complex sentences. The initial seed TSPP list is updated after each iteration where iteration refers to a new document. A detailed result analysis is given in Table 4.

Since our system relies upon the initial seed patterns, it underperforms when the considered document contains more new treatment-specific phrases compared to the listed seed values. Details of the generated seed patterns are depicted in Fig. 8.

The information retrieval efficiency of our system in terms of candidate sentence selection as well as relevant triple selection is given in Figs. 9 and 10. Result analysis clearly shows that compared to TSPP, Rel_TSPP improves the recall value though the precision value has decreased.

Table 4 Detailed result analyses of text documents

Name of herb		Basil	Amla	Neem	Aloe vera	Turmeric	Ginger	Curry leaves
Seed TSPPs		30	36	41	45	45	46	46
Analysis of candidate sentences extraction								
Actual no of sentence		10	4	6	10	7	7	6
No of extracted sentences	TSPP	5	2	3	2	1	2	2
	Rel_TSPP	10	3	6	5	2	4	3
Analysis of relevant triple extraction								
Actual no of triples		14	13	26	10	31	10	8
No of extracted triples	TSPP	7	9	12	2	17	2	3
	Rel_TSPP	13	11	14	3	24	2	4
Analysis of pattern extraction								
Existing patterns	TSPP	2	2	1	2	1	2	1
Newly retrieved patterns		4	4	2	Nil	Nil	Nil	1
Existing patterns	Rel_TSPP	6	6	3	2	1	2	2
Newly retrieved patterns		2	1	2	Nil	1	Nil	nil

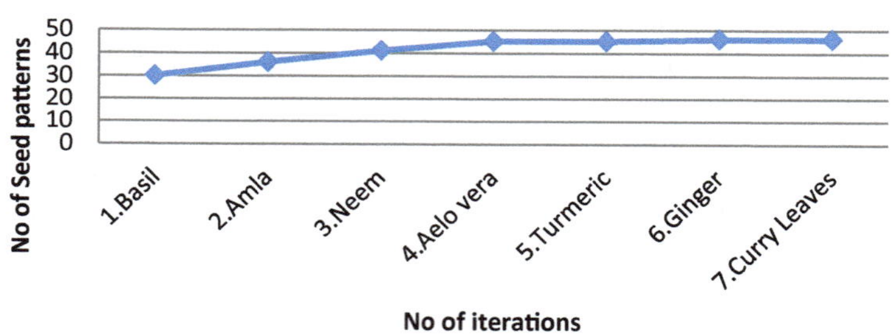

Fig. 8 Details of seed TSPP patterns used per iteration where every new document is considered as an iteration

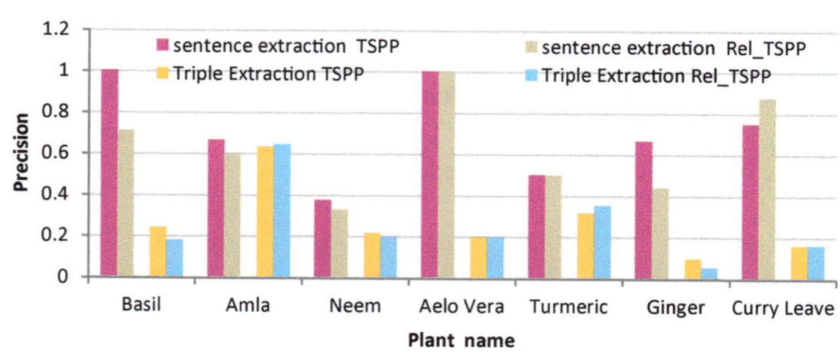

Fig. 9 Detailed precision value analysis of candidate sentence extraction as well as triple extraction from text documents

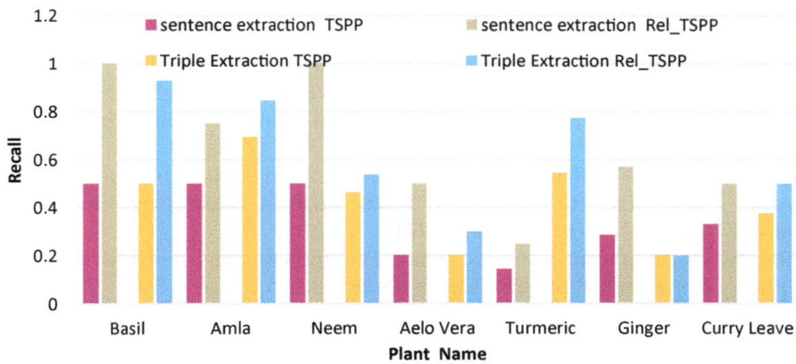

Fig. 10 Detailed recall value analysis of candidate sentence extraction as well as triple extraction from text documents

5 Conclusion

In this paper, we have proposed an automatic and domain-dependent approach to extract health benefits of known medicinal plants from unstructured text. This approach is capable of handling complex sentences which has been a major research topic in case of semantic relation extraction from text. Though the initial phase of the framework is domain dependent, the second phase, i.e., the clause analysis and semantic rules, can be used to extract relations for any domain. In future, we indent to extend our analysis part so that we can extract more information about medicinal plants. The proposed approach can be made more robust, if some unsupervised machine learning techniques can be incorporated in the treatment-specific phrase pattern selection process. In this paper, we have not considered the pattern type "disease/body part name TSPP medicinal plant name". In future, we plan to work on this reverse pattern and improve the performance.

References

1. "Indian Medicinal Plants For Diabetes Text Data Mining The Literature Of Different Electronic Databases For Future Therapeutics", Bhanumathi Selvaraj, Sakthivel Periyasamy, Biomedical Research Journal. 2016
2. "Herbal Plants Used For The Treatment Of Malaria- A Literature Review", Satish Bahekar, Ranjana Kale, Journal of Pharmacognosy and Phytochemistry, Volume 1 Issue 6, 2013
3. "Phytochemical Screening And In Vivo Antimalarial Activity Of Extracts From Three Medicinal Plants Used In Malaria Treatment In Nigeria", A. E. BankoleEmailauthorA. A. AdekunleA. A. Sowemimo C. E Umebese O AbiodunG. O. Gbotosho, Parasitology Research, January 2016, Volume 115, Issue 1, pp 299–305
4. "A Framework Of Protein-Drug Association For Malaria By Text Data Mining Of Biomedical Literature", Bhanumathi S, Sakthivel P. RJPBCS 2016; 7: 1493–1499.

5. “Natural Products For Chronic Cough: Text Mining The East Asian Historical Literature For Future Therapeutics”, Shergis JL, Wu L, May BH, Zhang AL, Guo X., Chron Respir Dis 2015; 12: 204–211.
6. “Medicinal Plants Used In Cancer Treatment: An Overview”, Babele, Sneha; Verma, Sachin; Dwivedi, Sumeet; Dubey, Raghvendra, International Journal of Pharmacy & Life Sciences. Sep 2016, Vol. 7 Issue 9
7. “Anti-Acne Activity Of Italian Medicinal Plants Used For Skin Infection”, Kate Nelson, James T. Lyles, Tracy Li, Alessandro Saitta, Eugenia Addie Noye, Paula Tyler, and Cassandra L. Quavel, Front Pharmacol. 2016; 7: 425
8. “A Review On Ayurvedic Medicinal Plants For Eye Disorders From Ancient To Modern Era”, International journal of pharmaceutical sciences and research, 2014; Vol. 5(12): 5088–5096.
9. “Large-scale extraction of accurate drug-disease treatment pairs from biomedical literature for drug repurposing”, Xu and Wang BMC Bioinformatics 2013, 14:181
10. ”An Automatic and Clause-Based Approach to Learn Relations for Ontologies”, D. Thenmozhi, Chandrabose Aravindan, The Computer Journal 59(6):bxv071 September 2015

Part III
Soft Computing Applications and Pattern Recognition

A Hybrid Machine Learning Technique for Fusing Fast *k*-NN and Training Set Reduction: Combining Both Improves the Effectiveness of Classification

Bhagirath Parshuram Prajapati and Dhaval R. Kathiriya

Abstract The primary dilemmas in nonparametric algorithms like *k*-nearest neighbor classification are the largest computational and storage requirements. Moreover, the effectiveness of classification decreases due to uneven distribution of training data. In this paper, we present three approaches to minimize computation time and storage requirements. In order to achieve the goal, we present three approaches: fast *k*-NN, training set reduction techniques, and a hybrid of the previous two approaches. We have compared three approaches to existing methods and results show that the effectiveness (in terms of execution time and storage requirement) of the three algorithms are significantly better than existing algorithms.

Keywords Machine learning · *k*-NN · Hybrid technic

1 Introduction

In simple *k*-nearest neighbor classifier, a training set (consists of input vectors and associated class labels) is given as an input to the machine learning algorithm without any changes in training set size. The algorithm calculates the distance between a new input test vector and each vector of the stored training set, and assigns a class label to the test vector. Hence, *k*-NN classifier requires a large amount of memory to store the training dataset and a large amount of time required to execute this algorithm. Hence, the requirement of space and time is higher

B. P. Prajapati (✉)
Department of Computer Engineering, A. D. Patel Institute of Technology, New V. V. Nagar, Anand, India
e-mail: bhagirath123@gmail.com

D. R. Kathiriya
Agricultural Information Technology, College of Agricultural Information Technology, Anand, India
e-mail: deanait@aau.in

C. R. Panigrahi et al. (eds.), *Progress in Advanced Computing and Intelligent Engineering*, Advances in Intelligent Systems and Computing 714,
https://doi.org/10.1007/978-981-13-0224-4_21

compared to other parametric classification algorithms where parameters are learned from the training set and algorithm uses these parameters to compute similarity measures required to compute the class label for an input vector. Since it stores all the training instances, these two problems motivate us to find key to reduce time and space of k-NN classifier. There are a few solutions to this problem which are feature selection, fast k-NN, and reducing the size of training dataset by removing noisy and unimportant training instances. In this paper, we have evaluated three training set reduction techniques. The evaluation of above three approaches is carried out on agriculture dataset. And results suggest that the effectiveness of above approaches is significant to existing methods.

This paper is organized as follows. Section 2 presents the background about the research topic. Section 3 presents the soil health card dataset. Section 4 covers simple k-NN and proposed algorithms. Section 5 is about comparisons of all methods and analysis. Section 6 concludes this papers and provides some future research direction.

2 Background

The area of machine learning is divided into three major areas. Supervised approach is where the algorithms are supervised by a labeled data, for example, classification. In unsupervised machine learning approach, no such labeled data are provided to the input rather than the learning algorithm learns from the data itself, for example, clustering. In semi-supervised approach, the combination of labeled and unlabeled records is provided as an input to the algorithm.

Figure 1 provides an overview of traditional classification where feature vectors are constructed from the training set (which is a combination of training record and class labels) after data cleaning and feature selections. The feature vectors are provided as an input to the machine learning algorithm, and learning model is created. The test instance where the goal is to find a class label is given as an input to the learning model and the learning model assigns a class label to it.

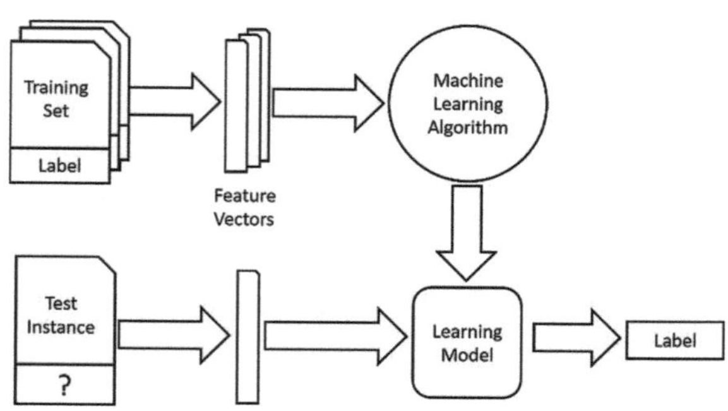

Fig. 1 An overview of classification task

2.1 k-Nearest Neighbors Algorithm

The *k*-nearest neighbor classification algorithm is a classical well-known method in machine learning [1, 2]. It is a well-established method in the area of pattern recognition and a lot of research has been done on *k*-NN [3–5]. For example, remote sensing [6, 7], image processing [8, 9], and so on. Raymer et al. [10] applied *k*-NN in combination with a genetic algorithm on medical datasets for knowledge discovery. Frigui et al. [11] used a *k*-NN classifier to perform detection of landmines; here, they adopted a possibilistic *k*-NN classifier. Yang et al. [9] adopted the local mean-based nearest neighbor algorithm to perform the discriminant analysis. Li et al. [12] adopted the *k*-NN classifier to the image classification of hyperspectral images. Bosch et al. [13] adopted *k*-NN classifier for classification of a scene.

2.2 Accelerating k-NN

To speed up the *k*-NN searching is an interesting area of research and it is mainly divided into two categories: template condensation and template reorganization [14]. Template condensation identifies the redundant patterns in template set and removes it [15–17]. While the restructuring of templates is done in the template reorganization algorithms [18–21], a lot of work has been done to find a new approach and in one such method, classification performance is not affected while reducing the storage and computation cost [22].

Hu et al. [23] applied sample weight learning on the nearest neighbor classifier. Parthasarathy and Chatterji [24] explored the way to use *k*-NN in case sample size is small. Some researchers have analyzed the data point's relationships to the nearest neighbor relationships, like centers of the classes and hyperplane data points. Gao et al. [25] have designed a nearest neighbor classifier based on the center called center base nearest neighbor classifier. Li et al. [26] used the local probabilistic centers of each class in the classification process. Vincent et al. [27] applied the *k*-local hyperplane NN technique.

3 Soil Health Card Data Set

This research work is concentrated on exploring the applicability of machine learning techniques on agricultural dataset of soil health card and to propose improved efficient machine learning algorithm to classify soil sample into the categories of the deficiencies of micro- and macronutrients (Fig. 2).

	A	B	C	D	E	F	G	H	I
1	SHC_POTA	SHC_SULP	SHC_MG	SHC_PHOS	SHC_IRON	SHC_MAN	SHC_ZINC	SHC_CU	lable
2	240	42	2	27	2	6.3	7.5	0.25	MaMi210
3	264	32.24	3.6	38	2	5.2	8	0.45	MaMi243
4	278	6.5	3.6	21	0.25	0.3	2.2	14.2	MaMi179
5	247	4.5	4.5	15	0.36	0.25	1.2	14.2	MaMi163
6	358	7.8	6.5	25	0.25	35	5	13.6	MaMi183
7	234	7.8	8.5	27	0.45	6	2.5	15	MaMi179
8	269	6.8	7.5	19	0.36	0.25	1.2	13	MaMi163
9	260	7.5	6.4	21	0.25	5	2.5	14.2	MaMi179
10	274	6.4	7.5	30	0.36	4	3.5	14.2	MaMi179
11	260	4.5	7.3	23	0.25	3	2.5	7	MaMi179
12	243	5.6	4.3	17	0.25	2	4.5	15.4	MaMi163
13	278	3.5	7.2	25	0.45	0.25	3.5	15	MaMi179
14	260	6.5	6.5	19	0.25	5	3.5	9.8	MaMi163

Fig. 2 Soil health card dataset

4 Proposed Work

4.1 Application of Classification Technique k-Nearest Neighbor

We have first applied *k*-nearest neighbor algorithm on dataset of district Kutch, which is having 14000 samples of soil parameters from SHCDB and calculated accuracy, precision, recall, F1 measures, and classification time in milliseconds.

Algorithm 1: ***k*-Nearest Neighbor (*k*-NN) Classifier**

Input: A set of agriculture records R = $\{R_1, R_2 \ldots R_n\}$, where n is the total number of agriculture records, training record set D.

Procedure:

Step 1: Divide the record data into one training set and test set as 50–50 split.

Step 2: For each test record, calculate similarity with each training record.

Step 3: Sort the training records in the descending order of the maximum cosine similarity and select the top k training records.

Step 4: Assign a class to test record which occurs maximum times in the top k training records.

Step 5: Construct a confusion matrix.

Step 6: Calculate performance measures from the confusion matrix.

4.2 Application of Classification Technique Fast k-Nearest (FKNN) Neighbor

The primary limitation of the simple *k*-NN algorithm is it needs to retain all the training data and prone to high computational cost. In order to reduce the computation cost of mentioned simple *k*-NN, we proposed and designed a fast *k*-nearest neighbor algorithm. The fast *k*-NN (FkNN) classifier first finds k clusters by employing *k*-means clustering algorithm. The class label of each cluster is the class whose maximum number of records are present in that particular cluster [28]. For each test data, we have calculated similarity with each cluster and assigned a class based on *k*-NN approach. The fast k-nearest neighbor algorithm is described as below.

Algorithm 2: Fast *k*-nearest neighbor algorithm (FkNN)

Input: A set of agriculture records $R = \{R_1, R_2 \ldots R_n\}$, where n is the total number of agriculture records, training record set D.

Procedure:

Step 1: Divide the record data into one training set and test set as 50–50 split.

Step 2: Construct k clusters using *k*-means clustering algorithm (validate k value for k-means clustering by elbow method or silhouette method) and assign a class label to each cluster based on maximum occurrences of a particular class in that cluster.

Step 3: For each test record, calculate similarity with each cluster's centroid.

Step 4: Sort the clusters in the descending order of the maximum cosine similarity and select the top k clusters.

Step 5: Assign a class to test record whose summation of similarity is maximum in the top k clusters.

Step 6: Construct a confusion matrix.

Step 7: Calculate performance measures from the confusion matrix.

In above algorithm in step 2, we have applied clustering validation techniques because for cluster analysis there is always a question of how to evaluate the goodness of clusters [29]. For *k*-means clustering, it is desirable to perform clustering with optimum k value to avoid finding patterns in noise and to compare it with other clustering algorithms. In this research work, we considered two methods of cluster validation: the first method computes the sum of square of error (SSE) [30, 31] in Eq. 1,

$$SSE = \sum_{i=1}^{k} \sum_{x \in c_i} dist(x, c_i)^2 \tag{1}$$

The SSE is defined as the sum of the squared distance between each member of the cluster and its centroid. It checks measure cohesion [32], which means how closely related are objects in a cluster.

4.3 Application of Classification Technique Training Set Reduction k-Nearest Neighbor

In this approach in the first phase, a training set is converted into a set of training vectors [33]. The training vectors are given as an input into training set reduction algorithm and the algorithm's output is the reduced training set. In the second phase, the reduced training set is employed by the classifier to classify a new test instance. We have applied shrink (subtractive) algorithm [34] to reduce the training set.

Algorithm 3: **Training set reduction *k*-NN (TRS-kNN)**

Phase I: Shrink (subtractive) algorithm.

Input: A set of training instances $T = \{T_1, T_2, \ldots T_n\}$ where n is the total number of agriculture records, training record set D.

Step 1: Assign all the training documents into S.

Step 2: Select randomly an instance P from S.

Step 3: Classify the instance P using remaining instances from S.

Step 4: Remove the instance P if it is correctly classified.

Phase II:

Input: A set of agriculture records $R = \{R_1, R_2 \ldots R_n\}$, where n is the total number of agriculture records reduced training record set D.

Procedure:

Step 1: Divide the record data into one training set and test set as 50–50 split.

Step 2: Construct k clusters using k-means clustering algorithm (validate k value for k-means clustering by elbow method or silhouette method) and assign a class label to each cluster based on maximum occurrences of a particular class in that cluster.

Step 3: Sort the training records in the descending order of the maximum cosine similarity and select the top k training records.

Step 4: Assign a class to test record which occurs maximum times in the top k training records.

Step 5: Construct a confusion matrix.

Step 6: Calculate performance measures from the confusion matrix.

4.4 Application of Classification Technique Training Set Reduction Fast k-Nearest Neighbor (TSR-FkNN)

This method is a hybrid method, where we have combined features of both fast *k*-NN and training set reduction. Figure 3 provided an overview of a proposed and designed hybrid approach of the previous two, i.e., FkNN and TSR-kNN.

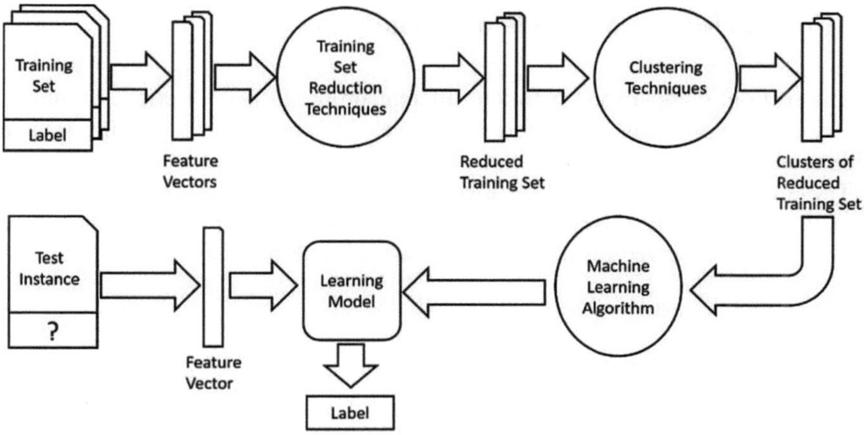

Fig. 3 An overview of hybrid machine learning technique TSR-FkNN

Algorithm 4: Training set reduction fast k-nearest neighbor
Phase I:
Input: A set of training instances T = {T_1, T_2, … T_n} where n is the total number of agriculture records, training record set D.

Step 1: Assign all the training documents into S.
Step 2: Select randomly an instance P from S.
Step 3: Classify the instance P using remaining instances from S.
Step 4: Remove the instance P if it is correctly classified.

Phase II:
Input: A set of agriculture records R = {R_1, R_2, …, R_n} where n is the total number of agriculture records, reduced training record set D.
Procedure:

Step 1: Divide the record data into one training set and test set as 50–50 split.
Step 2: Construct k clusters using k-means clustering algorithm (validate k value for k-means clustering by elbow method or silhouette method) and assign a class label to each cluster based on maximum occurrences of a particular class in that cluster
Step 3: For each test record, calculate similarity with each cluster's centroid.
Step 4: Sort the clusters in the descending order of the maximum cosine similarity and select the top k clusters.
Step 5: Assign a class to test record whose summation of similarity is maximum in the top k clusters.
Step 6: Construct a confusion matrix.
Step 7: Calculate performance measures from the confusion matrix.

5 Comparison of Methods and Analysis

In this section, comparisons between different proposed classification techniques are carried out in terms of performance measures and time of classification in milliseconds.

These results are performed on a computer with Intel i5 processor and 4 GB Ram, the software IDE is NetBeans 8.2. Depending on hardware some of the results may vary. The observed results are on average of multiple runs.

5.1 Accuracy Comparison

In Table 1, the accuracy of different *k*-NN classifiers is compared. Williams et al. [35] adopted accuracy as a measure to compare five machine learning algorithms and high accuracy algorithm is preferred. In our research work, it is found that proposed TSR-FkNN (applying SSE) classifier have the highest accuracy and all other classifiers also have accuracy less than the accuracy of TSR-FkNN (applying SSE).

5.2 Training Set Comparison

In Table 2, training instances of different *k*-NN classifiers are compared. Witten, Ian H. et al. [36] have dedicated a chapter on reduction techniques for instance-based learning algorithms. Here, the training set reduction machine learning algorithms are compared.

In our research, TSR-FkNN (applying SSE) has lowest training instances when the value of k is 33 and 35, respectively, All other classifiers have higher training instances, while *k*-NN has highest training instances. Here, training instances of all classifiers other than *k*-NN are reduced by applying novel techniques designed for this research.

Table 1 Comparison of accuracy for all *k*-NN classifiers

Sr. no	Value of k in *k*-NN	Accuracy of *k*-NN	Accuracy of FkNN	Accuracy of TSR-kNN	Accuracy of TSR-FkNN
1	31	89.21	88.64	88.21	89.92
2	33	88.85	87.92	89.71	90.42
3	35	88.41	88.85	89.57	**90.85**
4	37	89	88.78	89.07	89.85

Table 2 Comparison of training instances for all k-NN classifiers

Sr. no	Value of k in k-NN	Training instances of k-NN	Training instances of FkNN	Training instances of TSR-kNN	Training instances of TSR-FkNN
1	31	7000	141	3005	141
2	33	7000	131	2855	**61**
3	35	7000	131	2970	71
4	37	7000	181	3040	111

Table 3 Comparison of classification time for all k-NN classifiers

Sr. no	Value of k in k-NN	Classification time of k-NN	Classification time of FkNN	Classification time of TSR-kNN	Classification time of TSR-FkNN
1	31	5766	199	1021	248
2	33	5779	191	1017	**143**
3	35	5777	217	1039	180
4	37	5746	233	1343	217

5.3 Classification of Time Comparison in Milliseconds

In Table 3, comparison of all classifier is done in terms of classification time in milliseconds. Williams et al. [37] applied method of comparing the time of five classifiers. Bost, Raphael, et al. [38] applied machine learning algorithm on encrypted dataset and compared algorithm based on execution time. In our research, it is observed that TSR-FkNN (applying SSE) is having lowest classification time when the values of k are 33 and 35, respectively.

6 Conclusion and Future Directions

- **Storage Reduction**:
  - Storage requirement in *k*-NN is very high in comparison to other algorithms.
  - For TSR-FkNN (applying SSE), storage requirement is lowest when values of k are 33 and 35, respectively. Hence, in terms of storage, TSR-FkNN is efficient.
- **Execution Time**:
  - Execution time is highest in kNN followed by TSR-kNN as they store more number of instances for training purpose.

- Execution time is lowest in TSR-FkNN (applying SSE) followed by FkNN as they store less number of training instances.

- **Generalization Accuracy**:
  - Generalize accuracy of TSR-FkNN(applying SSE) is highest compared to other algorithms; hence, in terms of accuracy, TSR-FkNN (applying SSE) is recommended.

Association mining can be applied to the dataset to get important frequent association like association between crop and soil type. We can build recommended system from the above research to guide farmers to get important information about fertilizers, water supply, growing crop on a particular soil type, etc.

References

1. T. Cover, P. Hart, "Nearest neighbor pattern classification," IEEE Trans. Inf. Theory, vol. 13, no. 1, pp. 21–27, Jan. 1967.
2. T. Denoeux, "A *k*-nearest neighbor classification rule based on Dempster–Shafer theory," *IEEE Trans. Syst., Man, Cybern.*, vol. 25, no. 5, pp. 804–813, May 1995.
3. A. Bosch, A. Zisserman, and X. Muoz, "Scene classification using a hybrid generative/discriminative approach," *IEEE Tra. Pattern Anal. Mach. Intel.*, vol.30, no.4, pp. 712–727, Apr. 2008.
4. J. Yang, L. Zhang, J. Yang, and D. Zhang, "From classifiers to discriminators: A nearest neighbor rule induced discriminant analysis," *Pattern Recognit.*, vol. 44, no. 7, pp. 1387–1402, 2011.
5. J. Xu, J. Yang, and Z. Lai, "K-local hyperplane distance nearest neighbor classifier oriented local discriminant analysis," *Inf. Sci.*, vol. 232, pp. 11–26, May 2013.
6. H. Frigui and P. Gader, "Detection and discrimination of land mines in a ground-penetrating radar based on edge histogram descriptors and a possibilistic *K*-nearest neighbor classifier," *IEEE Trans. Fuzzy Syst.*, vol. 17, no. 1, pp. 185–199, Feb. 2009.
7. M. Li, M. M. Crawford, and J. Tian, "Local manifold learning-based *k*-nearest-neighbor for hyperspectral image classification," *IEEE Trans. Geosci. Remote Sens.*, vol. 48, no. 11, pp. 4099–4109, Nov. 2010.
8. T. Mensink, J. Verbeek, F. Perronnin, and G. Csurka, "Distance-based image classification: Generalizing to new classes at near-zero cost," *IEEE Trans. Pattern Anal. Mach. Intell.*, vol. 35, no. 11, pp. 2624–2637, Nov. 2013.
9. Acharya, Tinku, and Ajoy K. Ray. *Image processing: principles and applications*. John Wiley & Sons, 2005.
10. M. L. Raymer, T. E. Doom, L. A. Kuhn, and W. F. Punch, "Knowledge discovery in medical and biological datasets using a hybrid Bayes classifier/evolutionary algorithm," *IEEE Trans. Syst., Man, Cybern. B, Cybern.*, vol. 33, no. 5, pp. 802–813, Oct. 2003.
11. H. Frigui and P. Gader, "Detection and discrimination of land mines in a ground-penetrating radar based on edge histogram descriptors and a possibilistic *K*-nearest neighbor classifier," *IEEE Trans. Fuzzy Syst.*, vol. 17, no. 1, pp. 185–199, Feb. 2009.
12. J. Yang, L. Zhang, J. Yang, and D. Zhang, "From classifiers to discriminators: A nearest neighbor rule induced discriminant analysis," *Pattern Recognit.*, vol. 44, no. 7, pp. 1387–1402, 2011.

13. A. Bosch, A. Zisserman, and X. Muoz, "Scene classification using a hybrid generative/discriminative approach," *IEEE Trans. Pattern Anal. Mach. Intell.*, vol. 30, no. 4, pp. 712–727, Apr. 2008.
14. Zhang, Bin, and Sargur N. Srihari. "Fast k-nearest neighbor classification using cluster-based trees." IEEE Transactions on Pattern analysis and machine intelligence 26.4 (2004):
15. G.L. Ritter, H.B. Woodruff, S.R. Lowry, and T.L. Isenhour, "An Algorithm for a Selective Nearest Neighbor Decision Rule," IEEE Trans. Information Theory, vol. 21, pp. 665–669.
16. C.L. Chang, "Finding Prototypes for Nearest Neighbor Decision Rule," IEEE Trans. Computers, vol. 23, no. 11, pp. 1179–1184, Nov. 1974.
17. P.E. Hart, "Condensed Nearest Neighbor Rule," IEEE Trans. Information Theory, vol. 14, pp. 515–516, May 1968.
18. A.J. Broder, "Strategies for Efficient Incremental Nearest Neighbor Search," Pattern Recognition, vol. 23, nos. 1/2, pp. 171–178, Nov. 1986.
19. A. Farago, T. Linder, and G. Lugosi, "Fast Nearest-Neighbor Search in Dissimilarity Spaces," IEEE Trans. Pattern Analysis and Machine Intelligence, vol. 15, no. 9, pp. 957–962, Sept. 1993.
20. B.S. Kim and S.B. Park, "A Fast k Nearest Neighbor Finding Algorithm Based on the Ordered Partition," IEEE Trans. Pattern Analysis and Machine Intelligence, vol. 8, no. 6, pp. 761–766, Nov. 1986.
21. E. Vidal, "An Algorithm for Finding Nearest Neighbors in (Approximately) Constant Average Time," Pattern Recognition Letters, vol. 4, no. 3, pp. 145–157, July 1986.
22. Yu, Xiaopeng. "The Research on an adaptive k-nearest neighbors classifier." *Cognitive Informatics, 2006. ICCI 2006. 5th IEEE International Conference on*. Vol. 1. IEEE, 2006.
23. Q. Hu, P. Zhu, Y. Yang, and D. Yu, "Letters: Large-margin nearest neighbor classifiers via sample weight learning," Neurocomputing, vol. 74, no. 4, pp. 656–660, 2011.
24. G. Parthasarathy and B. N. Chatterji, "A class of new KNN methods for low sample problems," IEEE Trans. Syst., Man, Cybern., vol. 20, no. 3, pp. 715–718, May/Jun. 1990.
25. Q. Gao and Z. Wang, "Center-based nearest neighbor classifier," Pattern Recognit., vol. 40, no. 1, pp. 346–349, 2007.
26. B. Li, Y. W. Chen, and Y.-Q. Chen, "The nearest neighbor algorithm of local probability centers," IEEE Trans. Syst., Man, Cybern. B, Cybern., vol. 38, no. 1, pp. 141–154, Feb. 2008.
27. P. Vincent and Y. Bengio, "K-local hyperplane and convex distance nearest neighbour algorithms," in Proc. Adv. Neural Inf. Process. Syst. (NIPS), vol. 14. Vancouver, BC, Canada, 2002, pp. 985–992.
28. B. P. Prajapati, and D. R. Kathiriya. "Evaluation of Effectiveness of k-Means Cluster based Fast k-Nearest Neighbor classification applied on Agriculture Dataset." International Journal of Computer Science and Information Security 14.10 (2016): 800.
29. Hardy, André. "An examination of procedures for determining the number of clusters in a data set." New approaches in classification and data analysis. Springer, Berlin, Heidelberg, 1994.
30. Milligan, G. W., & Cooper, M. C. (1985). An examination of procedures for determining the number of clusters in a data set. Psychometrika, 50(2), 159–179.
31. Lee, Paul H., et al. "A cluster analysis of patterns of objectively measured physical activity in Hong Kong." Public health nutrition 16.8 (2013): 1436–1444.
32. Arbelaitz, Olatz, et al. "An extensive comparative study of cluster validity indices." Pattern Recognition 46.1 (2013): 243–256.
33. Prajapati, B.P. and Kathiriya, D.R., 2016. Reducing execution time of Machine Learning Techniques by Applying Greedy Algorithms for Training Set Reduction. International Journal of Computer Science and Information Security, 14(12), p. 705.
34. Wettschereck, D., Aha, D.W. and Mohri, T., 1997. A review and empirical evaluation of feature weighting methods for a class of lazy learning algorithms. In Lazy learning (pp. 273–314).

35. Williams, Nigel, Sebastian Zander, and Grenville Armitage. "A preliminary performance comparison of five machine learning algorithms for practical IP traffic flow classification." ACM SIGCOMM Computer Communication Review 36.5 (2006): 5–16.
36. Witten, Ian H., et al. Data Mining: Practical machine learning tools and techniques. Morgan Kaufmann, 2016.
37. Williams, Nigel, Sebastian Zander, and Grenville Armitage. "A preliminary performance comparison of five machine learning algorithms for practical IP traffic flow classification." ACM SIGCOMM Computer Communication Review 36.5 (2006): 5–16.
38. Bost, Raphael, et al. "Machine Learning Classification over Encrypted Data." NDSS. 2015.

An Improved Bio-inspired BAT Algorithm for Optimization

Gopal Purkait, Dharmpal Singh, Madhusmita Mishra, Amrut Ranjan Jena and Abhishek Banerjee

Abstract Metaheuristic algorithms are used today to solve many optimization-related problems. The firefly algorithm, particle swarm optimization, harmony search and BAT algorithm are used as metaheuristic searched algorithms to find the optimized solution of the problem domain. The BAT algorithm was developed by using the unique characteristics of BAT which used the advanced capability of echolocation to move in the dark, avoid obstacles or barrier and also find its food or pray. The main aim of this paper is to represent and translate the behaviour of BAT algorithm in the form of improved BAT algorithm. The paper also describes the implication, advantage, disadvantage and application of BAT algorithm in different areas used by diversified authors.

Keywords Metaheuristic algorithms • BAT algorithm • Optimization problem Echolocation • Pulse rate

G. Purkait (✉) · A. Banerjee
Department of Computer Science & Engineering, Pailan College of Management & Technology, Kolkata, West Bengal, India
e-mail: purkait.gopal@gmail.com

A. Banerjee
e-mail: abhishek.barrackpore@gmail.com

D. Singh · M. Mishra · A. R. Jena
Department of Computer Science & Engineering, JIS College of Engineering, Kalyani, West Bengal, India
e-mail: dharmpal1982@gmail.com

M. Mishra
e-mail: madhu.smita7@gmail.com

A. R. Jena
e-mail: amrut.ranjan7@gmail.com

C. R. Panigrahi et al. (eds.), *Progress in Advanced Computing and Intelligent Engineering*, Advances in Intelligent Systems and Computing 714,
https://doi.org/10.1007/978-981-13-0224-4_22

1 Introduction

Today's optimization becomes an important and crucial procedure in each and every field, especially in mathematics and engineering. Most of the optimization problems have various complex constraints to find a most favourable solution or even sub-optimal solution for the known problems. Metaheuristic algorithms are very useful techniques used in the intricate problem to find the most favourable solution. Numerous metaheuristic algorithms, e.g. BAT algorithm, ant colony optimization, harmony search, swarm intelligence and particle swarm optimization, have been used for finding optimal solution to a problem. Metaheuristic algorithms are classified into different categories based on the working methodology and the application of different types of optimization problems. In today's world, metaheuristic algorithms are used to resolve difficulties, e.g. decision-making, resource allocation and also in medical and engineering fields. Metaheuristic algorithms have been classified into many groups as depicted in Fig. 1.

The last two three decades of different kinds of metaheuristic searching algorithms are applied in different areas of engineering and others to optimize the solution state. BAT algorithm is one variety of metaheuristics searching algorithm to use the echolocation behaviour of BAT. Here, the detailed discussion about the working principle and the application of BAT algorithm has been analysed. Fuzzy logic bat algorithm (FLBA is based on fuzzy logic discussed by Khan et al. [2]. Multi-objective bat algorithm (MOBA) is based on multi-objective optimization discussed by Yang [3]. K-means bat algorithm (KMBA) is a mixture of K-means and BAT algorithm presented by Komarasamy and Wahi [4]. Lin et al. [5] have used the chaotic bat algorithm (CBA) with tic maps on parameter estimation in dynamic biological systems presented by Levy flights, whereas Nakamura et al. [6] used the chao and binary bat algorithm (BBA) which is the discrete version of BAT algorithm to solve classifications and feature selection.

The detection of hairline crack bone in medical X-ray images discussed by Das [7] using the BAT algorithm concept. The author has applied BAT algorithm in preprocessing stage to improve the image, self-organizing map (SOM) and K-means clustering. Furthermore, the authors used these techniques to produce the objective image and opined that BAT algorithm was effective for image enhancement.

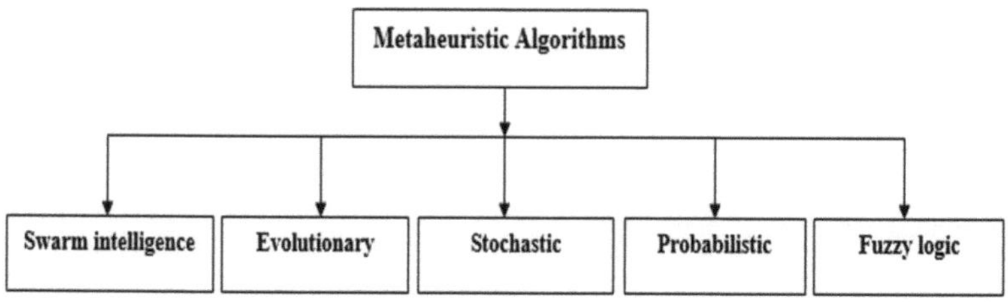

Fig. 1 Different types of metaheuristic algorithms

Moreover, the authors [8] proposed a new algorithm using bats echolocation behaviour to know the initial value to overcome the issues of K-medoids. The authors have used combination of K-medoids clustering algorithm and BAT algorithm to achieve better result. The authors showed the difference between K-medoid clustering technique with BAT algorithm and K-medoid itself.

A hybrid BAT algorithm with the Nelder–Mead method has been proposed by [9] to solve integer-programming problems. Furthermore, they opined that BAT algorithm has shown good performance for a wide exploration and a deep exploitation search, whereas the Nelder–Mead method performances potent for local search method to enhance the exploitation ability of the proposed ABATA algorithm. They have tested the performance of ABATA on seven integer-programming problems and compared against four benchmark algorithms. They have opined that the performance of ABATA was very good for global optimal solution.

The localization problem in WSN using BAT algorithm was proposed by [10] to solve the localization problem. The authors opined that BA can achieve higher accurate position estimation as compared to other existing algorithms for localization accuracy.

BAT algorithm-based approach is used by [11] to solve a variety of unit commitment problems. The authors opined that proposed method has performed superior with stable convergence characteristic to avoid premature convergence. They further opined that the proposed method outperformed the other algorithms for the optimum solution in less computational time.

A new optimization of BAT algorithm has proposed by [12] to solve CEED problem on three and six generating units, and they have further opined that proposed algorithm outperformed other algorithm in superior features, stable convergence and good computational efficiency.

A new swarm intelligence optimization algorithm named as DBA (Discrete BAT Algorithm) was proposed by authors [13] to identify the number of communities for global optimal solution to prevail over the shortcomings of conventional algorithms.

A comparative bacterial foraging optimization algorithm (BFO) with BAT algorithm (BA) has been proposed by authors [14] on twelve selected benchmark functions to get more accurate solution faster convergence rate as compared to BA. BAT algorithm [15] also used for handwritten digit recognition on standard MNIST dataset to achieve global accuracy. The authors have shown that the proposed method given 95.60% result as compared to other algorithm.

A hybrid method has proposed by [16] to improve the dynamic steadiness of the power system to optimize the maximum power loss and optimum capacity of UPFC with minimum cost, whereas a novel bat algorithm (NBA) [17] has proposed an algorithm to improve the optimization problems of original BAT algorithm on dataset for better performance of NBA as compared to BA.

Few authors also used a novel adaptive bat algorithm (NABA) [18], hybrid bat algorithm (BADE) [19] and improved bat algorithm (IBA) [20] to improve the explorative characteristics of BAT algorithm, enhance the performance of the conventional BAT Algorithm, increase utilization capabilities towards the end of the cycles, local search ability (exploitation) on dataset for more efficient and robust result as compared to other algorithms.

Metaheuristic search techniques for robotic path planning using the concept of cuckoo search and BAT algorithm for problem-solving was proposed by authors [21] along with the result of BAT algorithm which outperformed the cuckoo search.

Introduction has been furnished in Sect. 1 and methodology and modified BAT has been shown in Sect. 2. The conclusion has been given in Sect. 3.

2 Methodology of BAT Algorithm

Bats are animals with superior ability of echolocation. They have the diverse sizes with echolocation towards certain degree, but amongst micro-bats employ echolocation characteristics widely to sense prey, keep him away from obstacles. They produce an extremely loud sound pulse and wait to take note for the echo that returned from the nearby objects. They also employ the time delay from the production and recognition of the echo, to calculate the distance with orientation of the aim along the moving speed of the prey.

2.1 BAT Algorithm

BAT algorithm was introduced by Xin-She Yang [1] on the echolocation features of micro-bats to solve a variety of optimization problems based on the following furnished rule.

1. Bats uses its echolocation behaviours to calculate distance and should distinguish the variation among foodstuff/target and surrounding obstacles.
2. It has assumed that bats fly with arbitrarily velocity v_i at location x_i with a preset frequency f_{min}, with altering wavelength and loudness A_0 to look for food. It has also assumed that they should also adjust the wavelength (or frequency) of emitted pulses $r \in [0, 1]$;
3. Even though the loudness can diverge in lots of ways, it has assumed that loudness varies from a large (positive) value A_0 to a minimum constant value A_{min}.

The progress of virtual bats is based on the following formula where x_i is the position and v_i is the velocities in a d-dimensional search space. The new solutions x_i^t and velocities v_i^t at time step t are given by

$$f_i = f_{min} + (f_{max} - f_{min})\beta \tag{1}$$

$$v_i^t = v_i^{t-1} + (x_i^t - x_*)f_i \tag{2}$$

$$x_i^t = x_i^{t-1} + v_i^t \tag{3}$$

$\beta[0, 1]$ assumed as random vector and $x*$ is the present global best location (solution) after comparing the solutions of n bats.

2.2 Pseudocode of Improved BAT Algorithm

See Fig. 2.

```
Initialization: Objective function f(x) represent by percentage
error less than 2%, bat population  represent as xi (i = 1,
2,...., n), velocity  represent as  vi and  pulse frequency
represent as  fiat xi

Estimated  error=  ((Estimated  value-  actual  value)/actual
value))*100

Produce fresh solutions by adjusting frequency, and update the
velocities and solutions by the equations (1) to (3) repeatedly

  While (Estimated error <1%) do

Bat new frequency, velocity and position will be produced by
given three equations
```

$f_i = f_{min} + (f_{max} - f_{min})\beta$, -------- (1) where f_i is the frequency of bat x_i

$v_i^t = v_i^{t-1} + (x_i^t - x_*)f_i$, ---------(2)

Where v_i^t represent as the velocity of bat at moment t, v_i^{t-1}: represent as the velocity of bat at moment step t-1, x_i^t represent the position of bat at moment step t, x_*: represent as the global best location among n bats, β

$x_i^t = x_i^{t-1} + v_i^t$, ----------------(3)

Where x_i^t represent the situation of bat at moment step t is, x_i^{t-1} represent the situation of bat at time step t-1, v_i^t represent the velocity of bat at time step t,

```
    Rank the bats solution has been done based on the estimated
    error and current best solution.

  Termination of while loop

Analyse outcomes
```

Fig. 2 Pseudocode of the modified BAT algorithm

2.3 Flowchart of Improved BAT Algorithm

The traditional BAT algorithm has not offered the obvious idea about the selection of initial parameter of bat and movements of virtual bat in a d-dimensional space. It has been further observed that different authors get the different results based on the different choices of parameter's selection for the similar dataset. Therefore, here an endeavour has been set to choose the primary condition and progress of virtual bat in a d-dimensional space in the form of flowchart (furnished in Fig. 3) of the proposed algorithm.

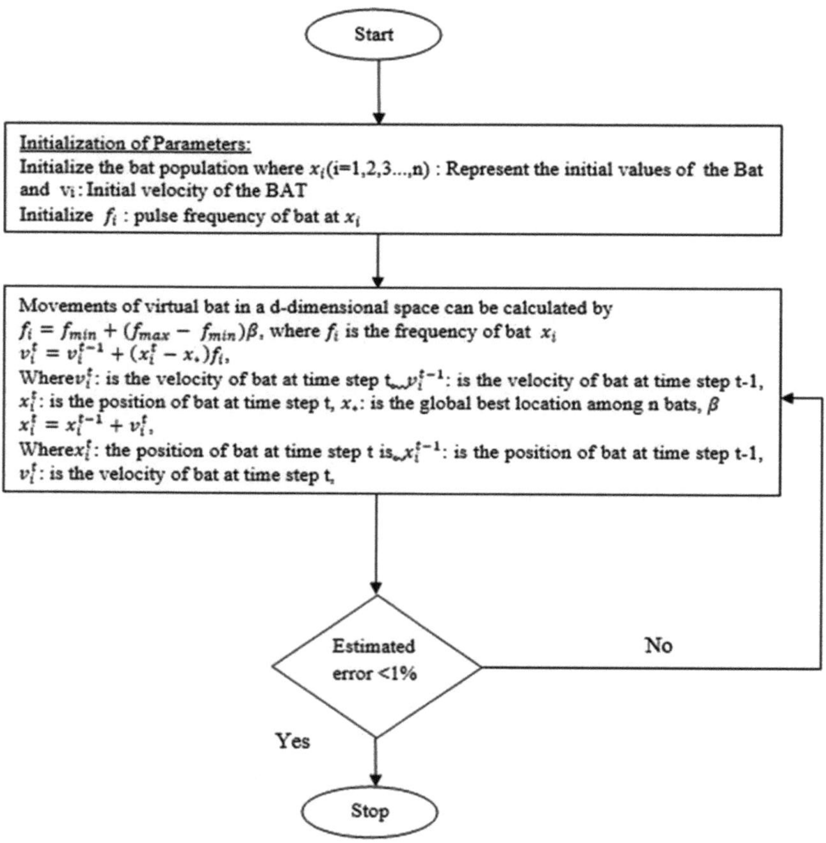

Fig. 3 Flowchart of the improved BAT algorithm

3 Conclusions and Future Proposed Work

It has been observed that lots of metaheuristic algorithms along with BAT algorithm show the simplicity and flexibility in problem domain. BAT algorithm frequency tuning, automatic zooming and parameter manipulate make it a totally simple algorithm in reality as compared to other algorithm. BA used its echolocation behaviours and rule to enhance their universal performance further as compared to other algorithms. But how to excite the convergence on the set of rules of the BAT algorithm is a completely tough question for the researchers. This was the main reason to provide the idea of modified BAT algorithm to resolve the problem in the different data domains. This paper made an effort to give the modified form of BAT algorithm to resolve the problem. The abstract concept of the algorithm has been furnished in the paper, and practical approach will be the proposed work of this paper. The proposed work is mainly based on design an integrated system to extract the knowledge based on the concept of BAT algorithm.

By analysing the significant blessings of upgrading equations, three essential points/features have been summarizing as follows:

How to perform frequency tuning: BA echolocation and frequencies tuning functionality is almost same as the essential feature of particle swarm optimization and harmony search. Therefore, BA owns its advantages on swarm intelligence-based algorithms.

How to do automatic zooming: BAT algorithm selects the gain over diverse metaheuristic algorithms with robotically zooming capability for a role for treatments. This zooming capability of bat is followed by six automatic replaces of explorative moves of nearby surroundings. Therefore, due to this reason, BA shows the short convergence rate as compared to other algorithms with minimum opening degrees of the iterations.

How to manipulate the parameters: Metaheuristic algorithms captivated to use the parameters used by several predefined sets of rules on structured tips. On the other hand, BA used the manipulate parameters to vary the values of barriers in the iterations. This behaviour of bat mechanically switches the BAT algorithm from exploration to exploitation on top-rated approaches. This behaviour of bat makes it differ from over various metaheuristic algorithms.

References

1. Xin-She Yang: Bat algorithm: literature review and applications, Int. J. Bio-Inspired Computation, Vol. 5, No. 3, (2013) 141–149.
2. Khan, K., Nikov, A., Sahai A: A fuzzy bat clustering method for ergonomic screening of office workplaces, S3T 2011, Advances in Intelligent and Soft Computing, 2011, Volume 101/2011, pp. 59–66.
3. Yang, X. S.: Bat algorithm for multi-objective optimisation, Int. J. Bio-Inspired Computation, Vol. 3, No. 5, (2011), 267–274.

4. Komarasamy, G., Wahi, A.: An optimized K-means clustering technique using bat algorithm, European J. Scientific Research, Vol. 84, No. 2, (2011)263–273.
5. Lin, J. H., Chou,C. W., Yang, C. H. Tsai, H. L.: A chaotic Levy flightbat algorithm for parameter estimation in nonlinear dynamic biological systems, J.Computer and Information Technology, Vol. 2, No. 2, (2012) 56–63.
6. Nakamura, R. Y. M., Pereira, L. A. M., Costa, K. A., Rodrigues, D., Papa, J. P. Yang, X. S.: BBA: A Binary bat algorithm for feature selection, in: 25th SIBGRAPI Conference on Graphics, Patterns and Images (SIBGRAPI), (2012) 291–297.
7. Goutam Das: Bat algorithm based Softcomputing Approach to Perceive Hairline Bone Fracture in Medical X-ray Images, International Journal of Computer Science & Engineering Technology (IJCSET) Vol. 4 No. 04, (2013) 435.
8. Monica Sood: K-Medoids Clustering Technique using Bat Algorithm" International Journal of Applied Information Systems (IJAIS),USA Volume 5, No. 8, (2013) 535–560.
9. Ahmed Fouad Ali: Accelerated Bat Algorithm for Solving Integer Programming Problems, Egyptian Computer Science Journal Vol. 39 No. 1 (2015).
10. Sonia Goyal; Manjeet Singh Patterh:Wireless Sensor Network Localization Based on BAT Algorithm, International Journal of Emerging Technologies in Computational and Applied Sciences (IJETCAS), (2013) 507–518.
11. Anand, R; A. Azeezur Rahman: Solution of Unit Commitment Problem Using BAT Algorithm" IJETI International Journal of Engineering & Technology Innovations, Vol. 1 Issue 2, (2014).
12. Bandi Ramesh; V Chandra Jagan Mohan; V C Veera Reddy: Application ofBat Algorithm For Combined Economic Load And Emission Dispatch, Int. J. Elec & Electr. Eng & Telecoms. (2013).
13. Anping Song; Mingbo Li; Xuehai Ding; Wei Cao; Ke Pu: Community Detection Using Discrete Bat Algorithm, IAENG International Journal of Computer Science, (2014)
14. Yazan A. Alsariera; Hammoudeh S. Alamri; Abdullah M. Nasser; Mazlina A. Majid; Kamal Z. Zamli: Comparative Performance Analysis of Bat Algorithm and Bacterial Foraging Optimization Algorithm using Standard Benchmark Functions, 8th. Malaysian Software Engineering Conference (MySEC), Langkawi, (2014) 295–300.
15. Eva Tuba; Milan Tuba; Dana Simian: Handwritten Digit Recognition by Support Vector Machine Optimized by Bat Algorithm, GECCO'17 Proceedings of the Genetic and Evolutionary Computation Conference Companion, (2017)125–126.
16. B. Vijay Kumar; N. V. Srikanth: Bat Algorithm and Firefly Algorithm for Improving Dynamic Stability of Power Systems Using UPFC, International Journal on Electrical Engineering and Informatics, Volume 8, Number 1, (2016).
17. Xian-Bing Meng; X. Z. Gao; Yu Liu, Hengzhen Zhang "A novel bat algorithm with habitat selection and Doppler effect in echoes for optimization"http://dx.doi.org/10.1016/j.eswa.2015.04.0260957-4174_2015 Elsevier Ltd.
18. Wasi Ul Kabir; Nazmus Sakib; Syed Mustafizur Rahman Chowdhury; Mohammad Shafiul Alam: A Novel Adaptive Bat Algorithm to Control Explorations and Exploitations for Continuous Optimization Problems, International Journal of Computer Applications (0975 – 8887) Volume 94 – No 13, May 2014
19. A Novel Hybrid Xianbing Meng; X. Z. Gao; Yu Liu: Bat Algorithm with Differential Evolution Strategy for Constrained Optimization, International Journal of Hybrid Information Technology Vol. 8, No. 1 (2015).
20. Selim Yilmaz; Ecir U. Kucuksille: Improved Bat Algorithm (IBA) on Continuous Optimization Problems, Lecture Notes on Software Engineering, Vol. 1, No. 3, August 2013 https://doi.org/10.7763/lnse.2013.v1.61 279.
21. Yogita Gigras; Kusum Gupta; Vandana, Kavita Choudhary: A Comparison between Bat Algorithm and Cuckoo Search for Path Planning, International Journal of Innovative Research in Computer and Communication Engineering Vol. 3, Issue 5, May 2015.

Primitive Feature-Based Optical Character Recognition of the Devanagari Script

Richa Sharma and Tarun Mudgal

Abstract The Devanagari script forms the backbone of the writing system of several Indian languages including Sanskrit and Hindi. This paper proposes a method to recognize a Devanagari character from a digital image using primitive feature information. The procedure involves representing each character in terms of the presence and location of primitive features like vertical lines, the frequency, and location of the intersections and the frequency of intersections of character body with *Shirorekha* (the top horizontal line of a Devanagari character). The classification of the character is done on the basis of the existence and (if present) the location of these features in the glyph (test character). The proposed method gave 93.33% accuracy with 21 fonts used for Hindi, Sanskrit, and Marathi and 72.72% accuracy for the handwritten character samples taken from 22 different people from varied age groups for the *Ka-Varga*—the first five consonants of the Devanagari script. The method worked better for handwritten samples of younger people (aged 20–25 years) than the older ones (aged 40–50 years).

Keywords Optical character recognition (OCR) · Devanagari
Hindi OCR · Primitive feature-based OCR

1 Introduction

Optical character recognition (OCR) is the technique of recognizing the glyph (test character) from its digital image. This area, belonging to "Computer Vision" domain, deals in recognizing the printed as well as the handwritten text.

R. Sharma · T. Mudgal (✉)
Department of Computer Science, Keshav Mahavidyalaya, University of Delhi, New Delhi, Delhi, India
e-mail: tarunmudgal@outlook.com

R. Sharma
e-mail: rsharma@cs.du.ac.in

C. R. Panigrahi et al. (eds.), *Progress in Advanced Computing and Intelligent Engineering*, Advances in Intelligent Systems and Computing 714,
https://doi.org/10.1007/978-981-13-0224-4_23

Since the introduction of computers, one of the fields that the researchers have been trying to explore is the imitation of the skills possessed by the human beings. One such skill is the ability to read. To bestow this ability to a computer is the major area of research these days. The motivation behind developing such systems is due to (a) the need to conserve human efforts in reading huge volumes of printed/handwritten text; (b) avoid problems arising due to inefficient and erroneous processing of data, and (c) its promising usage in several fields like education, business, trade, and banking where this technique finds its application in reading bank checks, commercial forms, government records, and postal address sorter, among others.

The Devanagari script provides a rich writing system to a number of Indian including Sanskrit, Hindi, Marathi, and Nepali among others [1]. Research on Devanagari OCR began a few decades ago and several OCR techniques for the Devanagari character recognition have surfaced since then. However, unlike the western counterparts, Devanagari OCR has not evolved enough to be widely accepted. The typical procedure of Devanagari OCR includes either correlating the test image with the images of the characters or by representing the test image (i.e., the glyph in the image) in terms of some features. These features may be boxes (block of pixels in the image) [2], primitive features like presence of D-curve, U-curve, etc. [3], moments [4], and gradient features [5], among others. Various classification techniques like fuzzy logic [2, 6, 7], decision tree [3], Kohonen neural networks [8], and combinational classifiers are then used to classify the character.

Unlike English alphabet in which characters are composed of some combination of vertical, horizontal, and slant lines along with a few curves, Devanagari characters are comparatively complex, comprising curves, strokes, bars, *matras,* and dots. However, a deep observation of the character set hints toward a potential method to recognize the characters based on primitive features like presence and location of the vertical bar, the number of times the character body touches *Shirorekha*, number and location of intersections, and slope of the ending stroke. The method proposed in this paper is based on the detection of these features inherent to a character and classification of the character accordingly and works on all sizes of images without the need of scaling.

The related work done by other researchers in this field is discussed in the next section. Section 3 elaborates on the unique features of the Devanagari characters. Section 4 puts forth the proposed methodology. Results are discussed in Sect. 5 followed by challenges and future work in Sect. 6. Section 7 concludes the paper.

2 Related Work

Work on OCR of Indian scripts began in the decade of 1970 [9]. Sethi and Chatterjee [10] demonstrated Devanagari numeral recognition. They characterized the numeral on the basis of occurrence of four basic primitives—horizontal line segment, vertical line segment, right-slanted stroke, and left-slanted stroke. The presence of these attributes and their interconnections were used to classify the numeral. Later, they used a similar approach [3] to recognize the Devanagari characters. Sinha and Mahabala [11] presented a knowledge-based recognition system for the Devanagari script where the system stored the structure of each valid character in terms of primitives and corresponding relationships. The recognition was performed by searching for the unknown character primitives based on the stored description.

Siromoney et al. [12] proposed a technique to recognize printed Brahmi—the ancestor of the Devanagari script. They devised a method to extract the features of the character by scanning the character matrix row-wise and column-wise and then perform the recognition. Sinha [13] also proposed a method of rule-based contextual checker in the form of finite state machine. Substitution rules were in the form of <condition, action> pairs. Each rule had a penalty associated with it and the total penalty value gave a confidence score to detect the character. Marudarajan et al. [14] presented the adaptive threshold logic for printed Hindi numeral recognition. Banashree [15] used a technique to extract features using 16-segment display concept from the halftoned image and fed those to neural network classifier to recognize the numeral. Mukherji and Rege [7] segmented strokes of the character and extracted the features using a customized algorithm which encoded the strokes using Freeman chain coding along with consideration of the angular factor. These features along with fuzzy features extracted on basis of the fuzzy membership functions are used to recognize the character using multi-class tree classifier. Ghosh and Roy [16] detected Devanagari characters by first extracting the features of the constituent strokes of the character zone-wise and feeding them to an SVM classifier to recognize the stroke and then checking the presence of the combination of the detected strokes which would determine the character. Kushwah and Joshi [17] presented an algorithm to recognize the isolated modifier (*matra*) of Devanagari characters using pixel relationship. Kant and Vyavahare [18] worked on Devanagari OCR with focus on Marathi language and classified the characters using SVM classifier and achieved 85% accuracy. Gupta et al. [19] found 345 frequently occurring characters and conjuncts, divided them into 16 categories, and classified them holistically on the basis of the location of vertical bar and the frequency of intersection with the header line. Chaudhuri et al. [20] performed feature-based classification of Hindi characters using soft computing techniques, namely rough fuzzy multilayer perceptron, fuzzy support vector machines, fuzzy rough support vector machines, and fuzzy Markov random fields.

The next section describes the characteristics of the Devanagari consonants. The uniqueness of some of these features is utilized in the proposed methodology to recognize the first group of the Devanagari characters.

3 Devanagari Character Features

All the characters in the Devanagari script are composed of curves, strokes, circular regions, *matras,* and dots. Figure 1 shows the consonant set of Devanagari.

The group of first five letters (first row) is called the *Ka-Varga*. A major distinguishing feature of Devanagari characters is the top horizontal line (called *Shirorekha*) which is used to segment the core body of the character from upper modifiers (*matras*). The processing is done on the core body after removing *Shirorekha* by computing the horizontal projection of the character image and removing the row containing the maximum number of white pixels [21].

A database of the feature sets of the alphabet is made on the basis of the following attributes:

1. Vertical Bar
   a. Absent
   b. Present—Right
   c. Present—Center
2. Intersections
   a. Present
      i. The corresponding positions
   b. Absent
3. *Shirorekha* Touch Frequency
   a. The number of times the character body touches *shirorekha*
4. Number of endpoints and slopes of ending curves

The next section chalks out the methodology proposed in this paper to recognize the Devanagari characters by extracting the abovementioned features of these characters.

क ख ग घ ङ
च छ ज झ ञ
ट ठ ड ढ ण
त थ द ध न
प फ ब भ म
य र ल व
श ष स ह

Fig. 1 Devanagari consonants

4 Proposed Methodology

The proposed methodology involves the stages shown in Fig. 2.

4.1 Input and Digitization

The input of the OCR system is an image of the character. Digitization includes scanning the image of the characters and converting it into digital format in form of pixels.

4.2 Preprocessing

The image is then preprocessed to remove noise and extract the region of interest in order to perform the recognition with a better accuracy rate. Preprocessing is done by performing the following operations (shown in Fig. 3).

Smoothening/Blurring. The image of the character is smoothened using Gaussian filter in order to remove salt-and-pepper noise as shown in Fig. 3a. Kernel size of 5 × 5 and the σ value as 3 gave the best results.

Binarization. The grayscale image is converted into a binary image with all pixels in the image with luminance greater than *level* with the value 1 (white) and replaces all other pixels with the value 0 (black) as shown in Fig. 3b. The *level* lies

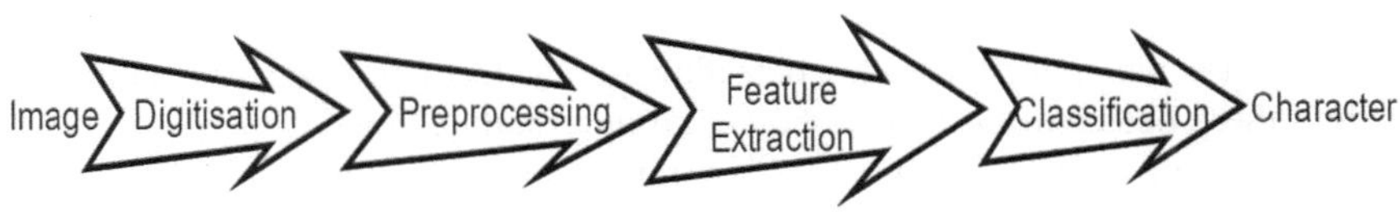

Fig. 2 Stages of OCR

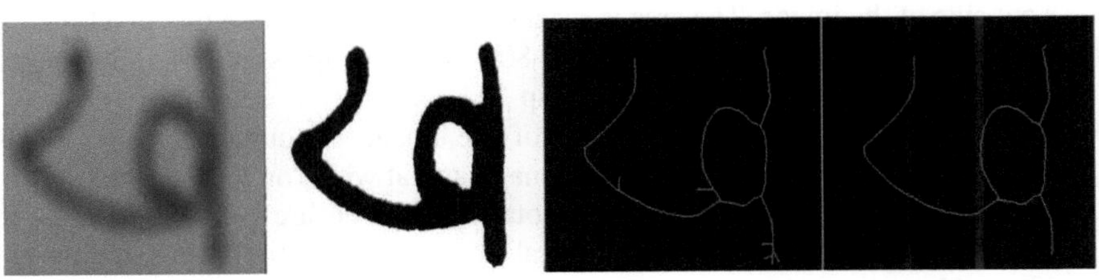

(a) Smoothening **(b)** Binarization **(c)** Thinning **(d)** Spur Removal

Fig. 3 Operations involved in preprocessing

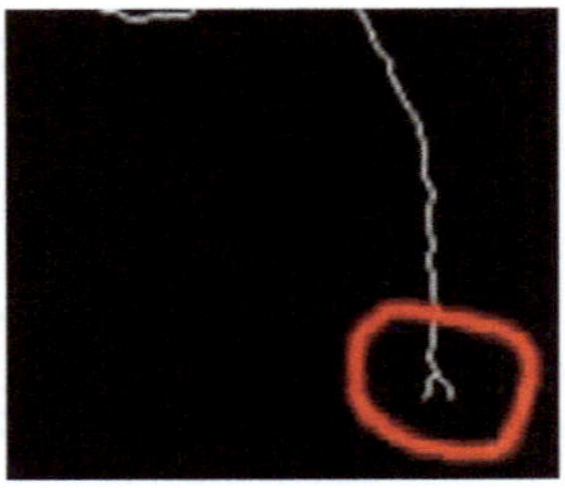

Fig. 4 Example of a spur

in the range [0, 1] and is computed using Otsu's algorithm [22]. The image is later inverted (black pixel converted into white and vice versa).

Thinning. The binarized image is thinned by removing pixels so that an object without holes shrinks to a minimally connected stroke, and an object with holes shrinks to a connected ring halfway between each hole and the outer boundary, as shown in Fig. 3c.

Spur Removal. Noise and/or thinning sometimes introduce some inadvertent small branches called spurs (shown in Fig. 4), which may affect the recognition procedure. These spurs which also are an integrated part of handwritten characters are removed by detecting the intersection point and the corresponding endpoint. The path between the two is then calculated, and if found more than 12% of the height of the image (or 5 pixels if image height is less than 40 pixels), it is removed from the original image. This is iteratively done for all the edges present in the image. The result of the same is shown in Fig. 3d.

4.3 Feature Extraction

Feature extraction phase includes detection of the features described above and feeding these detected features to the classifying module.

Vertical Bar Check. The aim of this test is to detect the presence of the vertical bar. In case the vertical bar is present, the test checks the corresponding location of the vertical bar.

- A region with the width equivalent to 12% of the total width is formed from the right side of the image. The area is scanned for the number of white pixels and if the number is greater than or equal to 80% of the image's height, right central vertical is considered as present (Group 1).
- The horizontally middle pixel—*mid* of the image is found and a region is formed by width equivalent to 9% of image's total width on both the sides from *mid*. This region is scanned and if the total white pixels are greater than equal to 80% of the image's height, the central vertical bar is considered as present (Group 2).
- If both the above checks fail, then the vertical bar is absent (Group 3).

Shirorekha Touch Frequency calculation. Padding of 10% of the total height is left from the top. A region of height 2% (*roiHeight*) of the total height or 1 pixel (if image height is less than 50 pixels) is formed and scanned. The total number of white pixels is counted and normalized by dividing by *roiHeight* which gives the touch frequency.

Intersections count. The number of intersections is calculated, and the coordinates of the intersection points are noted in an array. The intersection points are further analyzed on the basis of their location by counting the number of points on the left side and on the right side of the middle of the character body.

Figure 5 shows the flow of the character recognition process.

4.4 *Classification*

All the features extracted are then matched with the feature set database. First, the group is matched and then other attributes are matched to recognize the character. Some characters falling in the same group even after the aforementioned checks cannot hitherto be differentiated and are passed through a slope detection check.

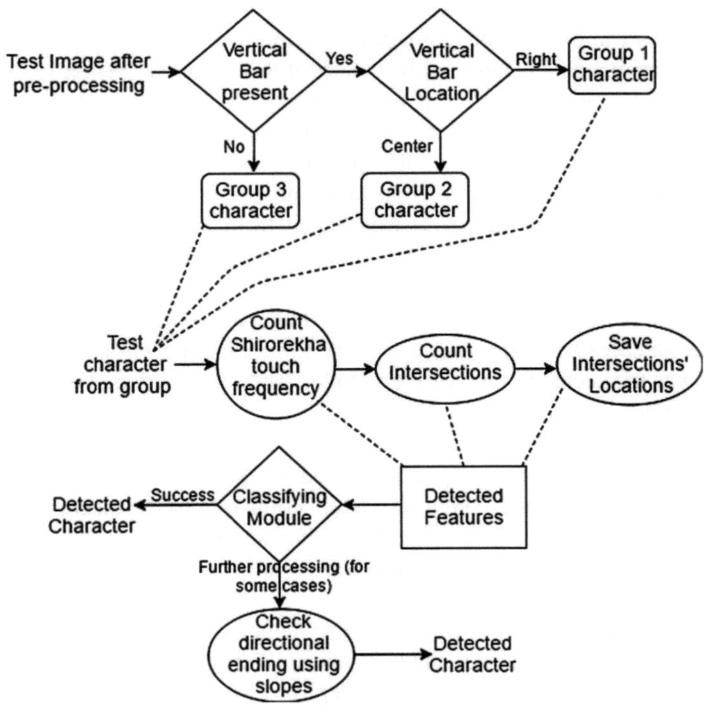

Fig. 5 Feature detection and flow of character recognition process

Fig. 6 Slope detection

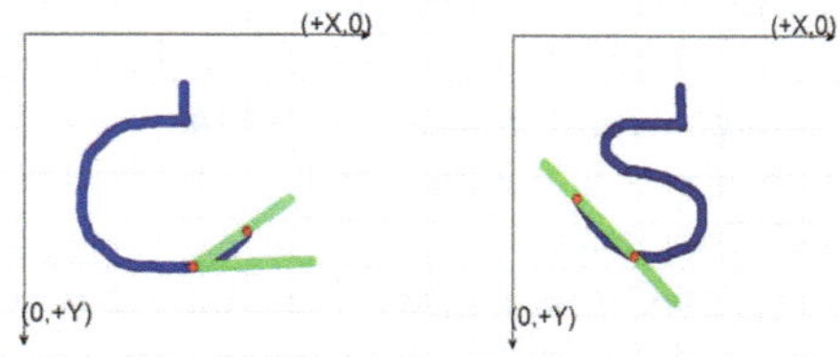

Slope Detection. To discern one character from the other, the slope is calculated taking the endpoint other than the top endpoint and another point which is 15% of image height pixels away from the endpoint (the points shown with red dots in Fig. 6) on the stroke path. The slope value shows the quadrant the character stroke lies in.

5 Results

Unlike other languages in which standard datasets are available to compare and evaluate the accuracy, there is no such standardized dataset available in the Devanagari script. Though an attempt has been made by Kumar et al. [23] to provide a dataset in order to benchmark the accuracy of Devanagari characters, the dataset site was unreachable during the course of this research. Hence, to implement and evaluate the method proposed in this paper, datasets were prepared by the authors. For the printed characters, various Devanagari fonts were searched for designing the inputs. Finally, the samples of *Ka-Varga* were taken from 21 different Devanagari fonts used for Hindi, Sanskrit, and Marathi languages. Samples of handwritten characters of *Ka-Varga* were taken from 22 people belonging to different age groups. There were 18 young students (aged 20–25) and 4 elders (aged 35–50).

The accuracy in case of recognizing printed characters was found to be 93.33%. Among the samples from the students, 78.89% were recognized correctly, whereas only 45% was the accuracy rate for the samples taken from elders. Overall accuracy level was found to be 72.72%.

Tables 1 and 2 show the result statistics in a concise manner.

The results show that the proposed method works best with printed characters and better on students' handwritten characters compared to the elders' handwritten characters. This may be attributed to the fact that students are usually in touch with writing more than the elders and tend to follow the character strokes pattern more strictly.

It was observed that the recognition rate depends on the complexity of the composition of the character, i.e., the more the strokes the character is composed of, the higher is the inaccuracy. This is also depicted by the incongruous recognition

Table 1 Accuracy rate of algorithm for printed characters

Character	Total number of characters	Number of correctly recognized characters	Accuracy (in %)
क	21	21	100
ख	21	16	76.19
ग	21	19	90.48
घ	21	21	100
ङ	21	21	100
Total	105	98	93.33

Table 2 Accuracy rate of algorithm for handwritten characters

Character	Recognition accuracy rate (in %)		
	Students sample	Old people sample	Overall
क	94.44	50	86.36
ख	44.44	25	40.91
ग	94.44	25	81.81
घ	88.89	50	81.81
ङ	72.22	75	72.72
Total	78.89	45	72.72

rate of character ख which is written in varied styles by different writers as well as in different fonts, the left and right halves of the character not being connected in some fonts.

6 Challenges and Future Work

Although the mentioned procedure has shown substantial accuracy for the tested characters, some characters may still pose difficulty. A character of the form M* (e.g., अः or ङ) is always recognized as M (अ or ड) as the dots are usually removed in the noise-removal phase.

The handwritten character recognition is comparatively more challenging because of variations in the slant, strokes, and connectivity. The algorithm will need some modifications including but not limited to consideration of slant or skewness in order to nullify the foregoing effects and to enhance the accuracy rate.

Future work aims to implement the algorithm for all the characters in the alphabet. The authors also aim to increase the accuracy rate further by implementing machine learning models and training the system instead of using hard-coded values.

7 Conclusion

In this paper, we have presented how recognition of Devanagari characters can be done on the basis of their primitive characteristics. The proposed method which worked the best on printed characters and significantly well on handwritten characters can be used in applications where the character formation rules are followed strictly, e.g., in applications which can be used to teach writing Hindi or any other language based on the Devanagari script.

References

1. Cardona, G.: Devanagari. (Encyclopædia Britannica, inc.), https://www.britannica.com/topic/Devanagari, last accessed 2017/05/17.
2. Hanmandlu, M., Murthy, O. R.: Fuzzy model based recognition of handwritten numerals. The journal of the pattern recognition society, 1840–1854 (2007).
3. Sethi, K., Chatterjee, B.: Machine Recognition of constrained handprinted Devanagari. Pattern Recognition 9, 69–75 (1977).
4. Bansal, V., Sinha, R.: A Complete OCR for Printed Hindi Text in Devanagari Script. Proceedings 6th conference on document analysis and recognition, 800–804 (2001).
5. Aggarwal, A., Rani, R., RenuDhir.: Handwritten Devanagari Character Recognition Using Gradient Features. International Journal of Advanced Research in Computer Science and Software Engineering 2, 85–90 (2012).
6. Sarkar, R., Sen, B., Das, N., Basu, S.: Handwritten Devanagari Script Segmentation: A Non-linear Fuzzy Approach. Proc. (CD) of IEEE Conference on AI Tools and Engineering (ICAITE-08) (2008).
7. Mukherji, P., Rege, P. P.: Shape Feature and Fuzzy Logic Based Offline Devnagari Handwritten Optical Character Recognition. Journal of Pattern Recognition Research, 52–68 (2009).
8. Goyal, P., Diwakar, S., Agrawal, A.: Devanagari Character Recognition towards natural Human-Computer Interaction. Proceedings India HCI No EPFL-CONF-168804 (2010).
9. Pal, U., Chaudhuri, B.: Indian script character recognition: a survey. The journal of pattern recognition, 1887–1899 (2004).
10. Sethi, I. K., Chatterjee, B.: Machine Recognition of handprinted Devanagari Numerals. J. Institute of Electrical Telecommunication Engineering (India) 22, 532–535 (1976).
11. Sinha, R. M., Mahabala, H. N.: Machine recognition of Devanagari script. IEEE Transactions on Systems, Man and Cybernetics, 435–441(1979).
12. Siromoney, G., Chandrasekaran, R., Chandrasekaran, M.: Machine recognition of Brahmi script. IEEE Transactions on Systems, Man and Cybernetics (1983).
13. Sinha, R. M.: Role of contextual postprocessing for Devanagari text recognition. Pattern Recognition, 475–485 (1987).
14. Marudarajan, A. R., Jayanthi, K., Rajeswari, M.: Extension of adaptive threshold logic to printed Hindi numeral recognition. Journal of Institute of Electricaland Telecommunication Engineering (India), 223–225 (1978).
15. Banashree., P. N., Dharani, A., Vasanta, R., Satyanarayana, P. S.: OCR for Script Identification of Hindi (Devnagari) Numerals using Error Diffusion Halftoning Algorithm with Neural Classifier. International Journal of Computer, Electrical, Automation, Control and Information Engineering 1, 307–311 (2007).

16. Ghosh, R., Roy, P. P.: Study of two zone-based features for online Bengali and Devanagari character recognition. 2015 13th Int. Conf. on Document Analysis and Recognition (ICDAR), 401–405 (2015).
17. Kushwah, K. K., Joshi, B. K.: Hindi modifier recognition based on pixel relationship. 2016 Int. Conf. on ICT in Business Industry & Government (ICTBIG) (2016).
18. Kant, Mr Akshay J., Mrs Arati J. Vyavahare.: Devanagari OCR Using Projection Profile Segmentation Method. International Research Journal of Engineering and Technology (IRJET) (2016).
19. Gupta, M.K., Lakshmi, C.V., Hanmandlu, M., Patvardhan, C.: An Exhaustive Font and Size Invariant Classification Scheme for OCR of Devanagari Character. International Journal on Natural Language Computing, 4(1), 1–21 (2014).
20. Chaudhuri, A., Mandaviya, K., Badelia, P., Ghosh, S. K.: Optical Character Recognition Systems for Hindi Language. In Optical Character Recognition Systems for Different Languages with Soft Computing, pp. 193–216. Springer International Publishing (2017).
21. Bansal, V., Sinha, R.M.K.: Segmentation of touching and fused Devanagari characters. Pattern Recognition 35, 875–893 (2002).
22. Otsu, N.: A threshold selection method from gray level histograms. IEEE Transactions on Systems, Man, and Cybernetics, 62–66 (1979).
23. Kumar, R., Kumar, A., Ahmed, P.: A benchmark dataset for Devanagari document recognition research. 6th International Conference on Visualization, Imaging and Simulation (VIS'13), 258–263 (2013).

Population Dynamics Indicators for Evolutionary Many-Objective Optimization

Raunak Sengupta, Monalisa Pal, Sriparna Saha and Sanghamitra Bandyopadhyay

Abstract Recent research on multi- and many-objective optimization has led to the development of various state-of-the-art algorithms which produce satisfactory results for various kinds of problems. However, in real life, the underlying objective functions or the characteristic landscape formed by the objectives may not be known beforehand. This makes it difficult for a user to choose the correct optimization algorithm. This paper proposes new indicators which attempt to summarize the population dynamics across iterations. The statistics of the population movement can help in identifying various features of the problem at hand and the capacity of an algorithm to deal with the challenges corresponding to the features. The analysis of population movement can enable further modifications of an existing algorithm according to the optimization problem. The indicators can also help in the development of adaptive optimization algorithms by providing feedback during the search for optimality.

Keywords Many-objective optimization · Population movement Pareto-optimality · Visualization

R. Sengupta
Department of Electrical Engineering, Indian Institute of Technology Patna, Patna, India
e-mail: raunaksengupta@gmail.com

S. Saha
Department of Computer Science and Technology,
Indian Institute of Technology Patna, Patna, India
e-mail: sriparna.saha@gmail.com

M. Pal (✉) · S. Bandyopadhyay
Machine Intelligence Unit, Indian Statistical Institute, Kolkata, India
e-mail: monalisap90@gmail.com

S. Bandyopadhyay
e-mail: sanghami@gmail.com

C. R. Panigrahi et al. (eds.), *Progress in Advanced Computing and Intelligent Engineering*, Advances in Intelligent Systems and Computing 714,
https://doi.org/10.1007/978-981-13-0224-4_24

1 Introduction

In the recent years, several many-objective optimization (MaOO) algorithms have been developed that make use of different kinds of strategies. Algorithms based on reference points such as θ-DEA [1], MOEA/D [2], NSGA-III [3], etc. have been shown to perform well for several benchmark test problems like DTLZ [4] and WFG [5] functions. NSGA-III uses the concept of non-dominated sorting for selecting the members of the new population [3], whereas θ-DEA introduces θ-dominance based on the argument that non-dominated sorting does not produce enough selection pressure [1]. MOEA/D, which has mating constraints, performs very well for problems like DTLZ2, but its performance is not very good for problems such as DTLZ4, which has a biased density of solutions [2]. Another major class of MaOO algorithms is indicator-based algorithms. Algorithms such as HypE [6], IBEA [7], and MOMBI-II [8] come under this class as they make use of indicator values to perform selection. The problems that need to be optimized can provide different kinds of challenges. The possible features of a problem at hand have been classified into five different types in literature [5]:

1. *Geometry*: Shape of Pareto-front can be convex, concave, linear, mixed, and degenerate.
2. *Parameter Dependencies*: Separable objectives, non-separable objectives (Capability to determine ideal points by considering only one objective at a time).
3. *Bias*: Presence of a bias while mapping solutions from decision space to fitness functions in objective space.
4. *Many-to-One mappings*: Pareto one-to-one, Pareto many-to-one, flat regions, and isolated optima.
5. *Modality*: Uni-modal, multimodal (presence of local optimal fronts).

All of the algorithms have their own advantages and disadvantages and give best results for problems with only certain kinds of features. In real life, the problems that need to be optimized often either have a too complicated mathematical expression or are in the form of a black box (simulation, physical device) and thus much is not known about the features. This makes it difficult to choose the correct optimization problem suited for the function. Vigorous work has also not been done on relating the various problem features with the strategies best suited to overcome the challenges corresponding to the problem features.

In this paper, we propose indicators that can be observed throughout the iterations of an algorithm and based on them, the user can understand the features of the problem at hand. Much work has not been done on visualizing the dynamics of the points during search for optima. The indicators attempt to summarize the overall movement of the points in a population for a particular problem and algorithm. This will help in identifying the shortcomings of an algorithm, enhancing an algorithm and even designing better algorithms.

2 Indicators and Visualization of Population Movement

Information about the population movement can give a great deal of insight about both the problem function and the nature of the algorithm being used. Several works in literature use the information obtained from the previous iterations to optimize a problem more efficiently. Neighborhood-based cross-generation mutation (NCG) [9] produces new solutions by performing vector differencing between points in the same neighborhood, appearing in consecutive generations. This strategy has been justified as an attempt to direct a point toward the optimal front. On the other hand, fitness improvement rate (FIR) [10] measures the improvement in solutions and uses this information as a heuristic to select a reproduction operator for the next generation.

However, much work has not been done on explicitly quantifying and visualizing the movement of a population. To achieve this, we need to design quantitative indicators that define different properties of a distribution at a particular iteration as well as the changes that occur through the iterations. The proposed indicators are inspired by the recently developed radial plot visualization technique [11]. The movement of the population on a radial plot is shown in Fig. 1 for various iterations when NSGA-III is used for optimizing the three-objective DTLZ1 problem. As shown in Fig. 1a, the population members gather around a point at a distance corresponding to radius 1 on the polar plot indicating that the algorithm is trying to overcome the local optimal front. The points then start to converge at this local optimum as shown in Fig. 1b. At this stage, the consecutive generations keep improving the diversity until some new points overcome the local optimal front (Fig. 1c). At this point, there is a relative deterioration in the diversity until the points finally converge to the global optimum at a radius of 0.5 (Fig. 1d).

Table 1 lists the hypervolume values for four algorithms, viz.,-MOEA/D-PBI, θ-DEA, HypE, and NSGA-III, which are obtained from [1]. The final Pareto-front obtained by MOEA/D-PBI after optimizing on DTLZ4 problem is shown in Fig. 2a. The plot clearly shows the poor distribution of solutions which results in the relatively poor hypervolume values as compared to the other algorithms (Table 1). The radial plots in Fig. 2b, c show that poor diversity is obtained by NSGA-III and θ-DEA, respectively, for WFG6 problem. However, these plots contradict with the

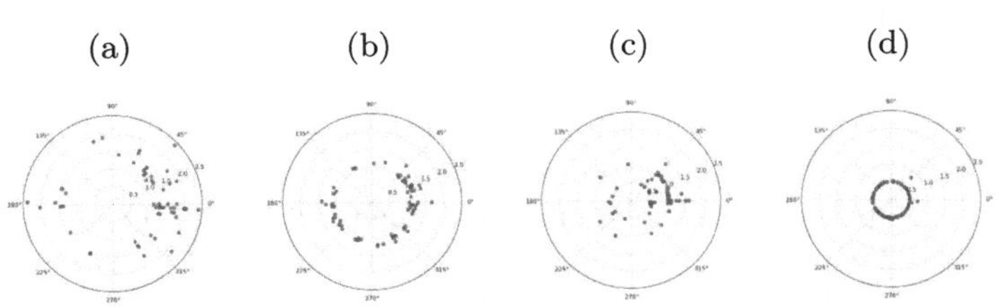

Fig. 1 Radial plots of NSGA-III run on DTLZ1 with three objectives **a** iteration 59, **b** iteration 87, **c** iteration 113, **d** iteration 170

Table 1 Mean hypervolumes on DTLZ4 and WFG6 (three objectives)

Problem	MOEA/D-PBI	NSGA-III	θ-DEA	HypE
DTLZ4	0.406020	0.744634	0.729265	0.549999
WFG6	0.654956	0.685939	0.690060	0.708633

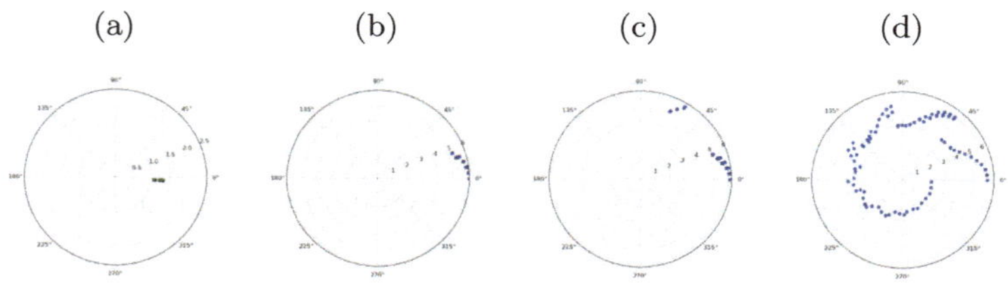

Fig. 2 Radial plots for three objective functions **a** MOEA/D on DTLZ4, **b** NSGA-III on WFG6, **c** θ-DEA on WFG6, **d** a good distribution for WFG6

corresponding hypervolume values in Table 1 which are relatively good and comparable with the other state-of-the-art algorithms. This implies that hypervolume indicator values may not always be enough to assess the efficacy of an algorithm.

To have an overall understanding of the movement, it is necessary to have several indicators each indicating different properties and statistics of the movement. The various properties of the movement that have been deemed necessary to have an overall summary of the response of an algorithm to a problem are as follows:

1. The tendency of a population distribution to diversify and spread over iterations needs to be captured. This indicator should not be affected by the tendency of the population members to improve their fitnesses.
2. An indicator is required for capturing the overall tendency of the population distribution to shrink in when plotted on radial coordinates, considering a minimization problem. This indicator should be as independent of the diversification nature of the population as possible.
3. An indicator capturing the overall rate of change in the fitnesses of the population would be able to tell about the speed of convergence.
4. An indicator for the consistency among the members could tell whether the tendency to improve is same for population members from different regions.

Based on the above discussions, this paper proposes several indicators to summarize the movement and performance of an algorithm. The indicators defined in this paper are inspired by the radial coordinate plots. At any generation (or iteration), g, each of the members of the population is associated with one of the reference lines closest to it, as done in [11]. Therefore, we have a vector called *Spread*, which is of the same length as that of the number of reference lines. Each member of this array represents the number of population members associated with it. Each member of the population also has an associated distance value, r, which is its Euclidean distance from the origin. Since a reference line might have several associated members, it also might have several corresponding values of r. We define vectors *r_inner* and *r_outer*. Each member of these vectors represents the smallest and largest value of r associated with each reference line, respectively. In case a reference line has only one value of r associated with it, the corresponding *r_inner* and *r_outer* values become equal. If a reference line does not have any associated member, its corresponding value of r is considered undefined and not used for calculations.

The indicators can be classified into two types—Independent and Dependent.

1. *Independent Indicators*: The values of the indicators under this class are independent of factors such as the range of values each objective can take, the scales of different objectives, and other factors which are dependent on the objective functions. Such indicators have a fixed range of values. Indicators under this class are as follows:

   (a) *D_metric*: This indicator is a measure of the diversity of the population at a certain iteration, g. To define this indicator, we first define the vector, *Spread*, as in Eq. (1), whose ith element is the number of points coupled with the ith reference line.

$$Spread^g = (s_1^g, s_2^g, \ldots, s_n^g)^T \tag{1}$$

   We also define another vector called *Ideal_Spread* such that $s_i^g = \frac{pop\_size}{n}$, where n is the number of reference lines being used, *pop_size* is the cardinality of the population, and $i = 1, 2, \ldots, n$. We have chosen all the members of *Ideal_Spread* to be equal. However, one is free to define a distribution according to requirements.
   Now, *D_metric* (as defined by Eq. (2)) at generation g is the euclidean distance of the current *Spread* vector from the *Ideal_Spread* vector. A value close to zero implies good diversity while a higher value implies poor diversity.

$$D\_metric^g = \frac{n}{pop\_size}\sqrt{\sum_{i=1}^{n}(Spread_i^g - Ideal\_Spread_i)^2} \tag{2}$$

   (b) *V_metric*: This indicator is a measure of the tendency of the population in general to move toward the origin of the radial plot, i.e., improve the member's fitnesses considering a minimization problem. This indicator is defined in Eq. (3).

$$V\_metric^g = \frac{\sum_{i=1}^{n}(1 - a_i^g)}{n^g} \tag{3}$$

where

$$a_i^g = \begin{cases} 1, & (r\_inner_i^{g-1} - r\_inner_i^g) > \epsilon \\ 0, & -\epsilon < (r\_inner_i^{g-1} - r\_inner_i^g) < \epsilon \\ -1, & \text{otherwise} \end{cases}$$

such that n^g is the number of points satisfying the conditions: $Spread_i^g \neq 0$ and $Spread_i^{g-1} \neq 0$. ϵ in the equation is just a small constant . Intuitively, V_metric is the fraction of reference lines with an associated point that cease to show improvement, i.e., move toward origin.

(c) *Consistency*: This indicator (given by Eq. 4) attempts to capture the consistency in the improvement of members associated with a reference line. A poor value of consistency in the initial iterations would imply the presence of certain regions where improvement is difficult for the algorithm.

$$Consistency^g = \sum_{i=1}^{n} \frac{a_i^g}{l_i^g} \tag{4}$$

where l_i^g is the total number of generations up to g with continued association of at least one point, for the ith reference line. If the number of points associated with the ith reference line becomes zero for any iteration, the corresponding value of l_i^g resets to zero and again starts counting when an associated point appears.

(d) *Velocity*: Velocity tries to measure the amount of change occurring per generation as shown in Eq. (5).

$$Velocity^g = \frac{\sum_{i=1}^{n} \tan^{-1}(r\_inner_i^{g-1} - r\_inner_i^g)}{n^g} \tag{5}$$

It captures the amount of change at each generation by calculating the average slope of the plot of the inner radii associated with each of the reference lines through iterations.

2. *Dependent Indicators*: The values of indicators under this class are dependent on the range of values that the objective function can take. These indicators are relatively simple and naïve. However, these are necessary and provide a lot of information. Indicators under this class are as follows:

(a) *Innermost Radius*: As the name suggests, the value of this indicator is given by the minimum value of the radius, r, among the population members at a particular generation. This is also equivalent to $\min_{i=1}^{n} r\_inner_i^g$.

(b) *Outermost Radius*: The value of this indicator is given by the maximum value of the radius, r, among the population members at generation g. This is also equivalent to $\max_{i=1}^{n} r\_outer_i^g$.

(c) *Average Inner Radius*: Average of all the inner radii associated with the reference lines at generation g. This is equivalent to $(\sum_{i=1}^{n} r\_inner_i^g)/n$.
(d) *Average Outer Radius*: Average of all the outer radii associated with the reference lines at generation g. This is equivalent to $(\sum_{i=1}^{n} r\_outer_i)/n$.
(e) *Inner Band*: Difference between the maximum inner radius and the minimum inner radius at generation g. This measure is an attempt to capture the variance in the inner radii of the members. This is also equivalent to $(\max_{i=1}^{n} r\_inner_i - \min_{i=1}^{n} r\_inner_i)$.

3 Results and Discussions

This section presents a few of the results that were obtained by employing the indicators. All of the experiments have been run using Python 2.7.6 on a system with Intel Core i7 processor @ 2.5 GHz and GTX 860 M GPU.

We have done our analysis on DTLZ1 to DTLZ4 problems with three and eight objectives. The algorithms that have been used for various comparisons are NSGA-II, NSGA-III, MOEA/D, and θ-DEA. Parameters of the algorithms have been set according to the recommendations in [1].

Due to constraint of space, only the plots which give a deeper insight have been shown and discussed over. Following are some of the inferences that have been enabled by observing the indicators:

1. As observed from plots in Fig. 3, problems with local optimal fronts like DTLZ1 and DTLZ3 usually tend to have a relatively more bumpy plot of diversity. Further, the plot of the innermost radius for such problems (Fig. 4) often tends to have flat regions. These regions correspond to the local optimal fronts where the points get stuck for some time. At these points, the diversity usually starts to improve until the points eventually overcome the optimal front.
2. It has been pointed out in literature that NSGA-III has a lower convergence speed than algorithms using PBI (penalty based boundary intersection) functions due to its lower selection pressure [1]. It can be observed from the plots in Fig. 4 that MOEA/D reaches global optimal front roughly by 70 iterations, while NSGA-III takes more than 90 iterations. NSGA-II on the other hand gets stuck at a

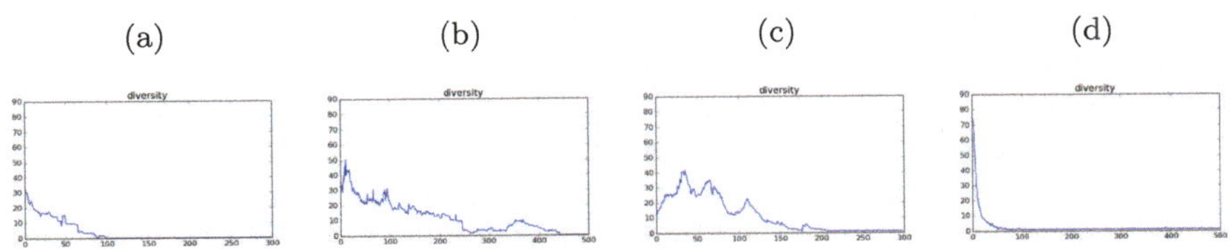

Fig. 3 Plots of *D_metric* **a** MOEA/D on DTLZ2 (8-obj), **b** MOEA/D on DTLZ3 (8-obj), **c** NSGA-III on DTLZ1 (3-obj), **d** NSGA-III on DTLZ4 (3-obj)

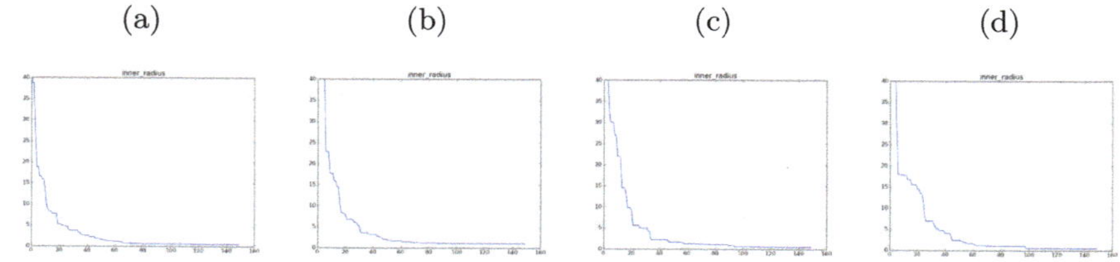

Fig. 4 Plots of innermost radius on three-objective problems **a** MOEA/D on DTLZ1, **b** NSGA-II on DTLZ1, **c** NSGA-III on DTLZ1, **d** θ-DEA on DTLZ1

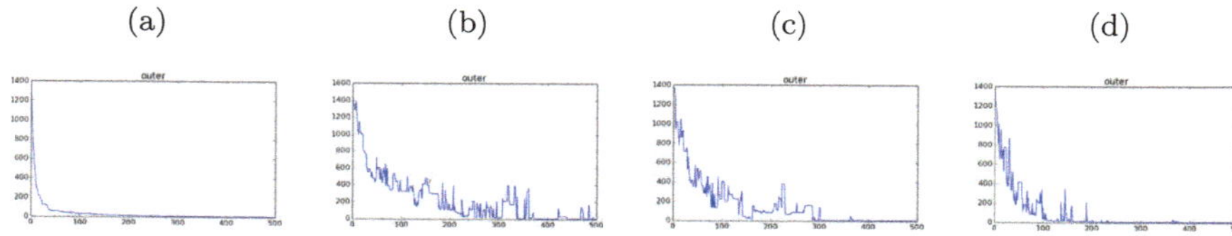

Fig. 5 Plots of outermost radius on three-objective problems **a** MOEA/D on DTLZ1, **b** NSGA-II on DTLZ1, **c** NSGA-III on DTLZ1, **d** θ-DEA on DTLZ1

local optimal front corresponding to radius equal to 1. However, NSGA-III and θ-DEA are comparable.

3. It can be observed from Fig. 5 that NSGA-II and NSGA-III in general have a higher tendency to have outliers, MOEA/D has the least tendency, while θ-DEA is in between. This implies that performing non-dominated sorting whether based on Pareto dominance or θ-dominance has a tendency to generate outliers. This behavior can be explained through the argument that non-dominated sorting enables relatively higher exploration and simultaneously reduces the selection pressure. Using mating constraints and PBI, such as done in MOEA/D, leads to an increased selection pressure and more exploitation which ultimately results in faster convergence rates and less outliers.
4. MOEA/D-PBI, however, suffers from a very poor diversity when running on problems with a biased density of solutions such as DTLZ4 as can be seen from the plots in Fig. 6. This happens because of the presence of mating constraints which decreases the ability of an algorithm to explore, which is required very much when there are regions with biased density of solutions.
5. From the diversity plots (Fig. 6), we also observe the improved capability of reference point-based techniques in obtaining a good diversity of solutions when compared to methods such as crowding distance.
6. The behavior of the reference point-based algorithms does not change much with an increase in the number of objectives. Only the required number of function evaluation increases. However, NSGA-II shows a significant difference in its behavior as can be observed from the plots for the eight-objective problems (Fig. 7). The value of the innermost radius does not decrease steadily like it does for the other algorithms. In fact, it increases and converges to a high value for all

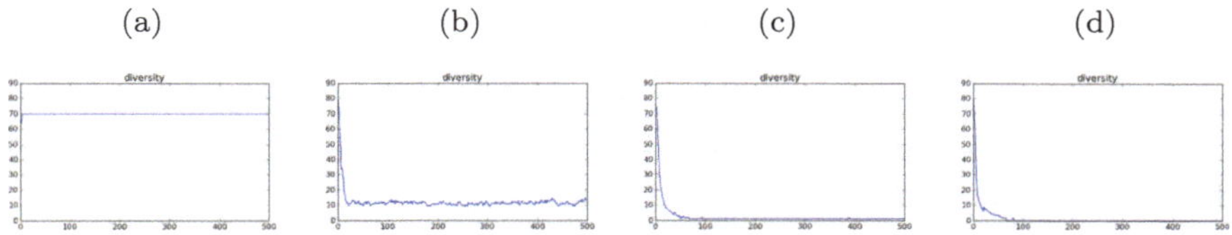

Fig. 6 Plots of *D_metric* on DTLZ4 (3-objective) functions **a** MOEA/D-PBI, **b** NSGA-II, **c** NSGA-III, **d** θ-DEA

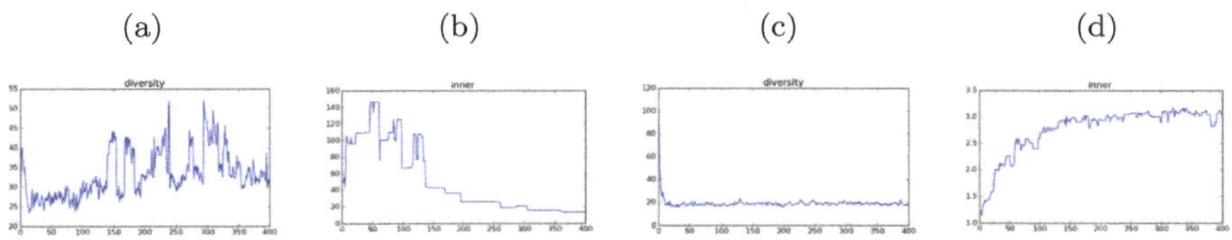

Fig. 7 NSGA-II plots for eight-objective problems **a** *D_metric* for DTLZ1, **b** innermost radius for DTLZ1, **c** *D_metric* for DTLZ4, **d** innermost radius for DTLZ4

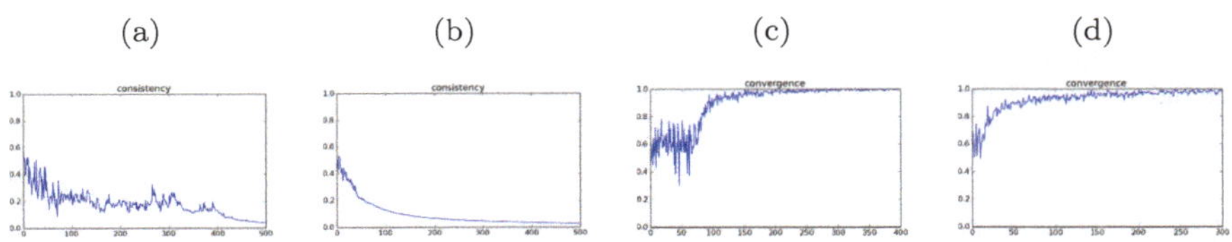

Fig. 8 Plots of consistency **a** NSGA-III on DTLZ3 (3-obj), **b** NSGA-III on DTLZ4 (3-obj); and *V_metric*, **c** θ-DEA on DTLZ1 (8-obj), **d** θ-DEA on DTLZ2 (8-obj)

the test problems except DTLZ1. This shows that crowding distance renders useless when the number of objectives is as high as 8. This combined with the low selection pressure of non-dominated sorting makes NSGA-II not very suitable for problems with number of objectives as high as 8.

7. It can be observed in general that the consistency plots and the *V-metric* plots (Fig. 8) associated with problems such as DTLZ1 and DTLZ3 which have local optimal fronts in general have more spikes and large variances.

4 Conclusion

This paper provides a direction toward explicitly quantifying the population movement of an evolutionary multi-objective optimization and making various inferences based on their analysis. This work is based on the argument that a single indicator fails to provide sufficient information about the performance of an algorithm. The indicators observed through iterations will be able to tell us about the response of an algorithm while optimizing a particular kind of a problem and in turn tell us about its advantages, disadvantages, and how well it is able to overcome various challenges posed by an algorithm. The indicators will also enable identification of various features that might be present in unknown objective functions and act as a heuristic in choosing the right strategies for optimization.

Future work includes developing indicators that take into account the shape and structure that emerges from the iterations. The variance among the innermost solutions at a particular iteration might potentially be informative. However, using the variance in its simplest form will not work if the shape is complicated. One might also consider deriving other indicators from the ones mentioned in this paper by simply performing various signal processing operations on them. A major direction that can be pursued from here is using the information obtained from the indicators as a feedback for the development of an adaptive and robust many-objective optimization algorithm and correlating various features of objective functions to the best suited strategies. Future work also includes performing a much more detailed survey incorporating various classes of algorithms and a large variety of challenging test problems.

Acknowledgements This work is funded by the project (DST-INRIA/2015-02/BIDEE/0978) of the Indo-French Centre for the Promotion of Advanced Research (IFCPAR).

References

1. Yuan, Y., Xu, H., Wang, B., Yao, X.: A new dominance relation-based evolutionary algorithm for many-objective optimization. IEEE Trans. Evolutionary Computation 20(1), 16–37 (2016), http://dx.doi.org/10.1109/TEVC.2015.2420112
2. Zhang, Q., Li, H.: MOEA/D: A multiobjective evolutionary algorithm based on decomposition. IEEE Trans. Evolutionary Computation 11(6), 712–731 (2007), http://dx.doi.org/10.1109/TEVC.2007.892759
3. Deb, K., Jain, H.: An evolutionary many-objective optimization algorithm using reference-point-based nondominated sorting approach, part i: solving problems with box constraints. Evolutionary Computation, IEEE Transactions on 18(4), 577–601 (2014)
4. Deb, K., Thiele, L., Laumanns, M., Zitzler, E.: Scalable test problems for evolutionary multi-objective optimization. Evolutionary Multiobjective Optimization pp. 105–145 (2005)
5. Huband, S., Hingston, P., Barone, L., While, R.L.: A review of multiobjective test problems and a scalable test problem toolkit. IEEE Trans. Evolutionary Computation 10(5), 477–506 (2006), http://dx.doi.org/10.1109/TEVC.2005.861417
6. Bader, J., Zitzler, E.: Hype: An algorithm for fast hypervolume-based many-objective optimization. Evolutionary Computation 19(1), 45–76 (2011), http://dx.doi.org/10.1162/EVCO_a_00009
7. Zitzler, E., Künzli, S.: Indicator-Based Selection in Multiobjective Search, pp. 832–842. Springer Berlin Heidelberg, Berlin, Heidelberg (2004), http://dx.doi.org/10.1007/978-3-540-30217-9_84
8. Hernández Gómez, R., Coello Coello, C.A.: Improved metaheuristic based on the r2 indicator for many-objective optimization. In: Proceedings of the 2015 Annual Conference on Genetic and Evolutionary Computation. pp. 679–686. GECCO '15, ACM, New York, NY, USA (2015), http://doi.acm.org/10.1145/2739480.2754776

9. Qiu, X., Xu, J.X., Tan, K.C., Abbass, H.A.: Adaptive cross-generation differential evolution operators for multiobjective optimization. IEEE Transactions on Evolutionary Computation 20(2), 232–244 (2016)
10. Li, K., Fialho, Kwong, S., Zhang, Q.: Adaptive operator selection with bandits for a multiobjective evolutionary algorithm based on decomposition. IEEE Transactions on Evolutionary Computation 18(1), 114–130 (2014)
11. He, Z., Yen, G.G.: Visualization and performance metric in many-objective optimization. IEEE Trans. Evolutionary Computation 20(3), 386–402 (2016), http://dx.doi.org/10.1109/TEVC.2015.2472283

Classification of Electrical Home Appliances Based on Harmonic Analysis Using ANN

Bighnaraj Panda, Madhusmita Mohanty and Bidyadhar Rout

Abstract This paper proposes the study of different electrical parameters to identify the demand characteristics of several electrical home appliances. Voltage and current signals of various loads are collected with respect to time using Digital Storage Oscilloscope (DSO) connected with computer. Significant parameters of the loads containing different harmonics are calculated using Fourier Series Analysis (FSA). These parameters are then used to characterize the loads. The different electrical load identification models are prepared by processing these parameters with the implementation of Backpropagation Artificial Neural Network (BP-ANN). This identification of loads will help the utility to properly manage the usage of the energy by putting Time of Use tariff (TOU) for the electrical home appliances. Thus, a better demand-side management of electrical energy can be obtained.

Keywords Fourier series analysis · BP-ANN and TOU

1 Introduction

In the present times, the energy consumption is of great concern to the whole world because of depletion of fossil fuel reservoirs. Further, energy generations from different renewable energy sources are growing by leaps and bounds. However, a small measure of managing energy consumption will not only augment the effective use of natural resources but will help in a large way to make the small-scale rooftop installations of distributed generations viable. This can be done by using demand-side management [1], which means managing the loads on the consumer

B. Panda (✉)
Gandhi Engineering College, Bhubaneswar, Odisha, India
e-mail: bighnarajpanda90@gmail.com

M. Mohanty
NIT Puducherry, Karaikal, Puducherry, India

B. Rout
VSSUT, Burla, Odisha, India

C. R. Panigrahi et al. (eds.), *Progress in Advanced Computing and Intelligent Engineering*, Advances in Intelligent Systems and Computing 714,
https://doi.org/10.1007/978-981-13-0224-4_25

side according to the availability of supply. For demand-side management different tariffs have to be defined for different loads at different times. Different tariffs for different loads can only be defined only when the loads are identified and classified into different groups of tariffs.

Studies on load identification were being conducted since 1970s and 1980s. In 1980s, Fred Schweppes and George hart [2] at MIT also developed some approaches for nonintrusive monitoring of load, which had its origins in load monitoring for residential buildings. Strategies for nonintrusive monitoring have been developed over the last 20 years [3]. However, many non-computational tools useful in practical and field-based Nonintrusive Load Monitoring (NILM) system have been developed due to advances in computing technology. The growing use of high-efficiency smart appliances necessitates review of the high-performance nonintrusive load and diagnostic monitoring techniques. Under Hart's scheme, the operating schedules of individual loads or groups of loads are determined by identifying times at which electrical power measurements change from one nearly constant (steady-state) value to another. The events which correspond to either turn ON or turn OFF of the load are characterized by the magnitude and sign of the real and reactive power as the steady-state changes. Pairing of events to establish the operating cycles and energy consumption of different electrical loads are done with the events having equal magnitudes and opposite signs. The commercial version of the Hart's work has been done in 1999 using the steady-state changes of the loads.

A recent review and classification of the forecasting methods has been given by Alfares and Nazeeruddin [4], where novel methods including fuzzy logic, genetic algorithms, and neural networks [5] have been included in addition to the conventional econometric models [6]. Load signatures have been developed and used to classify and identify the loads based on their V-I trajectories [7]. For proper classification and identification of electrical home appliances, modeling technique is of the essence in data collection, parameter extraction using their spectral coefficients [8] and training the neural network with these parameters [9]. In [10], the authors propose a model for scheduling of electrical loads in which the household electrical appliances are divided into two categories, that is, constant and one-shot electrical loads depending upon their power consumption period. The constant loads have their power consumption constant over a period of time, and one-shot loads consume power for a short period of time.

In this work, the authors acquired the voltage and current samples of various electrical household appliances using a pc interfaced digital storage oscilloscope in order to characterize model identification system using backpropagation-based Artificial Neural Network (ANN).

The paper organization is as follows: Sect. 2.1 of Sect. 2 describes the measurement methodology and its processing using Fourier series analysis. Different spectral coefficients are used to calculate the active power, reactive power, peak voltage, and peak current and total harmonic distortion for different harmonics of the signal. Section B discusses artificial neural network implementation for using the electrical parameters calculated from Sect. 2.1. Backpropagation algorithm is used to assign or adjust the weights while learning the neurons of the network. After the training is completed, the neurons are provided with test data as inputs to identify the load. The results and discussion are provided in Sect. 3.

2 Methodology

The process organization steps involve data acquisition and preprocessing, the ANN model formulation, training of the ANN model with load sample signatures, and testing of the random samples for load identification and storage for future coordination for an effective TOU tariff objective. The process flowchart is as follows (Fig. 1).

2.1 *Preprocessing of Data*

The loads to be studied under this research are the typical electrical household appliances. Most of the household electrical appliances are resistive and electronic type. Inductive loads such as ceiling fan and CRT monitor are also included. Desktop computer in standby mode, desktop computer with scanner, CFL lamp, audio system running, and audio system standby, and some other electronic loads are also taken. Voltage and current of the loads with respect to time are taken using digital storage oscilloscope (DSO). Two terminals of the load are connected as input to the digital storage oscilloscope and the output terminals of DSO are

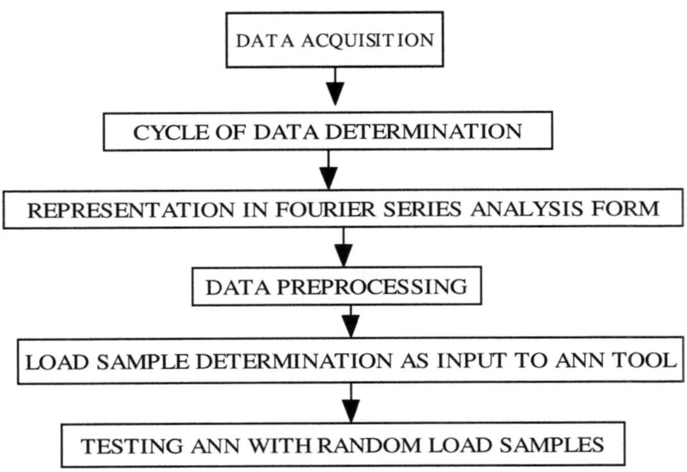

Fig. 1 Flowchart of the process

connected to a laptop for the storage of waveforms. These readings are preprocessed using MATLAB in order to be represented in Fourier analysis form as in Eqs. (1) and (2):

$$v(t) = \sum_{k=1}^{n} V_{mk} \cos(k\omega t + phi) \tag{1}$$

$$i(t) = \sum_{k=1}^{n} I_{mk} \cos(k\omega t + theta) \tag{2}$$

From the voltage and current signals as represented in Eqs. (1) and (2), all the spectral coefficients are calculated using Fourier series analysis as per Eqs. (3)–(8):

$$a_{vk} = \left(\frac{2}{T}\right) \int_{t-T}^{t} v(\tau) \cos(k\omega\tau) d\tau \tag{3}$$

$$a_{ik} = \left(\frac{2}{T}\right) \int_{t-T}^{t} i(\tau) \cos(k\omega\tau) d\tau \tag{4}$$

$$b_{vk} = (2/T) \int_{t-T}^{t} v(\tau) \sin(k\omega\tau) d\tau \tag{5}$$

$$b_{ik} = (2/T) \int_{t-T}^{t} i(\tau) \sin(k\omega\tau) d\tau \tag{6}$$

$$c_{vk} = \sqrt{a_{vk}^2 + b_{vk}^2} \tag{7}$$

$$c_{ik} = \sqrt{a_{ik}^2 + b_{ik}^2} \tag{8}$$

where

k	harmonic index
a_{ik} *and* a_{vk}	in-phase spectral coefficients
b_{ik} *and* b_{vk}	quadrature phase spectral coefficients
c_{ik} *and* c_{vk}	resultant coefficient

The spectral coefficients calculated from different voltage and current signals are used to determine the active power, reactive power, peak voltage, and peak current [11] of different harmonics of the given signal as in Eqs. (11)–(14):

$$phi = \tan^{-1}(b_{vk}/a_{vk}) \tag{9}$$

$$theta = \tan^{-1}(b_{ik}/a_{ik}) \tag{10}$$

$$If\ 0 < (phi - theta) < 90$$

$$P_k = V_{mk}{*}I_{mk}{*}\cos(phi - theta) \tag{11}$$

$$Q_k = V_{mk}{*}I_{mk}{*}\sin(phi - theta) \tag{12}$$

$$if\ (phi - theta) > 90$$

$$P_k = \left(\frac{V_{mk}{*}I_{mk}}{2}\right) * \cos(phi + theta) \tag{13}$$

$$Q_k = \left(\frac{V_{mk}{*}I_{mk}}{2}\right) * \sin(phi + theta) \tag{14}$$

The total harmonic distortions (THD) of voltage and current signals of different loads are also determined from Eqs. (15) and (16)

$$THD_V = \frac{V_k}{V_1} \tag{15}$$

$$THD_I = \frac{I_k}{I_1} \tag{16}$$

where

V_k Peak harmonic voltage
I_k Peak harmonics current
V_1 Fundamental component of voltage
I_1 Fundamental component of current.

2.2 *Artificial Neural Network Model*

The parameters of different loads calculated from Fourier series analysis are fed into artificial neural network. The network is trained with the load parameters. For training the network with input and target pairs, the parameter signature and the corresponding load type are provided as a vector-scalar combination pair, i.e., the names of electrical home appliance are set as target values and the parameters calculated from Fourier series analysis are the input values. There are three different methods to train the ANN in a multilayer perceptron network, viz., Backpropagation algorithm, Bayesian regularization, and scaled conjugate gradient. Backpropagation algorithm is preferred as it is simple and converges within less time. The number of hidden layers in the multilayer perceptron network is decided depending upon the performance and regression analysis. This is a hit-and-trial method, where arbitrary numbers of neurons are taken in the hidden layer and the network performance is studied; the regression curve should be nearer to one. The value of hidden layer for which all training, validations, and testing are satisfied is to be selected. While characterizing the hidden layer, the training process is undertaken with weight adjustment through the backpropagation algorithm.

For testing, an unknown parameter signature (dataset) is given as input. When this input file is processed with the trained weight matrix, the weighted sum passes through the threshold nonlinearity function and the output classifies into a particular load class depending upon the proximity of the processed output to the predefined class targets.

For each load, the training, validation, and testing data set are prepared by the value calculated from the MATLAB program. A neural network simulation program was designed using MATLAB. A laptop with Intel Core i3 central processing unit was used for both data acquisition and the MATLAB-based ANN implementation. For training purpose, 62 different neurons were used in the input layer, 21 neurons used in the hidden layer, and 8 neurons are used in the output layer. The input to the input layer neurons are the different parameters of the electrical loads calculated from the MATLAB program, and the initial weights for different neurons were randomly selected. Training for every electrical load was done according to the backpropagation algorithm till the mean square error (MSE) is less than 0.001 and the regression and the performance curves reach 1.

3 Results and Discussion

Experimental datasets were generated using Fourier series analysis of voltage and current waveform of the loads. Each sample consists of (62 * 1) parameters obtained for the particular load, which defines the load signature. To confirm the inferential power of the neural network, the full raw dataset creates (62 * N) matrix as training data and the test dataset, respectively. The full input dataset (62 * N) matrix includes the training, validation, and testing dataset. Here, 62 is the input parameters calculated from Fourier series analysis in the MATLAB program and N is the number of electrical loads. The parameters include the active power, reactive power, peak voltages, and peak currents of 15 harmonics and the total harmonics distortion of voltage and current signals.

Active powers of all the harmonics from first to fifteenth are P_1, P_2, P_3, P_4, P_5, P_6, P_7, P_8, P_9, P_{10}, P_{11}, P_{12}, P_{13}, P_{14}, and P_{15}; the reactive powers of all the harmonics from first to fifteenth are Q_1, Q_2, Q_3, Q_4, Q_5, Q_6, Q_7, Q_8, Q_9, Q_{10}, Q_{11}, Q_{12}, Q_{13}, Q_{14}, and Q_{15}; the peak harmonic voltages from first to fifteenth are V_1, V_2, V_3, V_4, V_5, V_6, V_7, V_8, V_9, V_{10}, V_{11}, V_{12}, V_{13}, V_{14}, and V_{15}; the peak currents

Table 1 Output and the target values in BP-ANN

Loads	Target value	Output value (x)
Audio system running	1	0.5 < x < 1.6
Audio system standby	2	1.6 < x < 2.2
CFL lamp	3	2.2 < x < 3.5
CRT monitor	4	3.5 < x < 4.6
Desktop computer	5	4.5 < x < 5.5
Desktop computer standby mode	6	5.5 < x < 6.5
Desktop computer with scanner	7	6.5 < x < 7.6
Ceiling fan	8	7.6 < x < 8.4

of all the harmonics from first to fifteenth are I_1, I_2, I_3, I_4, I_5, I_6, I_7, I_8, I_9, I_{10}, I_{11}, I_{12}, I_{13}, I_{14}, and I_{15}; and the total harmonic distortions of current and voltage which were calculated from the Fourier series analysis are given as input to the input layer neurons.

Some numerical values (i.e., number codes) should be assigned to different load signatures, so that the output can be compared numerically. We take audio system running as "1", audio system standby as "2", CFL lamp as "3", CRT monitor as "4", desktop computer as "5", desktop computer in standby mode as "6", desktop Computer with scanner as "7", and ceiling fan as "8". These numbers "1", "2", "3", "4", "5", "6", "7", and "8" are used as target values. The neural network has to learn that the parameters calculated from the Fourier series analysis (i.e., Active power, reactive power, peak voltage, peak current, and total harmonic distortions) of voltage and current are given as input samples for each load as the characteristics of load are named "1", "2", "3", "4", "5", "6", "7", and "8".

Now, in the test simulation process, the input parameters of a random load are processed with the developed weight matrix and the outcome is compared against the target vector of different load number codes. In some cases, the output calculated is not nearer to these target values, leading to errors or misclassification. For minimizing this error, the output is chosen within a range of values. The proposed algorithm in artificial neural network has better recognition accuracy for different household electrical appliances.

4 Conclusion

This paper has employed Fourier series analysis and backpropagation algorithm of artificial neural network to identify the household electrical loads. The proposed method has better recognition accuracy than the previously used methods either by using load signature method or by using voltage and current trajectory method. It is simpler and error encountered is less as it takes extensive (i.e., Sixty two) numbers of parameters of the load against only two to three parameters reported previously.

As more parameters of a load are taken, the identification accuracy is improved. The errors in this proposed method can only be attributed to voltage and current irregularities from the supply voltage and current. The results shown in Table 1 can further be coordinated toward characterizing time of use tariffs for demand-side management.

References

1. T. Logenthiran, D. Srinivasan, and T. Z. Shun: Demand side management in smart grid using heuristic optimization, IEEE Trans. on Smart Grid, Vol. 3, Sept 2012 pp. 1244–1252 (2012).
2. M. Akbar and Dr Z. A. Khan: Modified nonintrusive appliance load monitoring for nonlinear devices, IEEE International Multitopic Conf. (INMIC 2007), Lahore, Pakistan, pp. 28–30 Dec (2007).
3. C. Laughman, K. Lee, R. Cox, S. Shaw, S. Leeb, L. Norford and P. Armstrong: Power Signature Analysis, IEEE power & energy magazine,1540–7977/03 pp. 56-63.
4. Alfares and Nazeerudin: Electric load forecasting: literature survey and classification of methods, International Journal of Systems Science Vol. 33, Issue 1, 2002, pp. 23–34 (2002).
5. H. S. Hippert, C. E. Pedreira and R. C. Souza: Neural networks for short-term load forecasting: a review and evaluation, IEEE Trans. on Power Systems Vol. 16, Issue: 1 Feb 2001 pp. 44–55 (2001).
6. Pindyck, Robert, Rubinfeld and Daniel: Economic models and economic forecasts, McGraw-Hill/Irwin Book International Edition (1997).
7. H. Y. Lam, G. S. K. Fung and W. K. Lee: A Novel method to construct taxonomy of electrical appliances based on load signatures, IEEE Trans. on Consumer Electronics, Vol. 53, No. 2, MAY 2007, pp-653–660 (2007).
8. S. R. Shaw, S. B. Leeb, L. K. Norford and R. W. Cox: Nonintrusive load monitoring and diagnostics in power systems, IEEE Trans. on Instrumentation and Measurement, Vol. 57, No. 7, July 2008 pp. 1445–1454 (2008).
9. Yu-Hsiu Lin and Men-ShenTsai: Non-Intrusive load Monitoring by Novel Neuro Fuzzy Classification Considering Uncertainties, IEEE Trans. on Smart Grid, Vol. 5, No. 5, September (2014).
10. Tomas Lennvall, Larisa Rizvanovic and Pia Stoll: Scheduling of Electrical load in Home Automation Systems. IEEE International Conference on Automation Science and Engineering August 24–25, 2015, IEEE, Gothenburg, Sweden, (2015).
11. IEEE Standards Definitions for the Measurement of Electric Power Quantities under Sinusoidal, Non sinusoidal, Balanced, or Unbalanced conditions.

Multi-header Pulse Interval Modulation (MH-PIM) for Visible Light Communication System

Sandip Das and Soumitra Kumar Mandal

Abstract In this paper, a Multi-header Pulse Interval Modulation (MH-PIM) is presented for visible light communication. The paper discusses the fundamental principles of the symbol structure, its code properties, and slot error performance of the proposed scheme. Throughout the paper, the proposed scheme performance is analyzed and compared with PPM, DPIM, and DH-PIM in terms of average frame length, bandwidth requirement, and slot error performance. The system is simulated in MATLAB considering the channel adds white Gaussian noise and no interference is added from ambient light sources. It is verified from the result that MH-PIM has less average frame length, and requires less bandwidth and better slot error performance compared to other modulation schemes.

Keywords Visible light communication · Interval modulation
OOK · PPM

1 Introduction

The advancement and rapid growth of solid-state lighting technology have drawn attention of many researchers to research upon and develop various applications of Visible Light Communication (VLC) technologies and provide electromagnetic interference (EMI)-free and license-free communication [1–3]. As the wavelength

S. Das (✉)
Electronics & Communication Engineering, University of Engineering & Management, Jaipur, Rajasthan, India
e-mail: info.sandipec@gmail.com

S. K. Mandal
Electrical Engineering, NITTTR, Kolkata, West Bengal, India
e-mail: mandal_soumitra@yahoo.com

C. R. Panigrahi et al. (eds.), *Progress in Advanced Computing and Intelligent Engineering*, Advances in Intelligent Systems and Computing 714,
https://doi.org/10.1007/978-981-13-0224-4_26

decreases along the electromagnetic spectrum, the frequency as well as the energy of the wave increases. The visible band occupies a frequency range of 400–800 THz (wavelength of 380–780 nm) which is much larger compared to RF range used for present-day communication. With rapidly decreasing RF spectrum bandwidth of traditional RF communication below ~6 GHz for high data rate communication, VLC is an alternative short-range optical wireless communication technology and a solution to overcrowded RF spectrum.

In Visible Light Communication system (VLC), the information signal or its modulated version is used to directly modulate onto the optical carrier signal using intensity. Along with intensity modulation (IM), almost all practical optical communication systems employ electrical modulation [4–6]. Appropriate modulation schemes for visible light communication system depend on number of criteria such as power and bandwidth efficiencies. A few modulation schemes have been proposed in recent times which are adopted in VLC systems, such as On-Off Keying (OOK), Pulse Position Modulation (PPM), Digital Pulse Interval Modulation (DPIM), Dual Header Pulse Interval Modulation (DH-PIM), and Multilevel Digital Pulse Interval Modulation (MDPIM). OOK and PPM were first discussed in [7] for optical wireless communication, and it shows that OOK had minimum bandwidth requirement and PPM had better bit error rate performance. OOK is the simplest modulation scheme, where LEDs are turned ON or OFF depending on the data bits (1 or 0). In PPM, each symbol interval of duration T is divided into $L = 2^M$ sub-intervals or chips with duration of T/L, and only one pulse is transmitted [8]. In DPIM, the input sequence is encoded by varying the number of empty slots between the adjacent pulses, and thus symbol length is not fixed [9]. DH-PIM uses two different initial header pulses depending on the MSB of the input sequence and hence this scheme also does not have a fixed symbol length [10]. MDPIM also uses two different header pulses depending on the MSB of the input sequence but the amplitude of the pulse varies, for example, if MSB of the input sequence is "0", then the MDPIM symbol pulse starts with an amplitude "v"; if MSB of input sequence is "1", then MDPIM symbol pulse has an amplitude "2v" [11]. Among the modulation scheme discussed so far, DPIM, DH-PIM, and MDPIM do not have synchronization problem compared with OOK and PPM. Also in terms of bandwidth requirement and transmission efficiency, DPIM, DH-PIM, and MDPIM have better performance compared to OOK and PPM. Though MDPIM shows better performance compared to DPIM and DH-PIM in terms of bandwidth efficiency and transmission capacity, MDPIM may not be an appropriate choice because it may result in flicker problem.

In this paper, a Multi-header Pulse Interval Modulation (MH-PIM) is proposed where four different header pulses are used to represent the input sequence. The proposed modulation symbol structure, code properties, and its bandwidth requirement are given in Sect. 2, whereas error performance is discussed in Sect. 3, and finally the paper is concluded.

2 Proposed MH-PIM System Theory

At the transmitter, an input symbol of M bits OOK signal is mapped into an MH-PIM frame which starts with a header pulse. The header pulse is selected in a way that it represents the weight of the decimal value of two most significant bits (MSB) of the input symbol. The header pulse in a frame is followed by the number of empty time slots which represents the information carried by the frame. MH-PIM frame initiates with one of the four possible headers namely H1, H2, H3, and H4 and ends with a guard space. The pulse width duration of each header is given as follows:

$$\text{Pulse duration of Header, H1} - \frac{\alpha T_s}{4} \tag{1}$$

$$\text{Pulse duration of Header, H2} - \frac{\alpha T_s}{2} \tag{2}$$

$$\text{Pulse duration of Header, H3} - \frac{3\alpha T_s}{4} \tag{3}$$

$$\text{Pulse duration of Header, H4} - \alpha T_s \tag{4}$$

where $\alpha > 0$ and T_s is the slot duration. Thus, an MH-PIM starts with any of the four possible header and is followed by number of empty time slots $(d_n T_s)$, where $d_n \in \{0, 1, 2, \ldots (2^{(M-2)} - 1)\}$, M is the number of bits per symbol, and $\text{N} = 2^M$ symbols. MH-PIM frame with four different headers H_1, H_2, H_3, and H_4 is shown in Fig. 1.

a. Header one, H_1, b. Header two, H_2, c. Header three, H_3, d. Header four, H_4

The mapping of OOK symbol into MH-DPIM is implemented in the following ways:

1. If the decimal value of Most Significant Bit (MSB) and the bit next to MSB of the M bits OOK input signal is equal to "0", then header H_1 is used followed by information slot "d" (empty slot) which is equal to the decimal value of the remaining two least significant bits, i.e., if decimal value of the remaining two LSB is "1", then information slot d = 1, i.e., H_1 will be followed by only one zero (empty slot), if decimal value is "2", then information slot d = 2, i.e., H_1 will be followed by two zeros (empty slot) and so on. For example, in a 4-bit OOK symbol, "0001", MSB and bit next to MSB are "00", whose decimal value is "0" then header H_1 (10000) is used and decimal value of remaining two least significant bits "01" is "1"; hence, header H_1 is followed by only one zero, resulting in MH-PIM frame as "100000".
2. If the decimal value of most significant bit (MSB) and the bit next to MSB of the M bits OOK input signal is equal to "1", then H_2 is used followed by information slot "d" (empty slot) which is equal to the decimal value one's

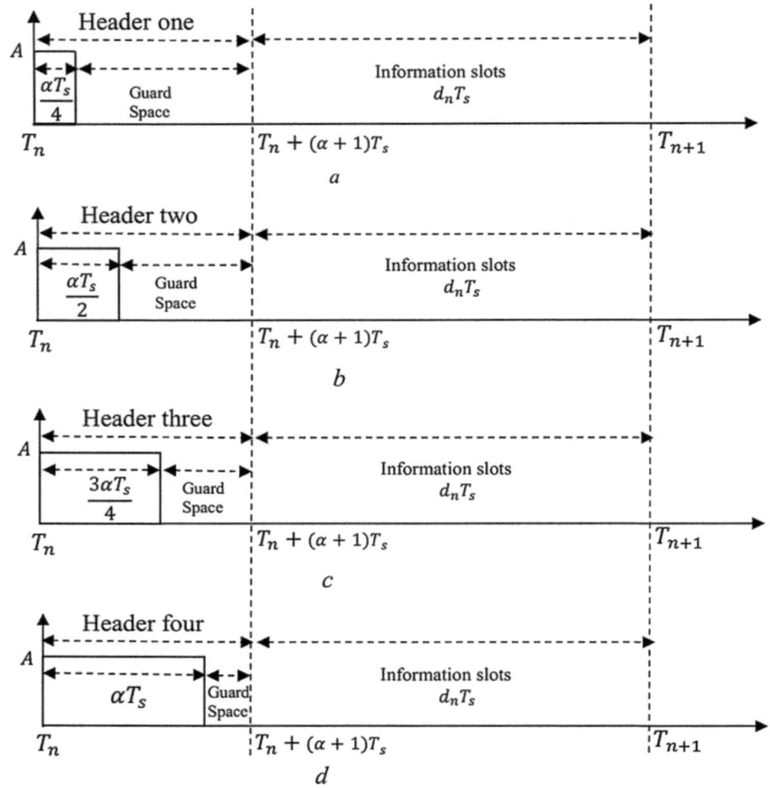

Fig. 1 The *n*th frame of MH-PIM with four different header pulses. Pulse width will vary for different values of α

complement of the remaining two least significant bits, i.e., if decimal value of the remaining two LSB is "1", then information slot d = 1, i.e., H_2 will be followed by only one zero (empty slot), if decimal value is "2", then information slot d = 2, i.e., H_2 will be followed by two zeros (empty slot), and so on. For example, in a 4-bit OOK symbol, 0110, MSB and bit next to MSB is "01", whose decimal value is "1", then header H_2 (11000) is used and decimal value of one's complement of remaining least significant bits "10" is "1"; hence, header H_2 is followed by only one zero, resulting in MH-PIM frame as "110000".

3. If the decimal value of most significant bit (MSB) and the bit next to MSB of the M bits OOK input signal is equal to "2", then H_3 is used followed by information slot "d" which is equal to the decimal value of the remaining two least significant bits, i.e., if decimal value of the remaining two LSB is "1", then information slot d = 1, i.e., H_3 will be followed by only one zero (empty slot), if decimal value is "2", then information slot d = 2, i.e., H_3 will be followed by

two zeros (empty slot), and so on. For example, in a 4-bit OOK symbol, 1001, MSB and bit next to MSB is "10", whose decimal value is "2", then header H_3 (11100) is used and decimal value of remaining two least significant bits "01" is "1"; hence, header H_3 is followed by only one zero, resulting in MH-PIM frame as "111000".

4. If the decimal value of most significant bit (MSB) and the bit next to MSB of the M bits OOK input signal is equal to "3", then H_4 is used followed by information slot "d" which is equal to the decimal value one's complement of the remaining two least significant bits, i.e., if decimal value is "1", then information slot d = 1, i.e., H_4 will be followed by only one zero, if decimal value is "2", then information slot d = 2, i.e., H_4 will be followed by two zeros, and so on. For example, in a 4-bit OOK symbol, 1101, MSB and bit next to MSB is "11", whose decimal value is "3", then header H_4 (11110) is used and decimal value of one's complement of remaining two least significant bits "10" is "2"; hence, header H_4 is followed by only two zero, resulting in MH-PIM frame as "1111000".

3 Code Properties of MH-PIM

The *n*th frame structure of MH-PIM shown in Fig. 1 is expressed using rectangular pulse function starting at $t=T_n$ and has a duration of $\tau=(1+h_n)\frac{\alpha T_s}{4}$, where $h_n \in \{0,1,2,3\}$ indicating header H_1, H_2, H_3, and H_4, respectively, and n represents the instantaneous frame number. Thus, MH-PIM signal can be expressed as

$$x(t)=A\sum\nolimits_{n=0}^{\infty} rect\left[\frac{4(t-T_n)}{\alpha T_s}-\frac{1}{2}\right]+h_n rect\left[\frac{4(t-T_n)}{\alpha T_s}-\frac{3}{2}\right] \tag{5}$$

where A is the amplitude of the pulse, and rectangular pulse function is defined as

$$rect(u)=\begin{cases} 1; & -0.5<u<0.5 \\ 0; & otherwise \end{cases} \tag{6}$$

The start of the *n*th MH-PIM frame is given by

$$T_n=T_0+T_s\left[n(\alpha+1)+\sum\nolimits_{k=0}^{n-1} d_k\right] \tag{7}$$

where $d_k \in \{0,1,2,\ldots,(2^{(M-2)}-1)\}$ represents the number of time slots in the *n*th frame of the *k*th symbol and is the start time of the first pulse at n = 0.

The mapping of all possible combinations of 4-bit OOK code word into 16-MH-PIM for $\alpha = 1$ is shown in Table 1.

Table 1 Mapping of OOK code word into 16-PPM, 16-DPIM, 16-DH-PIM, MH-PIM

Sl. No.	OOK	16-PPM	16-DPIM	16-DH-PIM$_2$	16-MH-PIM$_1$
1	0000	1000000000000000	10	100	10000
2	0001	0100000000000000	100	1000	100000
3	0010	0010000000000000	1000	10000	1000000
4	0011	0001000000000000	10000	100000	10000000
5	0100	0000100000000000	100000	1000000	11000000
6	0101	0000010000000000	1000000	10000000	1100000
7	0110	0000001000000000	10000000	100000000	110000
8	0111	0000000100000000	100000000	1000000000	11000
9	1000	0000000010000000	1000000000	1100000000	11100
10	1001	0000000001000000	10000000000	110000000	111000
11	1010	0000000000100000	100000000000	11000000	1110000
12	1011	0000000000010000	1000000000000	1100000	11100000
13	1100	0000000000001000	10000000000000	110000	11110000
14	1101	0000000000000100	100000000000000	11000	1111000
15	1110	0000000000000010	1000000000000000	1100	111100
16	1111	0000000000000001	10000000000000000	110	11110

It is seen that MH-PIM effectively removes the redundant time slots that follow the pulse in PPM. Clearly, the average symbol length in MH-PIM reduces when compared with DPIM; hence, it increases the data throughput. The minimum, maximum, and average frame length of MH-PIM is given as

$$L_{min} = \left(2^{\frac{M}{2}} + \alpha\right) T_s \tag{8}$$

$$L_{max} = \left(2^{\frac{M}{2}} + \alpha + 3\right) T_s \tag{9}$$

$$\bar{L} = \frac{\left[2\alpha + 3 + 2^{\left(\frac{M}{2}+1\right)}\right] T_s}{2} \tag{10}$$

The average frame length of PPM, DPIM, DH-PIM$_2$, and MH-PIM is tabulated in Table 2 and comparison of the average frame length is plotted in Fig. 2. It is observed from Fig. 2 that as the order of M increases, the average frame length increases for PPM, DPIM, and DH-PIM$_2$ but MH-PIM has minimum average length compared to all other modulation schemes.

Considering the time slot duration of OOK as τ_{OOK}, the average time slot durations of PPM, DPIM, and DH-PIM$_2$ are $\frac{M}{2^M}\tau_{OOK}$, $\frac{2M}{2^M+3}\tau_{OOK}$, and $\frac{2M}{2^{M-1}+2\alpha+1}\tau_{OOK}$, whereas, in case of MH-PIM, the average time slot duration is $\frac{2M}{2\alpha+3+2^{\left(\frac{M}{2}+1\right)}}\tau_{OOK}$, which is same as DH-PIM$_2$. Thus, using $B_{OOK} = R_b = \frac{1}{\tau_{OOK}}$ to

Table 2 Average frame length of PPM, DPIM, DH-PIM$_2$, and MH-PIM

Modulation schemes	Average frame length
PPM	2^M
DPIM	$\frac{2^M+3}{2}$
DH-PIM$_2$	$\frac{2^{M-1}+2\alpha+1}{2}$
MH-PIM	$\frac{2\alpha+3+2^{\left(\frac{M}{2}+1\right)}}{2}$

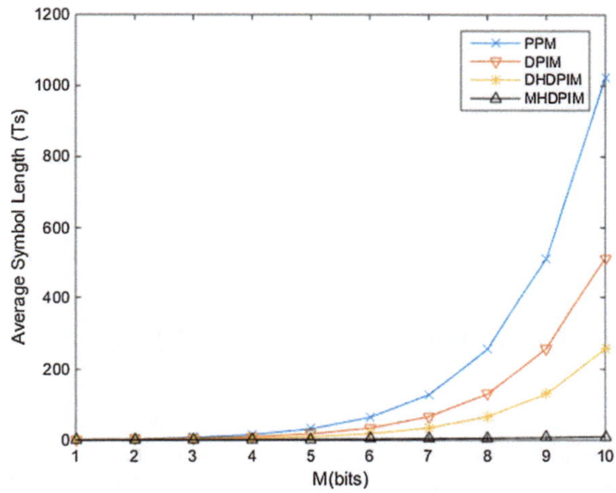

Fig. 2 Comparison of average frame length of PPM, DPIM, DH-PIM$_2$, and MH-PIM

Table 3 Bandwidths of PPM, DPIM, DH-PIM, and MH-PIM for a given R_b

Modulation schemes	Bandwidth
OOK	R_b
PPM	$\frac{2^M}{M}R_b$
DPIM	$\frac{2^M+3}{2M}R_b$
DH-PIM$_2$	$\frac{2^{M-1}+2\alpha+1}{2M}R_b$
MH-PIM	$\frac{2\alpha+3+2^{\left(\frac{M}{2}+1\right)}}{2M}R_b$

represent the bandwidth of OOK, the bandwidths of PPM, DPIM, DH-PIM$_2$, and MH-PIM are deduced and shown in Table 3.

Figure 3 shows the comparison of bandwidth requirement of PPM, DPIM, DH-PIM2, and MH-PIM and it is clear from Fig. 3 that for M > 2 the bandwidth requirement increases. The highest bandwidth requirement is for PPM followed by DPIM and DH-PIM$_2$, whereas MH-PIM requires very low bandwidth and, hence, has high data throughput compared to other modulation scheme discussed so far.

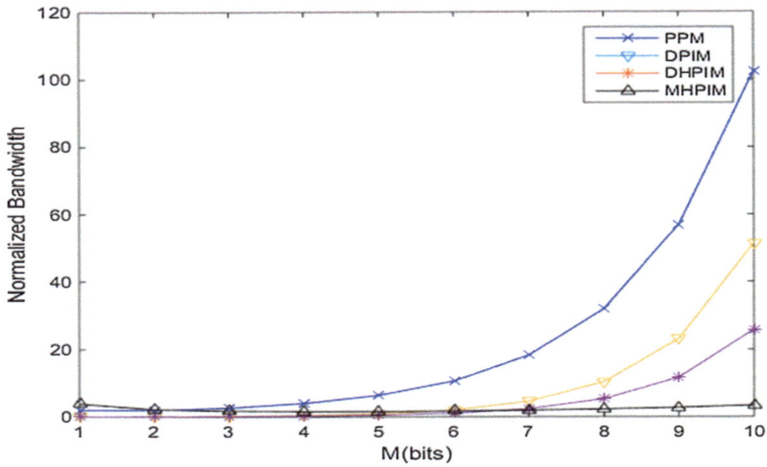

Fig. 3 Bandwidth requirement of PPM, DPIM, DH-PIM, and MH-PIM

4 Error Performance of MH-PIM

In Fig. 4, a block diagram of proposed MH-PIM system is shown where the input signal is composed of random binary bits of "1" and "0". The input signal is mapped into MH-PIM frame by the MH-PIM modulator. This MH-PIM signal is added with white noise and fed into a matched filter. The matched filter output signal is sampled at slot frequency $f_s = \frac{1}{T_s}$, followed by a decision circuit which interprets the received bits as "1" or "0" depending on the received optical power, which is fed into the MH-PIM demodulator to recover the original signal.

Assuming the channel as a distortion-free channel, no interference from ambient light and the dominant source of noise as background shot noise, error performance of proposed system is studied in this section. Considering the average received optical power as $\bar{P}$, photodetector responsivity as R, and an equiprobable occurrence of the header H_1, H_2, H_3, and H_4, the peak photocurrent can be given as

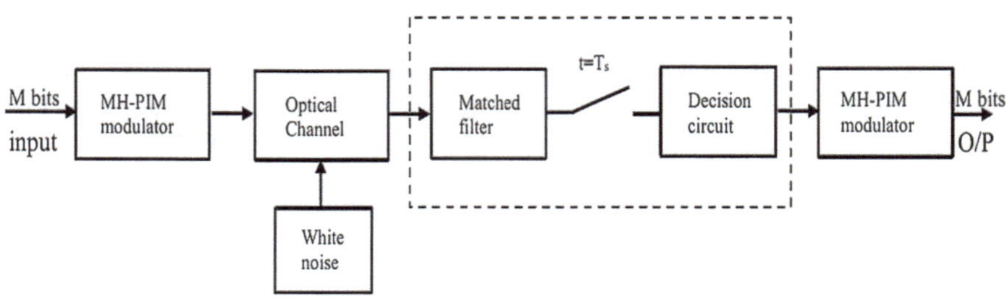

Fig. 4 Block diagram of MH-PIM system

$$i_{peak} = \frac{5\bar{L}}{4\alpha} R\bar{P} \tag{11}$$

Thus, energy received at the matched filter output is given by

$$E = \begin{cases} i^2_{peak} T_s; & \textit{if 1 is sent} \\ 0; & \textit{if 0 is sent} \end{cases}$$

$$E = \frac{25\bar{L}R^2\bar{P}^2 M}{16\alpha^2 R_b} \tag{12}$$

where R_b, is the bit transmission rate.

Now, considering η as the noise spectral density, the signal-to-noise ratio of OOK is $SNR_{OOK} = \frac{2R\bar{P}^2}{\eta R_b}$ and hence, the slot error probability of MH-PIM normalized to OOK can be given as

$$P_{se} = Q\left(\sqrt{\frac{25\bar{L}RM}{32\alpha^2} SNR_{OOK}}\right) \tag{13}$$

The slot error performance of MH-PIM is thus calculated and compared with PPM, DPIM, and DH-PIM2 as shown in Fig. 5. It is clear from the figure that MH-PIM has better performance compared to DPIM and DH-PIM2 but it is inferior to PPM.

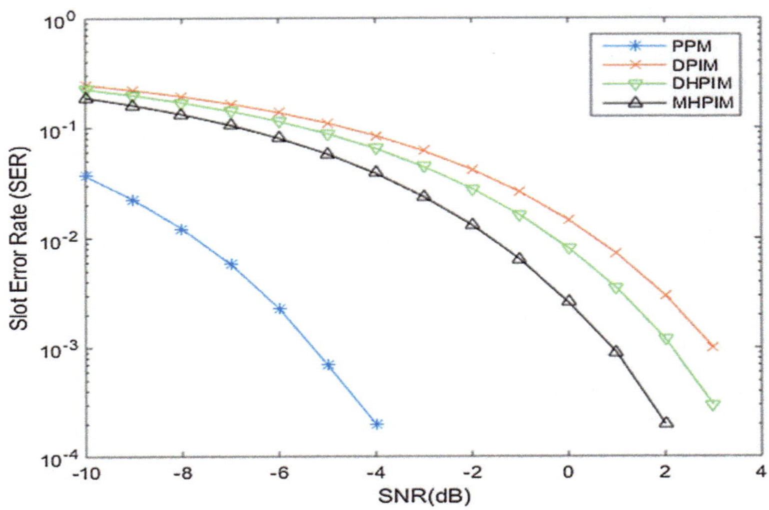

Fig. 5 Slot error performance of PPM, DPIM, DH-PIM2, and MH-PIM versus SNR (normalized to OOK)

5 Conclusion

In this paper, a new modulation scheme named MH-PIM is proposed for visible light communication. The symbol structure and ways to implement the mapping of OOK symbol into MH-PIM is discussed thoroughly. The proposed scheme's average frame length, bandwidth requirement, and slot error performance are compared and analyzed with PPM, DPIM, and DH-PIM. Simulation result shows that MH-PIM has better performance in terms of average frame length and bandwidth requirement compared to PPM, DPIM, and DH-PIM. MH-PIM has significantly reduced slot error rate compared with DPIM and DH-PIM. From the simulation result, it is seen that PPM has better slot error rate performance than DPIM, DH-PIM, and MH-PIM but lacks in average frame length and bandwidth requirement. Thus, analyzing all the categories, it is concluded that proposed MH-PIM is better than PPM, DPIM, and DH-PIM.

References

1. Komine T., Nakagawa M.: "Fundamental analysis for visible-light communication system using LED lights," IEEE Transactions on Consumer Electronics, 2014, 50(1), pp. 100–107.
2. Sewaiwar A., Tiwari S., Chung Y.H., "Novel user allocation scheme for full duplex multiuser bidirectional Li-Fi network," Optical Communication, 2015, 339, pp. 153–156.
3. Bandara A., Chung Y.H., "Novel color-clustered multiuser visible light communication," Trans. Emerging Telecommunication Technology, 2014, 25(6), pp. 579–590.
4. J.R. Barry, "Wireless Infrared Communications," Boston, MA: Kluwer, 1994.
5. F.R. Gfeller and U.H. Bapst, "Wireless in-house data communication via diffuse infrared radiation," Proc. IEEE, vol. 67, pp. 1474–1486, Nov. 1979.
6. D.J.T. Heatley, D.R. Wisely, I. Neild, and P. Cochrane, "Optical wireless: the story so far," IEEE Commun. Mag., vol. 36, pp. 72–74, 79-82, Dec. 1998.
7. Joseph M. Kahn, John R. Barry, "Wireless Infrared Communications", Proceedings of the IEEE, Vol. 85, No. 2, pp. 265–298, 1997.
8. Ma, Xiaoxue; Lee, Kyujin; Lee, Kyesan: 'Appropriate modulation scheme for visible light communication systems considering illumination', Electronics Letters, 2012, 48, (18), p. 1137–1139.
9. Z. Ghassemlooy, A. R. Hayes, N. L. Seed and E. D. Kaluarachchi, "Digital pulse interval modulation for optical communications," in IEEE Communications Magazine, vol. 36, no. 12, pp. 95–99, Dec 1998.
10. N.M. Aldibbiat, Z. Ghassemlooy and R. McLaughlin, "Dual header pulse interval modulation for dispersive indoor optical wireless communication systems," IEE Proceedings Circuits, Devices and Systems, vol. 148, no. 3, pp. 187–192, 2002.
11. Z. Ghassemlooy and N. M. Aldibbiat, "Multilevel Digital Pulse Interval Modulation Scheme for Optical Wireless Communications," 2006 International Conference on Transparent Optical Networks, Nottingham, 2006, pp. 149–153.

Text-to-Speech Synthesis System for Mymensinghiya Dialect of Bangla Language

Afruza Begum, S. Md. S. Askari and Utpal Sharma

Abstract Speech is the most popular form of communication medium in everyday life. Mymensingia dialect is a dialect of Bangla language. In this paper, we have developed a text-to-speech synthesis system (TTS) for the Mymensingia dialect of Bangla language using Festival speech synthesis toolkit. We have shown here how this Mymensingia dialect is phonologically and lexically different from standard Bangla Language. An attempt is made to convert the synthetic voice of Mymensingia dialect produced from the developed TTS to standard Bangla, covering some selected prosodic features of Mymensingia.

Keywords Text to speech · Prosody · Pitch · MFCC

1 Introduction

Speech is the most widely used and most popular form of communication medium in human society. A text-to-speech (TTS) system helps a large group of people mainly visually impaired and illiterate people to overcome human–computer interaction problem. It helps them to read online news, e-books, car navigation systems and enhancing other information systems. Typically, there are two main components in a TTS system [1, 2]: *(a) Text analysis. (b) Speech waveform Generation* these are

A. Begum
Department of Electronics and Communication Engineering, NERIST,
Itanagar 791109, India
e-mail: afruza.moon@gmail.com

S. Md. S. Askari (✉)
Department of Computer Science and Engineering, Rajiv Gandhi University,
Itanagar 791112, India
e-mail: askari.sikdar@gmail.com

U. Sharma
Department of Computer Science and Engineering, Tezpur University,
Tezpur 784028, India
e-mail: utpal@tezu.ernet.in

C. R. Panigrahi et al. (eds.), *Progress in Advanced Computing and Intelligent Engineering*, Advances in Intelligent Systems and Computing 714,
https://doi.org/10.1007/978-981-13-0224-4_27

also called front end and back end, respectively [3]. The input text is converted into a linguistic specification, which will consist components like phonemes, diphones, triphones, etc., and this is done in the text analysis component [4]. Speech waveforms are generated in the speech waveform generation component, from the produced linguistic specification [2, 4]. Speech synthesis systems are mainly divided into four categories [3]. Articulatory synthesis, formant synthesis, concatenative speech synthesis and statistical parametric speech synthesis which is also called hidden Markov model based synthesis [5].

Although attempts have been made for Bangla speech synthesis, the Mymensingia dialect of Bangla language is distinct from standard Bangla. Other dialects of Bangla language are Sylheti, Rongpori, Barisali, etc. These dialects are mutually distinct and sometimes would not be understood by a native speaker of standard Bangla. Also, there are differences in prosody (phrasing, intonation, and duration), pronunciation, vocabulary, and other aspects. Here is an example; it shows the difference between standard Bangla and Mymensingia dialect:

English Translation: "A man had two son"
Bengali Shadhubhasha: "aek bektir duiti putro asilo"
Standard Bangla: "aek jon loker duita chhele chhilo"
Mymensingia dialect: "aek jon mayensher duida pola asilo"

A lot of works have been done on standard Bangla language. N.P. Narendra et al. [6] developed a screen reader system for Bangla language using Festival TTS synthesizer engine for unrestricted domain. Shankar Mukherjee et al. [7] built a TTS system for Bengali language. They compared the output with the previously developed epoch synchronous non-overlap add (ES-NOLA) based concatenative speech synthesis technique [8]. The total average score for the original sentences was 4.66 and the ES-NOLA based synthesis sentence was 2.3 and in HTS they obtained 3.6 [7]. Divya Bansal1 et al. [9] built a HMM-based speech synthesis system for Punjabi language using HTK toolkit. So far it is found from our survey that no speech synthesis system is available in Mymensingia dialect, so we are motivated to discuss lexical and phonological variations of Mynensingia dialect with standard Bangla and build a text-to-speech synthesis system for Mymensingia dialect of Bangla language.We have not come across reports of synthesizing specific dialects of an Indian language. Our objective is to build a text-to-speech synthesis system to produce natural and intelligible sound for Mymensingia dialect of Bangla language.

In this paper, we have discussed various lexical and phonological variations of Mymensingia dialect with standard Bangla and built a text-to-speech synthesis system (TTS) for Mymensingia dialect of Bangla language and besides the above, and the conversion of synthesized voice produced from developed TTS of Mymensingia dialect to standard Bangla is also attempted. Festival TTS (text-to-speech synthesis) toolkit is used here to develop the text-to-speech synthesis system for Mymensingia dialect of Bangla. It is a very well documented, popular, and tested by many users. The Festival speech synthesis system is a general multilingual speech synthesis system. It offers a general frame work for building text-to-speech synthesis system. It was developed by Alan W. Black et al. [10] at Center Speech Technology Research at the University of Edinburgh. It is designed in such a way that it supports multiple languages like English, Japanese, Hindi, Bangla, Tamil, Marathi etc. Complete Festival speech synthesis tools consist of speech tools and Festvox [4]. Festival TTS synthesis system has mainly three modules, text analysis module, linguistic analysis module, and speech synthesis engine (waveform generation module). Here when the text is entered, according to the given text, prerecorded best speech sounds are selected from the speech database and those sounds are concatenated to produce the required output [11]. In text analysis phase, from the raw text basic utterances and words are identified, in linguistic analysis phase, pronunciation of the word and prosody structures (phrasing, intonation, and duration) are assigned to those words and in waveform generation phase, fully specified waveform is generated for the corresponding text [12].

We have organized this paper as follows, Sect. 2 describes the various phonological and lexical variations of Mymensingia dialect with Standard Bangla, Sect. 3 describes the architecture of Festival TTS; how it works and implementation details of TTS for Mymensingia dialect, Sect. 4 describes how the conversion of Mymensingia dialect to standard Bangla is carried out, Sect. 5 discusses the result and Sect. 6 concludes the paper.

2 Lexical and Phonological Variation of Mymensingia Dialect and Standard Bangla

All the language changes over time and changes according to the place, culture and religion. Sometimes a native speaker is not able to understand the dialects of the same language. Same thing happens with Bangla language, there are huge lexical variation in word and phrases of Mymensingia dialect with standard Bangla. Some examples of lexical alternation of Mymensingia dialect and standard Bangla are as follows:

1. Invitation: “nimontron/nimontrono” in standard Bangla corresponds to “jiapot” in Mymensingia dialect.
2. Meat: “mangsho” in standard Bangla corresponds to “gosto” in Mymensingia dialect.
3. Chili: “longka” in standard Bangla corresponds to “morich” in Mymensingia dialect.
4. Picture: “chobi” in Standard bangla corresponds to “phodo” in Mymensingia dialect.
5. Son: “chele” in Standard Bangla corresponds to “pola” in Mymensingia dialect.
6. Boat: “nowka” in standrard bangla corresponds to “naw” in Mymensingia dialect.
7. Persons: “loker” in Standard bangla corresponds to “mynsher” in Mymensingia dialect.
8. Say: “bolen/bolben” in standard Bangla corresponds to “kon/kow/koiben” in Mymensingia dialect.

9. Bath: "sunkora/gawdhowa" in standard Bangla corresponds to "dubdewa" in Mymensingia dialect.
10. Our: "amader" in standard Bangla corresponds to "ango" in Mymensingia dialect.
11. How: "kemon" in standard Bangla corresponds to "Kiba" in Mymensingia dialect.

The phoneme \t\is palato-alveolar, plosive, unaspirated & voiceless consonant in standard Bangla is altered to \d\palato-alveolar, plosive, unaspirated & voiced consonant in Mymensingia dialect [13]. For example, moda to mota (fat), ata to ada (flour), ati to adi (bundel). \bh\bilabial, plosive, aspirated & voiceless of standard Bangla phoneme altered to \b\bilabial, plosive, unaspirated & voiced consonant in Mymensingia dialect. \p\is the bilabial, plosive, voiceless & aspirated stop consonant, \ph\is altered to \s\alveolar, voiceless, aspirated & fricatives. For example Iftar to Istar (Iftar). Some phonological variations of Mymensingia dialect with standard bangla are as follows:

1. Rice: "dhan" in standard Bangla to "dan" in Mymensingia dialect.
2. Head: "matha" in standard Bangla "mata" in Mymensingia dialect.
3. Stick: "lathi" in standard Bangla to "ladi" in Mymensingia dialect.
4. Borrow: "dhar" in standard Bangla to "daar" in Mymensingia dialect.
5. Learn: "shikha" in standard Bangla to "hika" in Mymensingia dialect.
6. Profit: "labh" in standard Bangla to "lab" in Mymensingia dialect.
7. Iftar: "iftar" in standard Bangla to "istar" in Mynensingia dialect.
8. Money: "taka" in standard Bangla to "teha" in Mymensingia dialect.
9. Body: "shorir" in standard Bangla to "shoril" in Mymensingia dialect.
10. Flour: "ata" in standard Bangla to "ada" in Mymensingia dialect.

Beside the above words, there are many other words, here we have only shown some of them. We have found from our data analysis that in mymensingia dialect, pitch of the consonant is low and the pitch of the vowel is high, and the duration of pronunciation of a word in Mymensingia dialect is longer than standard Bangla.

3 Text-to-Speech Synthesis System for Mymensingia Dialect of Bangla Language Design and Implementation

3.1 Festival Architecture

In Festival, synthesized voice is produced in three phases as shown in Fig. 1. First text analysis phase, in this phase identification of words and basic utterances from the raw texts are done [12]. Next is the linguistic analysis phase, where finding the pronunciation of words and prosody structures (phrasing, intonation and duration) are assigned to those words [14]. The third and final phase is waveform generation

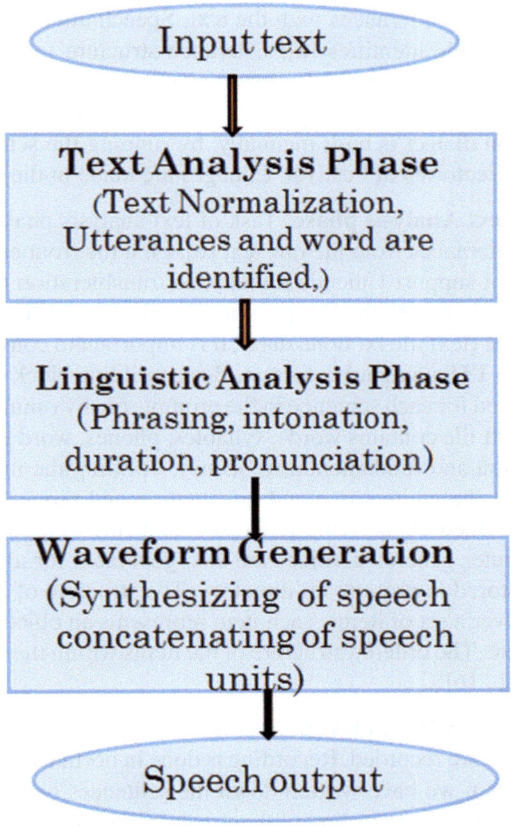

Fig. 1 Architecture of Festival TTS

phase; here fully specified waveform is generated for the corresponding text [12]. For waveform generation, database of real speeches is collected, then from the database, appropriate units are selected and wave generation is done by concatenating them [4].

3.2 A Synthesis System for Mymensingia Dialect

For building TTS system for Mymensingia dialect of Bangla first, the prompts file is created, it is the text transcription of spoken utterance and it is recorded in the next phase. The prompts file is kept in etc/ directory. Then the synthesizer front end is customized and generation of prompts is done; .utt files and computer-generated voice of all the sentences of prompts file is generated and stored in prompt_utt and prompt_wav respectively. All the sentences of prompts file is recorded and stored in wav/ directory. Autolabelling of the prompts file is done, i.e., aligning the speech utterances with the text. Speech utterances are segmented and labeled with syllable identities [6]. Utterance structure for recorded utterances is built [4] and pitchmarks and Mel cepstral coefficient are extracted. Cluster is built, after building cluster finally waveform synthesizer is built. This synthesis system for Mymensingia dialect is built manually, by running the scripts, which are available in various directories of Festival. Changes are made in the scripts according to our needs.

Text Analysis phase: Task of text analysis phase is to identify the words and basic utterances from the raw text [6]. First the front encoding is performed. Festival does not support Unicode directly, so transliteration of unicode text to ASCII code [15] according to the Bangla phone set is done. After completion of writing the prompts file (text file txt.done.data), it is important to convert the text of Mymensingia dialect to TTS compatible format. Basic building block of Festival utterance, .utt file is created for each sentence in the prompts file by running the build_idom.scm script. Each .utt file contains words, syllables, phones, word POS, syllable stress, phoneme duration and fundamental frequency. Splitting the input text (tokenizing) is done based on the white space and punctuation and pronunciation of each word in prompts is defined by applying letter-to-sound rule to the letter in the word [4]. Default computer generated voices are also generated for all the sentences of prompts file and stored in prompt_utt directory. The structure of .utt file has a collection of relations over a set of items, each item represents an object such as a word, segment, syllable, etc. The ordered structure of the items within the .utt file is defined by these Relations [1, 16].

Recording and Autolabeling: All the sentences of Mymensingia dialect in prompts file are recorded. Recording is done in normal room environment in absence of noise. First, we have written down the sentences needed to be recorded, so that it can be helpful in recording without pausing or breathing. Recording is done by sony voice recorder, then these recorded wav files are annotated into 5–8 sec frames in Praat and we have set the default sample rate as 48000Hz, default format as 16 bit mono input channel and it is saved as .wav (Microsoft signal 16 bit PCM) format.

The recorded speech utterances are labeled after segmentation with the syllable identities by running make_labs script. Here, align the speech utterance with the text is carried out [6, 12]. Alignment is done by Baulm Weltch algorithm [17]. The label files are extracted based on default generated wave files and speaker spoken wave files. Format of generated label file is shown in Table 1.

Linguistic Analysis phase: In this phase, both pronunciation and prosody are considered. Prosody means phrasing, intonation, and duration. Prosody model is built for Mymensingia dialect to predict the duration, phrase break at the time of synthesis, and the pitch of the syllables [4, 18]. CART-based approach is used to predict the phrase break. Earlier all the words are properly identified and their pronunciation is also defined by applying letter-to-sound rule to the letter in the word. Here modification of pronunciation of the previous phase is done to a standard form, when they appear in continuous speech. First the utterance structure for the recorded utterance is constructed by running build_idom.scm script.Then pitchmarks of all spoken

Table 1 Label file format

0.2200	100	pau
0.2889	100	t
0.3184	100	ax
0.3912	100	m
0.4656	100	aa
0.5035	100	r
0.6251	100	k
0.7728	100	ae
0.8403	100	m
0.9298	100	k
1.0682	100	ae
1.2017	100	k
1.3058	100	eh
1.3732	100	m
1.4073	100	ax
1.4654	100	n
1.6025	100	ch
1.7673	100	aa
1.8715	100	l
1.9923	100	s
2.2123	100	pau

utterances are extracted by running make_pm_wave script, and the pitchmarks are the short time energy peak of each pitch pulse in a speech signal [14], i.e., the beginning of the pitch period. After extracting pitchmarks, we have generated the pitch synchronous mel cepstral coefficient (MELCEP) parameterization of the speech by running make_mcep script, which is used in building cluster synthesizer.

Building the Cluster: Cluster is built by running the script build_idom.sc, which is already available in Festvox directory. For building clusters, all the utterances are loaded into the database and are sorted into segment type and name are assigned for each and every utterances. Then acoustic parameters are loaded and a distance table is built, which contains the calculated distance for every segment of the same type. This pre-calculation saves time, while calculating the acoustic distance between each segment, weights of parameters in the coefficient files are used [4]. Selection of features (phone context, prosody position and whatever others) are dumped into each unit type. Clusters are defined with respect to the acoustic distance between each unit type in the cluster and they are indexed by these features [12]. Finally, all the generated trees are combined into a single tree and dumped into the catalogue file [4, 16].

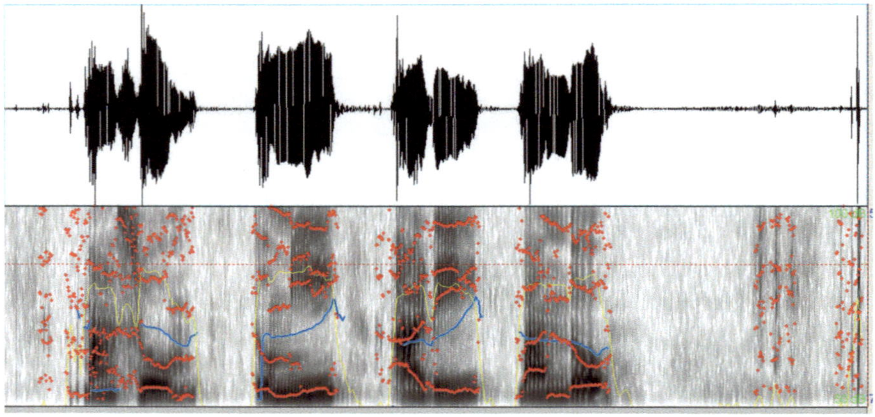

Fig. 2 Spectrogram of original voice of the sentence "masum tumi adi bandho"

Building Waveform Synthesizer: Waveform synthesizer for Mymensingia dialect is build by the script running tu_Mymensingia_afru_ldom.scm, which is available in festvox directory. At the time of synthesis, first, the given input text of Mymensingia dilect is converted to sequence of sound units (syllables) by applying letter to sound rule. Linguistic Analysis module generates phonetic, and contextual features related to each sound unit of input text and prosody module generates prosody features for each sound unit [12]. Finally unit selection algorithm is used to retrieve acoustic unit (speech) corresponding to the sound unit of input text from the stored inventory (clustered database) and to modify the retrieved acoustic unit, so that they match with the target prosody and these are concatenated (smooth) to form an output utterance [4]. SayText function is used to generate final output sound of Mymensingia dialect.

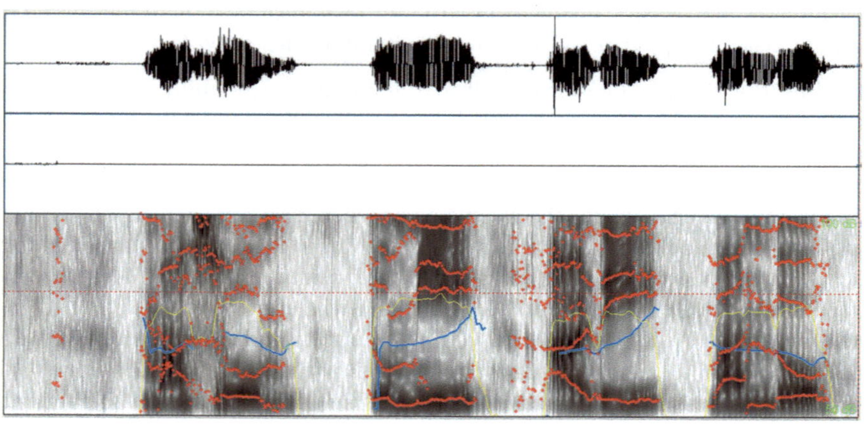

Fig. 3 Spectogram of synthesized voice of the same sentence "masum tumi adi bandho"

In this approach, produced a good quality, natural, and comprehensible synthetic voice of Mymensingia dialect. Figures 2, 3 respectively show the spectrogram of original wav file and synthesized wav files produced from our TTS for the same sentence “masum tumi adi bando” of Mymensingia dialect.

4 Conversion of Synthesized Mymensingia Dialect of Bangla to Standard Bangla

A speech synthesis system for Mymensingia dialect of bangla is built in this paper work. After getting synthesized voice of Mymensingia dialect from Festival TTS, the synthesized utterances are saved as wav files with the following command:

```
(utt.save.wave (utt.synth (Utterance Text “tomar dan balo”)) “test.wav”)
```

Here, we have taken the sentence “tomar dan balo” as example and saved as test.wav.

The objective is to convert these synthesized Mymensingia dialects to standard Bangla, Only some prosodic features are converted. Recording of the same sentence “tomar dan balo” is done, that was saved from Festival TTS by a standard Bangla speaker. Duration of the sentences of synthesized Mymensingia dialect and recorded standard Bangla are adjusted so that both the sentences have same duration. After that, the manipulated object of the standard Bangla files are created using praat, after that the pitch tier is extracted from it. Extracted pitch tier is shown in Fig. 4.

The same sentence of synthesized Mymensingia dialect is taken and also the manipulated object of it is created; then the the extracted pitch tier of standard Bangla file is replaced with the manipulated object of synthesized Mymensingia dialect. The manipulated object of mymensingia dialect before replacing pitch tier is shown in Fig. 5.

The manipulated object of mymensingia dialect after replacing pitch tier is shown in Fig. 6.

The pitch (fundamental frequency) of consonants is decreased and intensity of vowels is increased in the word of the resultant file, which is produced after replacing pitch tier. Finally, duration tier is added to the resultant file and duration of each word

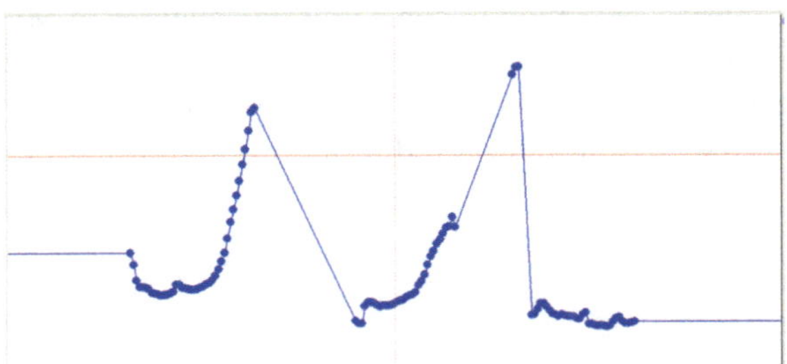

Fig. 4 Pitch tier of sentence “tomar dhan bhalo” of standard Bangla file

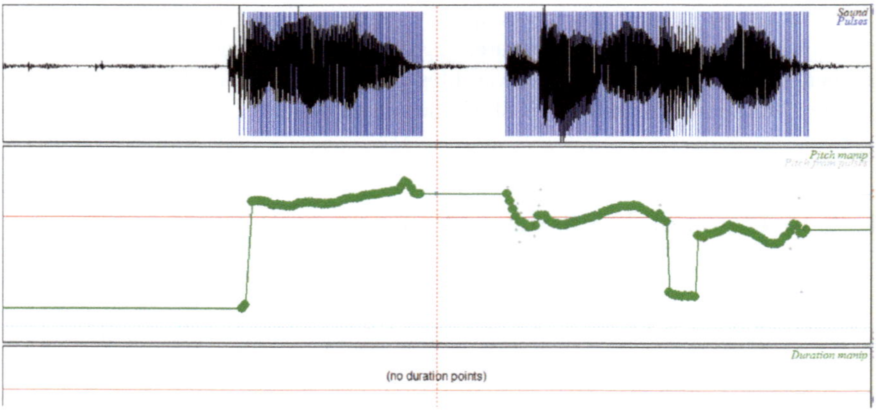

Fig. 5 Before replacing Pitch tier of sentence "tomar dan balo" of Mymensingia dialect

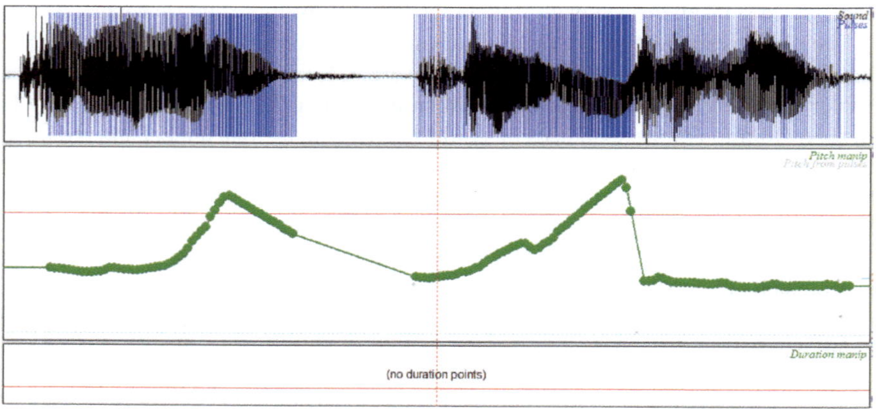

Fig. 6 After replacing Pitch tier of standard Bangla to the sentence "tomar dan balo" of Mymensingia dialect

in the resultant wave file is decreased very slightly. In Fig. 7, after adding duration point is also shown.

Here, conversion of some prosodic features of Mymensingia dialect to standard Bangla is done. We have experimented with only a small set of words, we have not got the the exact tone of standard Bangla, rather we get a sound which can be approximated (90%) to standard Bangla.

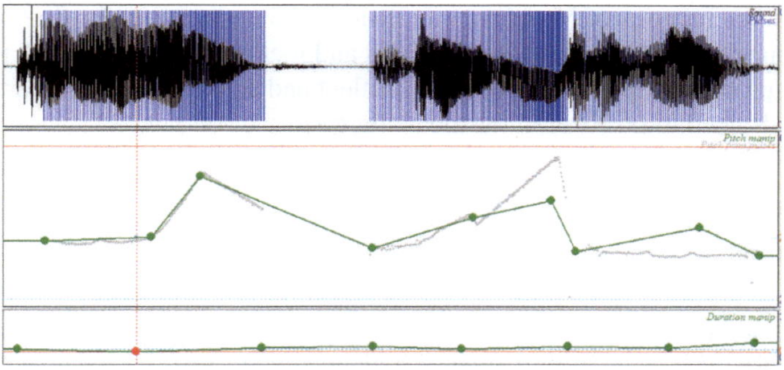

Fig. 7 After adding duration point

5 Discussion

A text-to-speech synthesis for Mymensingia dialect is built here, Fig. 2 shows the spectrogram of original sentence "masum tumi adi bandho" and Fig. 3 shows the spectrogram of synthesized voice of the same sentence produced from the TTS which we have built. From the developed synthesis system for Mymensingia dialect, we obtained a good quality, natural, and comprehensible synthetic voice. Training of our system is done by taking a small database, to get synthetic voice for all the word of Mymensingia dialect we need to increase our database. The drawback of this system is that if a very long sentence is taken, sound produced by the system lacks prosody. Another part of our work is to convert the synthetic voice of Mymensingia dialect to standard Bangla. Result shows partial success, we have not got the exact tone of standard Bangla, rather we get a sound which can be approximated (90%) as standard Bangla. We are still working on this particular problem.

6 Conclusion

A text-to-speech synthesis system for Mymensingia dialect of Bangla is built using Festival speech synthesizer engine. In text analysis and linguistic analysis module of Festival, pronunciation and prosody for each sound unit (syllables) are added and waveform synthesizer module selects the appropriate acoustic unit from speech database for each syllable and are concatenated to give desired speech output. We also try to convert the synthetic voice of Mymensingia dialect produced by our TTS to standard Bangla by replacing pitch tier, changing the fundamental frequency of vowels an consonants and adding duration tier using praat. We have tried to convert some prosodic features of Mymensingia dialect to Standard Bangla. So far we have experimented with only a small set of words and met with partial success in synthesizing the pronunciation in Mymensingia dialect and conversion to standard Bangla. More words need to be covered for the better quality of output.

References

1. Alan Black, Paul Taylor, Richard Caley, Rob Clark, Korin Richmond, Simon King, Volker Strom, and Heiga Zen. The festival speech synthesis system version 1.4.2. Software, Jun 2001.
2. Keiichi Tokuda, Yoshihiko Nankaku, Tomoki Toda, Heiga Zen, Junichi Yamagishi, and Keiichiro Oura. Speech synthesis based on hidden markov models. *Proceedings of the IEEE*, 101(5):1234–1252, 2013.
3. Youcef Tabet and Mohamed Boughazi. Speech synthesis techniques. a survey. In *Systems, Signal Processing and their Applications (WOSSPA), 2011 7th International Workshop on*, pages 67–70. IEEE, 2011.
4. Alan Black and Kevin Lenzo. Building voices in the festival speech synthesis system, 2000.
5. Keiichi Tokuda, Heiga Zen, and Alan W Black. An hmm-based speech synthesis system applied to english. In *IEEE Speech Synthesis Workshop*, pages 227–230, 2002.
6. NP Narendra, K Sreenivasa Rao, Krishnendu Ghosh, Vempada Ramu Reddy, and Sudhamay Maity. Development of bengali screen reader using festival speech synthesizer. In *India Conference (INDICON), 2011 Annual IEEE*, pages 1–4. IEEE, 2011.
7. Sankar Mukherjee and Shyamal Kumar Das Mandal. A bengali hmm based speech synthesis system. *arXiv preprint* arXiv:1406.3915, 2014.
8. Shyamal Kr Das Mandal and Asoke Kumar Datta. Epoch synchronous non-overlap-add (esnola) method-based concatenative speech synthesis system for bangla. In *SSW*, pages 351–355, 2007.
9. Divya Bansal, Ankita Goel, and Khushneet Jindal. Punjabi speech synthesis system using htk. *International Journal of Information Sciences and Techniques (IJIST) Vol*, 2, 2012.
10. Alan W Black, Rob Clark, Korin Richmond, Junichi Yamagishi, Keiichiro Oura, and Simon King. The center for speech technology research, university of edinberg. www.cstr.ed.ac.uk/projects/festival/, 2014. [Online; Accessed 21-March-2015].
11. Robert AJ Clark, Korin Richmond, and Simon King. Multisyn: Open-domain unit selection for the festival speech synthesis system. *Speech Communication*, 49(4):317–330, 2007.
12. Paul Taylor, Alan W Black, and Richard Caley. The architecture of the festival speech synthesis system. 1998.
13. Md Mostafa Rashel. Standard colloquial bengali and chatkhil dialect: a comparative phonological study. *Language in India*, 12(1), 2012.
14. Alan W Black and Paul A Taylor. Automatically clustering similar units for unit selection in speech synthesis. 1997.
15. Alexander Kain and Michael W Macon. Spectral voice conversion for text-to-speech synthesis. In *Acoustics, Speech and Signal Processing, 1998. Proceedings of the 1998 IEEE International Conference on*, volume 1, pages 285–288. IEEE, 1998.
16. CMU Speech Group. Festbox. Software, December 2014.

17. Paul M Baggenstoss. A modified baum-welch algorithm for hidden markov models with multiple observation spaces. *IEEE Transactions on speech and audio processing*, 9(4):411–416, 2001.
18. NP Narendra and K Sreenivasa Rao. Optimal weight tuning method for unit selection cost functions in syllable based text-to-speech synthesis. *Applied Soft Computing*, 13(2):773–781, 2013.

Order Reduction of Discrete System Models Employing Mixed Conventional Techniques and Evolutionary Techniques

Prabhakar Patnaik, Lini Mathew, Preeti Kumari, Seema Das and Ajit Kumar

Abstract A single-input single-output discrete system of high order is reduced to a second order in this paper employing two approaches: an indirect approach using conventional techniques and a direct approach using evolutionary techniques. In the indirect approach, the discrete system is transformed to a continuous system and reduced to a lower order by Padê approximation (by matching Time Moments or Markov parameters) combined with Routh approximation for ensuring stability and inverse transformed back to a lower order discrete system. In the evolutionary approach, the discrete system is reduced to a lower order discrete function and optimized based on minimization of ISE as the objective function using Genetic Algorithm (GA) and Particle Swarm Optimization (PSO) independently. The step responses of reduced order discrete systems obtained by the conventional and evolutionary approaches are compared to determine the best solution.

Keywords Order reduction · Discrete system · Continuous system
Padê approximation · Routh approximation · Stability · Genetic algorithm
Particle swarm optimization · ISE · Step response

1 Introduction

For analysis, synthesis, and to study the behavior of real-life systems, its high order complex mathematical model needs to be reduced to simpler lower order model whose behavior resembles that of original system as far as feasible. Various methods of model order reduction have been listed and described comprehensively

---

P. Patnaik (✉) · L. Mathew · P. Kumari
NITTTR, Sector 26, Chandigarh, India
e-mail: prabhakarpatnaik47@gmail.com

S. Das
IKGPTU, Jalandhar, Panjab, India

A. Kumar
Amity University, Patna, Bihar, India

C. R. Panigrahi et al. (eds.), *Progress in Advanced Computing and Intelligent Engineering*, Advances in Intelligent Systems and Computing 714,
https://doi.org/10.1007/978-981-13-0224-4_28

and comparatively by Genesio R. et al. and other authors [1–7]. Shamash Y. [7] showed that the models reduced from even the originally stable models by many methods are not always stable. This problem has been addressed by Hutton M and many others [8–26] by reduction methods using stability criterion like Routh approximation or Mihailov stability criterion and many without the aid of any stability criterion as well as using mixed techniques. Majority of various methods referred above are applicable for reduction of continuous systems only. The discrete systems can be reduced in discrete domain directly or indirectly using known conventional methods and their stability verified [27–33]. In the last two decades, bio-inspired evolutionary techniques such as Genetic Algorithm (GA) and Particle Swarm Optimization (PSO) method have emerged as modern tools for reduction and optimization [34–42]. In this paper, an *n*th-order single-input single-output discrete system has been reduced by two approaches. The first indirect approach consists of the transformation of the discrete system into a continuous system, order reduction of the continuous system using a combination of conventional techniques of Padê approximation (by matching Time Moments or Markov parameters) combined with Routh approximation for ensuring stability, and inverse transformation of the reduced continuous system back into a discrete system. The second approach consists of the direct method using evolutionary techniques of GA and PSO independently. This paper is organized into eight major sections with Sect. 1 already used for introduction. The problem statement is made in Sect. 2. Indirect approach of order reduction by a combination of conventional techniques has been presented in Sect. 3. In Sects. 4 and 5, order reduction and optimization of the solution by GA and PSO methods have been presented sequentially. All the above methods have been applied to an eighth-order discrete transfer function to obtain second-order reduced discrete transfer function in Sect. 6. The results obtained are compared in Sect. 7. Conclusions made are discussed in Sect. 8.

2 Problem Statement

An nth-order discrete system transfer function in z-transform is represented by:

$$G(z) = N(z)/D(z). \tag{1}$$

with numerator $N(z) = a_0 + a_1z + a_2z^2 + \cdots + a_{(n-1)}\, z^{(n-1)}$ and denominator $D(z) = b_0 + b_1z + b_2z^2 + \cdots + b_{(n-1)}\, z^{(n-1)} + b_n\, z^n$ where a_i $(0 \le i \le n - 1)$ and b_i $(0 \le i \le n)$ are the scalar coefficients of powers of 'z' in the expressions of numerator $N(z)$ and denominator $D(z)$, respectively.

The main objective is to derive a discrete system transfer function $R(z)$ of lower order 'r' $(r < n)$ using indirect (conventional) methods and direct methods (GA and PSO). The reduced transfer function is represented by

$$R(z) = N_r(z)/D_r(z). \quad (2)$$

with numerator $N_r(z) = c_0 + c_1 z + c_2 z^2 + \cdots + c_{(r-1)} z^{(r-1)}$ and denominator $D_r(z) = d_0 + d_1 z + d_2 z^2 + \cdots + d_{(r-1)} z^{(r-1)} + d_r z^r$ where c_i $(0 \le i \le r-1)$ and d_i $(0 \le i \le r)$ are the scalar coefficients of powers of 'z' in the expressions $N_r(z)$ and $D_r(z)$ respectively chosen such that the behavior and response of $R(z)$ should match as closely as possible to that of $G(z)$ for the same type of inputs.

3 Reduction by Conventional Method

Verify the stability of discrete system G(z) by applying Jury stability criterion. By applying bilinear transformation z = (1 + s)/(1 − s) to G(z) to Eq. (1), obtain an equivalent continuous system transfer function G(s) represented by Eq. (3) given below

$$G(s) = N(s)/D(s). \quad (3)$$

with numerator $N(s) = e_0 + e_1 s + e_2 s^2 + \cdots + e_{(n-1)} s^{(n-1)}$ and denominator $D(s) = f_0 + f_1 s + f_2 s^2 + \cdots + f_{(n-1)} s^{(n-1)} + f_n s^n$ where e_i $(0 \le i \le n-1)$ and f_i $(0 \le i \le n-1)$ are the scalar coefficients of powers of 's' in the expressions $N(s)$ and $D(s)$, respectively.

Construct Routh array from the denominator $D(s)$ by arranging the powers of s in the decreasing order. Verify the stability the $G(s)$ by applying Routh stability criterion. Using Routh approximation method proposed by Hutton and Friedland, modified by Shamas Y and illustrated by Panda S. et al. [8, 9, 40], obtain a reduced order denominator of desired order 'r' as shown below in Eq. (4):

$$D_r(s) = \left\{ \sum_{j=0}^{r} h_j s^j \right\} \text{ with } h_r = 1. \quad (4)$$

Following John S. et al. and Panda S. et al. [12, 14, 40], $G(s)$ is expanded about $s = 0$ (or $s = \infty$) into power series with time moments (or Markov Parameters) given below:

$$\text{Time Moment power series } G(s) = G_T(s) = \left(p_0 + p_1 s + p_2 s^2 + p_3 s^3 \ldots\right). \quad (5)$$

$$\text{and Markov Parameters power series } G(s) = G_M(s) = \left(m_1 s^{-1} + m_2 s^{-2} + m_3 s^{-3} \ldots\right). \quad (6)$$

Multiply the power series Eqs. (5) and (6) independently with the reduced denominators $D_r(s)$ Eq. (4) and limit the largest power of s in the products to one power less than that of the reduced denominator Eq. (4) to obtain the two reduced numerators as given below in Eq. (7):

$$N_r(s) = \left[\sum_{j=0}^{r-1} \left(q_j s^j\right)\right] \tag{7}$$

where j represents power of s and q_j represent the coefficient of jth power of s. By using reduced numerators and reduced denominator represented by Eqs. (7) and (4), obtain two reduced models of order 'r' in s domain, one model by matching Time Moments in Eq. (5) and one model by matching Markov parameters in Eq. (6), as given below in Eq. (8):

$$G_r(s) = N_r(s)/D_r(s) \tag{8}$$

Remove the steady state errors of the reduced functions by multiplying the reduced functions with respective gain correction factors $[G(s)/G_r(s)]_{s=0}$. Convert the reduced models in s domain to z domain by inverse bilinear transformation yielding four reduced discrete functions $G_r(z)$ by the indirect conventional approach. Once again steady-state errors are removed by applying suitable correction factors. Finally, verify the stability of each of the reduced discrete function $G_r(z)$ by determining that all the discrete system poles reside inside the area of circle of radius $|z| = 1$.

4 Reduction by GA Technique

The objective function chosen in this paper for the GA technique is minimization of the (ISE) or square of error between the transient state step responses of the original discrete system G(z) of high order and the reduced order discrete system R(z) integrated within the limits of time domain of transient state. The ISE computed by the integral I given by Eq. (9) is as follows:

$$I = \int_{0}^{T_s} [y(t) - y_r(t)]^2 d(t) \tag{9}$$

where y(t) and $y_r(t)$ represent the unit step responses of original and desired reduced discrete transfer functions and T_S represents the settling time of transient state response. The evolution in GA is initiated from a population of randomly generated variables. The reduced order transfer function R(z) is optimized through a number of iterations (generations). For a given optimization objective function, a number of solutions are possible.

An optimized reduced function of desired order is obtained using readily available MATLAB code for GA after defining the objective function, controlled variables, maximum number of generations, population size as well as the upper and lower bounds of each variable. The controlled variables are the scalar coefficients of powers of z in the numerator and denominator of the desired reduced discrete system transfer function arranged in a defined sequence. Their total number of controlled variables is five (5) for optimizing second-order reduced function. GA creates new solutions (akin to chromosomes of a living cell) using reproduction, crossover, and mutation.

5 Reduction by PSO Technique

In the PSO method, the reduced order transfer function R(z) is optimized through a number of iterations (generations) keeping an optimization objective. A number of feasible solutions (called as a particles) are produced in each generation by following the principles of fish schooling or flocking behavior of birds or social behavior of a flock of birds. The objective function of the PSO method is also the same as in the case of the GA method and is given by Eq. (9).

Each particle endeavors to improve its fitness in each successive generation by imitating the properties of more successful peers. It is capable of remembering its own best fitness position (referred to as the *p-best*) in the solution space so far visited by it. The overall best fitness position out of all the *p-best* positions of different particles in the population in a given generation is called as the group best or the *g-best* position. Each particle continuously makes effort to move toward the *g-best* position. Each particle flies in the solution space with a velocity determined by its own momentum which is modified dynamically according to its own flying experience (p-best position) (cognitive vector) as well as that of its peers or other particles (g-best position) (social vector). Various parameters are selected carefully according to past experience to guide the particle achieve an optimum value as fast as possible with suitable velocity and without resorting to excessive iterations. The PSO algorithm needs to be initialized with an initial swarm consisting of the coefficients of powers of z chosen from any known reduced discrete system of desired order. In each PSO run, the optimization ceases automatically after completing the preset number of generations (iterations). Barring the first PSO run, the initialization of the next PSO run is done using the optimized results achieved in the previous PSO run. The ISE will gradually reduce and get stabilized with the increase in the execution of a total number of iterations.

6 Numerical Example

6.1 Combined Routh Approximation and Padê Approximation Method

An eighth-order discrete system (converted from an eighth-order stable continuous system of Panda S. et al. [40]) is given in Eq. (10) as shown below:

$$G(z) = N(z)/D(z) \text{ with a sampling period of } 0.25\text{ s} \tag{10}$$

where $N(z) = 2.052\ z^7 - 5.461\ z^6 + 4.64\ z^5 - 0.04639\ z^4 - 2.228\ z^3 + 1.25\ z^2 - 0.1686\ z - 0.02627$ and $D(z) = z^8 - 3.044\ z^7 + 3.877\ z^6 - 2.697\ z^5 + 1.12\ z^4 - 0.2842\ z^3 + 0.04307\ z^2 - 0.003565\ z + 0.0001234$. The main objective is to derive a stable reduced second-order discrete system model which has a transient step response similar and as close as possible to that of the original stable discrete system of eighth order. The steps explained in Sect. 3 have been implemented on the eighth-order discrete transfer function (10) to obtain stable second-order reduced discrete transfer functions by the indirect conventional method as given below:

$$R_{TM}(z) = (1.632z - 0.5788)/(z^2 + 0.01711z + 0.0001234) \tag{11}$$

$$R_{MP}(z) = (1.142z - 0.0886)/(z^2 + 0.01711z + 0.0001234) \tag{12}$$

6.2 Genetic Algorithm Method

An optimized second-order discrete transfer function as shown below in Eq. (13) is obtained by running a GA program readily available on MATLAB after specifying various parameters, defining the objective function and entering the coefficients of powers of z of Eq. (10) in predefined sequence:

$$R_{GA}(z) = (1.3048z + 0.6109)/(1.0071z^2 + 0.2104z + 0.5314) \tag{13}$$

6.3 Particle Swarm Optimization Method

The coefficients of powers of z of the second-order reduced function $R_{T1}(z)$ Eq. (11) obtained by the conventional methods is used for initialization of PSO algorithm. An objective function is defined and various parameters are specified in Sect. 5. A program based on PSO algorithm is run on MATLAB after entering the coefficients of powers of z of Eq. (10) in the program. The reduced function obtained in a particular run is used for initialization of next PSO run and so on for subsequent runs. The second-order reduced discrete function obtained by PSO is given in Eq. (14) shown below.

$$R_{TMPSO}(z) = \frac{(-69.1165743z + 63.03683816)}{(-38.17950153z^2 + 43.06414420z - 10.78699324)} \quad (14)$$

The plot of ISE data generated during seven consecutive PSO runs versus the generation number in Fig. 1 shows optimization of solutions after 200 generations, the minimization, and the convergence of ISE data with an increase in the number of generations. In a similar manner, using the discrete second-order functions $R_{M1}(z)$ Eq. (12) for initialization of PSO, another second-order reduced optimized discrete transfer function is obtained as shown below in Eq. (15).

$$R_{MPPSO}(z) = \frac{(6.14176748z - 5.58841660)}{(3.65282698z^2 - 4.30474654z + 1.18888032)} \quad (15)$$

The step responses of second-order discrete system equations obtained by the second-order Eqs. (14) and (15) obtained by PSO, paired with Eqs. (11) and (12) obtained by indirect conventional method (by matching Time Moments or Markov parameters) combined with Routh approximation and Eq. (13) obtained by GA, along with original discrete eighth-order Eq. (10) are plotted in Figs. 2 and 3. The parameters of step responses, viz., settling time (T_S), rise time (T_R), peak time (T_P), and maximum overshoot (M_P) are measured from the plots of step responses. The values of poles for each transfer function are calculated. The ISE values are calculated for all the reduced second-order functions using Eq. (9). The results obtained are tabulated in Table 1 for comparison.

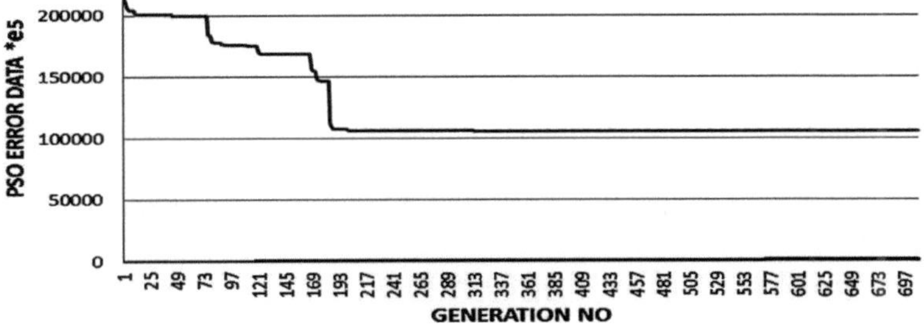

Fig. 1 ISE (between original transfer function and reduced function) versus PSO Generation No

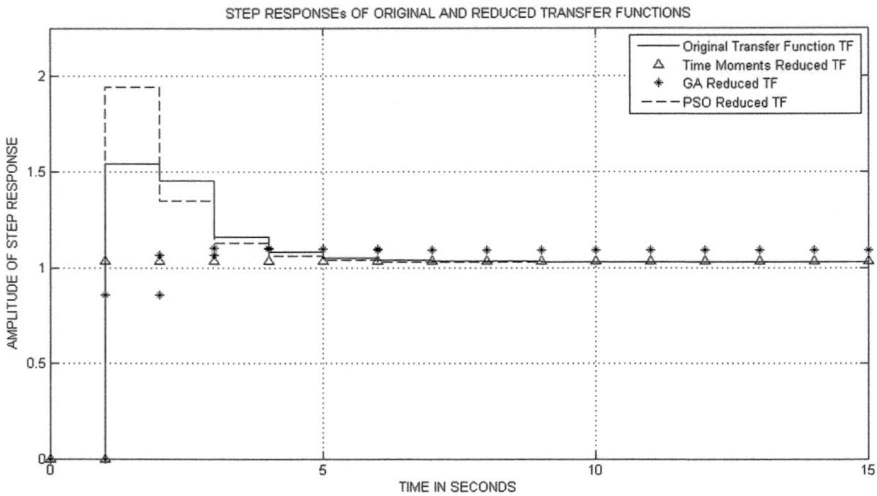

Fig. 2 Step responses of original discrete TF, reduced TF matched with time moments, GA reduced TF, and PSO reduced TF

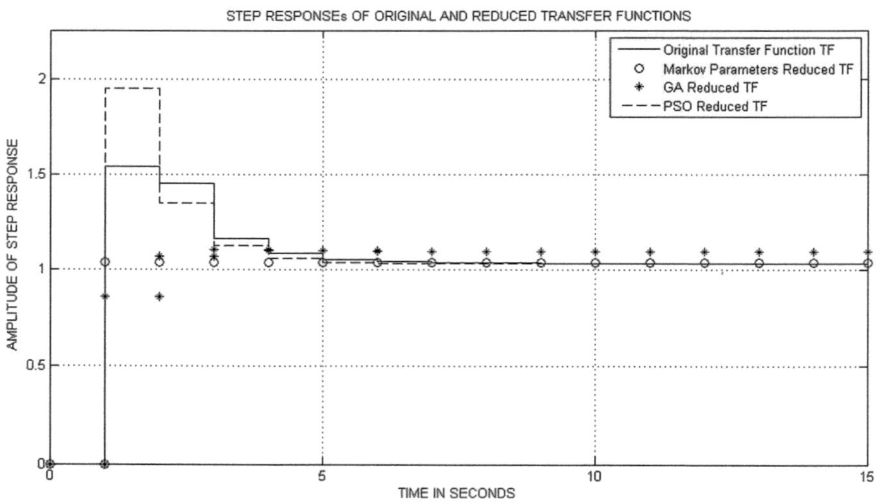

Fig. 3 Step responses of original discrete TF, reduced TF matched with Markov parameters, GA reduced TF and PSO reduced TF

Table 1 Comparison of step response parameters and poles of the original transfer function (OTF) and reduced functions TMTF, MPTF, GATF, and PSO RTF

Transfer function	T_R (s)	T_P (s)	M_P (%)	T_S (s)	ISE	Poles
Original eighth order Eq. (10) (OTF) G(z)	1.0	1.0	54	5.5	NA	0.7838, 0.6236, 0.4462 ± j0.1099, 0.2256 ± j0.0717, 0.1577, 0.1353
Reduced 2nd order $R_{TM}(z)$ Eq. (11)	1.0	1.0	3.54	1.0	0.3670	−0.0086 ± j0.0071
Reduced 2nd order $R_{MP}(z)$ Eq. (12)	1.0	1.0	3.56	1.0	0.6486	− 0.0086 ± j0.0071
Reduced 2nd order $R_{GA}(z)$ Eq. (13)	1.0	3.0	10	4	0.0261	− 0.1045 ± j0.7188
Reduced 2nd order $R_{TMPSO}(z)$ Eq. (14)	1.0	1.0	94	4.5	0.00067	0.7525 and 0.3755
Reduced 2nd order $R_{MPPSO}(z)$ Eq. (15)	1.0	1.0	95	4.5	0.0047	0.7367 and 0.4418

7 Discussion of Results and Comparison of Methods

It can be seen from Figs. 2, 3 and Table 1 that the magnitude of each of the poles of all functions is less than unity and lie within the unit circle. All the 2nd order Reduced Transfer Functions (RTFs) are stable like the 8th order Original Transfer Function (OTF). But the dominant poles of the PSO reduced RTFs are closer to that of the OTF. Though the step responses of all the RTFs have zero steady-state error, the amount of similarity and close resemblance of the shapes of step responses vary from one RTF to another. While the step response of OTF has a distinct large overshoot and the overshoots in step responses of RTFs due to GA and other conventional methods are insignificant, but the step responses of RTFs only due to PSO has got some significant values of overshoots comparable and even larger than that of 8th order OTF. On comparison of step response parameters settling time (T_s), maximum overshoot (M_p), peak time (T_p), and rise time (T_R) and ISE, it can be seen that the parameters of step responses of RTF obtained by PSO are the closest and best approximates of OTF, while those obtained by GA Padê and Routh approximations are quite inferior and poor approximates of 8th order OTF.

8 Conclusion

In this work, a discrete system transfer function of eighth order has been reduced to a discrete transfer function of second order by the indirect approach of combined conventional methods of Padê approximation (matching of Time Moments/Markov parameters) and Routh approximation and direct methods of Genetic Algorithm and Particle Swarm Optimization. A comparison of all parameters and the step responses shows that the results of the discrete functions reduced by PSO method are best and closest approximates of original model followed by that of GA method and the step responses of the reduced TF obtained by conventional techniques are poor approximates of the step response of original transfer function. This is also confirmed by the lowest ranges of ISE of the PSO reduced functions.

Therefore, it can be concluded that the PSO technique is the best of all the methods of reduction discussed above. Considering that the reduction is from a high eighth order to a low second order, the results are satisfactory and encouraging. Still better results and closer resemblance of step response of RTFs with that of OTF can be expected if the desired order of reduced function is increased to third order or higher.

References

1. Genesio R., Mlianese M.: A Note on Derivation and Use of Reduced-order Models, Technical Notes and Correspondence, IEEE Transactions on Automatic Control, Vol 21, pp 118–122, February 1976.
2. Bosley M.J., Lees F.P.: A Survey of Simple Transfer function derivations from higher order State variable models,Automotica, Vol 8, pp 765–775, 1978.
3. Fortuna L, Nunnari G, Gallo. A.: Model Order Reduction Techniques with applications in Electrical Engineering, Springer-Verlag, 1992.
4. Ali Eydgahi., Jalal Habibi., Behzad Moshiri.: A MATLAB Toolbox for Teaching Model Order reduction Techniques, Proceedings of International Conference on Engineering Education, Valencia, Spain, July 21–25, 2003.
5. Janardhanan S.: Model Order Reduction and Controller Design Techniques sp_topics_01.pdf.
6. Oliver P. D.: A Comparision of Reduced Order Model Techniques 41st South eastern Symposium on System Theory, University of Tennesses Space Institute, Tullahoma, TN, USA, TIA.3, pp 240–243, Mar 15-17, 2009.
7. Shamash Y.: Stable Reduced-Order Models using Padé-type approximations, Technical notes and Correspondence, IEEE Transactions on Automatic Control, pp 615–616, October 1974.
8. Hutton M F., Friedland B.: Routh Approximations for reducing Order of Linear Time Invariant System, IEEE Transactions on Automatic Control, Vol AC-20, No.3, pp 329–336, June 1975.
9. Shamash Y.: Model Reduction using the Routh Stability Criterion and the Padé approximation technique, International Journal of Control, Vol 21, No. 3, pp 475–484, 1975.
10. Appiah R.K.: Linear Model Reduction using Hurwitz polynomial approximation, International Journal, Vol. 20, pp 329–337, 1975.
11. Krishnamurty V., Seshadri V.: A Simple and Direct Method of Reducing Order of Linear Systems Using Routh Approximations in the Frequency Domain, Technical notes and Correspondence, IEEE Transactions on Automatic Control, pp 797–799, October 1976.
12. John Sarasu., Parthasarathy R.: System Reduction using Cauer Continued Fraction Expansion about $S = 0$ *and* $S =$ alternately, Electronic Letters Vol. 14 No. 8, pp 261–262, April 1978.
13. Krishnamurty V., Seshadri V.: Model Reduction Using the Routh Stability Criterion, IEEE Transactions on Automatic Control, Vol. AC-23, pp 729–731, Aug, 1978
14. John Sarasu., Parthasarathy R.: System Reduction by Routh approximation and modified Cauer Continued Fraction, Electronic Letters Vol. 5 No. 21, pp 691–692, 1979.
15. Shamash Y.: Failure of the Routh-Hurwitz Method of Reduction IEEE Transactions on Automatic Control, Vol. AC-25, No. 2, April 1980.

16. Bai-Wu Wan.: Linear Model reduction using Mihailov Stability criterion and Pade approximation technique, International Journal of Control, Vol. 33, pp 1073–1089,1981.
17. Singh V., Chandra D., Kar H.: Improved Routh-Padé Approximants: A Computer-Aided Approach IEEE Transactions on Automatic Control, Vol. 49, No. 2, February 2004.
18. Tomar S K., Prasad R.: Linear Model Reduction Using Mihailov Stability Criterion and Continued Fraction Expansions Proceedings of XXXII National Systems Conference, NSC 2008, pp 603–605, December 2008.
19. Chen T.C, Chang C.Y., Han K.W.: Reduction of transfer functions by the stability equation method, Journal of Franklin Institute, Vol. 308, pp 389–404, 1979
20. Pal J.: Improved Padé Approximations using Stability Equation Method Electronics Letters, Vol.19, No. 11, pp 426–427, May 1980.
21. Gutman P O., Carl F M., Molander P.: Contribution to the Model Reduction Problem IEEE Transactions on Automatic Control, Vol. AC-27, No.2, pp 454–455, April, 1982.
22. Lucas T N.: Factor Division: A useful Algorithm in Model Reduction IEE Proceedings, Vol. 130, Pt. D, No. 6, pp 362–364, November 1983.
23. Chen T.C, Chang C.Y., Han K.W.: Model reduction using the the stability equation method and Padé approximation technique, Journal of Franklin Institute, Vol. 309, pp 473–490,1980.
24. Habib N., Prasad R.: An Observation on the Differentiation and Modified Cauer Continued Fraction Expansion Approach of Model Reduction Technique XXXII National System Conference NSC 2008, pp 574–579, Dec 17-19,2008.
25. Vishwakarma C B., Prasad R.: Order Reduction using advantages of Differentiation method and Factor Division algorithm Indian Journal of Engineering & Material Sciences, Vol 15, pp 447–451, December 2008.
26. Vishwakarma C B., Prasad R.: System Reduction using Modified Pole Clustering and Padé Approximation X NSC 2008, pp 592–596, December 17-19, 2008.
27. Shamash Y., Continued fraction methods for reduction of discrete time dynamic systems, International Journal of Control, Vol. 20, pp. 267–268, 1974.
28. Rao A K., Naidu D S.: Singular Perturbation Method for Kalman Filter in Discrete Systems IEE Proceedings, Vol 131, Pt. D, No.1, January 1984.
29. Ismail O, Bandopadhay B, Gorez R.: Discrete Interval system reduction using Padé Approximation to Allow retention of Dominant Poles, IEEE Transactions on Circuits and Systems-1: Fundamental Theory and Applications, Vol. 49, No.11, November 1997.
30. Mittal S K., Chandra Dinesh.: Stable Optimal Model Reduction of Linear Discrete Time Systems Via Integral Squared Error Minimization: Computer-Aided Apporoach AMO-Advanced Modelling and Optimization, Vol. 11,pp 531–547, November 4, 2009.
31. Ramesh K., Nirmalkumar A., Guruswamy G.: Design of Digital IIR Filters with the Advantages of Model Order Reduction technique World Academy of Science, Engineering and Technology 52, pp 1229–1234, 2009.
32. M Gopal M.: Digital Control and State Variable Methods, Tata McGraw Hill, New Delhi, 2nd Edition, 2003.
33. Sivanandam S. N., Deepa S. N.: Control System Engg using MATLAB',Vikas Publishing House Pvt Ltd. Noida, 2009.
34. Kennedy J., Eberhart R.: Particle Swarm Optimization, Proceedings of the IEEE International Conference on Neural Networks, Perth, Australia, pp. 1942–1945, 1995.
35. Stron Rainer., Price Kenneth.: Differential Evolution- A Simple and Effective Adaptive Scheme for Global Optimization over Continuous Spaces, Journal of Global Optimization, Vol. 11, pp. 341–359, 1995.
36. Bipul L., Venayagamoorthy G K.: Differential Evolution Particle Swarm Optimization for Digital Filter Design 2008 IEEE Congress on Evolutionary Computation (CEC-2008), pp 3954–3961, 2008.
37. Chen-Chien H.: Chun-Hui G Digital Redesign of Interval Systems Via Particle Swarm Optimization World Academy of Science, Engineering and Technology 41, pp 582–587, 2008.

38. Tomar S K., Prasad R., Panda S., Ardil C.: Conventional and PSO based Approaches for Model Reduction of SISO Discrete Systems International Journal of Electrical and Electronics Engineering, pp 45–49, January 2009.
39. Yadav J S., Patidar N P., Singhai J.,Panda.: Differential Evolution Algorithm for Model Reduction of SISO discrete system Proceedings of World Science Congress on Nature & Biologically Inspired Computing (NABIC-2009), Coimbatore, India, pp 1053–1058, December, 2009.
40. Panda S., Tomar S.K., Prasad R., Ardil C.: Reduction of Linear Time-Invariant Systems Using Routh-Approximation and PSO International Journal of Applied Mathematics and Computer Sciences, pp 82–88, February, 2009.
41. Yadav J S., Patidar N P., Singhai J., Panda S., Ardil C.: A Combined Conventional and Differential Evolution Method for Model Orde Reduction International Journal of Information and Mathematical Sciences, pp 111–118, February, 2009.
42. Gupta Lipika., Mehra Rajesh.: Modified PSO based Adaptive IIR Filter Design for System Identification on FPGA International Journal of Computer Applicatrions (0975–8887), Vol. 22 No. 5, pp 1–7, May 2011.

Part IV
Big Data Applications, Internet of Things and Data Science

An Efficient Framework for Smart City Using Big Data Technologies and Internet of Things

Krishna Kumar Mohbey

Abstract The evolution of cloud computing, Internet of things (IoT) and big data has played vital role in the development of smart cities. IoT uses various types of embedded devices, such as sensors, actuators, Bluetooth, Wi-Fi, radio frequency identification (RFID), and ZigBee to collect data from different smart city applications. The huge amount of data, which are collected from different applications are known as big data. To perform real-time processing on data, data collected from smart city applications an efficient framework are required. This framework can combine big data technologies with IoT services toward smart city. In this paper, various IoT communication techniques are discussed with big data technologies. Then, a framework is proposed for handling big data generated from smart city applications. The proposed framework primarily focuses on problems related to smart city vision for real-time decision-making. In addition, this paper discusses the various principles and requirements of smart city for enhancing the life standard of people. The proposed framework can serve as a benchmark for authorities and policy makers in smart cities enhancement with the use of IoT concept, features, and big data technologies.

Keywords Smart city · Internet of things · Big data technologies · Cloud computing

1 Introduction

The concept of smart city includes a modern urban area that enhances the living standard of people through utilizing information technology, effective communication, and proper data management. Smart city development is a concept, which includes physical infrastructure enhancement as well as various factors and strategies related to citizenship and environment. With the enhancement of sophisticated computing

K. K. Mohbey (✉)
Department of Computer Science, Central University of Rajasthan, Ajmer, India
e-mail: kmohbey@gmail.com

C. R. Panigrahi et al. (eds.), *Progress in Advanced Computing and Intelligent Engineering*, Advances in Intelligent Systems and Computing 714,
https://doi.org/10.1007/978-981-13-0224-4_29

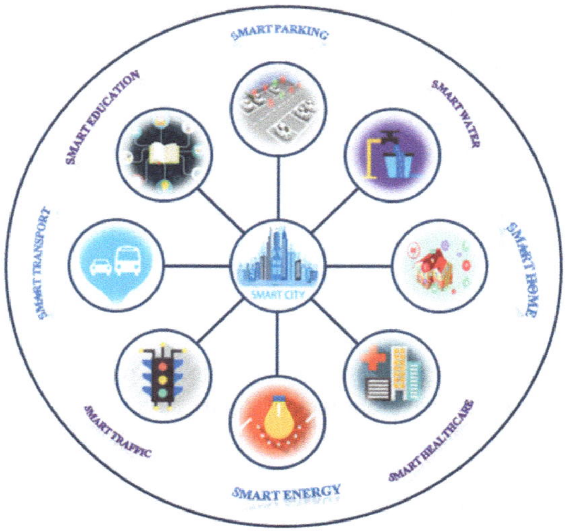

Fig. 1 Different applications of smart city

and technology [1], lots of devices are connected through sensors. In today's world, sensor technology and IoT have provided the solution for living in a smart environment, i.e., smart city. To make a city smarter, it is needed to manage environment in an efficient way that is required for natural resources, mobility, governance, transportation, traffic, health care, education, energy, etc. To fulfill these objectives, several services of smart city [2] have been introduced recently, including smart grid [3], smart transportation [4], smart home [5], smart health care [6], smart education, smart energy, and smart mobility [2, 7], and so on. Figure 1 shows the different applications of the smart city.

At present, smart city management mainly depends on the advancement of physical infrastructure and efficient communication for real-time decision-making [5]. The communication between various applications of smart city is possible through IoT. With the IoT technology, various objects of different applications are connected in the proper way to enhance the quality of communication and availability of real-time data.

Due to the rapid increase in the IoT technologies toward smart city applications, data is generating at high volume in every second. This data is generated in the form of high volume, velocity and variety [8, 9]. These collected data are useful for smart city enhancement since it describes the actual value and characteristics of smart city applications. The big data collected from various sources are mostly including unstructured features as compared to data collected by other means [10]. Figure 2 shows the relationship between various smart city applications, big data and cloud computing. This figure also demonstrates the exchange of information between various smart city applications using IoT sensors and cloud platform. This collected big data are unstructured in nature and stored in different data centers using a

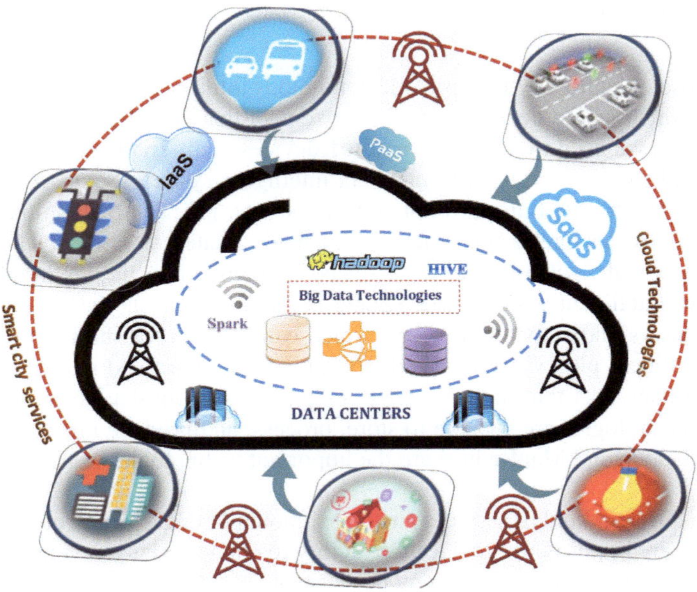

Fig. 2 The relationship between smart city applications, big data, and cloud infrastructure

distributed database such as NoSQL. The purpose of storing data in data center is to share among various applications of smart cities using cloud platform [11]. To process this big data, a programming model is also required with parallel algorithms to obtained valuable results. Smart city has enhanced the living standard of urban citizen through transmitting different areas of human life such as health, home, energy, transport, water, education, governance, pollution, etc. Various governments have already started smart city ideas in their country to fulfill requirements of the citizens [12].

With the development of various technologies toward smart city, applications can transform various sectors of the nations economy [13]. This transformation supports smart cities to collect various requirements for their enhancements. Improving the citizen lifestyle in smart city requires various technologies such as wireless sensor network (WSN), embedded devices, and IoT. In addition, big data analytics is also important for smart city applications [14]. Big data are generated from various sources like mobile phones, cameras, GPS, social media, sensors, computers, and so on. Therefore, efficient data storage and processing have become a challenging task to manage it. Proper management and effective analysis of data are the main tasks for the success of a smart city. One of the possible solutions for big data management is to collaborate cloud services with IoT. This study mainly focuses on conducting a survey of smart city applications with big data technologies toward enhancing the living standard of urban citizens.

2 Related Work

The purpose of smart city is to provide continuous information to make a real-time decision, which would be helpful to the citizens of that city. To develop such kind of smart city, we need to deploy several IoT devices at different places for different services. IoT devices are used to make an intelligent system for smart city which includes home, traffic, transportation, education, etc. [1]. The motivation of smart city is to make an intelligent system in all fields related to citizen. These fields included hospitals, schools, railways, roads, buildings, and the environment and so on [15]. Intelligent system development is possible with the interconnection of various sensors and actuating devices, which can be used for sharing information between different platforms. To develop such kind of smart and intelligent system requires frameworks that include cloud computing and big data technologies [1]. Big data technologies are capable to store, process, and produce information intelligently. Hadoop and MapReduce are the important technologies to handle big data [16].

Due to different kind of services and huge amount of data storing, cloud computing models are required. It provides the facility to connect many devices or clusters at real-time [17]. Cloud computing models are capable to handle complex and large-scale computing tasks [18]. Cloud computing services include platform as a service (PaaS), software as a service (SaaS), and infrastructure as a service (IaaS). Cloud computing also provides different services which can be used to manage data [19].

Real-time data storage and processing are the biggest task in smart city. It also includes streaming architecture and seamless communication between various sensors within the smart city services. A lot of research is going in this direction, but it still requires an efficient framework to enhance the efficiency and processing at real-time environment.

3 IoT Communication Technologies

An effective communication is required to make a city smart. This communication is possible through connecting various equipments to collect real-time data. This equipment includes smart home devices, sensors, smartphones, laptops, etc. The communication should be capable of transferring real-time collected data. In this section, various IoT communication technologies are highlighted, which are beneficial for smart cities.

3.1 WSN

A wireless sensor network (WSN) is used to connect various distributed and independent devices. It works on low power integrated circuits and wireless technology to connect devices. It is capable of monitoring physical as well as environmental conditions in real-time toward smart city services. It can monitor temperature, light, humidity, pressure, etc. In addition, radio transceiver is used in WSN for sending and receiving signals through wireless technology [20].

3.2 RFID

Radio frequency identification (RFID) is a technology which uses electromagnetic coupling in the radio frequency to uniquely identify an object. This technology is useful for smart city applications as well as IoT device communications because it is capable to identify any object. It can be used with any kind of objects such as person, car, animal, cloth, and so on. In addition, to make a city smarter, it can be applied in schools, hospitals, libraries, environments, and other places [7].

3.3 LTE, 4G, and 5G

The long-term evolution (LTE) technology is used to connect various devices with 4G wireless network. It supports hybrid data and voice communication. This technique is an efficient scheme which supports high data transfer. With this technique, multiple users can share a common channel. In addition, 5G supports high bandwidth up to 10 Gbit/s [7].

3.4 Wi-Fi

A smart city is fully connected with wireless communication because it is fast, flexible, and secure. It has various features like low cost, dynamic network improvement, and easy deployment. It is the replacement of the cable network and provides facility to access the Internet at broadband speed [11].

3.5 *ZigBee*

ZigBee is a wireless communication technology and generally used for short range communication between devices. It is capable for reliable, robust, secure, and low power consumption. In smart city services, it is widely used in a smart home for connecting devices, in smart lighting and in other places [7].

3.6 *Bluetooth*

Short range communication is possible through Bluetooth which uses wireless radio system. It replaced cable for computer peripherals such as keyboard, mouse, printer, joysticks, and so on. Due to low power consumption, it is useful for communicating various smart city objects [19].

4 The Role of Big Data in Smart City Services

Smart city services produce huge amount of data every day due to real-time data collection. To store, manage, and process, this data required advanced technologies for efficient data processing. Big data technologies provide various tools and methods which are able to collaborate with different services and enhance the smart city standards. These technologies are also capable to analyze data and predict decisions toward smart city enhancement. In this section, some important services of smart city are discussed.

4.1 *Smart Home*

Smart home is an important service of smart city in which different objects are connected through IoT sensors and controlled by smartphones or other computing devices. Home objects collect data from various sources by sensors, camera, Wi-Fi, Bluetooth, and so on. The collected data transfer to storage unit for further processing [21].

4.2 *Smart Health Care*

Smart healthcare systems manage health-related e-data. In smart city, various healthcare centers are connected through sensors and IoT devices to provide communication between hospitals, patients, doctors, and diagnosis machines. It includes online medical services like online appointment, digital record storage, remote home services, alarm system, and remote patient monitoring [6].

4.3 Smart Transportation

The purpose of smart transportation is to minimize traffic congestion. Number of accidents can be reducing by providing alternate routes to vehicles. With the use of various IoT sensors, cameras, smart vehicles and RFID techniques transportation can make effective. This system is capable to predict real-time traffic patterns which are useful for safe and secure traffic [7, 21].

4.4 Smart Grid

This system uses advanced meters, readers, and communication network to understand real-time power demand and consumption. In smart grid, real-time monitoring can be achieved through computer-based remote controls. These controllers used between power producers and consumers to increase efficiency [22].

4.5 Smart Governance

The government can easily analyze various results using big data technologies which are beneficial for citizens of a smart city. Big data techniques can help the government to make policies, implement, and monitor in real-time.

5 Proposed Framework

Smart city services produced data continuously in different formats. Managing such kind of data existing approaches and techniques is not sufficient due to limited processing speed and limited storage capability. To handle this problem, it is needed to develop an efficient framework with the help of big data technologies. The proposed framework is based on parallel processing on distributed data storage. The proposed framework for smart city data processing is shown in Fig. 3.

This framework is divided into multiple layers, where each layer is responsible for a specific task.

The first layer is responsible for communication and data generation. It consists of various objects and IoT embedded devices in smart city. These devices generate

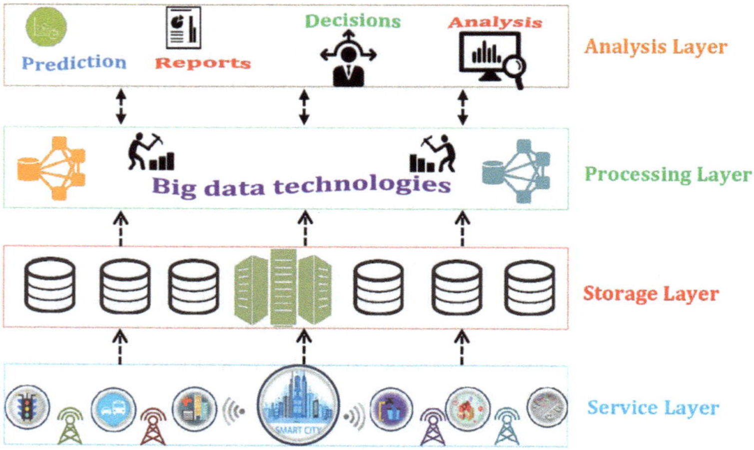

Fig. 3 Framework for smart city data processing using big data technologies

large amounts of heterogeneous data. Second layer is responsible for collecting and storing data in a distributed environment after applying preprocessing. The generated data are stored with the help of big data technologies such as Google cloud, Microsoft Azure, Amazon, and so on. In the third layer, the stored data will processed according to given queries using big data techniques such as MapReduce framework [23]. MapReduce is a high processing model for distributed and parallel processing. It makes various clusters of data for processing. Analysis layer is the last layer, which provides facility to interact people and devices directly to make real-time decisions. The result of analysis may be used for prediction, report generation, and recommendations for smart city.

6 Conclusion

In this paper, an efficient framework for smart city services has been proposed, which will be useful for managing real-time generated data. This large data is continuously generated by the various IoT embedded devices of smart city services. This paper also describes various IoT communication technologies toward smart city services. Finally, this paper concludes that big data and IoT devices are the building blocks of a smart city. The proposed framework is capable for decision-making and policies enhancement which can change the lifestyle of the citizens.

References

1. Gubbi, J., Buyya, R., Marusic, S., & Palaniswami, M.: Internet of Things (IoT): a vision, architectural elements, and future directions. Future Generation Computer Systems, 29(7), 1645–1660 (2013).
2. Chourabi, H., Nam, T., Walker, S., Gil-Garcia, J. R., Mellouli, S., Nahon, K., & Scholl, H. J.: Understanding smart cities: an integrative framework. Paper presented at the 45th Hawaii International Conference on System Science (HICSS), (2012).
3. Chen, S.-y., Song, S.-f, Li, L., & Shen, J.: Survey on smart grid technology. Power System Technology, 33(8), 17 (2009).
4. Adeli, H., & Jiang, X.: Intelligent infrastructure: neural networks wavelets, and chaos theory for intelligent transportation systems and smart structures. CRC press, (2009).
5. Caragliu, A., Del Bo, C., & Nijkamp, P.: Smart cities in Europe. Journal of Urban Technology, 18(2), 6582 (2011).
6. Demirkan, H.: A smart healthcare systems framework. It Professional, 15(5), 3845 (2013).
7. Hashem, I. A. T., Chang, V., Anuar, N. B., Adewole, K., Yaqoob, I., Gani, A., & Chiroma, H.: The role of big data in smart city. International Journal of Information Management, 36(5), 748–758 (2016).
8. Gani, A., Siddiqa, A., Shamshirband, S., & Hanum, F.: A survey on indexing techniques for big data: taxonomy and performance evaluation. Knowledge and Information Systems, 46(2), 241–284 (2016).
9. Khan, N., Yaqoob, I., Hashem, I. A. T., Inayat, Z., Mahmoud Ali, W. K., Alam, M., & Gani, A.: Big data: survey, technologies, opportunities, and challenges. The Scientific World Journal, (2014). https://doi.org/10.1155/2014/712826.
10. Chen, M., Mao, S., & Liu, Y.: Big data: a survey. Mobile Networks and Applications, 19(2), 171–209 (2014).
11. Borgia, E.: The internet of things vision: key features, applications and open issues. Computer Communications, 54, 131 (2014).
12. Jimenez, C. E., Solanas, A., & Falcone, F.: E-government interoperability: linking open and smart government. Computer, 47(10), 22–24 (2014).
13. Batty, M.: Big data, smart cities and city planning. Dialogues in Human Geography, 3(3), 274–279 (2013).
14. Al Nuaimi, E., Al Neyadi, H., Mohamed, N., & Al-Jaroodi, J.: Applications of big data to smart cities. Journal of Internet Services and Applications, 6(1), 115 (2015).
15. Su, K., Li, J., & Fu, H.: Smart city and the applications. Paper presented at the 2011 International Conference on Electronics, Communications and Control (ICECC) (2011).
16. Hashem, I. A. T., Anuar, N. B., Gani, A., Yaqoob, I., Xia, F., & Khan, S. U.: MapReduce: review and open challenges. Scientometrics, 134 (2016). https://doi.org/10.1007/s11192-016-1945-y.
17. Mell, P., & Grance, T.: The NIST definition of cloud computing (2011).
18. Chang, V., Bacigalupo, D., Wills, G., & Roure, D. D.: A categorisation of cloud computing business models. Paper presented at the Proceedings of the 2010, 10th IEEE/ACM International Conference on Cluster, Cloud and Grid Computing (2010).
19. Hashem, I. A. T., Yaqoob, I., Anuar, N. B., Mokhtar, S., Gani, A., & Khan, S. U.: The rise of big data on cloud computing: review and open research issues. Information Systems, 47, 98115 (2015). https://doi.org/10.1016/j.is.2014.07.006.
20. Dargie, W. W., & Poellabauer, C.: Fundamentals of wireless sensor networks: theory and practice. John Wiley & Sons (2010).

21. Mohbey, K.K.: The role of big data, cloud computing and IoT to make cities smarter, Int. J. Society Systems Science, Vol. 9, No. 1, pp. 75–88 (2017).
22. Lai, C. S., & McCulloch, M. D.: Big data analytics for smart grid. (2015). Retrieved from Accessed 23. 04. 16. http://smartgrid.ieee.org/newsletters/october-2015/big-data-analyticsfor-smart-grid.
23. Dean, J., & Ghemawat, S.: MapReduce: simplified data processing on large clusters. Communications of the ACM, 51(1), 107–113 (2008).

A Practical Implementation of Optimal Telecommunication Tower Placement Strategy Using Data Science

Harsh Agarwal, Bhaskar Tejaswi and Debika Bhattacharya

Abstract The exponential growth in the tele-density in India and around the world has put forth a lot of challenges for the network operators. The customers look for good signal reception, fast data speeds, and call quality while choosing their cell phone operators. The aim of this study is to obtain an optimum number of telecommunications towers using data science algorithms like mean shift, SVM classification, and K-means algorithm and practically implemented it using Android application. We propose a new method for optimizing the position of cell towers to get the coverage area of the widest service through three stages: Clustering, classification, and positioning. The proposed cell phone tower placement scheme involves data extraction from cell phone users through an Android application and the analysis of the data to obtain a set of possible candidate sites for establishing a base station.

Keywords Mobile communication · Data science · Tower placement

1 Introduction

In the recent times, a phenomenal growth has been seen in the mobile communication sector in India. The number of cell phone subscribers in India was 1127.37 million as of December 31, 2016 [1]. The technology used in cell phones has also evolved over the years. Penetration of smartphones in the Indian market has increased rapidly over the years, crossing 300 million mark in 2016 [2]. The increased usage of smartphones has provided a huge scope for network providers to

H. Agarwal (✉) · B. Tejaswi · D. Bhattacharya
Department of CSE, IEM Kolkata, Kolkata, India
e-mail: agarwal.harshnu@gmail.com

B. Tejaswi
e-mail: bhaskartejaswi2008@gmail.com

D. Bhattacharya
e-mail: bdebika@iemcal.com

C. R. Panigrahi et al. (eds.), *Progress in Advanced Computing and Intelligent Engineering*, Advances in Intelligent Systems and Computing 714,
https://doi.org/10.1007/978-981-13-0224-4_30

enhance their services by anonymously collecting usage data from several users and finding trends and patterns in cell phone usage.

Network operators face a lot of challenges due to dynamic expansion of cities and redistribution of human population. Mobile phone coverage is one of the most important parameters considered by customers when choosing a network provider [3]. The operators perform field survey to install new towers. Cost, number of subscribers in a locality, and the availability of space are some of the factors that are needed to be considered for base station installation. Further, companies also act upon complaints received by the consumers about poor call quality, low signal strengths, etc. Technology can become an enabler in this process and can help network operators make informed decisions driven by near real-time data.

The following are the specific contributions we intend to make through this proposed work:

- We propose a scheme for finding a set of possible candidate sites for placing new towers.
- We have designed and implemented a practical solution using an Android application. We collect data using the Android application and perform data analysis using Python and advanced data science algorithms.
- The proposed scheme has been rigorously examined and verified using data analysis on real-world data.

2 Related Work

Several studies have been conducted regarding data analysis on mobile phone data. Doug et al. propose a solution to analyze traveling patterns of people using mobile phone location and timestamp data for traffic planning [4]. Liao et al. propose a framework for sensing and logging a users' daily life and derive inferences on the users' mobility and behavior [5]. Swati et al. analyze location data to predict mobility patterns of a user [6].

All coverage models are based on an assumption that the customers do not get proper services beyond a given range. Set coverage problem (SCP) aims at determining the number of nodal centers required along with their prospective locations such that all nodes in a wireless network get an adequate level of coverage (signal level). In wireless communication, a balance is sought between the economic feasibility of the coverage and the overall benefits that result from setting up new nodal centers. Church and ReVelle propose this trade-off as the Maximal Covering Location Problem [7]. This problem considers budget as a constraint while locating the number of facilities. In our proposed scheme, we build on the MCLP model in the context of base station deployment.

Placement of new base stations is an optimization problem that involves several variables such as traffic density, channel allocation, number of base stations, channel interference, and other network related parameters. Pereira et al. use particle swarm optimization for placement of multiple base stations in a metropolitan area [8]. Komnakos et al. propose a solution with minimization of energy consumption in focus [9]. Recent studies have also explored the possibilities of utilization of big data platforms and data analytics by telecom operators [10, 11]. However, implementation of a practical solution for tower placement problem using real-world data is missing in the existing literature.

3 Proposed Scheme

In order to obtain a set of possible candidate sites for deploying base stations, we collect signal strengths at several locations from a number of users having our designed android application on their smartphone. We analyze the data collected from all users by performing the following steps:

1. **Clustering:** We perform clustering of the sampled locations using geographical coordinates provided by the device location of different users at different instances of time. We consider a particular area on a basic scale of 1–2 km. In order to perform location-based clustering, we prefer using mean shift algorithm [12] over K-means clustering as we do not know beforehand, the number of clusters that we will obtain after clustering. We now briefly discuss the working of Mean shift algorithm:

- Consider a dataset **Z**.
- Consider a point z $\in$ **Z**, in the sample and find the set of its neighboring points Np(z).
- Calculate the mean shift **m(x)**, which is the weighted average of the points in **Np(z)** with respect to the point z and update **z** $\leftarrow$ **m(x)**.
- Repeat the steps b and c, till the points are not moving or moving by negligible distances.

By performing the clustering of geolocation data, we obtain N_c number of clusters. For each cluster, we get a centroid which acts as the center of cluster. After this, we further classify the area on the basis of good and bad signal strength.

2. **Classification**

   Let us consider the following parameters:

- S_g: Minimum signal strength above which the signal strength is good for normal usage by a user.
- S_b: Maximum signal strength below which the signal strength is bad for normal usage by a user.
- $S_{i,j}$: Signal strength at the ***j*th** sample point in the ***i*th** cluster.
- N_i: Number of data points inside the ***i*th** cluster.

- N_{min}: Minimum number of users per base station that the network provider wants to have, so that the installation is economically feasible.
- R^{i}_{signal}: Cluster signal ratio for the ***i*th** cluster.
- R^{min}_{signal}: Minimum desirable cluster signal Ratio.

We define cluster signal ratio R^{i}_{signal} for ***i*th** cluster as the ratio of the number of points (N_g) in the cluster where $S_{i,j} > S_g$ to the number of points (N_b) in the cluster where $S_{i,j} < S_b$.

$$R^{i}_{signal} = \frac{Ng}{Nb}$$

From the N_c clusters, for further steps, we consider only those clusters where $N_i > N_{min}$ and $R^{i}_{signal} < R^{min}_{signal}$. The group of interest obtained is further divided into two types of zones, good signal strength zones, and bad signal strength zones. This is done using support vector machine (SVM) classification. SVM classifies a given set of data into two discrete sets. Classification helps in concentrating the focus completely on the region which actually requires a new tower placement and installation.

3. **Positioning:** We now concentrate on the weak signal zone obtained from classification. We then perform K-means algorithm to find the centroid which is the possible candidate site for tower placement.

   The K-means algorithm works in the following steps:

- Randomly select "**k**" number of cluster centers.
- Calculate the distance between each data point and each cluster center.
- To every data point, assign that cluster center which is closest to that point.
- Calculate the new cluster center. Again, calculate the distance between each data point and the newly obtained cluster center.

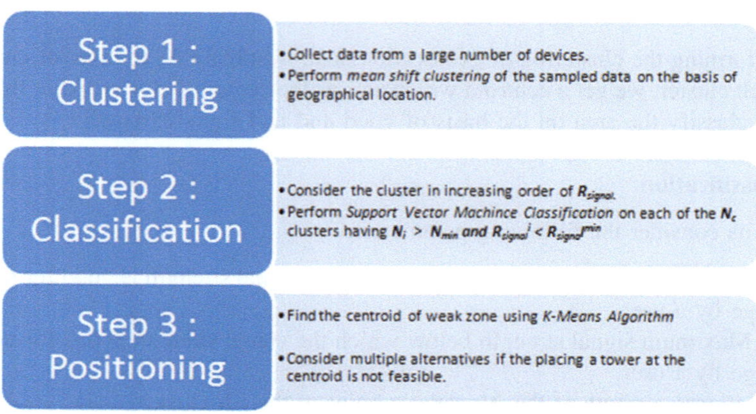

Fig. 1 Proposed tower placement scheme

- If none of the data points is reassigned to a new cluster center, stop, otherwise repeat steps (c) through (e).

We then consider the given candidate site so as to see the geographical suitability (Water bodies, hilly areas, etc., are neglected) and land availability. Figure 1 summarizes the proposed tower placement scheme and the steps involved in it.

4 Practical Implementation

We have designed an Android application for collecting data from the users. The application is compatible with Android 6.0 and below. Figure 2 contains a screenshot of the application. It collects the following data from a user:

- Network provider name
- Network type
- Signal strength
- Date and time
- Tower ID
- Location (GPS)

We use IMEI number to uniquely identify a user. The android application runs in the background and stores the aforementioned data in a file on the local storage, which is sent at regular intervals (a week) to a centralized server. We collect signal strength received by phone and the location from all users to increase the accuracy of signal strength used for analysis. Files of data collected from all users become the

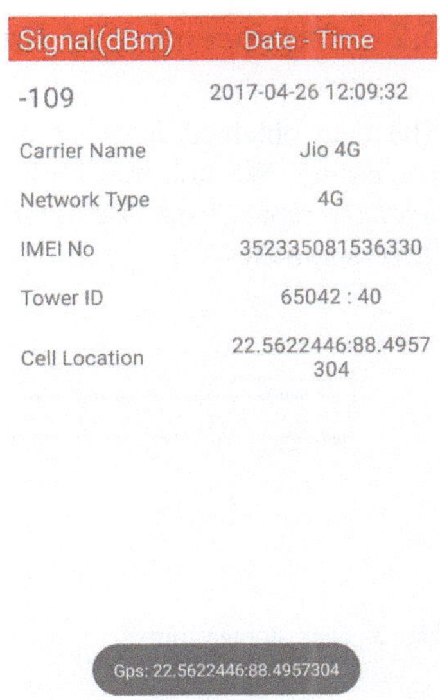

Fig. 2 Screenshot of the designed application

Table 1 Python modules used for practical implementation

Sl. No.	Module name	Purpose
1	Scikit-learn	Machine learning library with inbuilt support for classification
2	Matplotlib	Used for plotting the results obtained after clustering
3	Scipy	Used for plotting and results obtained after clustering
4	Pandas	Used for scientific and technical computation
5	Numpy	Provides support for large multidimensional arrays

inputs for a Python program, which reads the files, performs clustering and classification according to the proposed scheme and gives as output, the set of possible candidate sites for deploying base station. Table 1 contains information about the Python modules used for implementation purposes.

5 Experimental Setup

5.1 Data Collection

In order to collect the data required for verifying our proposed scheme, we install the application on 15 different devices, all present in an area of radius 2 km, in Lake Town, Kolkata. We collect the data for a period of 2 days, at random intervals. We use the same carrier on all 15 devices. Figure 3 shows the sample data format and Table 2 contains details about each component of the format.

5.2 Preprocessing of Data

The data obtained from all the devices is merged into a single file for further processing. We find that some of the samples have some components missing or wrongly represented. We filter the entries and accept those which are free from these anomalies:

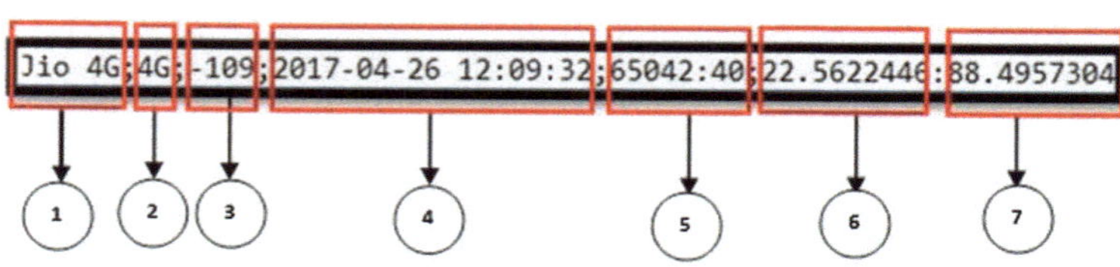

Fig. 3 Data storage format

Table 2 Fields

Index	Field name
1	Carrier name
2	Type of network
3	Signal strength in dBm
4	Date and time
5	Tower id
6	Latitude in radian
7	Longitude in radian

- The geographical location is missing from the entry.
- Redundant entry with the same timestamp and the same location value.
- If "Wifi" is detected as network type, we replace that with the last network type detected other than Wifi.

6 Results

We perform the experiment as explained in Sect. 5 of the paper and present our findings in this section. The values of the various parameters we use during the experimentation as mentioned in Table 3. In Fig. 4, we plot the result obtained after applying mean shift algorithm on the GPS location data collected with our android application. The black colored "X" marks plotted in Fig. 4 denotes the cluster centers obtained for several clusters.

We observe that we obtain several clusters, but only a few have a large number of sampled values. We ignore the smaller clusters, accepting those clusters that have cluster size greater than 10. This condition helps our proposed strategy concentrate on densely populated areas, and placing telecommunication tower in such areas would help improve services for a large consumer base.

In Fig. 5, we plot the signal strength quality (good signal strength and bad signal strength) at the respective locations where they were measured. A red colored point indicates that the signal strength received at that location is bad, and a green colored point indicates that the signal strength received at that location is good.

The distinction between good and bad signal strength depends on $\boldsymbol{S_g}$ and $\boldsymbol{S_b}$ parameters, as considered in Step 2 of our proposed scheme. The values of $\boldsymbol{S_g}$ and $\boldsymbol{S_b}$ that we consider during the experiment are mentioned in Table 3. Figure 5

Table 3 Values of parameters considered during experimentation

Sl. No	Field name	Value
1	S_g	−70 dBm
2	S_b	−100 dBm
3	N_{min}	10
4	R_{signal}^{min}	0.75

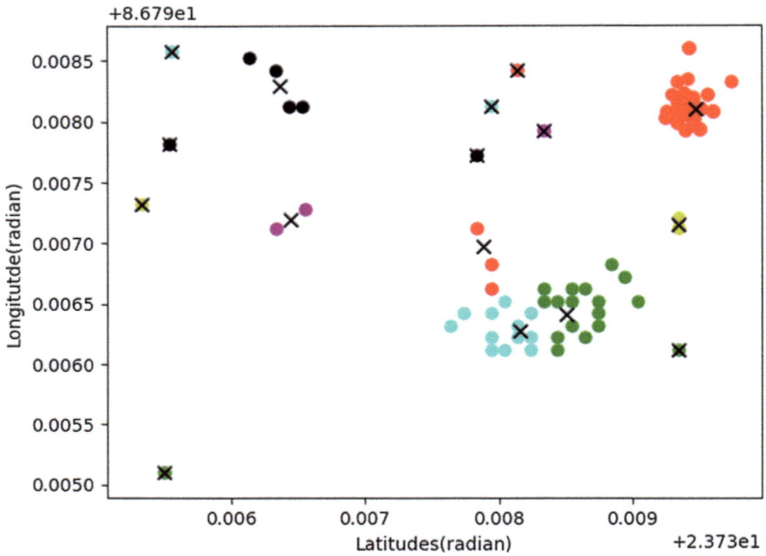

Fig. 4 Plot of location data obtained from the Android application. The "X"-marked points are centroids of their respective clusters which we obtain from Step 1 of the proposed scheme

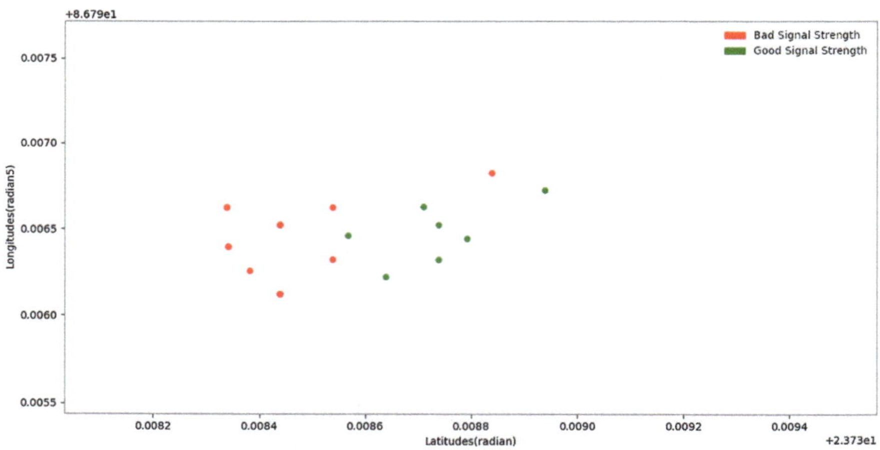

Fig. 5 Plot of signal strength received at sampled data points

shows the signal strengths received in one of the weak signal zones obtained during our experiment where $R^i_{signal} < R^{min}_{signal}$, which indicates that the coverage ratio is less than the minimum level set by the telecom operator, thus increasing its priority for tower installation among the sampled zones.

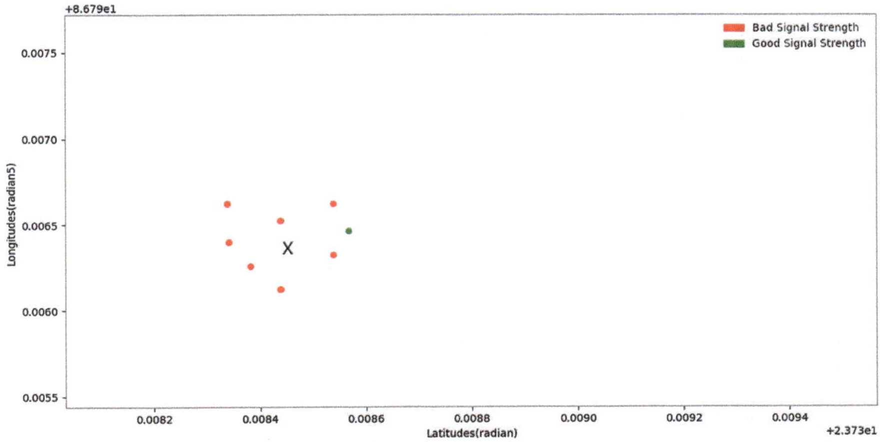

Fig. 6 The black colored X denotes the position of new tower suggested by the proposed scheme after K-means algorithm

We then plot the cluster center obtained for the weak signal zone using K-means algorithm. Figure 6 shows the location of tower installation finally proposed by the designed scheme, marked with a black colored X mark.

7 Conclusion

In this paper, we have presented a model for tower placement with practical implementation. We have designed an android application to facilitate data collection and the data collected has been analyzed using data science through Python Libraries. Our proposed scheme suggests a tower installation in an area with poor signal strength, where otherwise the network provider would have installed more than one tower. Thus, our optimal tower placement strategy would help the network providers make informed and data-driven decisions regarding new tower installation.

References

1. TRAI *Press Release on Telecom Subscription Data, 2016.*
2. GSMA Association report *The Mobile Economy, 2017.*
3. Communications Consumer Panel report *Mobile coverage: the consumer perspective, 2009.*
4. Honghui Dong *et al.*, "Urban residents travel analysis based on mobile communication data," *16th International IEEE Conference on Intelligent Transportation Systems (ITSC 2013)*, The Hague, pp. 1487–1492. 2013.

5. J. Liao, Z. Wang, L. Wan, Q. C. Cao and H. Qi, “Smart Diary: A Smartphone-Based Framework for Sensing, Inferring, and Logging Users’ Daily Life,” in *IEEE Sensors Journal*, vol. 15, no. 5, pp. 2761–2773, May 2015.
6. S. Rallapalli, W. Dong, G. M. Lee, Y. C. Chen and L. Qiu, “Analysis and applications of smartphone user mobility,” *2013 Proceedings IEEE INFOCOM*, Turin, pp. 3465–3470, 2013.
7. Church, R.L., ReVelle, C., The maximal covering location problem. Regional Science 30, 101–118, 1974.
8. M. B. Pereira, F. R. P. Cavalcanti and T. F. Maciel, “Particle Swarm Optimization for base station placement,” *2014 International Telecommunications Symposium (ITS)*, Sao Paulo, pp. 1–5, 2014.
9. D. Komnakos, A. Rouskas and A. Gotsis, “Energy Efficient Base Station Placement and Operation in Mobile Networks,” *European Wireless 2013; 19th European Wireless Conference*, Guildford, UK, pp. 1–5, 2013.
10. O. Celebi, E. Zeydan, O. Kurt, O. Dedeoglu, O. Iieri, B. Aykut Sungur, A. Akan, S. Ergut, On use of big data for enhancing network coverage analysis, in: *20th International Conference on Telecommunications (ICT)*, pp. 1– 5, 2013.
11. A. Karatepe, E. Zeydan, Anomaly detection in cellular network data using big data analytics, in: *Proceedings of 20th European Wireless Conference*, pp. 1–5, 2014.
12. D. Comaniciu and P. Meer. Mean shift: A robust approach toward feature space analysis. *IEEE Trans. Pattern Anal. Machine Intell.* 24: pp. 603–619, 2002.

Evaluation of IoT-Based Computational Intelligence Tools for DNA Sequence Analysis in Bioinformatics

Zainab Alansari, Nor Badrul Anuar, Amirrudin Kamsin, Safeeullah Soomro and Mohammad Riyaz Belgaum

Abstract In contemporary age, computational intelligence (CI) performs an essential role in the interpretation of big biological data considering that it could provide all of the molecular biology and DNA sequencing computations. For this purpose, many researchers have attempted to implement different tools in this field and have competed aggressively. Hence, determining the best of them among the enormous number of available tools is not an easy task, selecting the one which accomplishes big data in the concise time and with no error can significantly improve the scientist's contribution in the bioinformatics field. This study uses different analyses and methods such as fuzzy, Dempster–Shafer, Murphy, and entropy Shannon to provide the most significant and reliable evaluation of IoT-based computational intelligence tools for DNA sequence analysis. The outcomes of this study can be advantageous to the bioinformatics community, researchers, and experts in big biological data.

Keywords Internet of things · Computational intelligence · DNA sequence analysis · Bioinformatics · Big data · Entropy analysis

---

Z. Alansari (✉) · N. B. Anuar · A. Kamsin
College of Computer Science and Information Technology,
University of Malaya, Kuala Lumpur, Malaysia
e-mail: z.alansari@siswa.um.edu.my; zeinab@amaiu.edu.bh

N. B. Anuar
e-mail: badrul@um.edu.my

A. Kamsin
e-mail: amir@um.edu.my

Z. Alansari · S. Soomro · M. R. Belgaum
College of Computer Studies, AMA International University, Salmabad, Bahrain
e-mail: s.soomro@amaiu.edu.bh

M. R. Belgaum
e-mail: bmdriyaz@amaiu.edu.bh

C. R. Panigrahi et al. (eds.), *Progress in Advanced Computing and Intelligent Engineering*, Advances in Intelligent Systems and Computing 714,
https://doi.org/10.1007/978-981-13-0224-4_31

1 Introduction

Internet of things is a new revolution on the Internet. Objects make themselves recognizable and getting smarter by creating and providing relevant decisions. They can connect to each other and can have access to collected information by other objects or can be a part of a larger complex service. This development coincides with the emergence of cloud computing capabilities and the transition from the traditional Internet to the IPv6 unlimited addressing capacity.

Computational intelligence (CI) is one of the most significant sectors of AI which applies a variety of methods for the AI realization. The tools used in computational intelligence are often mathematical tools that somehow inspired by nature and the world around. The following are some of the most valuable tools and templates considered in computational intelligence:

- Evolutionary computation which is a set of methods that are known as evolutionary algorithms. The most popular algorithms are genetic algorithm inspired by the theory of evolution and genetics. This algorithm stimulated the evolution process that happened in nature over millions of years. The primary applications of evolutionary algorithms are solving optimization problems and mathematical planning [1].
- Swarm intelligence and the methods that fall into this category suggest another model for solving optimization problems. In this way, a significant number of very simple and low intelligence agents collaborate or compete to form a different type of swarm intelligence or collective intelligence. For example, an ant colony optimization algorithm which simulated by the collective presence of ants is one of the swarm intelligence algorithms [2].
- Artificial neural network (ANN) is also one of the most important CI algorithms. Almost all scientists are confident that the human brain is the most known complex structure in the entire universe. Mathematicians and engineers of artificial intelligence that inspired by the findings of neuroscientists (neurologist) introduced an ANN which uses a variety of information modeling and classification. Perhaps neural networks can be considered as the most valuable tool in machine learning field [3].
- Fuzzy systems are using concepts like high or low to describe an idea instead of using exact numbers. For example, in phrases such as high profit, the amount of profit is not exactly clear [4]. Today, fuzzy systems excess usage to design different smart appliances. In addition to the above, other mathematical tools are used to improve the overall performance of systems based on computational intelligence. The primary goal of researchers in the fields of artificial intelligence and computational intelligence is to create such tools that provide us a closer alliance with human intelligence [5].

In this paper, we focus on the evaluation of IoT-based computational intelligence tools for DNA sequence analysis and the most important challenges and open issues.

2 Literature Review

Bioinformatics handles the immature data collected from researchers daily to form the image, charts, and numbers. It also sorts the data gathered from a large variety of databases on the network. The meaning and significance of initial tests on data collection including experimental errors, principles or to data collection for statistical coincidence means need careful experimental design and multiplicity results [6]. Experiments in professional conditions, reactants, equipment, and time are costly. To conclude, biological data are always incomplete [7].

Driven large amounts of data from recent biological tests led to the creation of massive databases that contain genes, proteins, and genetic data and another data type's sprocket [8]. Introduced big data and reviewed related technologies, such as cloud computing, Internet of things, and Hadoop. Researchers insist on recovering data from some of the central databases specification such as nuclide or amino acid chain, organism, marginal genes, or proteins name [9]. To increase the production of experimental data, computer simulations based on CI play a fundamental part in biological methods [10]. Considered the IoT criteria in the health sector for sustainable development. Quick result and conclusion based on CI increased the biological information products [11]. Discussed the relationship between sequence database, IoT, and bioinformatics.

Genetic engineering refers to a set of methods which are used for isolation, purification and implying and expression of a particular gene in a host [12]. It ultimately causes a particular trait or produces the desired product in the host organism. Today, the technology and knowledge of genetic engineering and molecular biotechnology seem almost unlimited [13]. In recent years, development of tools for DNA sequencing provided recombinant revolutions in the treatment of many human diseases including all kind of cancers and most of the autoimmune diseases such as diabetes and the detection, prevention, and treatment of many congenital diseases [14].

With the development of technology and biologically advanced tools, researchers faced the massive amounts of big biological data which the analysis of them using experimental methods associated with some challenges [15]. Therefore, some new ways with high speed and accuracy are needed for this purpose. Due to the high speed and accuracy of computational intelligence methods, it can be said that they are a good alternative for being considered instead of laboratory procedures [16].

3 Research Methodology

This study used a practical descriptive survey given that it is based on the decision team to provide the needed data for determining the considerable sample size. The research's data collected from engineers, researchers, and experts by questionnaires and interviews.

3.1 Theory of Dempster–Shafer

In uncertainty time, data integration is imperative, and for this purpose, Bayesian theory, fuzzy logic, and the evidence theory are the known methods. The theory of Dempster–Shafer is considered as one of the most methods used for uncertainty reasoning, modeling, and accuracy of intelligent systems [17].

Dempster–Shafer theory is one of the primary methods for evaluating the uncertainty of unstructured data. It was founded by Dempster using the concept of upper and lower probabilities, and then Shaffer introduced it as a theory [18]. Moreover, measurement of uncertainty is one of the most important roles of entropy as one of a basic concept of big data and can be used as an uncertainty analysis tools in a particular situation.

The uncertainty decision is one of the most important research issues in computational systems and unstructured data. In recent years, the researchers and engineers provided useful definitions of uncertainty. The dual uncertainty nature is expressed by Helton [19] with the following definitions:

1. Aleatory uncertainty: as the fact that system can act randomly.
2. Epistemic uncertainty: happens when there is a shortage of data about the particular system, and it is a feature for performance analyzing.

Dempster combination rule is critical to combine evidence from different sources [20] and is a potential tool to evaluate the risk assessment and computational results. This method used in the impossibility of test's accurate measurement or inference knowledge of expert's opinion. One significant feature of this theory is the combination of evidence which extracted from different sources and modeling of conflict between them.

3.2 Reliability of Research Tools

According to the prepared questionnaires, it can be said that they measure all the criterions and options. In another word, most of the questions contain the desired structure considering that all the criterions investigated and the designer was not able to design an absolute orientation in questionnaire's design. Moreover, interviews were conducted to determine the security level of criterions and the overall rating level was calculated using the fuzzy and Dempster–Shafer method for which the obtained outputs were logically correct. The measurement's reliability in this study does not benefit from the quantitative methods thus the assessment credit of assessors is said to be considered as a criterion for reliability analysis. However, for paired comparison of the questionnaires which is based on the Saaty's scale [21], we can use the compatibility rating to evaluate the reliability. Therefore, the compatibility criterion which is used to measure the incompatibility in the paired comparison matrix, when we use the group AHP to combine the matrixes, is determined as follows:

$$CI = \frac{(\lambda_{max} - n)}{n}. \quad (1)$$

Then compatibility rating (CR) will be calculated by $= \frac{CI}{RI}$. If the result of CR is less than 0.1, the matrix compatibility is acceptable. The research process is shown in Fig. 1.

3.3 Research Process

(1) First Stage

At this stage of research, all the computation tools for DNA sequence analysis in bioinformatics are identified through library research, literature review, and all current researches in this area. Then they were placed in a cycle form with an appropriate structure which a total of 14 criteria constitute the main elements of the cycle. Finally, the obtained components and subcomponents were verified by bioinformatics experts. The criterion's structure is shown in Fig. 2.

(2) Second Stage

The aim of this phase is to determine the importance of criterions using entropy analysis of Shannon [22]. Therefore, to achieve the best result, a questionnaire was given to some experts of bioinformatics and the specific matrices to each were formed. Due to the formation of multiple matrices, we should use the geometric

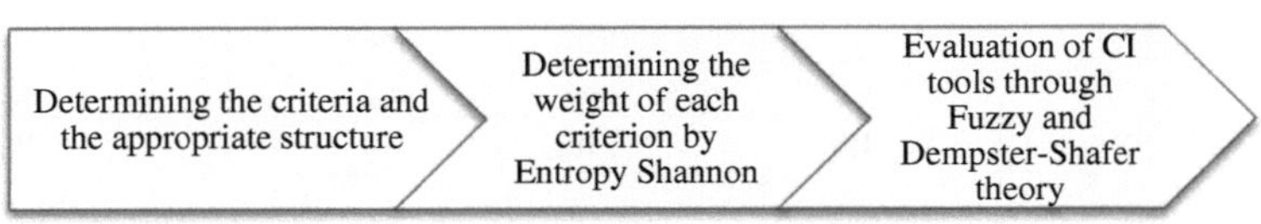

Fig. 1 The process of study

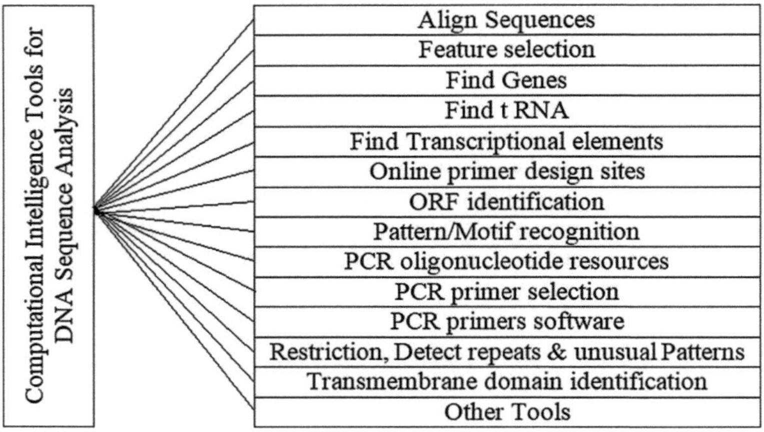

Fig. 2 Computational Intelligence tools for DNA sequence analysis

mean for the variable of each matrix $D = \left\| a_{ij} = \frac{w_i}{w_j} \right\|$ to obtain the final matrix. Equation 2 has been used to calculate the weights of each variable:

$$a'_{ij} = \left(\prod_{i=1}^{k} a_{ijI} \right)^{\frac{1}{k}}. \tag{2}$$

By getting the integrated matrix from Eq. 2, compatibility rating calculated which is used as an input for entropy Shannon. Finally, the final weight of each variable is computed using entropy Shannon noting that the input and output of each stage in this study is different from each other and each part is covering some specific requirements of this research. Therefore, each fuzzy and entropy method used in this study has different output. In this regard, to determine the variable's weight, the steps below are followed. The decision matrix contains some data which is used in entropy as an evaluation criterion. Suppose that the obtained decision matrix using paired comparison and combining geometric mean is as given in Table 1.

Table 1 Decision about indicators

Indicator	C_1	C_2		C_n
C_1	a_{11}	a_{12}	…	a_{1n}
C_2	a_{21}	a_{22}	…	a_{2n}
…	…	…	…	…
C_n	a_{m1}	a_{m2}	…	a_{mn}
W_j	W_1	W_2	…	W_n

Using this matrix, P_{ij} is calculated as Eq. 3:

$$P_{ij} = \frac{a_{ij}}{\sum_{i=1}^{m} a_{ij}} \; ; \forall_{i,j}. \tag{3}$$

The entropy's indicator E_j is obtained by Eq. 4:

$$E_j = -k \sum_{i=1}^{m} \left[P_{ij} \ln P_{ij} \right] ; \forall_j. \tag{4}$$

Uncertainty or deviation degree d_j which obtained for indicator j shows that the specific indicator of j, how much useful information is provided for the decision. The amount of W_j is obtained from Eq. 5:

$$d_j = 1 - E_j \, ; \forall_j. \tag{5}$$

Then the weight is calculated using Eq. 6:

$$W_j = \frac{d_j}{\sum_{j=1}^{n} d_j} \; ; \forall_j. \tag{6}$$

If a particular weight was considered earlier like λ_j for indicator j, the adjusted weight of $W_j^{'}$ is calculated as Eq. 7:

$$W_j^{'} = \frac{\lambda_j W_j}{\sum_{j=1}^{n} \lambda_j W_j} \; ; \forall_j. \tag{7}$$

(3) Third Stage

In this phase, the assessment of CI tools for DNA sequence analysis should be evaluated. By collecting the considered data, the next level starts which is data analysis and evaluation of final decision [23]. In order to determine the level of each tool, the collected data from four questionnaires used as input in fuzzy functions. The primary motivation for the fuzzy sets is uncertainty. A characterized function can define each subset of fuzzy A in the main set of X. These functions are called the membership function which for each x member, from the central set X, allocate a number in the range of 0.1 which represents the degree of x membership in the fuzzy set of A. Therefore, it is defined as A: X [1.0]. An example of a fuzzy set A in a defined set of X is $A = \{\langle x, \mu_A(x)\rangle | x \in X$ which $\mu_A : X \rightarrow [0.1]$ is the membership function of A. The real value of $\mu_A(x)$, describes the degree of $x \in X$ in A. For a finite set of A = $\{x_1, \ldots, x_i, \ldots, x_n\}$, the fuzzy set of (A, m), usually shown as $A = \left\{ \frac{\mu_A(x_1)}{x_1}, \ldots, \frac{\mu_A(x_i)}{x_i}, \ldots, \frac{\mu_A(x_n)}{x_n} \right\}$.

In this study, if X is a defined set, five different variables describe the degree of CI tools in DNA sequence analysis which are X = {(VL) Very Low, (L) Low, (M) Medium, (H) High, (VH) Very High}. If we assume that only two adjacent variable overlaps, the fuzzy functions are defined as follows:

$$
\begin{aligned}
&f_{very\,low}(x) = -0.4x+1, && 0 \leq x \leq 2.5.\\
&f_{low}(x) = -0.4x, && 0 \leq x \leq 2.5.\\
&f_{low}(x) = -0.4x+2, && 2.5 \leq x \leq 5.\\
&f_{medium}(x) = 0.4x-1, && 2.5 \leq x \leq 5.\\
&f_{medium}(x) = -0.4x+3, && 5 \leq x \leq 7.5.\\
&f_{high}(x) = 0.4x-2, && 5 \leq x \leq 7.5.\\
&f_{high}(x) = -0.4x+4, && 7.5 \leq x \leq 10.\\
&f_{veryhigh}(x) = 0.4x-3, && 7.5 \leq x \leq 10.
\end{aligned} \tag{8}
$$

where f_{VL}, f_L, f_M, f_H, and f_{VH} are the membership functions of fuzzy sets. After determining the degree of each indicator, it is time to combine the same level functions. For this purpose, we must lower the functions to increase the confidence given to each indicator. In fact, the lower rate is used when an information source provides a basic probability assignment ($BPAm$) which has the same reliability as $\propto$. Therefore, $(1-\propto)$ considers as a lowering rate and the new $BPAm^{\propto}$ is defined as:

$$
\begin{aligned}
m'(A) = \propto m(A), \quad \forall A \subset \theta, \quad A \neq \theta.\\
m'(\theta) = 1 - \propto + \propto\, lm(\theta).
\end{aligned} \tag{9}
$$

All the mass functions should be lowered using $\propto$ which is called the lower factor where m is a mass function of a witness, m^a represents the indicative allocation function of initial lower probability and the lower factor a $(0 \leq a \leq 1)$, determine the evidence reliability. Noting that before the final composition, the overlap value of indicators must be calculated by Eq. 10:

$$
\begin{aligned}
m'(A) = \propto m(A), \quad \forall A \in \theta, \quad A \neq \theta.\\
m'(\{Y, A\}) = \tfrac{S(Y \cap A)}{S(X \cap A)} \times (1 - \propto m(A)), \quad Y \neq A, \quad Y \in X, \quad X \subset \theta.
\end{aligned} \tag{10}
$$

Then the combination level began. Given that this study contains some conflicts, an averaging method of Murphy [24] is used to overcome the conflicts. As Murphy proposed, if all the evidences are available concurrently, the mass average can be calculated and the final mass by joining the averaged values several times can be found. This rule can combine the two $BPA(m)$ of m_1 and m_2 for the new $BPA(m)$. Noting that the Dempster combination rule combines the multiple belief functions through $BPA(m)$. Dempster–Shafer combination rule is shown as $m = m_1 \oplus m_2$ and specifically obtained from the combination of BPAs m_1 and m_2:

$A \neq \emptyset$ and $(\emptyset) = 0$ when

$$m_{12}(A) = \frac{\sum_{B \cap C = A} m_1(B) m_2(C)}{1 - k}. \quad (11)$$

$$k = \sum_{B \cap C = \emptyset} m_1(B) m_2(C).$$

4 Analysis and Result

According to the decision matrix of Table 1, to gain the indicator's weight, we have to follow the steps below:

Step 1: Calculating P_{ij}: after calculating P_{ij} and gaining its values we follow the other steps as below.

Step 2: Calculate the entropy amount E_j according to the calculated values of P_{ij} and Eq. 4, the amount of entropy can be obtained which is shown in Table 2.

Step 3: Calculating uncertainty value (d_j): the values of uncertainty are gained according to entropy's values and Eq. 5.

Step 4: Calculating the weight (W_j): the weight of each indicator is gained according to the uncertainty value and Eq. 6.

Step 5: Adjusted weight (W_j'): the adjusted weights are calculating according to indicator's weight and intellectual weight (λ_j) is calculated according to Eq. 7.

Based on Table 2 and calculating the mean of entropy and uncertainty values and the weight of indicators with intellectual and adjusted weight, we find that

Table 2 Gained values (step 2 to 5)

Indicators	Entropy value (E_j)	Uncertainty value (d_j)	Indicator's weight (W_j)	Intellectual weight (λ_j)	Adjusted weight (W_j')
B1	0.966	0.034	0.245	0.2333	0.313
B2	0.963	0.037	0.263	0.2333	0.336
B3	0.985	0.015	0.106	0.2333	0.135
B4	0.982	0.018	0.13	0.1834	0.13
B5	0.977	0.023	0.166	0.0667	0.061
B6	0.987	0.013	0.091	0.05	0.025
B7	0.856	0.144	0.0293	0.4	0.355
B8	0.735	0.265	0.54	0.3	0.492
B9	0.918	0.082	0.167	0.3	0.152
B10	0.996	0.004	0.024	0.1667	0.27
B11	0.965	0.035	0.184	0.1333	0.17
B12	0.975	0.025	0.13	0.15	0.136
B13	0.992	0.008	0.043	0.15	0.045
B14	0.945	0.055	0.129	0.1833	0.372

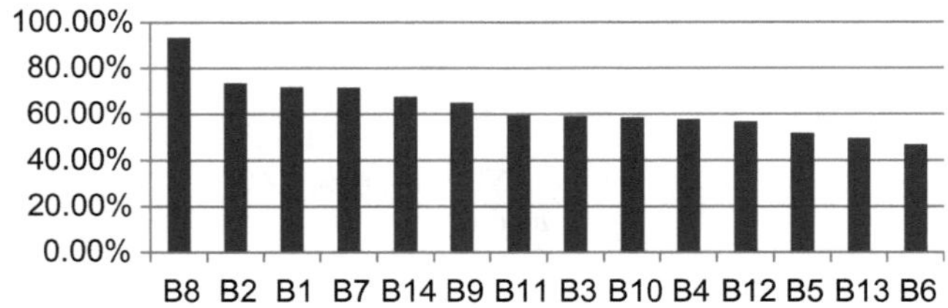

Fig. 3 Evaluation of IoT-Based CI Tools for DNA Sequence Analysis

Table 3 Indicator's rating

Indicators	Description	Rating
B1	Align sequences	M
B2	Feature selection	VH
B3	Find genes	H
B4	Find t RNA	M
B5	Find transcriptional elements	M
B6	Online primer design sites	VH
B7	ORF identification	H
B8	Pattern/Motif recognition	H
B9	PCR oligonucleotide resources	H
B10	PCR primer selection	VL
B11	PCR primers software	VL
B12	Restriction, detect repeats and unusual patterns	L
B13	Transmembrane domain identification	VL
B14	Other tools	H
B14	Other tools	H

almost all fourteen indicators have near rating and similarities which are shown in Fig. 3.

After calculating each indicator's weight, the second part of data which are usability, reliability, validity, and power of each indicator were collected from the experts and used as input for Eq. 8. Table 3 shows the calculated results:

By determining the indicator's score, the next step is to combine the indicators of each group. For this purpose, the following five diagnosis hypotheses are considered: $\theta = \{$(VL) Very Low, (L) Low, (M) Medium, (H) High, (VH) Very High$\}$

Each one of these is indicating the CI tools rating for DNA sequence analysis in bioinformatics and used as input in Dempster–Shafer theory. Noticing this evidence is preliminary and vague for combination, they need to lower first. Equation 9 is used for lowering the evidence, and the overlap between variables obtained using Eq. 10 and finally is composition stage turn. After synthesizing the evidence,

Table 4 Overall evaluation of IoT-Based CI Tools for DNA sequence analysis

Evidence combination	VL	L	M	H	VH	VL, L	L, M	M, H	H, VH
B1, B2, B3, B4	0	0	0	0	0	0	0.01	0.02	0.03
B3, B4, B5, B6	0	0	0	0	0	0	0.01	0.02	0.01
B5, B6, B7, B8	0	0	0	0.2	0	0	0.01	0.11	0.11
B7, B8, B9, B10	0	0	0	0.2	0	0	0	0.08	0.08
B9, B10, B11, B12	0	0	0	0.1	0	0.02	0.01	0.04	0.04
B11, B12, B13, B14	0	0	0	0.1	0	0.02	0.01	0.03	0.03
Average	0	0	0	0.1	0	0.01	0.01	0.05	0.05

maybe 100% assurance allocate to the particular focal element. Several ways have been introduced for facing such conflicts. This study used Murphy's proposed idea, and the calculations results are shown in Table 4.

5 Conclusion

As per the results of this study, it indicates that pattern/motif recognition tools have the highest ranking and the lowest ranking is given to online primer design sites. This study shows that the usability of all the DNS sequence analysis methods integrated with the computational intelligence tools are almost equal and it confirms the importance of CI in bioinformatics. This study uses distinctive analysis and approaches such as fuzzy system, Dempster–Shafer algorithm, a method of Murphy and Shannon's entropy to provide the most significant and reliable evaluation of computational intelligence tools for DNA sequence analysis. The CI tools play a fundamental role in bioinformatics DNA sequence analysis. Hence, the recommendation to the computer scientists, engineers, and researchers are to examine research, produce, and propose innovative tools and methods in this area using the presented consequences of this research. The findings of this study can be advantageous to the bioinformatics community, researchers, and experts in big biological data.

References

1. Sastry, Kumara, David E. Goldberg, and Graham Kendall. "Genetic algorithms." Search methodologies. Springer US, 2014. 93–117.
2. Alam, Shafiq, et al. "Research on particle swarm optimization based clustering: a systematic review of literature and techniques." Swarm and Evolutionary Computation 17 (2014): 1–13.
3. Graupe, Daniel. Principles of artificial neural networks. Vol. 7. World Scientific, 2013.
4. Alpaydin, Ethem. Introduction to machine learning. MIT press, 2014.

5. Shaikh, Pervez Hameed, et al. "A review on optimized control systems for building energy and comfort management of smart sustainable buildings." Renewable and Sustainable Energy Reviews 34 (2014): 409–429.
6. Mohammad, Riyaz, B., Safeeullah, S., & Zainab, A. "Cloud Service Ranking Using Checkpoint Based Load Balancing in Real Time Scheduling of Cloud Computing." International Conference on Advanced Computing and Intelligent Engineering. India: Springer, 2016.
7. Zainab Alansari, Safeeullah Soomro, Mohammad Riyaz Belgaum, Shahaboddin Shamshirband. "The Rise of Internet of Things (IoT) in Big Healthcare Data: Review and Open Research Issues". International Conference on Advanced Computing and Intelligent Engineering 2016, India. Springer. 2016.
8. Li, Songnian, et al. "Geospatial big data handling theory and methods: A review and research challenges." ISPRS Journal of Photogrammetry and Remote Sensing 115 (2016): 119–133.
9. Klipp, Edda, et al. Systems biology: a textbook. John Wiley & Sons, 2016.
10. Rong, Ke, et al. "Understanding business ecosystem using a 6C framework in Internet-of-Things-based sectors." International Journal of Production Economics 159 (2015): 41–55.
11. Acland, Abigail, et al. "Database resources of the national center for biotechnology information." Nucleic acids research 42. Database issue (2014): D7.
12. Nussbaum, Robert L., Roderick R. McInnes, and Huntington F. Willard. Thompson & Thompson Genetics in Medicine E-Book. Elsevier Health Sciences, 2015.
13. Lewis, Douglas R. Biotechnology: An Era of Hopes and Fears. Air Force Institute of Technology Wright Patterson AFB United States, 2016.
14. Zainab Alansari, Nor Badrul Anuar, Amirrudin Kamsin, Safeeullah Soomro and Mohammad Riyaz Belgaum. "The Internet of Things Adoption in Healthcare Applications". The IEEE 3rd International Conference on Engineering, Technologies and Social Sciences, 2017.
15. Chen, CL Philip, and Chun-Yang Zhang. "Data-intensive applications, challenges, techniques and technologies: A survey on Big Data." Information Sciences 275 (2014): 314–347.
16. Witten, Ian H., et al. Data Mining: Practical machine learning tools and techniques. Morgan Kaufmann, 2016.
17. Hall, David L., and James Llinas. "An introduction to multisensor data fusion." Proceedings of the IEEE 85.1 (1997): 6–23.
18. Yager, R. R. "On the Dempster-Shafer framework and new combination rules." Information sciences 41(2) (1987): 93–137.
19. Helton, J. C. "Uncertainty and sensitivity analysis in the presence of stochastic and subjective uncertainty." Journal of Statistical Computation and Simulation 57(1–4) (1997): 3–76.
20. Sentz, K., and S. Ferson. "Combination of evidence in Dempster-Shafer theory." Albuquerque: Sandia National Laboratories 4015 (2002).
21. Saaty, T. L. "What is the analytic hierarchy process?" Berlin Heidelberg: Springer, 1988. 109–121.
22. Shannon, C. E. "A mathematical theory of communication." ACM SIGMOBILE Mobile Computing and Communications Review 5(1) (2001): 3–55.
23. Ross, T. J. Fuzzy logic with engineering applications. John Wiley & Sons, 2009.
24. Murphy, C. K. "Combining belief functions when evidence conflicts." Decision support systems 29(1) (2000): 1–9.

Efficient Data Deduplication for Big Data Storage Systems

Naresh Kumar, Shobha and S. C. Jain

Abstract For efficient chunking, we propose Differential Evolution (DE) based approach which is optimized Two Thresholds Two Divisors (TTTD-P) Content Defined Chunking (CDC) to reduce the number of computing operations using single dynamic optimal parameter divisor D with optimal threshold value exploiting multi-operations nature of TTTD. To reduce chunk size variance, TTTD algorithm introduces an additional backup divisor D′ that has a higher probability of finding cut points, however, adding an additional divisor decreases chunking throughput. To this end, Asymmetric Extremum (AE) significantly improves chunking throughput by using local extreme value in a variable-sized asymmetric window to overcome Rabin and TTTD boundaries shift problem, while achieving nearby same deduplication ratio (DR). Therefore, we propose DE-based TTTD-P optimized chunking to maximize chunking throughput with increased DR; and scalable bucket indexing approach reduces hash values judgment time to identify and declare redundant chunks about 16 times than Rabin CDC, 5 times than AE CDC, 1.6 times than FAST CDC on Hadoop Distributed File System (HDFS).

Keywords Data deduplication · Content defined chunking · TTTD
HDFS

N. Kumar (✉) · Shobha
Department UIET, Computer Science and Engineering, Kurukshetra University, Kurukshetra 136119, India
e-mail: naresh_duhan@rediffmail.com

Shobha
e-mail: shobha.antwal@gmail.com

S. C. Jain
Department Computer Science and Engineering, Rajasthan Technical University, Kota 324010, India
e-mail: scjain1@yahoo.com

C. R. Panigrahi et al. (eds.), *Progress in Advanced Computing and Intelligent Engineering*, Advances in Intelligent Systems and Computing 714,
https://doi.org/10.1007/978-981-13-0224-4_32

1 Introduction

In the recent year, the exponential growth in big data makes it the next big thing in the IT world. Big data is similar to smaller data but bigger in size that includes large dataset made up of a variety of structured, semi-structured and unstructured data. Structured data is highly organized and easily searchable data that resides within a relational database SQL format. Semi-structured data has a self-describing structured that includes CSV, XML, JSON, and NoSQL formatted data. Unstructured data is difficult to analyze and search often includes text and multimedia content. Example of unstructured data includes message, word processing document, PDF, presentation, photographs, audio files, video files, E-mail message, web page, and many more other kinds of business documents.

In order to explore the enormous amount of data, big data was introduced to the computing world by Roger Magoulas from O'Reilly media in 2005. The challenges of big data include analyzing, searching, transfer, storing, and redundancy reduction [1, 2]. The typical software tool ability is not enough to store the vast amount of information so some tools or techniques are required to solve storage problem of big data. The automatically and human-generated information [3, 4] has brought greater pressure to traditional techniques. How to store the vast amount of information is the big task.

Data deduplication has increased attention in large-scale storage systems to data reduction by eliminating redundant data at chunk level and identify duplicate contents by cryptographically secure hash signatures like SHA-1 fingerprint. According to deduplication studies conducted by IBM, Microsoft, Google, Intel, and Motorola, approximately 80% of the data in distributed storage systems are redundant and can be eliminated by efficient deduplication technologies.

According to a report from International Data Corporation (IDC), the amount of digital information created and copied in the whole world is about 1.8 ZB ($\sim 10^{21}$ Bytes) in 2012 and the volume will reach 40 ZB in 2020 [5, 6]. By observing the problem, there are some techniques required that efficiently process and store all types of data. Cloud storage [7, 8–14] is one of the solutions but the management of various storage nodes is difficult in the cloud network. As a result, the complexity of network increases and the performance of cloud network degrade. Deduplication is another most important technology in storage that used by various users. Deduplication is the process of eliminating duplicate copies of data through a deduplication scanning process so that unique copy is stored and will then serve all authorized user.

Based on the location, deduplication can be performed on the client side (source side) where the redundancy removed from the file at the source side before transmission to the backup. It reduces the bandwidth utilization but it is slower than target based deduplication especially for large files. Target-based deduplication perform the redundancy removal from the backup server upon which the backup software resides. It reduces the amount of storage requirement but does not reduce the amount of data that must be sent across local area network during the process of backup. Another categorization of deduplication is based on the ownership. In single user, the process of deduplication is applied on a single user, the compression techniques are applied to reduce the data size. The redundant data is not removed in this, so the cross client is the better option in which the client and server match the duplicity and then remove the redundant data.

The hird categorization is based on the granularity, deduplication can be performed over the whole file where the complete file is taken as a single chunk and the hash value is computed using the hash function. Hash function is a part of cryptography but does not require any special key for coding. It is a function that maps any value into a fixed size hash value called a hash code. A single bit change in the value generates the different hash code. If the file hash value is matched with the previously stored hash value then only reference pointer is given, otherwise, the file hash value is stored in the storage device. Chunk level deduplication is performed within a file where the file content is first broken into smaller chunks. The chunk may be of fixed size or variable size. We have performed our experimental results with analysis on Hadoop Distributed File System. Hadoop is an open source framework for storing data and running applications. It includes the HDFS, Hadoop common, Hadoop YARN, and Hadoop MapReduce. The main components of Hadoop are HDFS and MapReduce. HDFS is a master–slave architecture in which master contains information of name node and job tracker. The slave contains the data node, task tracker, and MapReduce information.

The contribution of this paper is to present an optimized TTTD-P variable size data deduplication scheme for HDFS environment, in order to achieve higher deduplication ratio with low deduplication time. The rest of this paper is organized as follows: Sect. 2 presents the related review and background concept in deduplication. Section 3 describes our proposed variable size TTTD-P algorithm with optimal parameters. Section 4 describes the experimental results with bucket indexing. Finally, Sect. 5 concludes this paper.

2 Background and Motivation

A low bandwidth network file system (LBFS) was proposed by Athicha Muthitacharoen to exploit the interfile and cross file similarity [15]. It avoids transmitting the redundant data over the network so that the less bandwidth consume in the transmission. LBFS breaks the file into variable size chunk based on the content and then indexes each chunk by using hash value. The hash value of chunk was computed by hashing function like MD5 or SHA1. The hash functions are a combination of cryptographic application like digital signature, random value generator, one-way function, and integrity protection. It maps the arbitrary input into fixed size output bit-the hash value. Message Digest 5 (MD5) is most widely used algorithm in the list of hashing function [16]. It processed the variable length message into fixed size 128-bit output. The process of MD5 includes five steps: appending, padding bits, appending length, initialize MD5 buffer, process message in 16-words and output. SHA-1 is another most commonly used hash function that processes the input 512-bit block and produces 160-bit message digest [17]. The algorithm of SHA-1 includes padding, appending length, initializing the SHA-1 buffer, process message in 16-word block, and final output.

Deduplication algorithms detect the duplicate content by using the hashing function. The duplication can be found at the file level or at the block level. In 2009, Deepavali et al. [18] proposed a scalable deduplication technique for the backup storage that works at file level. It detects the duplicate chunk with high accuracy by breaking the chunk into two tiers. One resides in the primary index (RAM) and another resides in the secondary index (BIN). When deduplication is performed at block level, the data stream is partition into blocks that may be of fixed size or variable size. For backup application and large-scale file systems, many of organizations use fixed size blocks. But there is a limitation in this method; for every insertion or deletion in the original file it may generate a set of chunks that are entirely different from the original ones. The boundary of the newly formed chunk is totally different from the previous chunks. It creates lots of metadata with some minor changes that increase the storage data and also increase the CPU overhead. Frequency-based chunking, byte index, and multi-byte index are some examples of fixed size chunking. Frequency-based chunking is a two-stage algorithm [19] that identifies the high frequent chunk. At first stage, it identifies the fixed size high-frequency chunk. Then at second stage, it consists of coarse-grained and fine-grained based chunking. In coarse-grained chunking, content defined chunking algorithm is used to partition the data stream into large size chunks and then fine-grained chunking scan each coarse-grained chunk to find the frequent fix size chunk.

Byte index chunking [20] finds the duplicity at byte level by searching the high probability duplicate chunk byte by byte in the file. Index matrix of size 256 $\times$ 256 is used to find the high probable duplicate chunk in less time. After that hash function is applied to confirm that the chunk is definitely duplicate or not. After that multi-byte index chunking [21] was proposed that use two index matrix. First 32 KB index matrix is used for files whose size less than 5 GB. If the file size is more than 5 GB then second 4 MB index matrix will be used.

Variable size chunking solves the problem of "boundary shift problem" that comes in fixed size chunking. It partitions the file according to the contents. Some of the variable sizes chunking algorithms are leap-based chunking, bimodal chunking, multimodal chunking, and basic sliding window CDC. Leap-based CDC algorithm [22] improves the deduplication performance by adding another judgment function. The pseudorandom transformation is used to define whether a window is qualified or not. This is the replacement of rolling hash function that is used in the sliding window CDC. Bimodal chunking [23] introduced as opposed to the unimodal baseline CDC approach. This is the improved version of CDC that mixes different average size chunks together. The algorithm first chunks the data stream into large chunks and then split parts of them into small chunks. A Multimodal Content Defined Chunking (MCDC) was proposed as a new enhancement in

Bimodal Content Defined Chunking. MCDC [24] determines the optimal chunk size according to data size and compressibility.

A new content defined chunking algorithm Asymmetric Extremum (AE) was presented that mainly focuses to improve the chunking throughput and the chunk size variance. The Rabin fingerprint based CDC and MAXP CDC algorithm limitations are removed by AE algorithm [25]. A variable size window is used that finds the maximum value without going in the reverse direction as opposed to fixed size window in Rabin-based CDC and MAXP. The AE algorithm requires one comparison and two conditional branch operations per byte. Basic Sliding Window (BSW) algorithm is used in variable size chunking. A signature is created for each chunk if the signature matches the predefined bit pattern, the algorithm sets the chunk boundary at the end of the window. After each comparison, the window slides one byte position and compute hash function. To find the duplicated data, the content defined chunking proposed a Low Bandwidth Network File System (LBFS) that determines the file similarity and saves bandwidth. Content defined chunking [26] breaks the input data stream into variable-sized chunks according to data contents.

The BSW algorithm establishes a window of byte stream starting from the first content to last content of a file as shown in Fig. 1. It performs file chunking, fingerprint generation, and redundancy detection. It avoids the boundary shift problem by making chunk boundaries depend on the local content of the file. There are three parameters in BSW: a fixed size sliding window W, an integer divisor D and an integer remainder R, where R < D. Parameter R must lie between 0 and D − 1, and usually taken as D − 1.

There are two problems in BSW algorithm. First, it may determine the breakpoint in each shift if the file contains lot of continuous repeating string like "aaaaaaaa". Due to this, the metadata size is equal to or larger than original file. The second problem is that after scanning complete file if no breakpoint detected then whole file as a chunk will be treated.

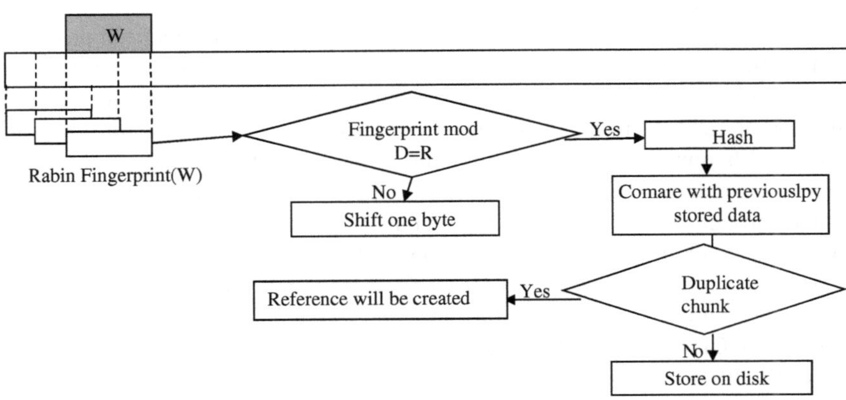

Fig. 1 Design of basic sliding window (BSW)

To resolve these problems, Two Thresholds Two Divisors (TTTD) algorithm [27] was introduced that takes second divisor D′, where D′ < D. It has a greater chance of searching the chunk boundary. In this, the entire stream is to be scanned and at every position, the primary and secondary divisor both compute the chunk boundary. If chunk boundary found by primary divisor D before going up to the T_{max} (Maximum Threshold) then it declare a breakpoint, otherwise, the breakpoint is determined by secondary divisor D′.

Some of the CDC algorithms help to find the more redundant data in less time. Ddelta [28] is one of the approaches that based on the Gear based CDC technique to eliminate the redundancy among identical data chunks. Merging the spooking-based fingerprint and gear-based chunking, it accelerates and improves the delta encoding and decoding processes. It also utilizes content locality of redundant data to search more duplicate. The wide research is the field acknowledges that 60% CPU overhead used during fingerprinting. Sample byte [29] is another solution to eliminate the redundant data in the end system services. With equivalent gain, it gives fast fingerprinting. To synchronize the commercial system services the novel approach Quick synchronization [30] is presented to shorten the synchronization time.

A fast and efficient approach is developed by Wen Xia et al. [31] to implement the gear based content defined chunking. Fast CDC is approximately 10 times faster than Rabin CDC (BSW) and 3 times faster than AE, but deduplication ratio of fast CDC is same as of Rabin [32] CDC means redundant data detection in both fast CDC and Rabin CDC is same; only computation overhead decreases in fast CDC. This new approach is the combination of fast gear CDC and the cut point skipping CDC. The first step is to perform the hash judgment by finding the hash value and then compare with previously stored value to find out the chunk cut point. After finding the sub-minimum chunk next step is to accelerate the speed of Gear based CDC by skipping the sub-minimum chunk. The final step is to normalize the chunk size distribution to a specified region that is larger than the minimum chunk size. Rabin CDC is time-consuming because it computes and judges the Rabin fingerprint [33–35] of the data byte by byte.

In order to speed up the CDC process, other hash algorithms had been proposed to replace the Rabin algorithm for CDC such as sample Byte, Gear, and AE. But gear-based CDC and AE CDC deduplication have the potential problem of low deduplication ratio means data redundancy detection is less, therefore, we are proposing differential evolution optimizing using bucket indexing to fast hash judgment by reducing computing operations time with increasing deduplication ratio and throughput. So we have Differential Evolution based data deduplication using bucket indexing which will be 16 times faster than Rabin CDC, 5 times faster than AE and 1.6 times faster than Fast CDC. Even more, DE-based approach has higher deduplication ratio by detecting maximum redundancy from data streams. Whereas fast CDC and AE have approximately same deduplication ratio as of Rabin CDC (BSW). So our new proposed DE-based data deduplication approach is faster as well as with more DR and throughput when analyzed with existing deduplication techniques.

3 Proposed Work

In this work, we have proposed an Optimized Two Thresholds, Two Divisors (TTTD-P) algorithm with optimal parameter values using the "Differential Evolution" function. The main function of this algorithm is to detect more duplicate content in the data stream. The TTTD algorithm uses the minimum and maximum threshold values to detect the duplicate content. An enhancement in TTTD is TTTD-S, it has a switch value that takes the average of the minimum and maximum threshold so that the duplicity detected more efficiently. Based on TTTD and TTTD-S algorithm, we have proposed an improvement in the TTTD algorithm that detects the more duplicity than the TTTD and TTTD-S. In this paper, we have done the comparative analysis of Rabin CDC, TTTD, AE, FAST CDC and TTTD-P in the HDFS environment. The variable size chunk hash value is computed by SHA1 [17].

A. *Two Thresholds, Two Divisors (TTTD) Algorithm*

In BSW and TD algorithm, the maximum threshold value causes the chunk to vary greatly in size. The small-sized chunks increase the quantity of chunks that results to be memory overhead. Two Thresholds Two Divisors is a combination of Small Chunk Merge (SCM) and Two Divisors (TD) algorithm.

The TTTD algorithm [26] uses four parameters D (Primary Divisor), D′ (Second Divisor), T_{max}(Maximum Threshold), T_{min}(Minimum Threshold). To control the variance in chunk size, the minimum and maximum threshold is to be set. The second divisor is half of the primary divisor as shown in Fig. 2.

B. *Two Thresholds, Two Divisors with Switch (TTTD-S) Algorithm*

An improvement in TTTD algorithm is TTTD-S that eliminates the disadvantage of TTTD algorithm [27]. A new parameter switch is added that takes an average of the minimum and maximum threshold. It switches the value at the specific position. The new parameter is known as average parameter. When the algorithm points to average position, the value of main and second divisor is halved. After that, it points to original value once the breakpoint is found. It reduces the computation time and avoids unnecessary comparisons and calculations.

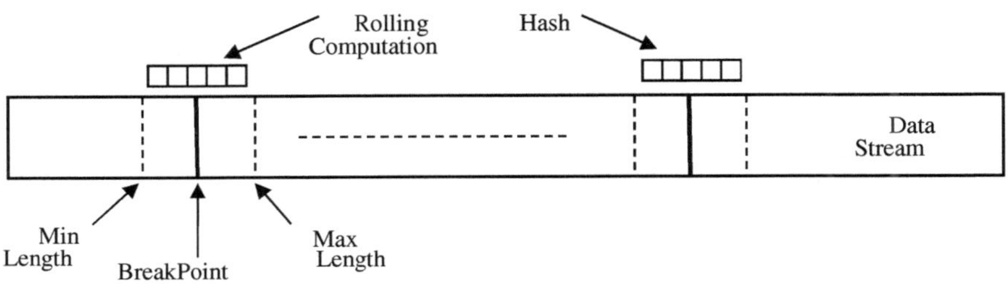

Fig. 2 Rolling hash computation in TTTD

C. *Two Thresholds, Two Divisors with Optimal Parameter (TTTD-P) Algorithm*

To detect more duplicity in the data we have proposed an improved of TTTD and TTTD-S algorithms. It maximally detects the duplicate content by using the optimal parameter. The optimal parameters are found by differential evolution. Differential evolution (DE) is proposed by Rainer Storn and Kenneth Price for optimization problem [36]. It is considered as one of the most powerful evolutionary algorithms for the real number function optimization nowadays.

An initial mutant parameter vector **Z** is created by choosing three members of the population, **a, b** and **c** at random. Then **Z** is generated as

$$z = a + F \times (b - c)$$

where F is a positive scale factor, effective values of which are typically less than one. The difference between two population members (a, b) is added to a third population member (c). The result (Z) is subject to mutate with the candidate for replacement to obtain a proposal. The basic algorithm of "Differential Evolution" contains the following procedure as shown in Fig. 3.

To optimize the function with D parameter, the size of population is initially declared. The step-by-step procedure of differential evolution is given in Fig. 3. Differential evolution is the four-step procedure [37, 38]:

$$X_{i,G} = [X_{1,i,G}, X_{2,i,G} \ldots X_{D,i,G}]$$
$$I = 1, 2 \ldots N \text{ and } G \text{ is generation number.}$$

(1) *Initialization*: A random vector is generated in the initialization phase in the interval $[X_i^L, X_j^U]$

$$X_j^L \le X_{j,i,1} \le X_j^U$$

where $X_{r1,G}$, $X_{r2,G}$, and $X_{r3,G}$ are three distinct candidate solution picked randomly among the population.

(2) *Mutation*: Elaborate the search space for the given parameter vector $X_{i,G}$ by adding three vectors $X_{r1,G}$, $X_{r2,G}$, and $X_{r3,G}$ such that the indices I, r1, r2, and r3 are distinct. Add the difference of weight of two vector to the third vector.

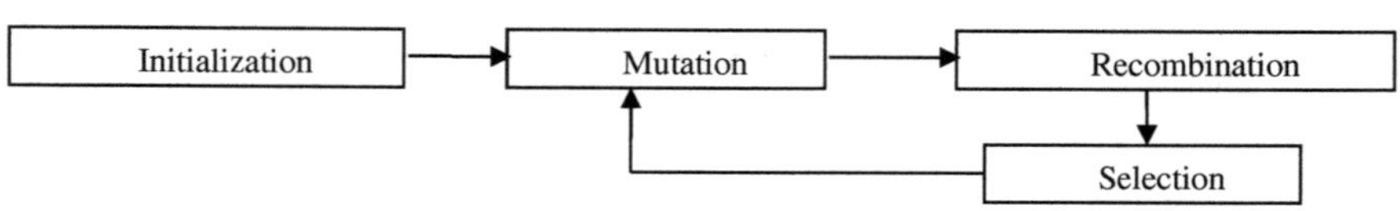

Fig. 3 Step by step procedure of differential evolution (DE)

$$V_{i,G+1} = X_{r1,G} + F(X_{r2,G} - X_{r3,G})$$

where $V_{i,G+1}$ is called donor vector.

DE/rand/1: $V_i = X_{r1} + F_1(X_{r2} - X_{r3})$

DE/rand/1: $V_i = X_{best} + F_1(X_{r2} - X_{r3})$

DE/rand to best/1: $V_i = X_{r1} + F_1(X_{r2} - X_{r3}) + F_2(X_{best} - X_{r1})$

DE/curr. to best/1: $V_i = X_i + F_1(X_{r2} - X_{r3}) + F_2(X_{best} - X_i)$

DE/rand/2: $V_i = X_{r1} + F_1(X_{r2} - X_{r3} + X_{r4} \ldots X_{r5})$

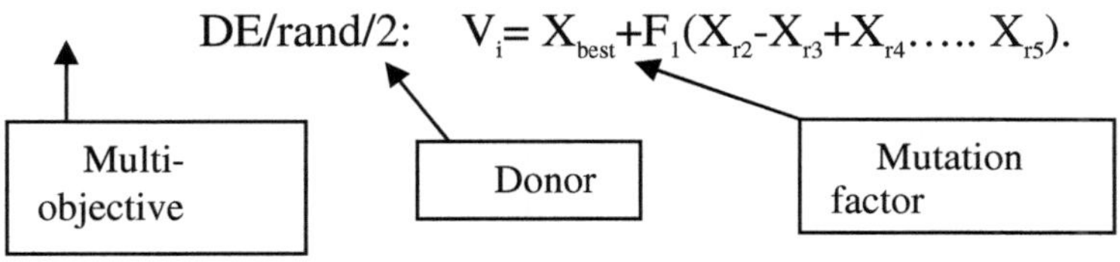

(3) *Recombination*: It pursuit successful solution from the previous generation with current donor. The $V_{i,G+1}$ finds from the target vector, $X_{i,G}$ and the element of donor vector $V_{i,G+1}$ with the probability CR.

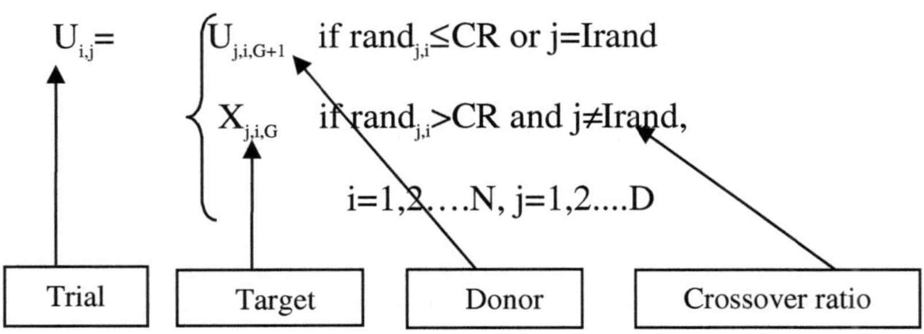

(4) *Selection*: In the selection phase the target vector $X_{i,G}$ is compared with the trial vector $V_{i,G+1}$ and the one with the lowest function value is permitted to the next generation. Greedy approach is key idea for fast convergence of DE. Mutation, recombination, and selection continue until some stopping criterion is reached.

$$X_i^{k+1} = \begin{cases} U_{i,G+1} & \text{if } F(U_{i,G+1}) < F(X_{i,G}), \\ X_{i,G} & \text{otherwise}, \\ & i = 1, 2 \ldots.. N \end{cases}$$

Multi-Objective

The differential evolution algorithm [39] involves loops. The outer loop mentions the stop criteria and the inner cycle point out each individual in a generation with probability CR. The pseudo code of DE algorithm is given as below:

```
Begin
G=0
Create a random initial population
For i=1 to NP do
 For j= 1to D do
   x_j,i^(G=D)=x_j^Min+rand_j[0,1].(x_j^max-x_j^min)
End for
End for
Evaluate fitness function for each
individual of population
For i= 1to NP do
F(x_i^(G=D))
End for
Test vector generation
For G=1 to Maxgen do
For i= 1to NP do
Select Randomly r1,r2,r3ϵ [1,NP],
r1≠r2≠r3≠i
Mutation and crossover process
 jrand=randInt[1:D]
For j= 1to D do
If(rand[0,1]<CR or j==jrand)
then
V_i,j^(G+1)=x_i,r1^(G)+F*(x_i,r2^(G)-X_i,r3^(G))
Else
V_i,j^(G+1)=x_i,j^(G)
Endif
End for
End for
Selection
If(f(v^(G+1))≤f(x_i^(G)) then
X_i^(G+1)=v_i,j^(G+1)
else
X_i^(G+1)=x_i^(G)
endif
endfor
endFor
End
```

The basic framework of proposed work is shown in Fig. 4. The input data is given to the deduplication system. The working of the whole process is divided into following steps:

(1) *Chunking*: The input file is divided into variable size chunks using the two thresholds two divisors with optimal parameters. These parameters are searched by the differential evolution. With differential evolution, the values of divisor and threshold are carried out as optimized values.
(2) *Hash Value Generation*: To calculate the hash value of each chunk SHA-1 is used. In some cases, the MD5 generates the same hash value of different chunks that creates the confusion and effects the performance. SHA-1 is used to create secure collision-free unique hash values of data chunks.
(3) *Redundancy Detection and Elimination*: After generating the hash values, store hash values into the corresponding bucket from bucket 0 to bucket 9 and bucket A to bucket F by measuring leftmost digit of hash value, i.e., 9aca34 … ef,

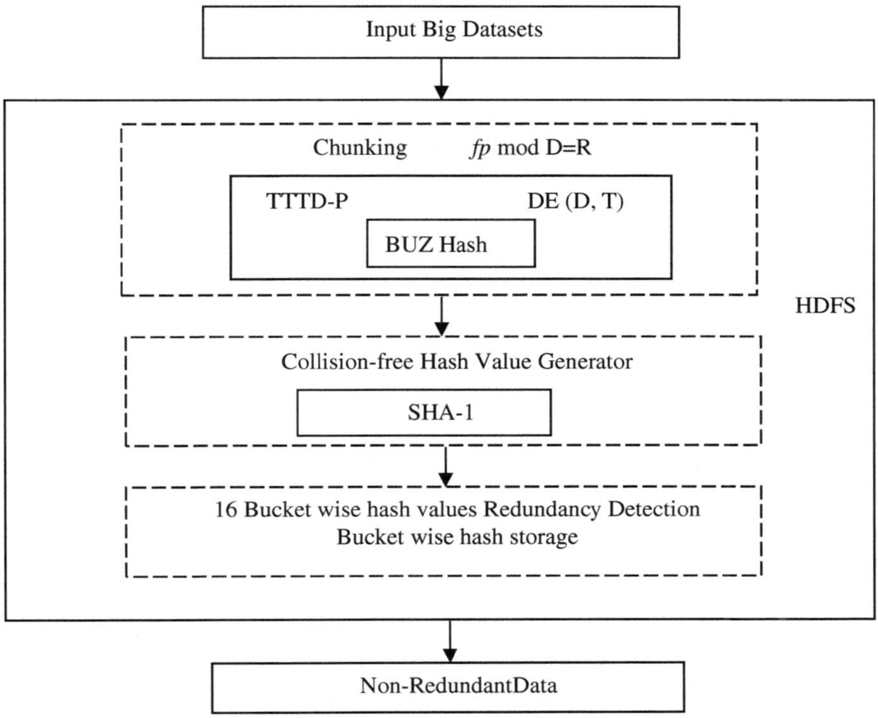

Fig. 4 Framework of proposed work in the HDFS environment

4ade923 … cf, b23a3ece … cd1 hash value in bucket 9, bucket 4, bucket B, respectively, based on bucket indexing [40]. Duplicate hash values are to be detected and then eliminated by Map and Reduce function. During a MapReduce [41] job, Hadoop sends Map and Reduce tasks to appropriate servers in the cluster.

Map Function

1. Input data stream is read by the mapping function.
2. With the help of "Two Thresholds, Two divisors with Optimal Parameters (TTTD-P)" the input data stream is broken into variable size chunks in chunking stage using fast BUZ rolling hash function [42] due to its lighter CPU overhead than Rabin hash in the sliding window based CDC.
3. Store the chunks and generate the hash values of each chunk by using SHA-1 hash values generator.

Reduce Function

1. Read the hash values of all chunks.
2. Identify unique hash values which give nonredundant chunks and store all hash values of complete Big Data streams chunks in 16 buckets based on leftmost bit of hash value (0–9 and A–F).
3. Eliminate duplicate chunks by comparing the new hash values with the previously stored hash values in buckets just by comparing with concerned bucket on basis of leftmost bit of hash value of a chunk.
4. Store only nonredundant data in the distributed storage systems.

4 Experimental Results and Analysis

In this section, the experiments are performed to evaluate the performance of Rabin CDC, TTTD, AE, FAST CDC and TTTD-P as per experimental setup given in Appendix A. Proposed algorithm DE-based Bucket Indexed data deduplication is implemented in Python [43] and Java [44] language to reduce computing operations in less time. Three big datasets from freedb [45] have been taken to perform the experimental evolution. Tables 1 and 2 show experimental results of CDC with three big datasets. The experiments are tested on HDFS environments with the following configuration:

- UBUNTU 14.04 (64-bit version)
- 64 GB Installed memory
- HADOOP 1.2.1 [46] and HIVE [40] installed on UBUNTU
- Eclipse Indigo installed on UBUNTU
- BUZ fast rolling hash function code [47], Differential Evolution code [42] and Bucket Indexing [48].

Matrices are used to evaluate the performance of Data Size after Deduplication, Duplicate Data Size, Deduplication Ratio, Hash Judgment Time and Throughput as shown in Table 2.

Table 1 Minimum chunk size, maximum chunk size and optimal parameter for big datasets by TTTD-P

Big datasets (GB)	Minimum chunk size (Bytes)	Maximum chunk size (Bytes)	Optimal parameter (threshold, divisor)
Dataset 1	461	2800	(1900, 280)
Dataset 2	420	2800	(1800, 270)
Dataset 3	191	2800	(1800, 270)

Table 2 Big datasets (Text, PDF, Audio, Video, TAR, LNX, WEB) three real-world datasets

Big datasets	Before deduplication input data size (GB)	Data deduplication algorithms	Redundant data size (GB)	After deduplication output data size (GB)	Deduplication ratio (input size/output size)	Hash judgment time (ms)	Throughput (MB/s)	Duplicate data chunks
Dataset 1	174.096609570	Rabin CDC	24.746572643	149.35003692	1.1656951224	26,451,963	339	19,731,533
		TTTD	39.900983962	134.19562560	1.2973344607	22,006,660	443	27,634,094
		AE	41.001354672	133.09525489	1.3080602288	11,500,854	1023	27,979,055
		FAST CDC	39.901023712	134.19558585	1.2973348450	2,641,785	3324	29,725,050
		DE bucket	49.901027316	124.19558225	1.4017938997	1,635,083	5313	39,053,222
Dataset 2	226.992897156	Rabin CDC	83.64454814	143.34834901	1.5835054866	38,236,973	437	62,299,656
		TTTD	100.46684505	126.52605210	1.7940407796	37,124,804	543	69,548,084
		AE	103.64761135	123.34528580	1.8403046025	12,745,658	1029	69,839,657
		FAST CDC	100.46711672	126.52578043	1.7940446317	3,821,978	3337	74,808,440
		DE bucket	118.46725039	108.52564676	2.0916060298	2,252,592	5343	93,486,302
Dataset 3	156.027040026	Rabin CDC	26.238750875	129.78828915	1.2021657812	24,874,984	331	19,353,753
		TTTD	38.823370578	117.20366944	1.3312470571	24,067,724	423	26,930,890
		AE	41.333531642	114.69350838	1.3603824857	10,815,211	1001	27,161,197
		FAST CDC	38.825812606	117.20122742	1.3312747951	2,488,509	3318	28,963,864
		DE bucket	58.826066794	97.200973232	1.6052003888	1,537,053	5311	48,449,552

A. *Deduplication Elimination Ratio (DER) or Deduplication Ratio (DR)*: The overall deduplication ratio is defined as input data size before deduplication divided by output data size after deduplication.

$$DR = \frac{Input\ data\ size\ before\ deduplication}{Output\ data\ size\ after\ deduplication}$$

B. *Deduplication Time*: Deduplication time is the time required by deduplication technique to give output response. First, chunking step (1) *Hashing* in which fingerprints of the data contents are generated using BUZ fast rolling hash function instead of heavier CPU overhead of Rabin hash and (2) *hash judgment* in which fingerprints are compared against a given value to identify and declare chunk cut points. Second, chunks hash values—*Bucket Indexing* in which SHA-1 hash values of chunks stored bucket-wise in 16 buckets based on left most digit of hash values. Third, data redundancy elimination—new data chunk hash value is compared based on left most digit of hash value with stored hashes in the corresponding bucket (Bucket 0–Bucket 9 and Bucket A–Bucket F). If it is found in bucket then declared as redundant chunk; otherwise store data chunks in storage systems.
C. *Throughput*: Throughput is a measure of how many units of information a system can process in a given amount of time. In data deduplication, throughput is the amount of data deduplicated in a given period and typically measure in bits per second (bps), megabits per second (Mbps) or gigabits per seconds (Gbps).

$$Throughput = \frac{Total\ Data\ Size}{Deduplication\ Time}$$

A good hash function BUZ [47] has a uniform distribution of hash values regardless of the hashed content best for bucket indexing by providing probably equal chance to all hash values with left most digit from 0 to 9 and A to F and uses far fewer calculation operations than Rabin, Adler, Gear, and Fast CDC [31]. Table 1 shows the datasets used in our experiments. Dataset 1, Dataset 2, and Dataset 3 are real datasets [45], whose sources are the web servers and mail servers.

TTTD-P: DE Based Bucket Indexed Data Deduplication

Chunk Number—Chunk Size (Bytes)—Chunk Hash Value

97c: 1—676—3ae22186413cc736b4c420223de515b2e1252b26

98c: 2—1636—f59430417084ef91008299282263fc50a3966d46

97c: 3—886—f26e3062370dc0fa30cb50a7ab5ea9e201b9f2f7

and so on.

Bucket Indexing: The deduplication technology is mainly on disk-based permanent storage to enhance space efficiency. But traditional approaches are facing two major problems. The first technical challenge is the duplicate-lookup disk bottleneck and second major challenge is storage node island effect. Traditional approaches store a complete index of data chunks. Therefore, the index becomes too large to store entire hash values while big data volume is increasing. In this scenario, it is very difficult by deduplicate process to lookup fingerprints in an on-disk index and hence degrades the overall performance of the system. The second issue is of storage node island effect to remove duplicates within primary or backup storage but not in distributed multiple storage nodes. Two recent famous papers on indexing issues for data deduplication DDFS [49] and Sparse Indexing [50], proposed novel strategies to eradicate the duplicate-lookup disk bottleneck by exploiting hash values localities in disk-to-disk I/O operations. However, traditional approaches are not in speedup, scaleup, and size up performance.

To overcome the drawbacks of the traditional deduplication indexing approaches, we propose a scalable very high throughput bucket indexing approach for distributed storage systems. Suppose there are 96 hash values generated for 96 chunks of a given data stream. A good hash function must have a uniform distribution of hash values regardless of the hashed content. Bucket indexing approach has scalable 16 buckets, i.e., B-0 to B-9 and B-a to B-f which store hash values by measuring leftmost digit of hash value to the corresponding bucket such as bucket B-a will store only hash values whose leftmost digit is "a". During hash, judgment hash values are compared against stored hash values to identify and declare redundant chunks. So a new chunk hash value will be compared only with stored hash values in a particular bucket instead of with all stored hash values.

For example, a hash function generates uniformly equally distributed 96 hash values for 96 variable size chunks means approximately 6 values for each having left most digit "0"–"9" and "a"–"f". During bucket indexing, a new chunk hash value will be compared only with stored approximately 6 hash values in a particular bucket instead of with all 96 hash values as in earlier techniques of hash judgment to identify and declare redundant chunk. Thus, it reduces hash judgment comparison 16 times approximately than previous Rabin CDC resolve the issues efficiently for big data storage systems. Proposed bucket indexing logic requires same storage space without extra cost as earlier by all generated hash values without bucket-wise, but drastically reduces hash judgment time 16 times significantly to identify and detect redundant chunks. It is clear from Table 3, the Summary Vector, Locality

Table 3 Index and Locality Reads

Data segments	Disk I/Os	% of total
No summary vector and no locality preserved caching	328,613,503	100.00
Summary vector (DDFS)	274,364,788	83.49
Locality preserved caching (LPC)	57,725,844	17.57
Proposed bucket indexing (BI)	20,538,344	6.25

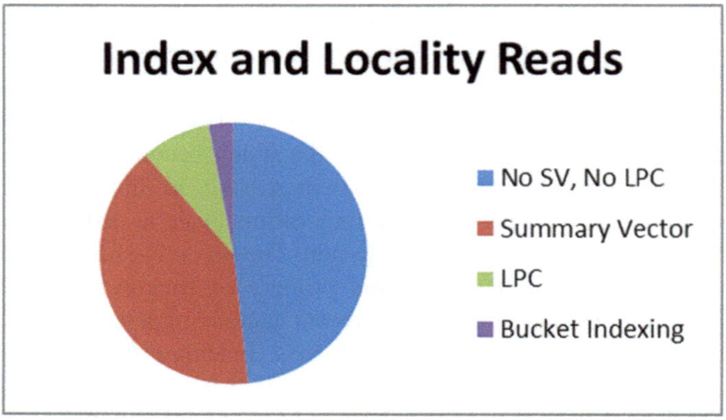

Fig. 5 Disk reads for index lookups for SV, LPC and BI

Preserved Caching, and Bucket Indexing have an astounding reduction in disk reads. Summary Vector (SV) [49] has disk I/Os reduction approximately 16.51% for 328,613,503 data chunk segments. The Locality Preserved Caching (LPC) [50] has disk I/Os reduction nearby 82.43% for 328,613,503 data segments. Proposed Scalable Bucket Indexing (BI) with high throughput alone reduces about 93.75% of the index lookup disk I/Os for distributed storage nodes for 328,613,503 data segments as shown in Fig. 5.

Bucket	Chunk number—Chunk size (Bytes)—Hash value (SHA-1)	Chunks	Data size (Byte)
B-0	8c: 38—895— 0aa63255bb51e1b3c5ffcf64b725008918bd8226 9c: 391—2460— 09dd19eb6b15adff08de995b2cc2d1783ade2ef5 9c: 402—1024— 02fa1ea993afd20db07dc910bde288aebee47d43 8c: 411—641— 0fa1f634a5a2711257e31ccabdf5686799007dcc 8c: 442—612— 0ff525fa5b4a38ebf019f172cccc215976c18d6f 8c: 543—473— 0e6e8153d9185edab8b2d35978445ef93df3171d 9c: 644—1349— 0757e9b9cbf8abdcc4f70f180a96fbea32d83216 8c: 745—538— 0d0ddd43de00e343732be7f48c0532c1e14cb3fa 8c: 749—548— 03f870fbf7d3720deb1ce2f0073511f20ff556fb 8c: 847—606— 05905ffc5e28d9854175768c152db9f5db538fab	1 2 4 1 1 7 3 1 1 6 Chunks = **27**	895 4920 4096 641 612 3311 4047 538 548 3636 Data size = **23,244**

(continued)

(continued)

Bucket	Chunk number—Chunk size (Bytes)—Hash value (SHA-1)	Chunks	Data size (Byte)
	Storage Used = 9146 **No. of Unique Chunks = 10** **Storage Space** **Saved = 23,244 − 9146 = 14,098**		
B-1			
…	…		
B-f			

5 Conclusion and Future Work

Deduplication techniques are mainly used to removes the redundancy from the data streams so that a large amount of space will be available to store the essential data. DE-based TTTD-P algorithm takes the optimal parameter to detect more duplicate content. The average chunk size and deduplication ratio are also maintained by using TTTD-P algorithm. The optimized chunking to maximize chunking throughput with increased DR; and bucket based fingerprint indexing approach reduces computation hash judgment maintained by using DE-based TTTD-P algorithm. The experiments on three big data sets has focused to reduce the storage space on remote servers. Furthermore, experimental result shows that bucket based fingerprint indexing approach reduces computation hash judgment time about 16 times than Rabin CDC, 5 times than AE CDC, 1.6 times than FAST CDC. DE-based bucket indexing TTTD-P algorithm not only successfully achieves the significant improvements in data deduplication and average chunk size but also obtains the better controls on the variations of chunk size by reducing the large-sized chunks in less computation time.

In our future work, we plan to incorporate bucket indexing architecture for big data storage systems on HDFS platform to explore the potentials and benefits of fast DE-based optimized data deduplication.

Acknowledgements For this research, we would like to show gratitude to Prof. Rohitashwa Shringi (and specially his wife Dr. Pramila) Mechanical Engineering department of Rajasthan Technical University, KOTA for their support, valuable advice, motivation, and encouragement. We would also like to thank Prof. S. C Jain Rajasthan Technical University, Kota for guiding this research work with valuable suggestions time to time.

Appendix: Experimental Setup

Chunking	Main divisor D	Backup Divisor D′	Fingerprinting *fp* mod D = R	Speed
BSW	1000	No	Rabin	Slow
BFS	1000	No	Rabin	Slow
TD	1200	600	Rabin	Slow
SCM	540	No	Rabin	Slow
TTTD	540	270	Rabin 16-bit incremental	Slow
TTTD-S	1600	800	Rabin 16-bit incremental	Slow
TTTD-P	Optimal DE (D, T)	No	BUZ 32-bit rolling hash	Very fast

References

1. Min Chen, Shiwen Mao, Yunhao Liu: Big Data: A Survey Mobile Networks and Applications Journal, Springer, Vol. 19, Issue 2, (2014) 171–209.
2. James Manyika, Michael Chui, Brad Brown, Jacques Bughin, Richard Dobbs, Charles Roxburgh, Angela Hung Byers: Big data: The next frontier for innovation, competition, and productivity. McKinsey Company, (2011) 1–156.
3. R. Storn: Differential Evolution: A simple and efficient heuristic strategy for global optimization over continuous spaces. Journal of Global Optimization, Vol. 11, (1997) 341–359.
4. Jaehong Min, Daeyoung Yoon, and Youjip Won: Efficient Deduplication Techniques for Modern Backup Operation. IEEE Transactions on Computers, Vol. 60, No. 6, (2011) 824–840.
5. J. Gantz, and D. Reinsel: The digital universe decade are you ready? IDC White Paper, May (2011).
6. H. Biggar: Experiencing data deduplication: Improving efficiency and reducing capacity requirements. White Paper, the Enterprise Strategy Group, Feb (2012).
7. Tin-Yu Wu, Jeng-Shyang Pan, and Chia-Fan Lin: Improving Accessing Efficiency of Cloud Storage Using Deduplication and Feedback Schemes. IEEE Systems Journal, Volume 8, No.1, (2014) 208–218.
8. Yinjin Fu, Lei Tian, Fang Liu, Hong Jiang, Nong Xiao: AA-Dedupe: An Application-Aware Source Deduplication Approach for Cloud Backup Services in the Personal Computing Environment. IEEE International Conference on Cluster Computing (CLUSTER), (2013) 112–120.
9. Ahmed El-Shimi, Ran Kalach, Ankit Kumar Adi Oltean Jin Li, Sudipta Sengupta: Primary Data Deduplication – Large Scale Study and System Design USENIX federated conference Week. June 12–15, (2012) 285–296.
10. Ross Neil Williams: Method for partitioning a block of data into sub-blocks and for storing and communicating such sub-blocks. Patent US5990810 A, 23 Nov (1999).

11. Purushottam Kulkarni, Fred Douglis Jason LaVoie, John M. Tracey, Redundancy Elimination Within Large Collections of Files Proceedings of the annual conference on USENIX Annual Technical Conference, General Track, (2004) 59–72.
12. Dirk Meister, Jürgen Kaiser, Andre Brinkmann, Michael Kuhn, Julian Martin Kunkel, Toni Cortes: A Study on Data Deduplication in HPC Storage Systems. Conference Proceedings of the ACM/IEEE Conference on High Performance Computing (2012).
13. Guanlin Lu: An Efficient Data Deduplication Design with Flash-Memory Based Solid State Drive. A Dissertation Submitted to The Faculty of the Graduate School of the University of Minnesota, (2012) 1–114.
14. Nikolaj Bjorner, Andreas Blass, Yuri Gurevich: Content-Dependent Chunking for Differential Compression, The Local Maximum Approach. Journal of Computer and System Sciences, Elsevier, Vol. 79, No. 3, (2010) 154–203.
15. Athicha Muthitacharoen, Benjie Chen and David Mazieres: A Low-bandwidth Network File System, proceeding of the 18th ACM Symposium on operating System principle (Sosp '01), Chateau lake louise, Banff, Canada, (2001) 174–187.
16. Janaka Deepakumara, Howard M. Heys and R. Venkatesan: FPGA Implementation of MD5 Hash Algorithm. IEEE Canadian Conference on Electrical and Computer Engineering, (2001) 919–924.
17. Dai Zibin Zhou Ning: FPGA Implementation of SHA-1 Algorithm. IEEE 5th international conference on ASIC, (2003) 1321–1324.
18. Deepavali Bhagwat, Kave Eshghi, Darrell D.E. Long and Mark Lillibridge: Extreme Binning: Scalable, Parallel De-duplication for Chunk-based File Backup. Proceeding of 17th IEEE/ ACM International Symposium on Modeling, Analysis and Simulation of Computer and Telecommunication systems (2009) 1–9.
19. Guanlin Lu, Yu Jin, and David H.C. Du,: Frequency Based Chunking for Data De-Duplication. 18th Annual IEEE/ACM International Symposium on Modeling, Analysis and Simulation of Computer and Telecommunication Systems, (2010) 287–296.
20. IderLkhagvasuren, Jung Min So, Jeong Gun Lee, Chuck Yoo and Young WoongKo: Byte-index Chunking Algorithm for Data Deduplication System. International Journal of Security and Its Applications, SERSC, Vol. 7, No. 5, (2013) 415–424.
21. IderLkhagvasuren, Jung Min So, Jeong Gun Lee, Jin Kim and Young WoongKo, Multi-level Byte Index Chunking Mechanism for File Synchronization. International Journal of Software Engineering and Applications Vol. 8, No. 3, (2014) 339–350.
22. Chuanshuai Yu, Chengwei Zhang, Yiping Mao, Fulu Li: Leap Based Content Defined Chunking- Theory and Implementation, IEEE 31st Symposium on Mass Storage Systems and Technologies (2015) 1–12.
23. Erik Kruus, Christian Ungureanu, Cezary Dubnicki: Bimodal Content Defined Chunking for Backup Streams, FAST Proceedings of the 8th USENIX Conference on file and Storage Technologies, USENIX, (2010) 239–252.
24. Jiansheng Wei, Junhua Zhu, Yong Li, "Multimodal Content Defined Chunking for Data Deduplication", https://www.researchgate.net/publication/261286019, Research gate, (2014).
25. Yucheng Zhang, Hong Jiang, Dan Feng, Wen Xia, Min Fu, Fangting Huang, Yukun Zhou: AE An Asymmetric Extremum Content Defined Chunking Algorithm for Fast and Bandwidth-Efficient Data Deduplication. IEEE Conference on Computer Communications (INFOCOM), (2015) 1337–1345.
26. Kave Eshghi, Hsiu Khuern Tang: A framework for analyzing and improving content based chunking Algorithms. Technical Report TR 2005–30, Hewlett-Packard Development Company, http://www.hpl.hp.com/techreports/2005/HPL-2005-30R1.html (2005).
27. Teng-Sheng Moh, Bing Chun Chang: A Running Time Improvement for the Two Thresholds Two Divisors Algorithm. *ACMSE '10*, April 15–17, (2010).

28. Xia, W., Jiang, H., Feng, D., Tian, L., Fu, M., and Zhou, Y.: Ddelta: A deduplication-inspired fast delta compression approach Performance Evaluation. 15 Proceedings of the 7th USENIX Conference on Hot Topics in Storage and File Systems, (2015) 258–272.
29. Aggarwal, B., Akella, A., Anand, A., et al. EndRE: an end-system redundancy elimination service for enterprises. In Proceedings of the 7th USENIX conference on Networked Systems Design and Implementation (NSDI '10) (San Jose, CA, USA), USENIX Association, April (2010) 14–28.
30. Cui, Y., Lal, Z., Wang, X., Dai, N., and Miao, C., "QuickSync: Improving Synchronization Efficiency for Mobile Cloud Storage Services" In Proceedings of the 21st Annual International Conference on Mobile Computing and Networking (Paris, France), ACM, Sept. (2015) 592–603.
31. Wen Xia, Yukun Zhou, Hong Jiang, Dan Feng, Yu Hua, Yuchong Hu, Yucheng Zhang, Qing Liu: FastCDC: a Fast and Efficient Content-Defined Chunking Approach for Data Deduplication. USENIX Open access to the Proceedings of USENIX Annual Technical Conference (USENIX ATC '16), (2016) 101–114.
32. Andrei Z. Broder: Some applications of Rabin's fingerprinting method. Sequences II: Methods in Communications, Security and Computer Science, Springer, (1993)143–152.
33. Michael O, Rabin: Fingerprinting by random polynomials. Center for Research in Computing Technology. Aiken Computation Laboratory, Univ., (1981).
34. Dubnicki, C., Kruus, E., Lichota, K., and Ungureanu, C.: Methods and systems for data management using multiple selection criteria. US Patent App. 11/566,122, Dec 1, (2006).
35. Min, J., Yoon, D., and Won, Y.: Efficient deduplication techniques for modern backup operation. IEEE Transactions on Computers Vol. 60, Issue 6, (2011) 824–840.
36. Derviskaraboga, Selcukokdem: A simple and Global optimization algorithm for engineering problem: Differential evolution algorithm. Turk Journal Electrical Engineering, Vol. 12, No. 1, (2004).
37. An Introduction to Differential Evolution, http://www.maths.uq.edu.au/MASCOS/MultiAgent04/Fleetwood.pdf.
38. Differential Evolution (DE): http://www.dii.unipd.it/alotto/didattica/corsi/Elettrotecnical computazionale/DE.pdf.
39. Prometeo Cortes-Antonio, Josue Rangel-Gonzalez, Luis A. Villa-Vargas, Marco Antonio Ramirez-Salinas, Heron Molina Lozano, Idar Batyrshin,: Design and Implementation of Differential Evolution Algorithm on FPGA for Double Precision Floating-Point Representation. Acta Polytechnic Hungarica, Vol. 11, No. 4, (2014).
40. Naresh Kumar and Ruchika Kumar, "Improved Join Operations Using ORC in HIVE", Springer Transactions on Information Communication Technology (ICT), Springer Vol. 4, Issue 2, pp. 209–215, Dec. (2016).
41. Jeffrey Dean and Sanjay Ghemawat, "Map-reduce: Simplified Data Processing on Large Clusters", To appear in OSDI, (2004).
42. http://www.icsi.berkeley.edu/storn/code.html.
43. http://www.python.org/.
44. http://www.java.org/.
45. Freedb.org, "http://freedb.org/pub/freedb/".
46. Apache hadoop 1.2.1, "http://hadoop.apache.org/".
47. http://www.serve.net/buz/hash.adt/java.000.html.
48. Naresh Kumar, Rahul Rawat and S.C Jain, "Bucket Based Data Deduplication Technique for Big Data Storage System", IEEE 5th International Conference on Reliability, Infocom Tech. and Optimization, Dec. (2016).

49. B. Zhu, K. Li, and H. Patterson, "Avoiding the Disk Bottleneck in the Data Domain Deduplication File System," in Proceedings of the 7th USENIX Conference on File and Storage Technologies (FAST) San Jose, CA, USA, pp. 269–282, Feb. (2008).
50. M. Lillibridge, K. Eshghi, D. Bhagwat, V. Deolalikar G. Trezise, and P. Campbell, "Sparse Indexing: Large Scale, Inline Deduplication Using Sampling and Locality," in Proceedings of the 7th USENIX Conference on File and Storage Technologies (FAST) San Jose, CA, USA, pp. 111–123, Feb. (2008).

Strategies for Inducing Intelligent Technologies to Enhance Last Mile Connectivity for Smart Mobility in Indian Cities

Moushila De, Shailja Sikarwar and Vijay Kumar

Abstract The rapid growth of India's urban population has put enormous strains on urban transport systems. During the last few decades, more cars, congestion, and the related urban transport problem arise. There are many issues related to safety and security especially for women, children, disabled, and senior citizens. People often have a problem in starting their trip, i.e., from their home. These difficulties do not lie in the main transport network but in the available options that a person has beyond his residence to reach the station located at main transport network. This paper is an attempt to identify the issues related to last mile connectivity at various metro stations of Delhi, to examine the constraints in ensuring last mile connectivity and to suggests measures to improve last mile connectivity in the selected case study stations with the help of various information- and communication-based solutions. Various ICT-based solutions have been suggested in this paper which will help to improve the last mile connectivity for smart mobility in Indian cities.

Keywords Smart urban transportation system · Automatic fare collection system
Integrated parking management system · Transi-oriented development
Passenger information system

M. De (✉)
MURP, Faculty of Architecture and Urban Planning, DCRUST, Murthal, India
e-mail: moushilade5@gmail.com

S. Sikarwar · V. Kumar
Faculty of Architecture and Urban Planning, DCRUST, Murthal, India
e-mail: sikarwarshailja@gmail.com

V. Kumar
e-mail: skvijayarch@gmail.com

C. R. Panigrahi et al. (eds.), *Progress in Advanced Computing and Intelligent Engineering*, Advances in Intelligent Systems and Computing 714,
https://doi.org/10.1007/978-981-13-0224-4_33

1 Introduction

Intelligent technologies play an important role to improve last mile connectivity for smart mobility in India's cities. People often have a problem in starting their trip, i.e., from their home. This defines the problem associated with first and last mile connectivity from home and from transit to destination. Though cities also develop into certain structure with regard to distribution of activities and the type of public transport network it has, in spite of which there are certain gaps and loopholes that fails to connect people directly to the main network. This is the problem of the Last Mile which starts at a user's residence till points where public transport network ends. Providing best last mile connectivity options can reduce the cases of various crimes and also helps in the prevention of crimes. Cities are more or less becoming car-centric cities to avoid the first and last mile problem. But still, people face problems related to traffic congestions. No one is ready to walk and use bicycle due to poor conditions of road infrastructure. Therefore, it is necessary to improve last mile connectivity with the help of various ICT based solutions.

2 About the Study

The study is research based for enhancing knowledge on the last mile connectivity for mass transit and evolving an approach to last mile connectivity. These studies evolve an approach to improve last mile connectivity of metro stations, which further results in the increase in number of users walking to metro.

3 Present Scenario of Last Mile Connectivity in India: Case Study New Delhi

In Indian cities, the conditions of last mile connectivity are very poor. Most of the cities do not have proper last mile connectivity facilities. To understand the existing conditions of last mile connectivity in Delhi, a survey was conducted in New Delhi to understand its existing scenario. It is inferred from the survey that presently only 41 out of 138 stations have proper feeder bus facility which has been connected to various MRTS stations. For understanding the present scenario of last mile connectivity, a field survey was conducted and four stations were chosen based on ridership data up to January 2017, last mile connectivity modes quality assessment through Reconnaissance survey, abutting land use around yellow line metro stations, activity intensity around the stations and typology of the stations. Based on the abovementioned criteria, four stations, i.e., Saket, Vishwavidyalaya, INA, and Sultanpur had been chosen and further existing scenario had been studied to improve the last mile connectivity through intelligent systems (Table 1).

Table 1 Case study identified parameters stations

Quality	Name of station	Ridership	Typology	Adjacent land uses	Density	Options for interchange
Good	Saket	57,000	Mid-block, underground	Residential and commercial	Medium	Feeder bus, auto, Grameen seva
Good	Vishwavidyalaya	23,182	Mid-block, underground	Commercial, institutional and residential	Medium	Rickshaw, auto, bus, E—rickshaw, bicycle
Average	I.N.A	30,590	Mid-block, underground	Commercial, residential and inner ring roads	Medium	Bus, auto
Average	Sultanpur	6377	Mid-block, elevated	Informal residential	High	Bus, E—rickshaw, auto

Figure 1 explains the calculated landuse data of four case study stations, i.e., Saket, INA, Sultanpur, and Vishwavidyalaya metro stations which have been calculated using ArcGIS 10.2. Figure 2 explains the comparative analysis of footfalls data. It is depicted from the figure that Saket has the highest footfalls, followed by

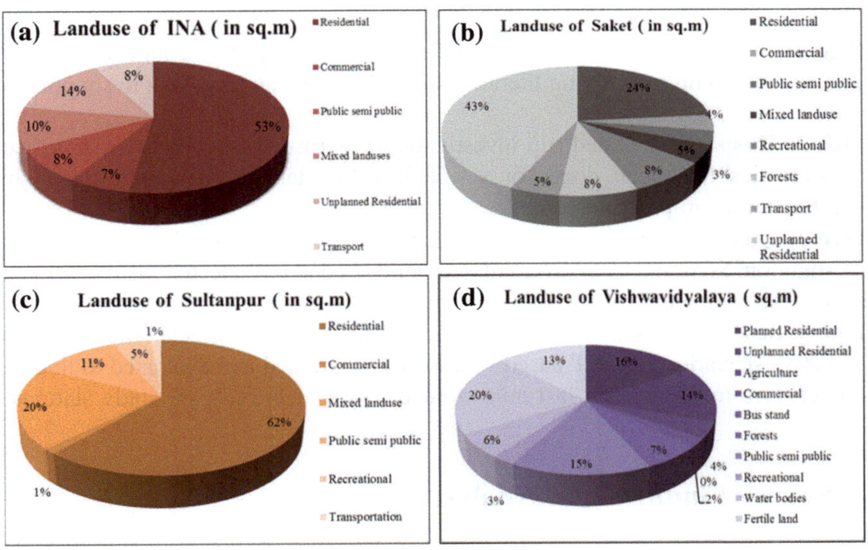

Fig. 1 **a–d** INA, Saket, Sultanpur and Vishwavidyalaya metro station landuse 2017 (calculated using ArcGIS 10.2)

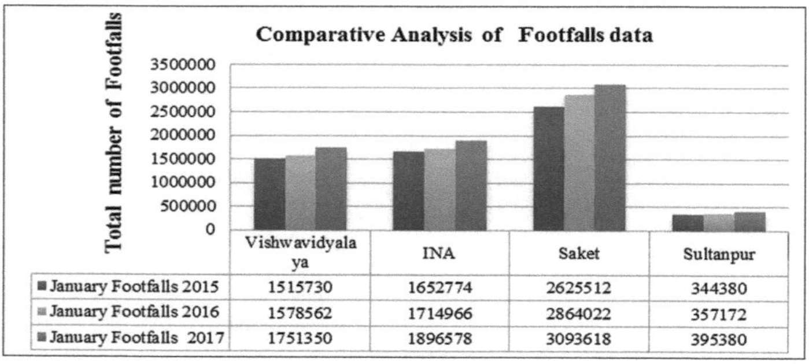

	Vishwavidyalaya	INA	Saket	Sultanpur
■ January Footfalls 2015	1515730	1652774	2625512	344380
■ January Footfalls 2016	1578562	1714966	2864022	357172
■ January Footfalls 2017	1751350	1896578	3093618	395380

Fig. 2 Comparative analysis of footfalls data. *Source* Retrieved from DMRC on February 2017

Vishwavidyalaya and INA has the mid footfalls and Sultanpur has the lowest footfalls. The main reason behind these difference in ridership is the availability of abutting landuse around the selected case study stations.

4 Methodology Adopted to Improve Last Mile Connectivity for Smart Mobility in Selected Case Study Stations

The study has been divided into four stages:

Stage 1: It is a preliminary stage establishing aims, objectives, and need of the study. It also involves studying literature and other facts and findings to support the study.
Stage 2: This stage included trip investigation in terms of trip characteristics, user characteristics had been done for selected metro stations. It was followed by identification of improvement areas.
Stage 3: Qualitative evaluation of trip characteristics was done based on the trip investigation. Station area assessment was conducted at the selected metro station to support the evaluation.
Stage 4: Based on the existing analysis and evaluation of the selected metro station, alternative scenarios and ICT based solutions were framed and evaluated to find the best scenario for improving last mile connectivity at selected case study stations.

5 Survey Findings and Analysis

(a) **IPT and NMT survey analysis**

For IPT (Intermediate Para Transit) survey operators within a kilometer radius were interviewed within chosen metro stations. For these, 50 surveys were

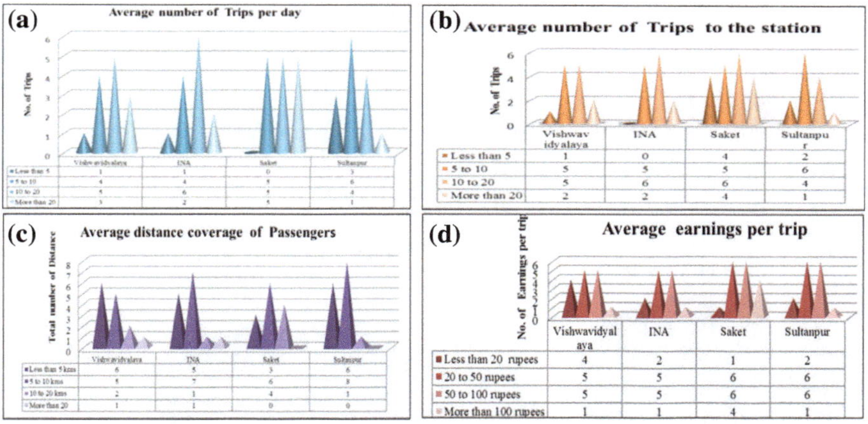

Fig. 3 **a–d** IPT and NMT survey analysis

conducted and interviews were taken from various NMT and IPT operators. A brief assessment was also conducted on the personal characteristics of NMT and IPT operators across stations to understand the average number of trips from the station, average number of trips to the station, average distance coverage of passengers, average earning per day of operators, issues regarding the operation of IPT and NMT operators, etc. The analysis of IPT and NMT operators survey is as follows (Fig. 3).

(b) **Bicycle facilities at selected case study stations**

In Saket metro station, total 36 cycles are available on both sides. In one side, almost 12–17 cycles have been rented by the people. The monthly cycle package is Rs. 300 per month. Almost 500–600 customers have taken the monthly package in Saket metro station. No other cycle stands near Saket metro station.

In Vishwavidyalaya metro station, total 26 cycles are available for rental purpose. The rental of cycle basically depends on the number of students. Sometimes, it goes 30, 50 or 70.

(c) **Private vehicle users survey analysis**

For private vehicle users survey was conducted near parking stations and within a kilometer radius of various offices, educational institution, etc.; 50% were car users along with 50% two-wheeler commuters in these study were taken for survey. 50 samples were collected near the four metro stations of the private vehicle users (Fig. 4).

(d) **Metro users survey analysis**

A brief assessment was also completed on the personal characteristics of users across station at trip producing and trip attracting area to understand the travel behavior and impact of ridership on metro. Users were assessed on basis of gender, age, income, occupation, frequency of trips, vehicle ownership, etc. (Fig. 5).

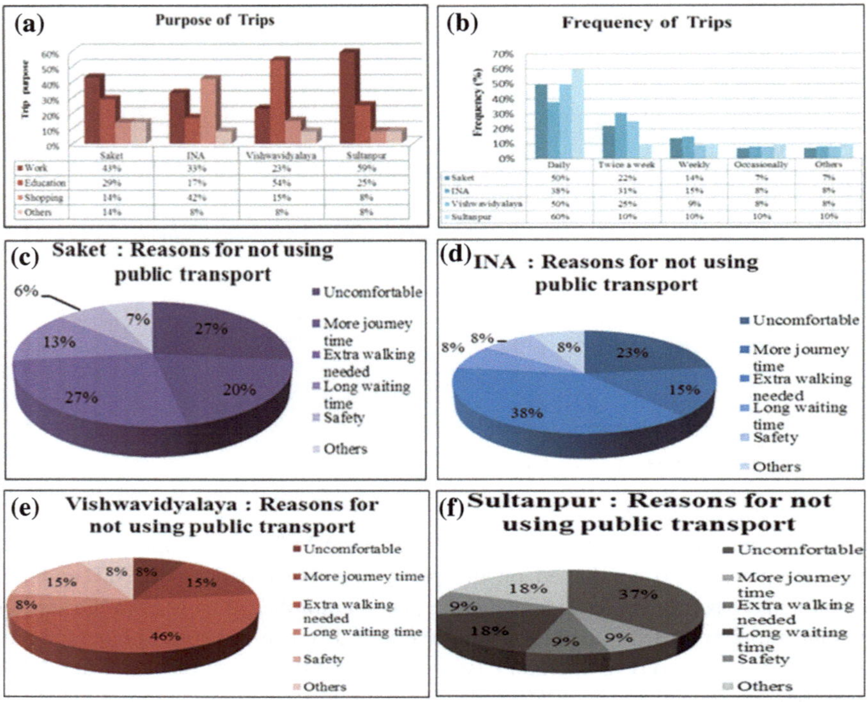

Fig. 4 a–f Private vehicle users survey analysis

(e) **Metro users and intermediate paratransit/nonmotorized transport operators suggestions and preferences**

(i) *Saket metro station*:

- Improvement and strengthening of road from SangamVihar to Saket (Near SBI) through Sainik Farms is needed to avoid traffic congestion. Pedestrilization needs to be developed near Saket metro station. The focus should be on people who are on foot.
- The road from Neb Sarai to Western Marg connecting Saket metro needs to be strengthened. Feeder bus service may be required from Saket metro station to pushpvihar.

(ii) *INA metro station*:

- New feeder bus services needs to be started connecting INA metro station to Sarojini Nagar and R.K. Puram Colony.
- Nonmotorized transport options need to be enhanced especially from Sarojini Nagar to INA metro station market or from Sarojini Nagar railway station to INA market, Sarojini Market.

Fig. 5 a–f Metro user survey analysis

- Inner roads of Lakshmi Bai Nagar and Sarojini Nagar needs to be strengthened, so that maximum people can use inner roads for cycling, biking, and walking directly to the metro stations.

(iii) *Vishwavidyalaya metro station*:

- Foot over bridge or subway needed near Vishwavidyalaya metro stations connecting both sides of the stations.
- There is scope for improvement in terms of more and better quality provisions for "bicycle on rent" as present supply runs short of huge demand.

(iv) *Sultanpur metro station*:

- Roads conditions are very bad, it needs to be strengthening. Feeder bus service is required.
- Roads from Gurudwara to Sultanpur metro station needs to be strengthened. Drainage problem needs to be solved. The focus should be given on pedestrilization specially the road connecting senior secondary school to Sultanpur metro stations.

6 Strategies for Inducing Intelligent Technologies to Enhance Last Mile Connectivity for Smart Mobility in Indian Cities

(i) **Smart mobility systems**

It is necessary to have smart mobility system and there is a need to integrate all modes with the main mode, i.e., multimodal transportation system. There should be designated parking for NMT vehicles such as E-Rickshaws and paratransit, TSR zone is also needed in every metro. Charging station should be provided near metro stations especially for battery operated vehicles such as E-rickshaws, etc. It is necessary to abolish the manual cycle rickshaw. In every metro station, total number of 100 parking spaces should be provided for E-Rickshaws and other multimodal environmental friendly modes (Fig. 6).

(ii) **Passenger information systems**

To reduce passenger waiting anxiety, Passengers Information Systems (PIS) boards need to be installed at all bus shelters, metro stations, etc. (Fig. 7).

(iii) **Multipurpose mobility card, automatic fare collection system, and automatic ticket vending machines**

It is necessary to use multiuse mobility card for last mile connectivity. The multiuse mobility card will help bring the various modes of transport even closer AFCS will reduce journey times by ensuring quicker boarding and alighting (Fig. 8).

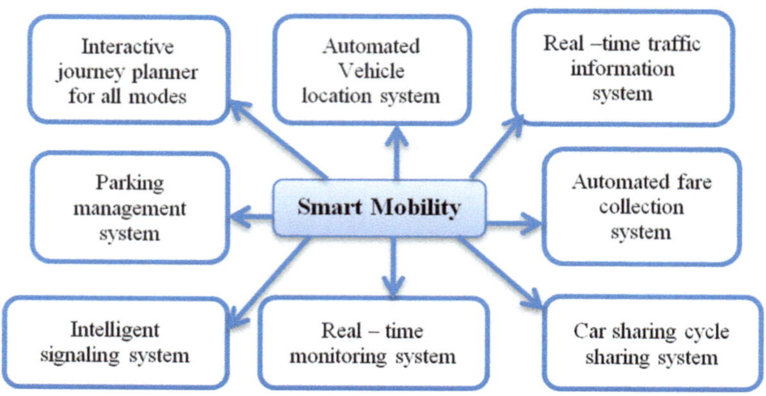

Fig. 6 Components of smart mobility system

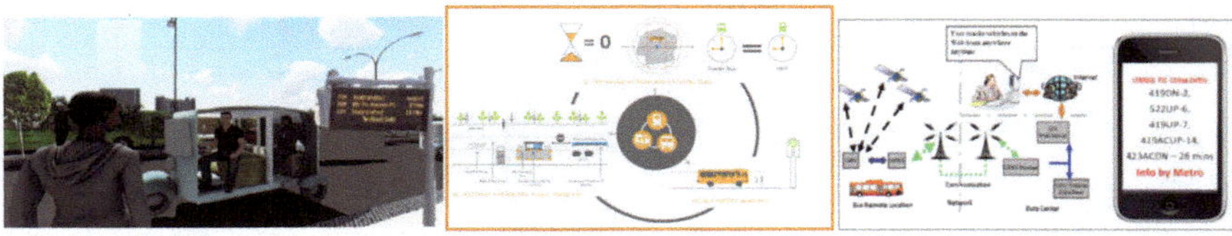

Fig. 7 Passenger information system. *Source* Safe access Manual, EMBARQ India

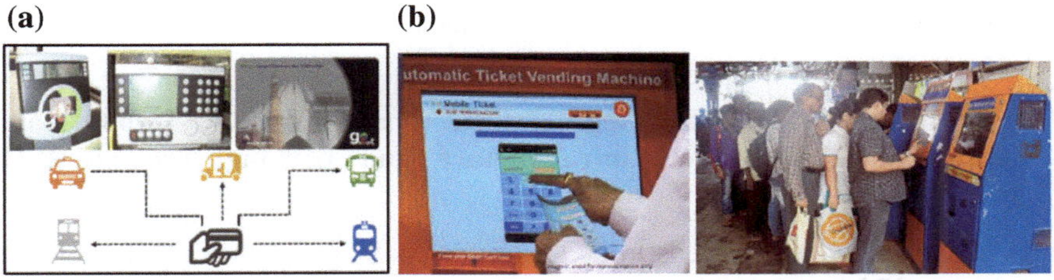

Fig. 8 **a**, **b** Multipurpose mobility card, and automatic ticket vending machine. *Source* Smart and Connected Transport—A Case Study of Delhi

(iv) **Operation control center, dynamic messaging system, smart traffic signaling system, and electronic road pricing system**

see Fig. 9.

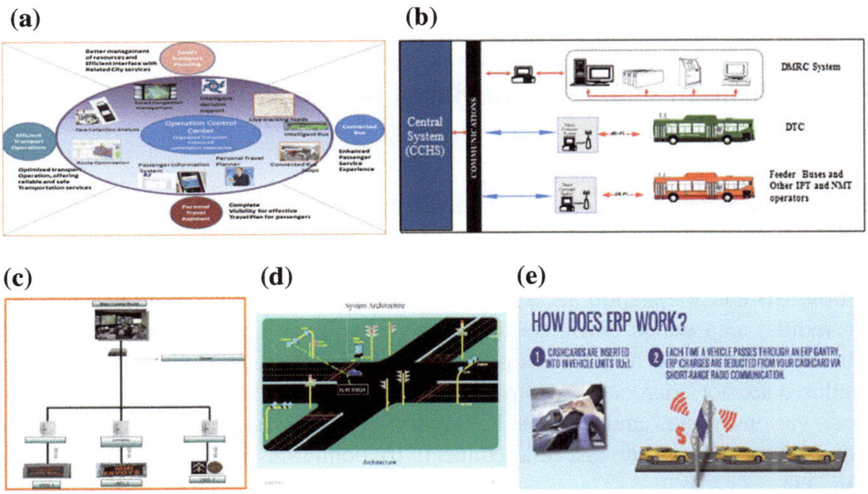

Fig. 9 **a–e** Operation control center, central systems, dynamic messaging system, smart traffic signaling system, electronic road pricing. *Source* Smart and Connected Transport—A Case Study of Delhi

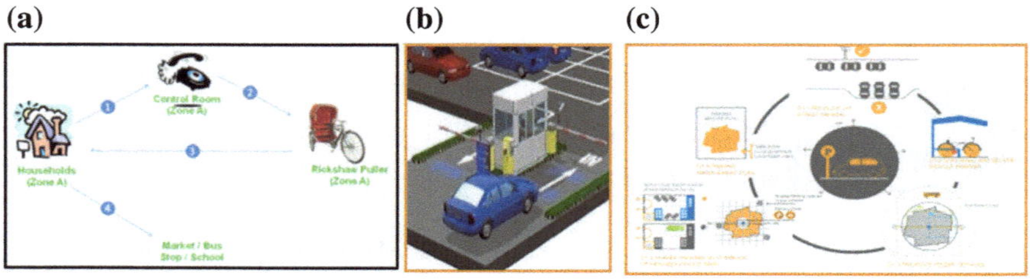

Fig. 10 **a–c** Dial-a-rickshaw facility, RFID based parking management system and other steps in parking management. *Source* Smart and Connected Transport—A Case Study of Delhi

(v) **GPS-based vehicle tracking system, dial-a-rickshaw facility, and improvement of bicycle infrastructure and parking, use of RFID technologies**

A separate call center needs to be set up for booking and the despatch of auto rickshaws. GPS-based Vehicle Tracking System needs to be installed in all auto rickshaw in the city. It is also necessary to introduce dial-a-rickshaw facility "GreenCAB" as a feeder service for the commuters who will use it as the last mile connectivity modes. There is a need for fully automated locking system that allows users to check bicycle easily or out of bike share systems. There is a need for wireless tracking system such as radio frequency identification devices (RFIDs) that locate within a bicycle and car are picked up, returned and identifies the users. Parking policy for metro stations and other important destinations should be formed to discourage increased private vehicle stations. For example, initiatives as high parking rates (Fig. 10).

7 Application of Intelligent Based Solutions in Saket Metro Station

The catchment area of Saket has been divided into three zones: Primary, Secondary and Tertiary zones, i.e., 500 m, 1 and 1.5 km (walking, cycling and other feeder modes). For station area improvement, it is divided into two parts i.e. (i) within station, (ii) Outside station access area of station i.e. the immediate influence (1.5 km radius) area which considers the access and transit area of a station. A multi-utility zone of 200 m has been proposed near Saket metro station, so that street vendor does not encroached footpaths and parking of NMT and IPT. Other than these various routes and various ICT based solutions have been suggested and proposed to improve the last mile connectivity conditions in that particular station (Fig. 11).

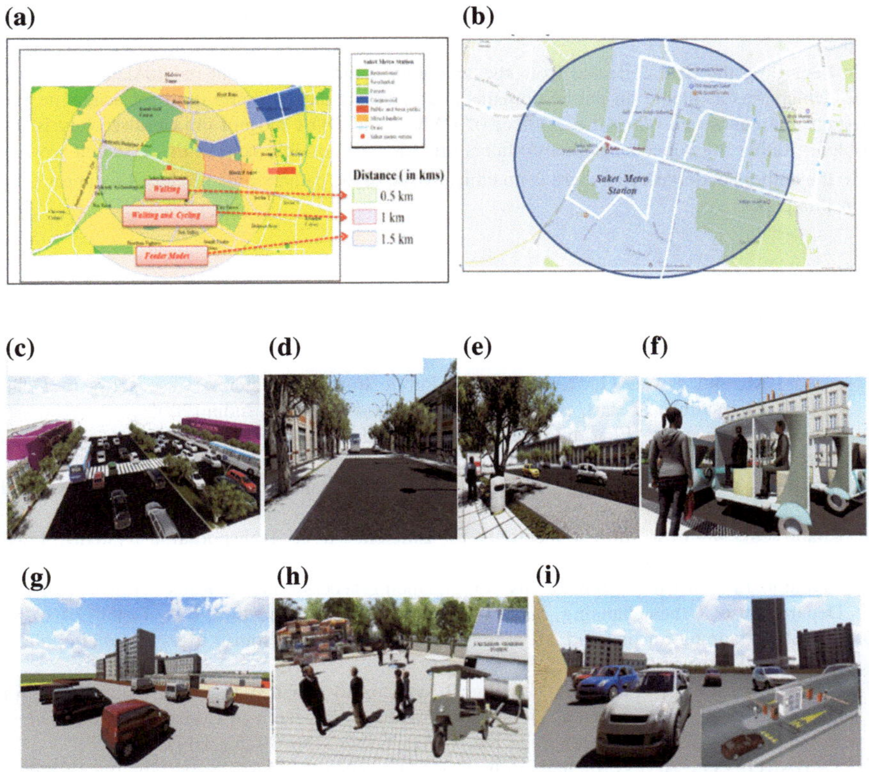

Fig. 11 Catchment area, improving pedestrian network and Application of Information and Communication based solutions in Saket metro stations

8 Conclusion

People often have problem in starting their trip, i.e., from their home. These difficulties do not lie in the main transport network but in available options that a person has beyond his residence to reach the station located at main transport network. During the last few decades, more cars, congestion and related urban transport problem and safety and security related problem arise. Thus, it is necessary that last mile connectivity should not be considered just an option at metro ends but be treated as an opportunity from a user's home. With physical integration and fare integration and with the help of intelligent systems, last mile connectivity can be improved.

Ethical Information

1. The paper contains the data collected and synthesized personally (through primary survey) with the users (with the prior-proper verbal consent) of the Metro and private vehicle users and NMT IPT operators, and they submitted absolutely no issue/problem in providing their opinions/

statements/observations regarding this data, as these data were collected for only academic purpose.

2. Regarding the data collected from the secondary sources, only that record/data have been used, which was available on public domains and since the data is being used for academic and research, the concerned office, i.e., Officer of DMRC and DDA did not object/raised any issue/ problem. Rather the concerned offices/officers provided all possible help in this research work.

So the authors undertake that there is no ethical issue in the data/records produced/mentioned by us in this paper.

References

1. Advani M. and Tiwari G. "Evaluation of Public transport systems: Case study of Delhi metro", Proceedings of start, conference held at IIT Kharagpur, India (2005).
2. Advocacy Advance—"First mile, last mile: How federal transit funds can improve access to transit for people who walk and bike" (2014).
3. BizLogics Technologies Pvt. Ltd, RFID based Parking Management System, article assessed on (22 April 2017).
4. Chaturvedi. N—Last-Mile Connectivity for Efficient Public Transportation Systems, TERI University, Internship report (2015).
5. Chidambara—Last mile connectivity for Enhancing accessibility of Rapid Transit Systems, Department of Urban Planning, School of Planning and Architecture, New Delhi (2016).
6. Das. A—Planning for first and last mile connectivity in a mass transit users in a metropolitan city, SPA TP thesis (2015).
7. De M, Sikarwar S, Kumar V, "Intelligent systems to enhance last mile connectivity for upcoming smart cities in India", Journal of Advanced Research in Construction and Urban Architecture. Volume 2 (3&4): Page No.16–31. ISSN: 2456–9925 (2017).
8. EMBARQ—Safe Access Manual, Volume I: Safe Access to Mass Transit Stations in Indian Cities, WRI India (2015).
9. Monika—First/Last Mile Connectivity Delhi: Case study Pitampura metro station, TCPO training report (2016).
10. Rawal. T, Devdas. V, Kumar. N—Integrated Multi Modal Transportation in India, IIT Roorkee, Changing Spectrum of Human Settlements and Planning Education, ISBN 978-93-5053-361-1 (May 2015).
11. Sahai N. S.—Smart and Connected Transport—A case study of Delhi, Delhi Integrated Multi —Modal Transit System Limited (2010).
12. Sanagapalli "Enhancement of Transit Ridership—Delhi Metro", SPA TP thesis (2012).
13. Sharma. A, "Planning and Design for Urban Neighbourhood based on Non-Motorised transport case study Delhi", IIT Roorkee thesis (2015).
14. Yadav. A, "Feasibility study of bicycle infrastructure in Delhi"—case study Delhi, SPA BPP thesis (2015).

A Proposed System for the Prediction of Coronary Heart Disease Using Raspberry Pi 3

Sahas Parimoo, Chaitali Chandankhede, Pulkit Jain, Shreya Patankar and Aishwarya Bogam

Abstract Technology is being used everywhere in our daily life to fulfill our requirements and aid our lives in every sphere including communication, traveling, entertainment, etc. But it has not been used up to its full potential in the field of health care. Real-time monitoring of patients and doctors is the primary concern of any healthcare facility and the conventional methods used in the hospitals are not being able to address these concerns efficiently. Our proposed system aims to provide a better service to the patients as compared to the conventional methods. It comprises of sensors for measuring temperature and pulse rate. A processor working in synchronization monitors the vitals of the patient and sends the captured data to the personal digital assistant (PDA) of head nurse and the doctors. In the PDA, an intelligent application gives statistical data of the vitals of patient in real-time manner. This data is used to analyze the health condition of patient over a period of time as well as monitor the patient in a real-time environment, remotely. We are also incorporating a coronary heart disease (CHD) predicting mechanism to determine whether a patient is susceptible to CHD. This mechanism is based on the concept of Traditional Chinese Medicine (TCM), which states that a large number of diseases (immaterial of the amount of severity they pose) can be diagnosed by sensing the pulse variations. This methodology is the foundation on which support vector machine (SVM) is predicting CHD in patients.

S. Parimoo · C. Chandankhede (✉) · P. Jain · S. Patankar · A. Bogam
M.I.T, Pune 411038, Maharashtra, India
e-mail: chaitalipb@gmail.com

S. Parimoo
e-mail: sahas.parimoo@gmail.com

P. Jain
e-mail: pulkitjain24.pulkit@gmail.com

S. Patankar
e-mail: shreyapatankar23@gmail.com

A. Bogam
e-mail: ashbogam@gmail.com

C. R. Panigrahi et al. (eds.), *Progress in Advanced Computing and Intelligent Engineering*, Advances in Intelligent Systems and Computing 714,
https://doi.org/10.1007/978-981-13-0224-4_34

Keywords Raspberry Pi 3 (RPi 3) · Coronary heart disease (CHD) Personal digital assistants (PDA) · Support vector machine (SVM) Pulse diagnosis theory (PDT) · Traditional Chinese medicine (TCM) Wireless sensor network (WSN) · Pulse waveform velocity (PWV) K-nearest neighbor (KNN) · Signal to noise ratio (SNR) · Artificial neural network (ANN) · Electrocardiograph (ECG) · Electromyographic (EMG) Oxygen saturation (SPO2) · Negative temperature coefficient (NTC)

1 Introduction

Health and fitness are among the top priorities of people from all walks of life. Technically sound hospitals are sought after and tend to be more competent. We are attempting to build a system which will help these hospitals in overcoming problems faced with the conventional methods. Using high accuracy sensors to capture the health parameters, a Raspberry Pi to process the collected data, and a PDA to access the same, remote monitoring can be done efficiently. It is an automation of the conventional methods providing constant vigilance over the patients and their care. Traditional Chinese Medicine entails that the pulse rate can be used as an important factor in predicting heart diseases [1]. Integrating IT with the healthcare sector results in ingenious solutions and breakthroughs. Heart rate along with the medical history of the patients can be used to predict the susceptibility of CHD [2]. We have used the classifier, SVM to classify patients as CHD or non-CHD [3]. The features used for classification are age, weight, gender, smoking habits, pulse rate, etc.

2 Problem Definition

Many patients require continuous monitoring and analysis of various parameters, which can be expensive and cumbersome. Thus, remote yet effective monitoring of patients is a major concern for the healthcare sector. Often the most important medical history of the patient is not readily available or compiled. Coronary heart disease (CHD) is a very critical heart disease resulting in enormous amount of deaths every year. Many lives can be saved if CHD is predicted using its early symptoms. We have proposed a system which will attempt to provide effective solutions to the abovementioned problems. Sensors can be used to effectively capture the vitals of the patients. The database server stores the captured parameters along with the medical history of the patients, through which they can be remotely accessed. We expect that using these vitals along with the medical history, and a machine learning algorithm, i.e., support vector machine (SVM), learning mechanism will efficiently try to predict the deadly coronary heart disease (CHD), which will increase its effectiveness over time. The proposed system can save time and effort, improving the health care of patients.

3 Research Methodology

3.1 Support Vector Machine

These are supervised learning models with the highly effective learning procedures that eventually will be used to carry out an effective analysis on the data which is based on classification. To classify patients as CHD and non-CHD using SVM, we have used the following features: age, gender, weight, pulse rate, height, smoking habits, and blood pressure. The pulse used is captured real time from the heartbeat sensor used.

3.2 Traditional Chinese Medicine

The pulse-based diagnosis of a patient is one of the major diagnoses out of the four examinations carried out, called the inspection, "auscultation and olfaction", inquiry and finally palpation. Since ages, the practitioner of Traditional Chinese Medicine used thigmesthesia to comprehend information of patients' pulse which is represented by arteriopalmus, so as to find when they discriminate data based on pulse attribute. Doctors diagnose the subject/patient by checking pulse beats at the measuring point on the forearm, on the radial artery, which requires a long experience and a high level of skill of the doctor involved. Pulse rate can be used as an indicator of heart diseases [1]. Heart rate of patients susceptible to heart diseases is usually very high as compared to healthy patients. This is because the heart needs to beat more to pump the same amount of blood. Inconsistencies in the pulse form may indicate blockage in the arteries causing heart diseases like CHD [4].

4 Implementation Methodology

Figure 1 highlights the working mechanism of the system. It is a flowchart representing the actual working of the system. It depicts the tasks that will be performed in the system.

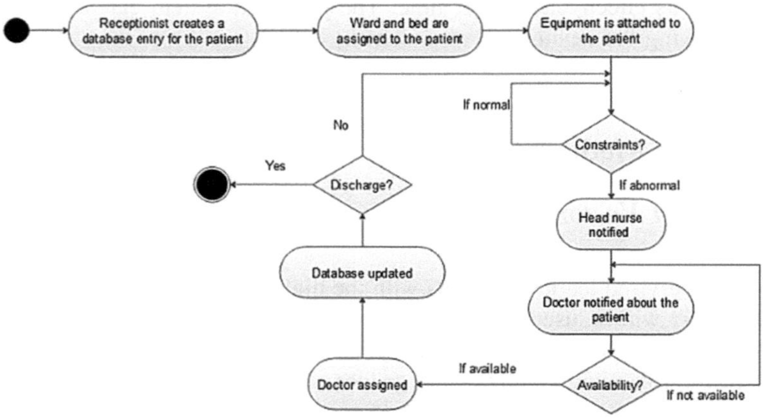

Fig. 1 Activity diagram

5 Application

Remote monitoring of patients is one of the state-of-the-art requirements and our system addresses it. According to our survey, this system has wide scope of application in today's scenario in the hospitals in our country. CHD Prediction allows us to extend our scope of real-time monitoring to cater to the hospitals providing cardiovascular care. To provide scalability [5], Arduino can be used along with Raspberry Pi, to monitor multiple patients at a time, as Arduino can gather the data from 128 sensors. Our system can be customized to be used individually for patients at home. In cases of lack of hospital staff or caretakers at home, this system will be applicable to remotely monitor the patient at home, from anywhere else. Patients with heart problems can be evaluated for CHD prediction, at an individual basis, also at home, to enhance further treatment for the same.

6 System Design

Our proposed system aims to ease the job of the medical staff by automating the conventional methods. This is done in four phases: data capturing, data collection and processing, data monitoring, and data prediction (Refer to Fig. 2).

Data capturing is achieved by using wearable body sensors like NTC Thermistor and pulse rate sensors are used for sensing the temperature and pulse rate of the patient, respectively [6]. These sensed vitals will be processed by the Raspberry Pi. This takes place in the data collection and processing phase.

The most significant phase of the entire process is the data monitoring phase. The Raspberry Pi 3 is an on-chip computer which comprises a quad-core based A53

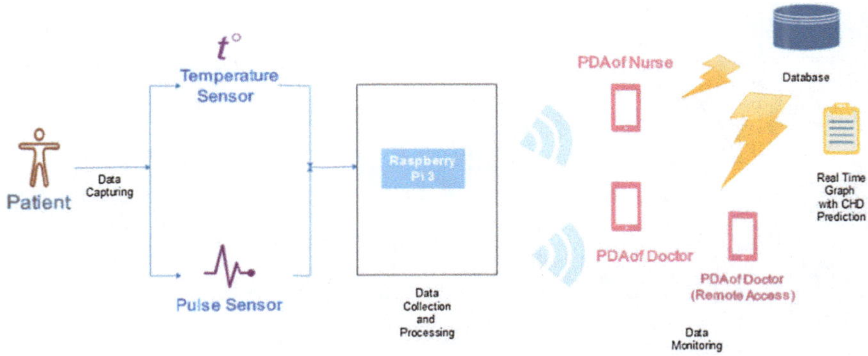

Fig. 2 System design

Cortex processor, reportedly 10 times more efficient than the Raspberry Pi 1. This is due to the two mechanisms: task threading and the instruction set use. The Raspberry Pi has an inbuilt Wi-Fi module, which will send the processed data to the PDAs of the nurse and doctor available in the premises. This real-time data will also be stored in the database for the prediction of CHD. Also, the action taken by the nurse or the medications given to him will be stored in the database for future references. In data prediction phase, the pulse rate and the medical history of the patient will be used to the predict CHD in patients, using SVM.

7 Experimental Setup

The basic experimental setup consisted of Raspberry Pi 3B, sensors, and a laptop to display the output of the sensors and an HDMI cable. The Raspberry Pi 3B was

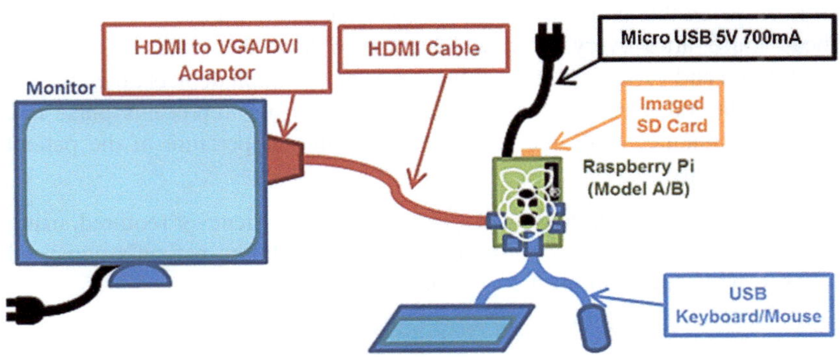

Fig. 3 Hardware setup

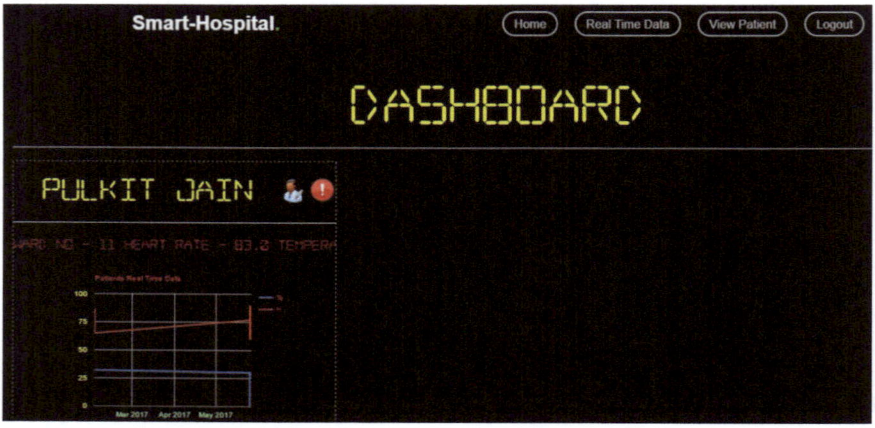

Fig. 4 Dashboard

connected to a power source and a monitor through the HDMI cable for the Raspbian OS to run. The sensors were connected to the Raspberry Pi 3B using a node head. The sensors successfully captured the vitals and sent them to the laptop, which was in the same network as that of the Raspberry Pi 3B via the Wi-Fi module of the Raspberry Pi 3B. Dynamic graphs were generated after every 30 s. An Internet-based SMS service was used to send the notifications remotely to the doctors. As for the CHD prediction part, the predictions were shown along with the vitals (Fig. 3).

8 Experimental Analysis

As a result of the above experiment performed, we observed the following (Fig. 4):

1. It is observed that graphs were generated dynamically on the PDA/laptop of the head nurse after every 30 s, which showed how the vitals of the patients monitored and varied with time. The X-axis of the graph depicts time and Y-axis depicts the value of the vitals captured. The red line represents pulse rate in beats per minute, while the blue line represents temperature of the patient in degree Celsius.

2. Notifications were sent to the doctors by the nurse, whenever required, using an Internet-based SMS service. The doctor logo beside the patient's name, when clicked will send a notification to the doctors.

3. CHD Prediction was done successfully. The red exclamation logo beside the doctor logo depicts high susceptibility of CHD, while a green tick logo depicts low susceptibility of CHD.
4. The view patient tab on the top right side of the page leads to the comprehensive database of the patients in tabular form.

9 Conclusion and Future Scope

In this project, we aim to monitor the health parameters of the patients efficiently and use the monitored data combined with their medical history to predict whether the patient may suffer from CHD or not. Also, the medical history of the patients will be updated regularly and stored in a systematic manner for quick references. We hope this proposed system will prove useful in saving time as well as lives.

This system can be expanded by using an Arduino along with the Raspberry Pi to accommodate up to 120 sensors [6]. Also, the patient data can be stored on the cloud to access remotely. Security provisions can be made to protect the privacy of the patients and the integrity of the data [7]. With the above modifications, the system will be more scalable, efficient, and secure.

References

1. Wenjie Xu, Haixia Yan, Jin Xu, Yiqin Wang, Zhaoxia Xu, Youwen Wang, and Rui Guo: Objective Study for Pulse Diagnosis of Traditional Chinese Medicine: Pulse Signal Analysis of Patients with Coronary Heart Disease, Conference on Control Automation (ICCA) June 18–20, 2014, Taichung, Taiwan, https://doi.org/10.1109/hic.2014.7038922
2. Kalia Orphanou, Arianna Dagliati, Lucia Sacchi, Athena Stas-sopoulou, Elpida Keravnou and Riccardo Bellazzi: Combining Na-ive Bayes Classifiers with Temporal Association Rules for Coro-nary Heart Disease Diagnosis Kalia, 2016 IEEE International Con-ference on Healthcare Informatics, https://doi.org/10.1109/ichi.2016.15
3. Yanwei Xing, Jie Wang and Zhihong Zhao Yonghong Gao: Combination data mining methods with new medical data to predicting outcome of Coronary Heart Disease, 2007, https://doi.org/10.1109/iccit.2007.204
4. Hai Xia Yan, Yi Qin Wang, Rui Guo, Zhao Rong Liu, Fu Feng Li, Feng Ying Run, Yu Jian Hong and Jian Jun Yan: Feature Extraction and Recognition for pulse waveform in Traditional Chinese Medi-cine based on Hemodynamics Principle, 2010 8th IEEE International Conference on Control and Automation Xiamen, China, June 9–11, 2010, https://doi.org/10.1109/icca.2010.5524147
5. Ahmad Mohawish, Ragini Rathi, Vibhanshu Abhishek, Thomas Lauritzen and Rema Padman: Predicting Coronary Heart Disease Risk Using Health Risk Assessment Data, SSH 2015: The 3rd International Workshop on Service Science for e-Health, IEEE, 2015, https://doi.org/10.1109/healthcom.2015.7454479

6. Vega Pradana Rachim and Wan-Young Chung: Wearable Non-Contact Armband for Mobile ECG Monitoring System, December 21, 2015, https://doi.org/10.1109/tbcas.2016.2519523
7. Ting Zhang, Jiang Lu, Fei Hu, Member, IEEE and Qi Hao: Bluetooth Low Energy for Wearable Sensor-based Healthcare Systems, 2014 Health Innovations and Point-of-Care Technologies Conference Seattle, Washington USA, October 8–10, 2014, https://doi.org/10.1109/hic.2014.7038922
8. A. Kampouraki, D. Vassis, P. Belsis, C. Skourlas: e-Doctor: A Web Based Support Vector Machine for Automatic Medical Diag-nosis, 2013, https://doi.org/10.1016/j.sbspro.2013.02.078
9. Jibing Gong, Shilong Lu, Rui Wang: PDhms: Pulse Diagnosis via Wearable Healthcare Sensor Network, 2011, https://doi.org/10.1109/icc.2011.5963341
10. Argyro Kampouraki, George Manis, and Christophoros Nikou: Heartbeat Time Series Classification With Support Vector Machines, July 2009, https://doi.org/10.1109/titb.2008.2003323

Simulation of Analytical Chemistry Experiments on Augmented Reality Platform

Ishan R. Dave, Vikas Chaudhary and Kishor P. Upla

Abstract The experiments of analytical chemistry are required to perform under controlled conditions. Also, they need a lot of safety precautions. In addition to this, in these experiments, there are many costly chemicals needed which may not be useful after those experiments. Augmented reality is the emerging field for training and education purpose nowadays. In the proposed work, we use augmented reality platform to perform analytical chemistry experiments which can eliminate the risks during experiments and also useful to prevent the waste of chemicals. The proposed algorithm deals with different computer vision techniques such as marker-based augmented reality, adaptive hand segmentation, gesture recognition, and hand pose estimation to manipulate virtual objects and perform experiment virtually. The algorithm is proposed for ego-centric videos to give real experience of experiments to the user and it is implemented on Android smartphone with virtual reality (VR) glasses. Such algorithm can be useful for smart educational environments in future.

Keywords Augmented reality · Chemistry education · Marker-less object manipulation

1 Introduction

Analytical chemistry is the discipline of finding, processing, and communicating compositional and structural data of a matter. Since it deals with wet chemicals, the experiment requires a lot of safety and it should be done under controlled conditions [1]. Some of the chemicals used in experiments are costly, which may

I. R. Dave (✉) · V. Chaudhary · K. P. Upla
Department of Electronics Engineering, Sardar Vallabhbhai National Institute of Technology (SVNIT), Surat, India
e-mail: ishandave95@gmail.com

V. Chaudhary
e-mail: vrnvikas1994@gmail.com

K. P. Upla
e-mail: kishorupla@gmail.com

C. R. Panigrahi et al. (eds.), *Progress in Advanced Computing and Intelligent Engineering*, Advances in Intelligent Systems and Computing 714,
https://doi.org/10.1007/978-981-13-0224-4_35

not be useful after those experiments. It is also crucial to minimize toxicity and disposal of chemical waste properly [1, 2]. Generally, analytical chemistry is introduced to the students of high schools. It is not feasible for low resources and rural schools to afford such costly chemicals and the expert lab persons to take care of safety requirements. This problem can be solved using augmented reality (AR). It is an enhancement in vision by combining real scene and virtual scene generated by computers [3]. Since past two decades, AR has been an emerging technology, which allows the user interacting with virtual three-dimensional objects in the real world [4, 5].

AR is very useful in the different applications such as entertainment, advertising, interior designing, industrial manufacturing, health care, training and education. Applications of AR and its systems are developed and used in various fields of education, for example, mathematics, chemistry, mechanical designing, and biology education. Students taking control of their own learning if an aspect which endears AR to the burgeoning concept of education. Augmented opportunity provides in terms of authentic education and training style, and further increases its appeal. Chang et al. [6] and Kerawalla et al. [7] state that the suggestion of virtual and augmented reality motivates not only for learning but also for real-time educational practices. One of its perks is that it offers a safe work environment in which mistakes made during skill training will not result in any real repercussion [8]. So AR can be a very beneficial tool in an educational environment to make it more interactive.

Over the last two decades, a plenty of research has been taken place in the field of AR. In spite of these efforts, integrating AR in the field of education and training has met with resistance. There are various factors for this resistance, for example, development cost, maintenance cost, and general reluctance to accept and adapt to new learning methods. Shelton [9] states that there is still uncertainty about the practice of AR in the domain of education and training. Problems such as cost cutting and performance between AR systems and traditional ways are the reasons for lingering uncertainty.

In literature, there are some attempts made in augmented chemistry education. Fjeld et al. [10] propose a setup for learning physical chemistry in AR environment. In the setup of the system, components name and printed picture are displayed by a booklet. A gripper is used to move atoms and a button is supposed to press to connect atoms to other molecular structure in virtual environment. Chen [11] also makes use of AR in physical chemistry education to visualize amino acid structure in three dimensions (3D). The result of his work shows that students like to manipulate virtual object by rotating makers. Cai et al. [12] perform a case study of AR in chemistry education on junior high school students. The study is targeted on "composition of structure", a topic of physical chemistry. They conclude that AR is effective tool in chemistry education. In augmented chemistry, most of the approaches are done in physical chemistry education. The proposed work is an attempt of using augmented reality in analytical chemistry domain. In this work, we propose an algorithm that allows the user to perform analytical chemistry experiments virtually using AR. The complete block schematic of the proposed algorithm is depicted in Fig. 1. It is primarily divided into three stages frame processing, pose estimation, and interaction with the virtual world.

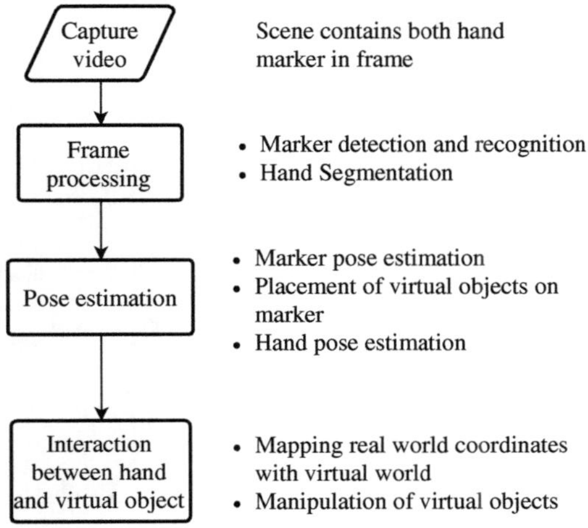

Fig. 1 Block diagram of proposed method

The rest of the paper is organized as follows. In Sect. 2, the different materials used in the proposed methods are discussed. The frame processing steps such as marker detection and segmentation of user's hand are elaborated in Sect. 3. Next to that in Sect. 4, the pose estimation techniques are discussed. The result section shows output of interaction of user's hand with the virtual world which is shown in Sect. 5. Finally, the conclusion is drawn in Sect. 6.

2 Materials and Setup

The experiment setup is shown in Fig. 2 (Human user is not directly involved in this study). Apparatus for this experiment includes VR glasses, Android smartphone, and fiducial marker. The user needs to wear VR glasses on which Android smartphone is attached. The smartphone is used for all purposes including video acquisition, video processing, and display purpose.

Fiducial marker is used as a benchmark for virtual experiment setup. The fiducial marker is 7×7 matrix that consists of binary colors (Fig. 3). Using such dimension of marker allows assignment of unique marker to each student, which prevents false marker detection in the classroom of a large number of students. Another advantage of this method of marker generation is that it requires only 49 bits to store a marker which is very less than storing a marker as an image. The virtual apparatus is created using unity software package. To give a real experience of chemistry experiments to the user, the system is proposed for ego-centric (first person) videos, which are captured by the smartphone camera.

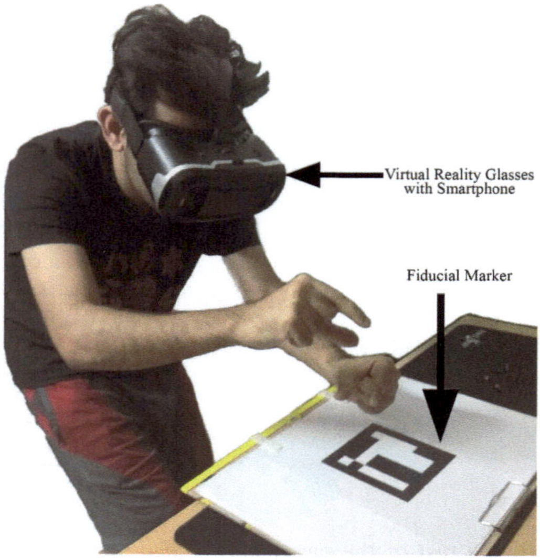

Fig. 2 System setup

Fig. 3 Fiducial marker

Fig. 4 Captured frame from camera

3 Frame Processing

A frame from the captured video is shown in Fig. 4. There are mainly two tasks carried out in frame processing simultaneously: marker detection and hand segmentation

(a) Marker in grey scale (b) Output of Canny edge Detector (c) Output of morphological operartions

Fig. 5 Marker detection

3.1 Marker Detection

Marker detection is an essential step to set up virtual experiment apparatus. In a captured frame, marker is detected using gradient based approach [13]. In the first step, the captured frame is converted into grayscale image (see Fig. 5a). Contours of the grayscale images are detected using Canny edge detection [14] (see Fig. 5b). Using mathematical morphological operations, the detected contours are linked to form segments, which finally result in quadrilaterals (see Fig. 5c). The marker pattern is identified from the marker database by matching detected quadrilateral pattern. After detecting marker, it is tracked in every 25th frame (called extended tracking) to reduce the computational complexity of the algorithm.

3.2 Hand Segmentation

For the sake of detection of the user's hand, each pixel of the captured frame is segmented as skin pixel or non-skin pixel. Here, we use adaptive hand segmentation technique, introduced by Lee et al. [15]. This technique utilizes combined probability from Gaussian mixture models and learned histograms of skin and non-skin models. The output of hand segmentation is shown in Fig. 6.

4 Pose Estimation

Considering pose estimation from the captured scene from the camera, the camera must be distortion free. So before pose estimation, camera's intrinsic parameters (A) are estimated by calibrating the camera with 10×7 checkerboard pattern [16]. Pose estimation is explained in the following sections.

Fig. 6 Output of adaptive hand segmentation

4.1 Pose Estimation of Fiducial Marker

Pose of fiducial marker with respect to camera is estimated by computing projective transformation matrix. The projective transformation matrix (B) is estimated by matching quadrilaterals from captured marker image and database marker image. The 3D model of virtual experiment setup is placed on the marker using intrinsic parameter matrix (A) and proactive transformation matrix (B) as,

$$[w] = [A][B][W], \tag{1}$$

where w and W are virtual and real-world coordinate, respectively. The placement of virtual experiment setup from different camera's view is shown in Fig. 7.

4.2 Pose Estimation of Hand

Since it is very difficult to estimate hand pose from a monocular camera, the proposed method works only for limited gestures of hand. The block diagram of the proposed method for hand pose estimation is shown in Fig. 8. A gesture from the segmented hand image (Fig. 6) can be considered as a fixed shape with variation in scale, rotation, and translation. So a gesture can be stated in terms of Hu's invariant moments [17]. First, the gesture of hand is continuously observed by finding the invariant moments from the segmented hand image. As soon as the invariant moments fall in the range of a gesture, hand pose is estimated from the training database. The algorithm requires one-time training for pose estimation for each gesture of hand. We are using fingertips as the features of hand. Fingertips are distinguished from hand contours by a curvature-based algorithm [18]. The system is trained for each gesture by

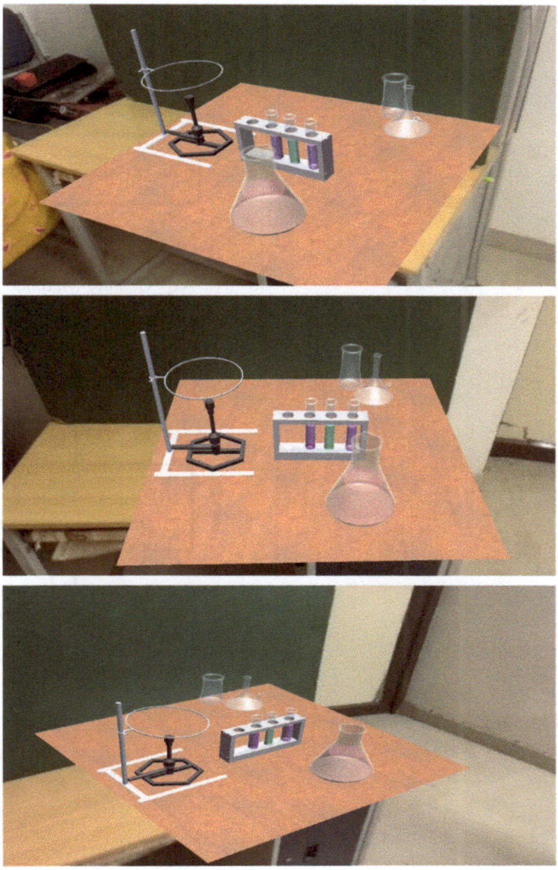

Fig. 7 Placement of virtual experiment setup on marker from different camera's views

calibrating with a checkerboard pattern as shown in Fig. 9. Hand pose is determined by estimating projective transformation matrix (B) from the curvature of fingertips detected from segmentation output image and database images as shown in Eq. (1).

5 Results

Using estimated pose of marker and hand, the user is able to manipulate objects of the virtual experiment. The simulation of analytical chemistry experiments using virtual object manipulation is shown in Fig. 10. Events take place in the experiment require knowledge of chemicals, which is stored in the system memory. The algorithm is implemented on Xiaomi Redmi Note 4 smartphone (CPU: Octa-core 2.0 GHz Cortex-A53) [19].

Fig. 8 Block diagram of hand pose estimation

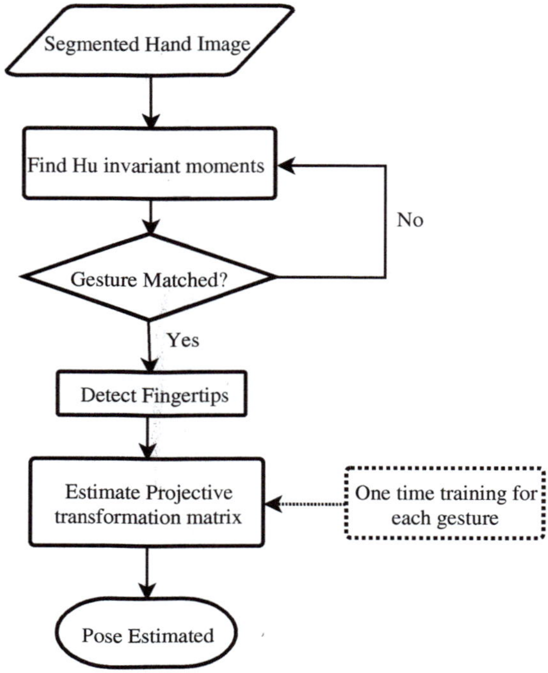

Fig. 9 One-time training for gripping gesture

6 Conclusion

Recently, attempts have been made to use augmented reality as a tool for learning about physical chemistry education. We have developed an algorithm which allows the user to simulate experiments virtually of analytical chemistry using augmented reality platform. The present algorithm is a collective problem of numerous computer vision techniques like marker-based augmented reality, adaptive hand segmentation, gesture recognition, and hand pose estimation. Marker-less hand interaction with the virtual experiment setup and ego-centric camera view give a real experience of analytical chemistry experiment to the user. Virtual experiment setup is created using unity software package and placed on fiducial marker using edge detection based

Fig. 10 Simulation of analytical chemistry experiments

method. Fiducial marker is generated by a predefined algorithm. This provides a unique marker and saves the space complexity of an algorithm. The hand is adaptively segmented using Gaussian mixture models and learned histograms of skin and non-skin models. Pose of marker and hand is estimated by finding projective transformation matrix from captured frame and training database. The system is trained for each defined gesture and results of each chemical events are prestored in the system. Marker-less hand interaction with the virtual experiment setup and ego-centric camera view give a real experience of analytical chemistry experiment to the user.

In the proposed work, analytical chemistry experiments are performed using AR platform in order to exclude the risks and chemical wastage during chemistry experiments. Smartphone implementation of the algorithm allows individual performance to learner, which results in a positive attitude of learning the subject. The algorithm has potential scope in future smart educational environments. It is worth to note that one can extend this system by training for more number of hand gestures and chemical reactions. Adaptive hand segmentation used in the proposed method works well in proper lighting conditions only, so one can improve system robustness by removing this constraint. Difficulty in hand pose estimation using monocular vision can be solved by using a stereo camera or leap motion controller as video acquisition unit.

Acknowledgements The authors would like to appreciate help supported by Dr. J. N. Sarvaiya (Head, ECED, SVNIT) and Mr. Vivek Bhargav (SMIS, Surat). The authors would also like to thank Ms. Therattil Anitta Saju for language amelioration in this paper.

References

1. Brundage, P., Palassis, J.: School chemistry laboratory safety guide. Centers for Disease Control and Prevention (2006)
2. Sales, M.G.F., Delerue-Matos, C., Martins, I., Serra, I., Silva, M., Morais, S.: A waste management school approach towards sustainability. Resources, conservation and recycling 48(2), 197–207 (2006)
3. Azuma, R.T.: A survey of augmented reality. Presence: Teleoperators and virtual environments 6(4), 355–385 (1997)
4. Fjeld, M., Juchli, P., Voegtli, B.M.: Chemistry education: a tangible interaction approach. In: Proc. INTERACT. pp. 287–294 (2003)
5. Feiner, S.K.: Augmented reality: A new way of seeing. Scientific American 286(4), 48–55 (2002)
6. Chang, G., Morreale, P., Medicherla, P.: Applications of augmented reality systems in education. In: Proceedings of society for information technology & teacher education international conference. vol. 1, pp. 1380–1385. AACE Chesapeake, VA (2010)
7. Kerawalla, L., Luckin, R., Seljeot, S., Woolard, A.: making it real: exploring the potential of augmented reality for teaching primary school science. Virtual Reality 10(3-4), 163–174 (2006)
8. Lee, K.: Augmented reality in education and training. TechTrends 56(2), 13–21 (2012)
9. Shelton, B.E.: Augmented reality and education: Current projects and the potential for classroom learning. New Horizons for Learning 9(1) (2002)
10. Fjeld, M., Voegtli, B.M.: Augmented chemistry: An interactive educational workbench. In: Mixed and Augmented Reality, 2002. ISMAR 2002. Proceedings. International Symposium on. pp. 259–321. IEEE (2002)
11. Chen, Y.C.: A study of comparing the use of augmented reality and physical models in chemistry education. In: Proceedings of the 2006 ACM international conference on Virtual reality continuum and its applications. pp. 369–372. ACM (2006)
12. Cai, S., Wang, X., Chiang, F.K.: A case study of augmented reality simulation system application in a chemistry course. Computers in Human Behavior 37, 31– 40 (2014)
13. Fiala, M.: ARtag, a fiducial marker system using digital techniques. In: Computer Vision and Pattern Recognition, 2005. CVPR 2005. IEEE Computer Society Conference on. vol. 2, pp. 590–596. IEEE (2005)
14. Canny, J.: A computational approach to edge detection. IEEE Transactions on pattern analysis and machine intelligence (6), 679–698 (1986)
15. Lee, T., Hollerer, T.: Handy AR: Markerless inspection of augmented reality objects using fingertip tracking. In: Wearable Computers, 2007 11th IEEE International Symposium on. pp. 83–90. IEEE (2007)
16. Zhang, Z.: A flexible new technique for camera calibration. IEEE Transactions on pattern analysis and machine intelligence 22(11), 1330–1334 (2000)

17. Hu, M.K.: Visual pattern recognition by moment invariants. IRE transactions on information theory 8(2), 179–187 (1962)
18. Argyros, A.A., Lourakis, M.I.: Vision-based interpretation of hand gestures for remote control of a computer mouse. In: European Conference on Computer Vision. pp. 40–51. Springer (2006)
19. Xiaomi Redmi Note 4, http://www.mi.com/in/note4/specs/

MCC and Big Data Integration for Various Technological Frameworks

Praveen Kumar Singh, Rajesh Kumar Verma and Joy Lal Sarkar

Abstract In the world of big data and Internet of things (IoT), data grows exponentially in terms of petabytes, and subsequent processing in large scale needs commodity-based clusters to run these applications. Mobile users cannot get the commodity cluster to run parallel, complex, and scientific applications. The portable devices can form a cloudlet and use the cloud based on available resources and required resources which can be taken care by our proposed scheduler engine. We are aware that the mobile devices have resource limitations, however the combinations of several component such as portable devices and cloud computing will help to fulfill the limitation in terms of resources. To encounter the problem of performing data intensive jobs using the mobile devices and also achieve interoperability with the cloudlet and different vendors of the cloud, we have proposed various architectures to integrate IoT and big data along with MCC. The proposed architectures use middleware for the integration of MCC and big data for various technological frameworks; in our approach, we use various appliances, sensors, and portable devices having efficient utilization of the resources on the cloud and cloudlets.

Keywords Cloud computing · Big data · Mobile cloud computing
Middleware · Computation offloading · Internet of things · Hadoop
Spark · Twister · Iterative MapReduce

P. K. Singh (✉)
Tata Consultancy Services Research, Mumbai, India
e-mail: praveenhelp78@gmail.com

R. K. Verma
Biju Patnaik University of Technology, Rourkela, Odisha, India
e-mail: rajeshverma_chicago2004@yahoo.com

J. L. Sarkar
Central University of Rajasthan, Ajmer, India
e-mail: joylalsarkar@gmail.com

C. R. Panigrahi et al. (eds.), *Progress in Advanced Computing and Intelligent Engineering*, Advances in Intelligent Systems and Computing 714,
https://doi.org/10.1007/978-981-13-0224-4_36

1 Introduction

In the world of big data, enormous data is generated from the various sources like mobile applications, social networks, scientific data, and through search engines. To analyze this vast amount of data and extract useful information from this data, people have come up with large data processing frameworks and all these frameworks are deployed on commodity cluster as well as on the cloud from different cloud providers. Cloud computing provides us the highly scalable resources as a service through the Internet and these services are provided as network as a services (NaaS), infrastructure as a services (IaaS), platform as a services (PaaS), software as a services (Saas) and storage as a service (STaaS). There are various open source frameworks which help to implement complex scientific application and execute them in a parallel manner which increases the speed of execution. These frameworks are parallel and distributed in nature. Users can distribute their data across the portable devices and execute their task in parallel over the data present in these distributed devices.

The frameworks like Hadoop and Twister use disk Input/Output (I/O)s to process huge amount whereas, Spark uses the in-memory concepts and reduces the I/O activity significantly during processing. The I/O activity does not always depend on the framework but it also depends on the application, whether the applications are I/O intensive and CPU intensive. The I/O intensive applications which require more number of disk access or local storage access and are CPU intensive indicates that the applications are very hungry for processing resources.

The paper aims to guide users to understand the tradeoff between resource optimization and proper utilization of available resources, with respect to the application stack.

This paper provides suitable architectures that enable the various user applications using the smart mobile devices and integrating with IOT as well as big data to process large data and come up with useful information for better decision-making through the architecture of IoT and big data with MCC, integration of MCC and big data with middleware and applications big data in MCC. Users can optimize the cloud and the cloudlet configuration and can also smartly utilize the computing resources properly and will only pay for the amount of resources that they have used (pay per use).

The real-world challenge for users is to perform CPU intensive jobs using the mobile devices with proper interoperability between different cloud vendors with cloudlet. From the perspective of users, we are stressing upon the scenarios where MCC and big data are integrated and also when they are used independently. In order to help users to choose the appropriate architecture and services with middleware, and which can be used to process their real world problem. We are categorizing the big data application in MCC cloud and cloudlet services based on their suitability and best resource utilization. In the categorization, we will focus on the different mobile application classes, different programming models, various storages used in processing different cloud vendors with cloudlet and some of the open source big data frameworks like Spark [19], Twister [10], Hadoop [3], and Flink [1].

The rest of the paper is organized as follows. In the next section, we will discuss about the related work, the architecture to integrate the MCC and big data using Middleware, applications of big data in MCC, and services architecture for MCC using middleware.

2 Related Work

Middleware is a software utility used to interact with the cloudlet and cloud vendors platforms. For the mobile devices to interact with the available technology frameworks, the middleware serves as a broker.

Akherfi et al. [7] proposed a middleware architecture which allows the users to use unrestricted cloud computation and cloud storage services. The proposed model improves the quality of services, availability, and lightweight responses for mobile cloud services. In the paper, Yin et al. [18] introduces middleware for low latency offloading of big data, which works on the Tasks Scheduler and Instance Manager. Offloading tasks are dispatched by the Tasks Scheduler to execute on the instances reserved by Instance Manager.

Computation offloading is essential during the execution of the big data jobs which are initiated by the user from their mobile devices. There are several challenges that are faced during the offloading process using the current frameworks and techniques. Chun et al. [9] proposed the "CloneCloud", in which application can be partitioned automatically using static analysis and dynamic profiling. The execution thread is migrated from the mobile device to the clone in the cloud at runtime. Ou et al. [14] proposed a service in which some of the tasks of an application are offloaded from mobile devices to the cloud as it has immense resources. However, in his approach, the application needs to be implemented in Java language.

The current architectures and frameworks present do not have the feasibility to connect with large datasets and also have language constraints.

3 Big Data and MCC Integration Using Middleware

Big data is used in various applications such as knowledge discovery in database (KDD), device-generated data analytics, fraud detection and prevention, recommendation based analytics, and many more.

3.1 Big Data and Associated Popular Frameworks

To process large-scale data, we need big data frameworks to be deployed in the cloudlet and the cloud. The various frameworks related to big data are as follows.

Hadoop: is used for batch processing in a parallel manner, it is generally deployed on the commodity cluster, cloud, and cloudlet. The earliest version [3] used in a single cluster had scalability problem which was resolved in the updated version Hadoop YARN [3]. Different studies that have been carried out in the past indicate that Hadoop is not suitable for long running jobs. However, Hadoop is very useful for running the job via the mobile devices.

Haloop [8]: represents a new programming model which is built on top of Hadoop. It is also one of iterative framework which has the capability of data caching through the loop-invariant mechanism and also controls the iterative jobs via the scheduler. It does not support fault tolerance for long running jobs. When a job failure occurs it needs to re-execute the MapReduce (MR) tasks.

Twister [6]: is an enhanced MR runtime which supports iterative MR computations efficiently. It was developed by Jaliya Ekanayake's and is currently supported by the SALSA Team at Indiana University [10]. Due to the availability of static and variable data, the computation cost is less. As the static data will be cached and the variable data size will keep on changing. Both the static and variable data can be reused in every iteration. The performance of Twister can be significantly improved by configuring the MR tasks as a cacheable. Twister can be easily used to implement iterative MR based algorithms [15, 16] for feature selection while building machine learning model in the cloud and the cloudlet.

Spark [19]: is used for batch, stream and graph processing. It has widely deployed frameworks for data analytics. The in-memory and lazy evaluation features enhanced the performance of the spark programs. It works on the resilient distributed datasets (RDDs). There are two main operations, transformations and actions which can be applied on RDDs. It is easily configurable and can be deployed on the cloudlet or cloud using a commodity cluster.

GraphX [11]: is available on spark and used for graph processing. Apart from GraphX for Spark, there are other available frameworks such as GraphLab [13], Giraph [2] and Pregel [17]. However, these other frameworks have compatibility issues with different cloud vendors.

Spark Streaming [20]: is used for streaming application and micro-batch processing. It is known for consistency, scalability, parallel execution, and fault recovery. Apart from Spark Streaming, Apache Storm [4] from Twitter, IBM InfoSphere Streams [5] are also some of the frameworks used for streaming different kinds of applications.

3.2 MCC

MCC can be used to overcome the limitations of computing resources and processing requirements of a SMDs by integrating it with the cloud. SMDs target the issues of security, performance, and environment once the integration of SMDs and the cloud takes place [12].

3.3 Architectural Components Used

The architecture is master–slave in nature and mainly used for building cluster on the cloud using both the local and remote resources. Figure 1 depicts the main components which are used in different layers.

- Message Broker: Used for sending the messages from the source to the target components. It is very important for the transfer of the messages (using message-oriented middleware).
- Technology Daemons: These are the services that are running on all devices. Various available frameworks provide different type of daemons to accomplish certain tasks.
- Scheduler Engine: It helps to schedule the tasks either in the cloudlet or on the cloud. This decision of running the tasks depends on the availability of the resources in the cloudlet.

The sequence of processing a particular tasks is as follows:

User interacts with the user application layer and provides all the required input. The input is processed by the MapReduce/Iterative MR program and sent across to the network layer. Using the network, the data is transmitted either to the cloud or to the cloudlet. The scheduler engine is responsible for deciding the place of execution of the tasks at hand. The scheduler engine has a scheduler script which will keep running on the backend at the user end.

The scheduler engine jobs are to get the application type, data size required by an application, estimate the resource required by an application through cost-based approach or prediction approach. Based on the requirement, scheduler engine will trigger and give signal whether to use cloudlet and perform mobile cloud computing or use cloud to perform cloud computing.

Once the required information is present in the cloudlet, data preprocessing and analyzing are performed locally. If the submitted application requires more amount of resources then what is present in the cloudlet the scheduler automatically routes the tasks to the cloud and subsequently each tasks is handled. It also helps to save money as well as energy.

The cluster formation can be done on the cloudlet as well as on the cloud. Mobile devices are connected in a master–slave manner, where the MR Jobs are present at Master node. Master node is connected with the Message Broker, which is the middleware component. The Message Broker is the publish/subscribe broker which is used for network and communication services. It is directly connected to the slave devices.

Slave devices, where technology daemons are running from the different technology stacks. Each slave will process MR jobs from the pool of various jobs which are

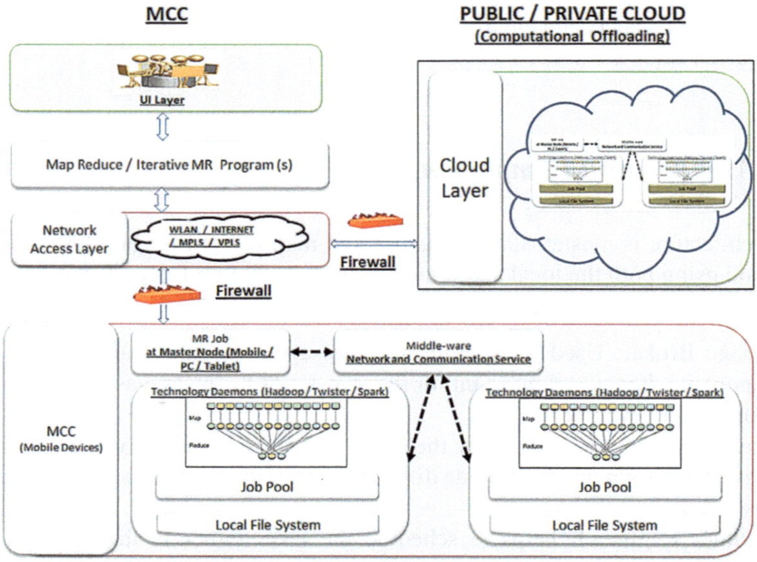

Fig. 1 Architecture of big data and MCC integration using middleware

waiting. Once the reduce phase is completed, the output is merged and sent across to the master node. If the required output needs to be stored data locally than the local file system is used on each of the devices.

If the submitted application requires a vast amount of resources, the cloud cluster will come into the picture, which has various machine instances available for processing of the MR job in a master–slave mechanism.

4 Proposed Layering Used in Big Data and MCC Context

The different layers that are used for any applications of big data using MCC are depicted in Fig. 2. The different layers and their functionalities are detailed as follows:

- Mobile Applications Layers: This layer hosts the various applications that require data analytics and mining. Several applications in the various domains, for example, healthcare applications, traffic applications, simulation-based applications, object captures, image classification, and detection based come under this category.
- Programming Model Layer: This layer contains different types of programming framework models which are used to process the Big Data. Several models such as M/R, RDD, high level language, workflow come under this category.

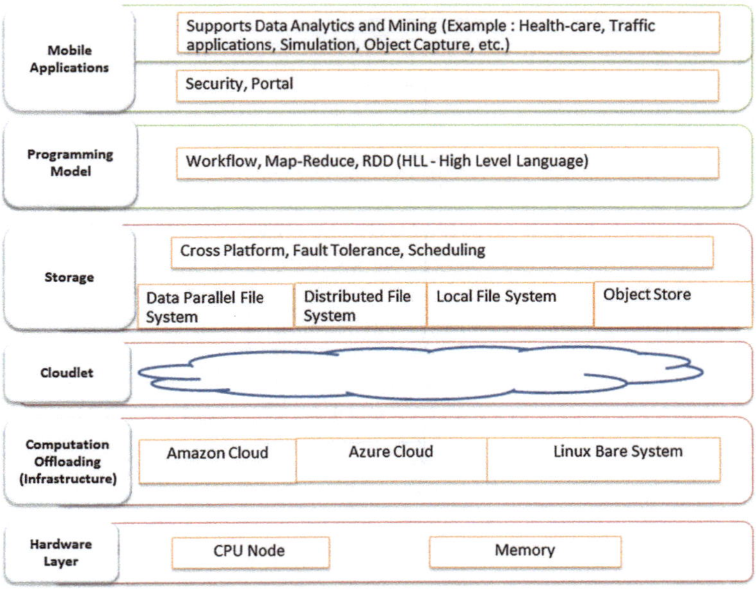

Fig. 2 Layering used in big data and MCC

- Storage Layer: This layer is useful for cross platform, fault tolerance, and scheduling. Data parallel file system, distributed file system, local file system, and object store are the file systems which are used in this layer.
- Cloudlet Layer: This layer contains various mobile devices which are connected together in a single-hop proximity to perform the computation.
- Computation Offloading Layer: Offloading is a mechanism which helps transfer the particular application which is computationally resource intensive to the cloud. Hence, it can be used to handle the problem of limited resource capabilities by offloading the computation and execution to the remote resources. Amazon cloud, Azure cloud, IBM cloud, and Linux bare systems fall under this category. This mechanism ensures better performance while running a complex applications. The flow chart shown in Fig. 3 depicts the logic used in case of offloading a particular application to the cloud.
- Hardware Layer: This layer consists of the CPU, memory, storage disk, and other peripheral devices which are used for computational purposes. The hardware presenting in this layer is generally heterogeneous (for example, the servers used for computations can be from different vendors.) in nature. Different types of hardware can be used here both the cloudlet as well as the cloud hardware infrastructure is part of the bottom-most layer.

Fig. 3 Flow chart of MCC execution

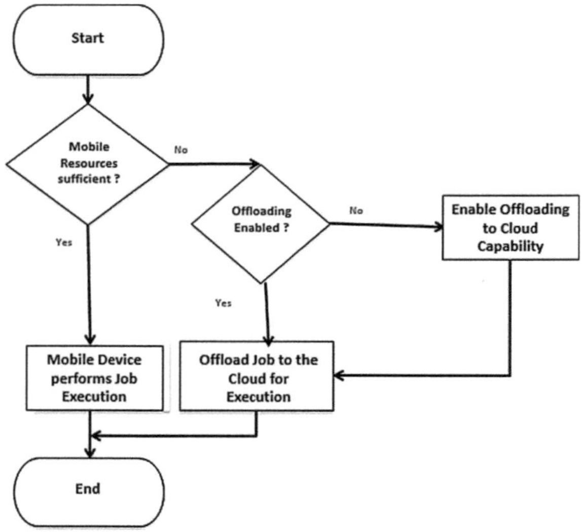

5 Proposed Service Architecture for MCC Using MiddleWare

Figure 4 depicts the service architecture for MCC using middleware, which contains the five modules as described below.

- User Module (Mobile devices such as Tablets, SMD etc.): This module contains the services provided to the users through the various applications which are smartly interconnected with each other. Data discovery and data processing can be done for various applications in the given environment, and smartly presented to the users via this interface.
- IoT Module: Smartly interconnected devices like cameras, cars, refrigerators, lights, and many more devices are connected through the Internet fall under this module. Each of the devices gathers processed data which is used for knowledge discovery and better decision making using the technology frameworks which are running on the cloudlet and the cloud.
- Sensor Module: This module is used for sensing the data of the environment (for example different type of sensors such as temperature, vibrations, light, etc.) continuously and also performs scientific calculations.
- Cloudlet Module: This module is used for computation using the cloudlet. The MR jobs which are running on the cloudlet helping computation. However, if the jobs are computationally intensive then they are subsequently offloaded to the cloud using the middleware services.
- Cloud Module: This module is hosted in the cloud (public cloud such as AWS, Google Cloud, Microsoft Azure, etc., or private data centers) and performs the

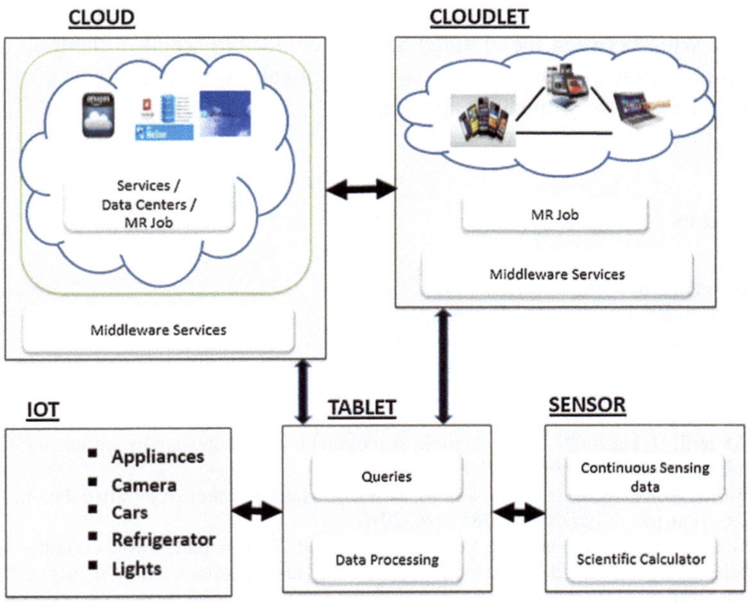

Fig. 4 Service architecture for MCC using middleware

computations. MR jobs present in this module perform the computationally complex jobs which requires lot resources (CPU, memory, and network).

Data is collected using various types of sensors from the different IOT devices. Processing and analysis can be done locally in the cloudlet. If the collected data size is huge and the processing requires rich remote resources, then it is required to offload the jobs to the cloud where subsequent processing and analysis are performed and results are subsequently transmitted to user devices using middleware services.

6 Conclusions and Future Work

The integration of MCC, big data, and cloud computing helps to create different IOT applications which are at the cutting edge technology today. User can smartly run parallel and complex applications which are easily controlled via the mobile devices. In this paper, our proposed architecture using middleware addresses the issues related to scalability of the application and huge data storage and processing. Heterogeneous environment provides the necessary computational and storage resources.

Also, we look forward to improve the proposed architecture by incorporating the security requirements, simultaneously dealing with multiple QoS, and fulfilling the clients service level agreement (SLA)'s, so that it can be easily adaptable in the near future. Also we look forward to coming up with an efficient offloading mechanism for load sharing between the cloudlet and the cloud. We are also planning to integrate this architecture with the Smart World applications such as Smart City, Smart Homes, and Smart recommender applications.

References

1. Apache flink. http://flink.apache.org/.
2. Apache giraph. http://giraph.apache.org/.
3. Apache hadoop. http://hadoop.apache.org/.
4. Apache storm. http://storm.apache.org/.
5. Ibm infosphere streams. http://www-03.ibm.com/software/products/en/ibm-streams.
6. Twister. http://www.iterativemapreduce.org/.
7. K. Akherfi, H. Harroud, and M. Gerndt. A mobile cloud middleware to support mobility and cloud interoperability. *IJARAS*, 7(1):41–58, 2016.
8. Y. Bu, B. Howe, M. Balazinska, and M. D. Ernst. Haloop: Efficient iterative data processing on large clusters. *PVLDB*, 3(1):285–296, 2010.
9. B.-G. Chun, S. Ihm, P. Maniatis, M. Naik, and A. Patti. Clonecloud: elastic execution between mobile device and cloud. In C. M. Kirsch and G. Heiser, editors, *EuroSys*, pages 301–314. ACM, 2011.
10. J. Ekanayake, H. Li, B. Zhang, T. Gunarathne, S.-H. Bae, J. Qiu, and G. Fox. Twister: a runtime for iterative mapreduce. In S. Hariri and K. Keahey, editors, *HPDC*, pages 810–818. ACM, 2010.
11. J. E. Gonzalez, R. S. Xin, A. Dave, D. Crankshaw, M. J. Franklin, and I. Stoica. Graphx: Graph processing in a distributed dataflow framework. In J. Flinn and H. Levy, editors, *OSDI*, pages 599–613. USENIX Association, 2014.
12. D. T. Hoang, C. Lee, D. Niyato, and P. Wang. A survey of mobile cloud computing: architecture, applications, and approaches. *Wireless Communications and Mobile Computing*, 13(18):1587–1611, 2013.
13. Y. Low, J. Gonzalez, A. Kyrola, D. Bickson, C. Guestrin, and J. M. Hellerstein. Distributed graphlab: A framework for machine learning in the cloud. *PVLDB*, 5(8):716–727, 2012.
14. S. Ou, K. Yang, and J. Zhang. An effective offloading middleware for pervasive services on mobile devices. *Pervasive and Mobile Computing*, 3(4):362–385, 2007.
15. P. S. V. S. S. Prasad, H. B. Subrahmanyam, and P. K. Singh. Scalable iqra_ig algorithm: An iterative mapreduce approach for reduct computation. In P. Krishnan, P. R. Krishna, and L. Parida, editors, *ICDCIT*, volume 10109 of *Lecture Notes in Computer Science*, pages 58–69. Springer, 2017.
16. P. K. Singh and P. S. V. S. S. Prasad. Scalable quick reduct algorithm: Iterative mapreduce approach. In *CODS*, 2016.
17. C. E. Tsourakakis. Pegasus: A system for large-scale graph processing. In S. Sakr and M. M. Gaber, editors, *Large Scale and Big Data*, pages 255–286. Auerbach Publications, 2014.
18. B. Yin, W. Shen, L. X. Cai, and Y. Cheng. A mobile cloud computing middleware for low latency offloading of big data. In Q. Li and D. Xuan, editors, *Mobidata@MobiHoc*, pages 31–35. ACM, 2015.
19. M. Zaharia, M. Chowdhury, M. J. Franklin, S. Shenker, and I. Stoica. Spark: Cluster computing with working sets. In E. M. Nahum and D. Xu, editors, *HotCloud*. USENIX Association, 2010.
20. M. Zaharia, T. Das, H. Li, S. Shenker, and I. Stoica. Discretized streams: An efficient and fault-tolerant model for stream processing on large clusters. In R. Fonseca and D. A. Maltz, editors, *HotCloud*. USENIX Association, 2012.

Smart HIV/AIDS Digital System Using Big Data Analytics

V. Ramasamy, B. Gomathy and Rajesh Kumar Verma

Abstract Smart HIV/AIDS digital system is a collection HIV/AIDS relevant electronic data integrated into a single location of the various data sources. This system will help to extract the useful information for various kinds of users like HIV/AIDS patients, doctors, researchers and government, etc., in a fast and flexible manner. Due to the huge amount of data collection in smart HIV/AIDS digital system, it needs to be processed with the help of big data technologies. So, the objective of this paper is to explain about the architecture of smart HIV/AIDS digital system. Various big data analytic techniques and its relevant models, algorithms, and tools to extract the useful information from the smart HIV/AIDS digital system with efficiently have also been discussed.

Keywords HIV/AIDS · Big data · Smart HIV/AIDS digital system

1 Introduction

The current electronic world produces the lot amount of electronic data in each and every day in all the fields like internet, social media, (Radio Frequency Identification) RFID sensors, educational research, business, space research, government activities, healthcare sector, weather control system and personal, etc. [1]. Also the electronic

V. Ramasamy (✉)
Department of Computer Science and Engineering, Park College of Engineering and Technology, Coimbatore, Tamil Nadu, India
e-mail: researchrams@gmail.com

B. Gomathy
Department of Computer Science and Engineering, Bannari Amman Institute of Technology, Coimbatore, Tamil Nadu, India
e-mail: bgomramesh@gmail.com

R. K. Verma
Department of Computer Science and Engineering, Biju Patnaik University of Technology, Rourkela, Odisha, India
e-mail: rajeshverma_chicago2004@yahoo.com

C. R. Panigrahi et al. (eds.), *Progress in Advanced Computing and Intelligent Engineering*, Advances in Intelligent Systems and Computing 714,
https://doi.org/10.1007/978-981-13-0224-4_37

data are generated at very fast manner in the form of structured, semi-structured and unstructured data formats like text, image, audio, video, etc. [1].

Nowadays, big data are represented as 7-Vs, which are namely Volume, Velocity, Variety, Variety, Veracity, Visualization, and Value [2]. To extract useful information from big data, specific techniques are required, which is normally known as big data analytics [3]. Traditional data processing techniques are not suitable for extracting useful hidden information from big data. However, big data analytics technology provides various techniques, algorithms, concepts, and models to handle all kinds (structured, semi-structured and unstructured) of big data [2].

One of the very dangerous disease for the human lives is Human Immunodeficiency Virus/Acquired Immune Deficiency Syndrome (HIV/AIDS). When the human affected by HIV/AIDS, probably he will face death [4]. The first HIV/AIDS affected human was popularly identified in the year 1981 in United States of America [5]. With the help of various advanced medications and treatment, 45% of the HIV/AIDS deaths had been reduced since from the year 2004. Even today, the number of human beings affected by HIV/AIDS all over the world is nearly 37 million [6].

HIV is the root cause of AIDS. HIV destroys or damages the white blood (CD4+) T cells in the human body severely [5]. By that immune system could not fight against even with normal viruses and bacteria due its inability. So in all the ways, human body gets damaged and it leads to death. HIV is transmitted to the human body through unprotected sex, blood sharing, mother to baby while birth and infected syringe and needle sharing, etc., and not transmitted through mosquito bites, handshaking or kissing, sharing of the toilet seat and food utensils [5]. HIV/AIDS cannot be cured fully but mortality rate could be reduced with the help of advanced treatments [7].

In the healthcare sector, HIV/AIDS data, which is generated from each of the patients in huge falls under the big data category. The gathered big data from patients are subsequently processed to get the necessary insights and inference which helps to reduce and control the HIV/AIDS relevant spreads or activities with the help of modern big data analysis techniques and tools.

2 Related Work

Arulananthan et al. [8] proposed the concept of SMART HEALTH which demonstrates the healthiest living environment. It was the combination of Information and Communication Technology (ICT) components like Cloud Computing (CC) [9, 10], smart sensing devices, Internet of Things (IoT) and big data technologies. It also consists of various electronic health records like blood pressure, heartbeat rate, ECG, etc.

Jokonya [3] explained the use of big data technologies in health care, which helps to combine the scattered small data sets into a single unit and offers to extract useful data or decision-making based on correlations and healthcare problem domain.

Patel et al. [4] discussed the various healthcare cost reduction concepts and explained how to get financial profit in the healthcare domain. They also discussed other concepts like personalized medicine, preventive care, health trend analysis, tracking of patients, and drug efficacy analysis.

Raghupathi et al. [7] described the importance of Hadoop [1] ecosystem components like HDFS, MapReduce [11], Pig, Hive [12] etc., in the healthcare domain for the purpose of distributed storage, processing, scheduling and data access, storage, serialization, intelligence, integration, visualization, management, monitoring, and orchestration.

3 Proposed Architecture

In this section, we have proposed an architecture of Smart HIV/AIDS digital system that is explained briefly and is shown as in Fig. 1. In the proposed architecture, the HIV/AIDS related data are collected from various data sources like doctors' clinic, Antiretroviral therapy (ART) units, HIV/AIDS scientist, insurance policy agencies of HIV/AIDS, medicine manufacturing companies of HIV/AIDS, and HIV/AIDS

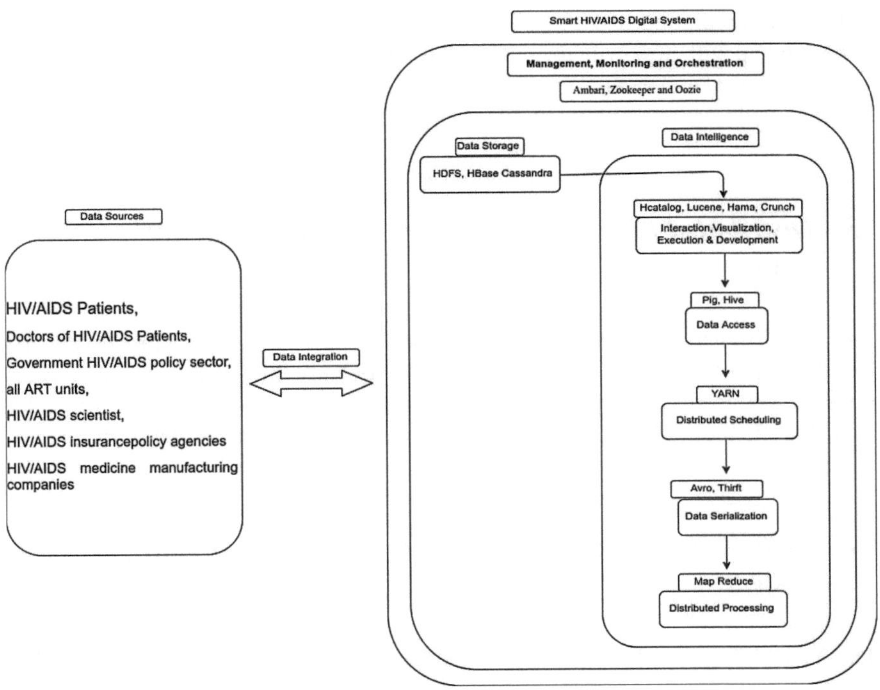

Fig. 1 Architecture of smart HIV/AIDS digital system

policy sector of government. Also HIV/AIDS patients' daily electronic health data like blood pressure, sugar, heartbeat rate, etc., are collected from wearable IoT-based smart watches, smart shoes, smart bracelet band, handheld gadgets, etc.

3.1 Distributed Storage in Smart HIV/AIDS Digital System

This data are stored in a safe and secure manner in the distributed cloud [13, 14] which has many data storage units, which are spread across the different Smart HIV/AIDS digital system. Here, for distributed storage, HDFS, HBase, and Cassandra [15] technologies are used.

3.2 Data Intelligence in Smart HIV/AIDS Digital System

In the data intelligence section, the important/necessary data only to be extracted from the completed data storage for useful decision-making of any HIV/AIDS related activities. For example, doctors required data for treatment of their patients, researchers required data for their research purpose, government needs data for controlling and predicting of HIV/AIDS outbreak and medical companies needs data for coming up with new drug formulae. Here, Apache Drill [16] and Apache Mahout [17, 18] are used and it can be processed and analyze the data very fast manner, even within a single second with many file system data in different locations.

3.3 Distributed Processing in Smart HIV/AIDS Digital System

Due to distributed data storage, there is a need of distributed data processing. By this, the requested query can be executed very quickly based on distributed query processing concepts. The query is executed in different data storage places parallel. Here, MapReduce [11, 19] is used for the distributed data processing purpose.

3.4 Distributed Scheduling in Smart HIV/AIDS Digital System

Here YARN (Yet Another Resource Negotiator) is used to schedule the users query in a distributed manner to perform the data processing task in very fast and quick manner [20]. Scheduling is useful for sequence of query execution to avoid unwanted and slow processing.

3.5 Data Access in Smart HIV/AIDS Digital System

Useful data can be accessed from the data storage with the help of any one of the user interface (data access) tool. Here, Apache Pig and Hive [12] are used to access the data very fast and user-friendly manner from the data storage.

3.6 Interaction, Visualization in Smart HIV/AIDS Digital System

To efficiently interact with the smart HIV/AIDS digital system data storage and extracting the useful data and visualize/understand the output data in very easiest/friendly manner by the user, the Apache Hcatalog, Lucene [21, 22], Hama [23, 24] and Crunch are used here. These tools are performing searching, indexing, joining, and aggregating related tasks on the HIV/AIDS data.

3.7 Data Serialization in Smart HIV/AIDS Digital System

Different format of HIV/AIDS data could not be transported to the different location in an easy manner. Here, the Apache Avro and Thrift are used for converting the different formats of HIV/AIDS data to binary format [25]. By this, the data can be very easily transportable rather than SOAP technologies on the HDFS cluster systems.

3.8 Data Integration in Smart HIV/AIDS Digital System

All the HIV/AIDS data are stored and processed in Hadoops HDFS [2]or HBase [19] or Casandra [15]. Most of the raw data in the form relational database format data. These data cannot be processed efficiently based on big data analytic processes. Here, Apache Sqoop, Flume, and Chukwa [26] are used for integrating related data from various data sources to HDFS clusters and vice versa.

3.9 Management, Monitoring, and Orchestration in Smart HIV/AIDS Digital System

Here all the data are stored in distributed HDFS cluster computers. It should be monitored, maintained, configured, offloaded [27], and administered in a proper manner. These activities in smart HIV/AIDS digital system are performed by Apache Ambari, Zookeeper, and Oozie [26].

4 Conclusion

The objectives of the proposed architecture of smart HIV/AIDS digital system presented in this paper are mainly focused on integrating the HIV/AIDS digital data collected from different data sources location into single location data storage concept but in a distributed manner in across many locations. Controlling and maintaining of this data are done in the form of single person administration point of view. By this, all HIV/AIDS relevant data are used by various users in a single location with IoT- and cloud-based storage. All useful inference and insights on the data are drawn with the help of big data analytics technology concepts. Our proposed architecture is very much helpful to doctors of HIV/AIDS patients, government agencies, research scientists, medicine manufacturing companies, and insurance policy agencies.

References

1. Shvachko, K., Kuang, H., Radia, S., and Ch ansler, R.: The Hadoop distributed file system. 2010 IEEE 26th Symposium on Mass Storage Systems and Technologies, MSST2010, pp. 1–10 (2010).
2. Dhyani, B., and Barthwal, A.: Big Data Analytics using Hadoop. International Journal of Computer Applications, 108(12) PP. 1–5, (2014).
3. Jokonya, O.: Towards a Big Data Framework for the prevention and control of HIV/AIDS, TB and Silicosis in the mining industry. International Conference on Health and Social Care Information Systems and Technologies, 16 pp. 1533–1541 (2014).
4. Patel, S., and Patel, A.: A Big Data Revolution in Health Care Sector: Opportunities, Challenges and Technological Advancements. International Journal of Information Sciences and Techniques (IJIST), 62(1), pp. 155–162, (2016).
5. http://www.amfar.org/About-HIV-and-AIDS/Basic-Facts-About-HIV/.
6. http://www.who.int/mediacentre/factsheets/fs360/en/.
7. Raghupathi, W., and Raghupathi, V.: Big data analytics in healthcare: promise and potential. Health Information Science and Systems, 2(1) pp. 1–10 (2014).
8. Arulananthan, C., and Hanifa, S.M.: SMART HEALTH POTENTIAL and PATHWAYS: A SURVEY. International Conference on Advanced Material Technologies (ICAMT), (2016).
9. Sarkar, J. L., Panigrahi, C. R., Pati, B., and Prasath, R.: MiW: An MCC-WMSNs Integration Approach for Performing Multimedia Applications. In Proc. of 4th International Conference on Mining Intelligence and Knowledge Exploration, pp. 83–92 (2016).

10. Panigrahi, C. R., Sarkar, J. L., Pati, B., and Das, H.: S2S: A Novel Approach for Source to Sink Node Communication in Wireless Sensor Networks. The 3rd International Conference on Mining Intelligence and Knowledge Exploration (MIKE-2015), pp. 406–414 (2015).
11. Wang, L., Tao, J., Ranjan, R., Marten, H., Streit, A., Chen, J., and Chen, D.: G-Hadoop: MapReduce across distributed data centers for data-intensive computing. Future Generation Computer Systems, 29(3), pp. 739–750, (2013).
12. Fuad, A., Erwin, A., and Ipung, H.P.: Processing performance on Apache Pig, Apache Hive and MySQL cluster. Proceedings of International Conference on Information, Communication Technology and System (ICTS) 2014, pp. 297–302 (2014).
13. Pati, B., Sarkar, J.L., Panigrahi, C.R., Debbarma S.: eCloud: An Efficient Transmission Policy for Mobile Cloud Computing in Emergency Areas. Progress in Intelligent Computing Techniques: Theory, Practice, and Applications. Advances in Intelligent Systems and Computing, 519, pp. 43–49 (2018).
14. Panigrahi, C.R., Sarkar, J.L., Pati, B., and Bakshi, S.: E^3M: An Energy Efficient Emergency Management System using mobile cloud computing. IEEE International Conference on Advanced Networks and Telecommunications Systems, pp. 1–6 (2016).
15. Chebotko, A., Kashlev, A., and Lu, S.: A Big Data Modeling Methodology for Apache Cassandra. 2015 IEEE International Congress on Big Data, pp. 238–245 (2015).
16. Hausenblas, M., and Nadeau, J.: Apache Drill: Interactive Ad-Hoc Analysis at Scale. Big Data, 1(2), pp. 100–104 (2013).
17. Thangavel, S. K., Thampi, N. S., and Johnpaul, C. I.: Performance Analysis of Various Recommendation Algorithms Using Apache Hadoop and Mahout. International Journal of Scientific and Engineering Research, 4(2), pp. 279–287 (2013).
18. Manu, M.N., and Ramesh, B.: Single-criteria Collaborative Filter Implementation using Apache Mahout in Big data. International Journal of Computer Sciences and Engineering Open Access, 5(1), pp. 7–13 (2017).
19. Taylor, R.C.: An overview of the Hadoop/MapReduce/HBase framework and its current applications in bioinformatics. Proceedings of the 11th Annual Bioinformatics Open Source Conference (BOSC) 2010, 11(12) pp. 1–6 (2010).
20. Kumar, Rajneesh., and Govindarajan, S.: Scheduling Techniques for Workload Distribution in YARN Containers. International Journal of Engineering Development and Research (IJEDR), 3(2) pp. 66–70 (2015).
21. Balipa, M., and Balasubramani, R.: Search Engine using Apache Lucene. International Journal of Computer Applications, 127(9) pp. 27–30, (2015).
22. Gao, R., Li, D., Li, W., and Dong, Y.: Application of Full Text Search Engine Based on Lucene. Advances in Internet of Things, 2(4), pp. 106–109 (2012).
23. Siddique, K., Akhtar, Z., Kim, Y.: Researching Apache Hama: A Pure BSP Computing Framework. Lecture Notes in Electrical Engineering, 393, Springer, Singapore (2016).
24. Siddique, K., Akhtar, Z., Yoon, E.J., Jeong, Y.S., Dasgupta, D., and Kim, Y.: Apache Hama: An emerging bulk synchronous parallel computing framework for big data applications. IEEE Access, 4 pp. 8879–8887 (2016).
25. Kanthi, A.M., and Patil, A. P.: Analytics on Command Centre Data in Healthcare Systems: A Case Study Implemented using Apache Hadoop, Avro and Crunch. International Journal of Innovative Research in Computer and Communication Engineering, 4(7) pp. 13674–13680 (2016).
26. Mehta, S., and Mehta, V.: Hadoop Ecosystem: An Introduction. International Journal of Science and Research (IJSR), 5(6) pp. 557–562 (2016).
27. Panigrahi, C. R., Pati, B., Tiwary, M., and Sarkar, J. L.: EEOA: Improving energy efficiency of mobile cloudlets using efficient offloading approach. Advanced Networks and Telecommuncations Systems (ANTS), pp. 1–6 (2016).

Applications of Smart HIV/AIDS Digital System Using Hadoop Ecosystem Components

V. Ramasamy, B. Gomathy and Rajesh Kumar Verma

Abstract Smart HIV/AIDS digital system is a collection of HIV/AIDS relevant electronic data integrated into a single place from the various data sources. After the successful storage of the data, there is a need to extract the necessary details of which will provide useful insight to the users. The main users of smart HIV/AIDS digital system are patients, doctors, researchers, government, etc. Due to the huge amount of data collection, normal data processing techniques are not sufficient and viable. Hence, there is a need of advanced technologies to extract the data as well as to view it in an effective, quick, user friendly, and convenient way. Hadoop ecosystem components are used to perform the user application related activities. In this paper, we have focused on explaining the different Hadoop ecosystem components and its intended uses to extract useful information from smart HIV/AIDS digital system.

Keywords HIV/AIDS · Big data · Digital system

1 Introduction

The current electronic world produces the lot amount of electronic data in each and every day in all the fields like Internet, social media, Radio Frequency Identification (RFID) sensors, educational research, business, space research, government activities, health care sector, weather control system and personal, etc. [1]. Also

V. Ramasamy (✉)
Department of Computer Science and Engineering, Park College of Engineering and Technology, Coimbatore, Tamil Nadu, India
e-mail: researchrams@gmail.com

B. Gomathy
Department of Computer Science and Engineering, Bannari Amman Institute of Technology, Coimbatore, Tamil Nadu, India
e-mail: bgomramesh@gmail.com

R. K. Verma
Department of Computer Science and Engineering, Biju Patnaik University of Technology, Rourkela, Odisha, India
e-mail: rajeshverma_chicago2004@yahoo.com

C. R. Panigrahi et al. (eds.), *Progress in Advanced Computing and Intelligent Engineering*, Advances in Intelligent Systems and Computing 714,
https://doi.org/10.1007/978-981-13-0224-4_38

the electronic data generated at very fast manner in the form of structured, semi-structured and unstructured data formats like text, image, audio, video, etc. [1].

Nowadays, big data are represented as 7-Vs, which are namely Volume, Velocity, Variety, Variety, Veracity, Visualization, and Value [2]. To extract useful information from big data, specific techniques are required, which is normally known as big data analytics [3]. Traditional data processing techniques are not suitable for extracting useful hidden information from big data. However big data analytics technology provides various techniques, algorithms, concepts, and models to handle all kinds (structured, semi-structured and unstructured) of big data [2].

One of the very dangerous disease for the human lives is Human Immunodeficiency Virus/Acquired Immune Deficiency Syndrome (HIV/AIDS). When the human affected by HIV/AIDS, probably he will face the death [4]. The first HIV/AIDS affected human was popularly identified in the year 1981 in United States of America [5]. With the help of various advanced medications and treatment, 45% of the HIV/AIDS deaths had been reduced since from the year 2004. Even today, the number of human beings affected by HIV/AIDS all over the world is nearly 37 million [6].

HIV is the root cause of AIDS. HIV destroys or damages the white blood (CD4+) T cells in the human body severely [5]. By that, immune system could not fight against even with normal viruses and bacteria due its inability. So in all the ways human body gets damaged and it leads to death. HIV transmitted to the human body through unprotected sex, blood sharing, mother to baby while birth and infected syringe and needle sharing, etc., and not transmitted through mosquito bites, handshaking or kissing, sharing of the toilet seat and food utensils [5]. HIV/AIDS cannot be cured fully but the mortality rate could be reduced with the help of advanced treatments [7].

In the healthcare sector, HIV/AIDS data, which is generated from each of the patients in huge falls under the big data category. The gathered big data from patients is subsequently processed to get the necessary insights and inference which helps to reduce and control the HIV/AIDS relevant spreads or activities with the help of modern big data analysis techniques and tools.

2 Related Work

Arulananthan et al. [8] proposed the concept of SMART HEALTH which demonstrates the healthiest living environment. It was the combination of Information and Communication Technology (ICT) components like Cloud Computing (CC) [9, 10], smart sensing devices, Internet of Things (IoT), and big data technologies. It also consists of various electronic health records like blood pressure, heartbeat rate and ECG, etc.

Jokonya [3] explained the use of big data technologies in health care, which helps to combine the scattered small data sets into a single unit and offers to extract useful data or decision making based on correlations and healthcare problem domain.

Patel et al. [4] discussed the various healthcare cost reduction concepts and explained how to get financial profit in the healthcare domain. They also discussed other concepts like personalized medicine, preventive care, health trend analysis, tracking of patients, and drug efficacy analysis.

Raghupathi et al. [7] described the importance of Hadoop [1] ecosystem components like HDFS, MapReduce [11], Pig, Hive [12], etc., in the healthcare domain for the purpose of distributed storage, processing, scheduling and data access, storage, serialization, intelligence, integration, visualization, management, monitoring, and orchestration.

3 Applications of Smart HIV/AIDS Digital System

In this section, the applications of smart HIV/AIDS digital system using Hadoop ecosystem components have been discussed briefly. In this application, the HIV/AIDS related data is collected from various data sources like the doctors clinic, Antiretroviral Therapy (ART) units, HIV/AIDS scientist, insurance policy agencies, medicine manufacturing companies, and HIV/AIDS policy sector of government. Also, health data like blood pressure, sugar and heartbeat rate, etc., is collected on a daily basis from wearable IoT-based smart watches, smart shoes, smart bracelet band, and handheld gadgets, etc., and is stored in smart HIV/AIDS digital system storage. Once the data has been stored there is a need to extract useful information and facts about the patients for subsequent treatment. The different Hadoop ecosystem components depicted Fig. 1 are used to process big data which are more advanced than the traditional data processing techniques.

3.1 Hadoop Distributed File System (HDFS)

In HDFS, one highly configured centralized main server (namenode/JobTracker) and a lot of of low cost commodity hardware clients (DataNode/TaskTracker) are grouped together (Hadoop cluster) to store and process the huge amount of data [1]. The server has full control over clients and its relevant activities. It is highly fault tolerant. Smart HIV/AIDS digital system uses the HDFS for its large amount of data storage.

3.2 Map Reduce

It is a master and slave based parallel distributed computer programming architecture on a Hadoop cluster. Here, mapper program runs on the server and divides the big input task into many small tasks and gives to clients. Reducer program runs on all

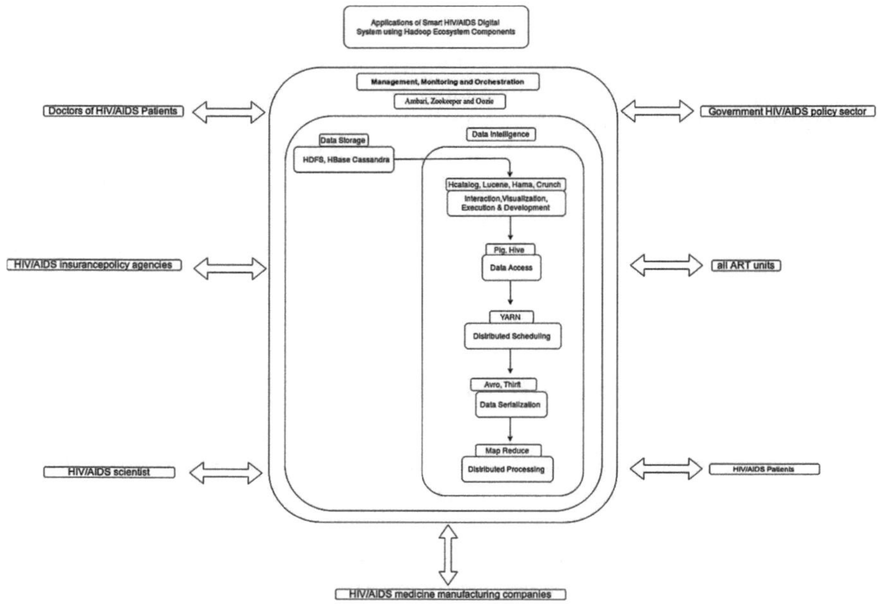

Fig. 1 The Hadoop ecosystem components for smart HIV/AIDS digital system application

clients and performs the original task in parallel given by the mapper and produce the combined single output. The cloud [13–15] based smart HIV/AIDS digital system queries obtained from users is executed based on MapReduce.

3.3 *Yet Another Resource Negotiator (YARN)*

It is a backbone for Hadoop MapReduce parallel distributed programming model on a commodity cluster and has the main job of providing the resources like communication network, CPU, storage memory and manage and monitor it in a distributed manner with the help of job submitter, resource manager, node manager, resource scheduler, container, and application master [16]. It is very much useful in smart HIV/AIDS digital system to schedule the distributed way of users queries for quick processing and avoid the unwanted activities.

3.4 *Pig and Hive*

These are the high level procedural language to easily analyze and process the big data, with a script-based MapReduce distributed programming along with parallelization and fault tolerance mainly for structured, semi-structured, and unstructured data [12]. This will be very much helpful for smart HIV/AIDS digital system users to access the important data from the huge volume of data collection.

3.5 HBase and Cassandra

HBase is a column-oriented distributed database technology with very good fault tolerant, low latency of I/O operations [17]. In HDFS, it takes more time to access to small data due to its sequential access. But using HBase, the small data from billions of data collection can be accessed very quickly due its random access capability. Cassandra [18] offers no single point of failure due its decentralized data deployments in many locations with replication. This is much helpful for storing and maintaining huge amount of data very easily in smart HIV/AIDS digital system.

3.6 Hcatalog, Lucene, Hama, and Crunch

In smart HIV/AIDS digital system, integration and visualization related issues are there due to most of the data stored in HDFS/HBase [17] format. HDFS and MapReduce integration related issues solved by Hcatalog. Searching and indexing related issues are overcome by Lucene [19, 20]. Hama [21, 22] performs the Bulk Synchronous Parallel (BSP) computing activities as well as join and aggregate related tasks are done by the crunch [23].

3.7 Avro, Thrift

Avro is a language neutral data serialization tool which helps to convert the structured data to binary or text format, in very fast manner with small size for easy transportation on the computer network [24]. Also Apache Thrift is useful for scalable cross language service support with many languages with less overhead and an alternative for SOAP in the clustering of Smart HIV/AIDS Digital System.

3.8 Drill and Mahout

In smart HIV/AIDS digital system, Apache Drill [25] analyzes a huge amount of HIV/AIDS data that is available in different locations in a single second. Apache Mahout [26, 27] creates necessary machine learning algorithms for collaborative filtering, clustering of HIV/AIDS data, classification of HIV/AIDS datasets and its recommendations with very fast manner.

3.9 Sqoop, Flume and Chukwa

By using Apache Sqoop, the data has been collected from different locations and moved to Smart HIV/AIDS Digital System [23]. As well as lots of log files are generated in smart HIV/AIDS digital system. These log files are collected, integrated, transported, and monitored by Apache Flume and Chukwa efficiently.

3.10 Ambari, Zookeeper, and Oozie

These are all very much useful in the smart HIV/AIDS digital system for the purpose of distributed cluster management and monitoring kind of activities with easily and efficiently [23].

4 Conclusion

In this paper, we have highlighted the applications of smart HIV/AIDS digital system using Hadoop ecosystem components, which mainly focuses on extracting the useful HIV/AIDS data from the big data storage, and is better than traditional data processing techniques. Hadoop ecosystem components help in data processing and extracting insights in an accurate, quick, convenient, fast, and efficient way. This provides great help to the doctors of the HIV/AIDS patients, governments, research scientists, medicine manufacturing companies, and insurance policy agencies for accessing the required data. Smart HIV/AIDS digital system provides a very user-friendly and intuitive graphical interface having different types of visualization such as graphs and charts.

References

1. Shvachko, K., Kuang, H., Radia, S., and Chansler, R.: The Hadoop distributed file system. 2010 IEEE 26th Symposium on Mass Storage Systems and Technologies, MSST2010, pp. 1–10 (2010).
2. Dhyani, B., and Barthwal, A.: Big Data Analytics using Hadoop. International Journal of Computer Applications, 108(12) PP. 1–5, (2014).

3. Jokonya, O.: Towards a Big Data Framework for the prevention and control of HIV/AIDS, TB and Silicosis in the mining industry. International Conference on Health and Social Care Information Systems and Technologies, 16 pp. 1533–1541 (2014).
4. Patel, S., and Patel, A.: A Big Data Revolution in Health Care Sector: Opportunities, Challenges and Technological Advancements. International Journal of Information Sciences and Techniques (IJIST), 62(1), pp. 155–162, (2016).
5. http://www.amfar.org/About-HIV-and-AIDS/Basic-Facts-About-HIV/.
6. http://www.who.int/mediacentre/factsheets/fs360/en/.
7. Raghupathi, W., and Raghupathi, V.: Big data analytics in healthcare: promise and potential. Health Information Science and Systems, 2(1) pp. 1–10 (2014).
8. Arulananthan, C., and Hanifa, S.M.: SMART HEALTH POTENTIAL and PATHWAYS: A SURVEY. International Conference on Advanced Material Technologies (ICAMT), (2016).
9. Sarkar, J. L., Panigrahi, C. R., Pati, B., and Prasath, R.: MiW: An MCC-WMSNs Integration Approach for Performing Multimedia Applications. In Proc. of 4th International Conference on Mining Intelligence and Knowledge Exploration, pp. 83–92 (2016).
10. Panigrahi, C. R., Sarkar, J. L., Pati, B., and Das, H.: S2S: A Novel Approach for Source to Sink Node Communication in Wireless Sensor Networks. The 3rd International Conference on Mining Intelligence and Knowledge Exploration (MIKE-2015), pp. 406–414 (2015).
11. Wang, L., Tao, J., Ranjan, R., Marten, H., Streit, A., Chen, J., and Chen, D.: G-Hadoop: MapReduce across distributed data centers for data-intensive computing. Future Generation Computer Systems, 29(3), pp. 739–750, (2013).
12. Fuad, A., Erwin, A., and Ipung, H.P.: Processing performance on Apache Pig, Apache Hive and MySQL cluster. Proceedings of International Conference on Information, Communication Technology and System (ICTS) 2014, pp. 297–302 (2014).
13. Pati, B., Sarkar, J.L., Panigrahi, C.R., Debbarma S.: eCloud: An Efficient Transmission Policy for Mobile Cloud Computing in Emergency Areas. Progress in Intelligent Computing Techniques: Theory, Practice, and Applications. Advances in Intelligent Systems and Computing, 519, pp. 43–49 (2018).
14. Panigrahi, C.R., Sarkar, J.L., Pati, B., and Bakshi, S.: E^3M: An Energy Efficient Emergency Management System using mobile cloud computing. IEEE International Conference on Advanced Networks and Telecommunications Systems, pp. 1–6 (2016).
15. Panigrahi, C. R., Pati, B., Tiwary, M., and Sarkar, J. L.: EEOA: Improving energy efficiency of mobile cloudlets using efficient offloading approach. Advanced Networks and Telecommunications Systems (ANTS), pp. 1–6 (2016).
16. Kumar, Rajneesh., and Govindarajan, S.: Scheduling Techniques for Workload Distribution in YARN Containers. International Journal of Engineering Development and Research (IJEDR), 3(2) pp. 66–70 (2015).
17. Taylor, R.C.: An overview of the Hadoop/MapReduce/HBase framework and its current applications in bioinformatics. Proceedings of the 11th Annual Bioinformatics Open Source Conference (BOSC) 2010, 11(12) pp. 1–6 (2010).
18. Chebotko, A., Kashlev, A., and Lu, S.: A Big Data Modeling Methodology for Apache Cassandra. 2015 IEEE International Congress on Big Data, pp. 238–245 (2015).
19. Balipa, M., and Balasubramani, R.: Search Engine using Apache Lucene. International Journal of Computer Applications, 127(9) pp. 27–30, (2015).
20. Gao, R., Li, D., Li, W., and Dong, Y.: Application of Full Text Search Engine Based on Lucene. Advances in Internet of Things, 2(4), pp. 106–109 (2012).
21. Siddique, K., Akhtar, Z., Kim, Y.: Researching Apache Hama: A Pure BSP Computing Framework. Lecture Notes in Electrical Engineering, 393, Springer, Singapore (2016).
22. Siddique, K., Akhtar, Z., Yoon, E.J., Jeong, Y.S., Dasgupta, D., and Kim, Y.: Apache Hama: An emerging bulk synchronous parallel computing framework for big data applications. IEEE Access, 4 pp. 8879–8887 (2016).
23. Mehta, S., and Mehta, V.: Hadoop Ecosystem: An Introduction. International Journal of Science and Research (IJSR), 5(6) pp. 557–562 (2016).

24. Kanthi, A.M., and Patil, A. P.: Analytics on Command Centre Data in Healthcare Systems: A Case Study Implemented using Apache Hadoop, Avro and Crunch. International Journal of Innovative Research in Computer and Communication Engineering, 4(7) pp. 13674–13680 (2016).
25. Hausenblas, M., and Nadeau, J.: Apache Drill: Interactive Ad-Hoc Analysis at Scale. Big Data, 1(2), pp. 100–104 (2013).
26. Thangavel, S. K., Thampi, N. S., and Johnpaul, C. I. : Performance Analysis of Various Recommendation Algorithms Using Apache Hadoop and Mahout. International Journal of Scientific and Engineering Research, 4(2), pp. 279–287 (2013).
27. Manu, M.N., and Ramesh, B.: Single-criteria Collaborative Filter Implementation using Apache Mahout in Big data. International Journal of Computer Sciences and Engineering Open Access, 5(1), pp. 7–13 (2017).

Part V
Advanced Networks, Software Defined Networks, and Robotics

Design and Implementation of Autonomous UAV Tracking System Using GPS and GPRS

Devang Thakkar, Pruthvish Rajput, Rahul Dubey and Rutu Parekh

Abstract This work presents a tracking system for autonomous unmanned aerial vehicle (UAV) and any ground vehicle using global positioning system (GPS) and general packet radio service (GPRS). A Google form is created and a Google spreadsheet is set as destination location to store Google form responses on ground station. Whenever UAV is flying autonomously, GPS finds its current location (latitude and longitude). Thereafter, GPRS sends hypertext transfer protocol (HTTP) request which consists of the current location of a UAV and submits a response in Google form and Google spreadsheet. A custom function is added in a Google spreadsheet to monitor the path of an UAV on static Google map using Google Apps Script (GAS).

Keywords Autonomous UAV · Tracking · Google apps script (GAS) · Global positioning system (GPS) · General packet radio service (GPRS)

1 Introduction

The tracking system is designed to track and monitor the path of any vehicle (e.g., UAV, car) in this paper. Now, we are referring to the tracking of a UAV. Research and development of UAV are increasing nowadays because of its applications like surveillance coordinating, reconnaissance operations, delivering medical supplies to remote or inaccessible regions, etc. The degree of autonomy is defined as the extent

D. Thakkar · P. Rajput (✉) · R. Dubey · R. Parekh
VLSI and Embedded Systems Research Group, Dhirubhai Ambani Institute of Information and Communication Technology, Gandhinagar 382007, Gujarat, India
e-mail: pruthvishrajput@gmail.com

D. Thakkar
e-mail: devangthakkar005@gmail.com

R. Dubey
e-mail: rahuldee@gmail.com

R. Parekh
e-mail: rutu_parekh@daiict.ac.in

C. R. Panigrahi et al. (eds.), *Progress in Advanced Computing and Intelligent Engineering*, Advances in Intelligent Systems and Computing 714,
https://doi.org/10.1007/978-981-13-0224-4_39

to which an UAV is free from our control. If the flight of an UAV is controlled using a remote control from ground station then it has 0% degree of autonomy. The fully autonomous UAV using single board computers (e.g., Beaglebone Black, Raspberry Pi) has 100% degree of autonomy.

UAV being autonomous, the tracking of its current location is more important for safety purposes. If certain unconditional failure of the system occurs, the last location of it can be seen by tracking its path. A wireless connection is required between a UAV and the ground station for tracking purpose. UAV sends its current location to ground station using wireless connectivity so that, its path can be monitored on Google map. Due to the limitation on connectivity range of Bluetooth (10 m), Wi-Fi (100 m) and ZigBee (10–100 m) [1], they are not advisable for autonomous UAV tracking which leads us to use internet connectivity.

There are modern embedded and communication technologies which are used for remote tracking and monitoring. A GPS sensor provides time and location information anywhere on the earth. The location information provided by GPS sensors can be seen using Google map. Global system for mobile communication (GSM) is a standard protocol for second generation (2G) digital cellular networks used by mobile phones [2]. GPRS is a 2.5G system based on GSM [3]. GSM and short message service (SMS) technology are used for vehicle tracking in [4, 5] which requires another GSM or cellular phone connected to a PC or laptop on the ground station. GSM short message service has low transmission speed and long-time transmission delay [6]. A web-based GPS-GPRS vehicle tracking system in [7] requires a server which receives the information and stores it in a database. The proposed tracking system in this paper does not require a GSM or a webserver on the ground station as shown in Fig. 1. This system takes less time to show the path of a vehicle and requires lesser hardware as compared to the system designed in [4, 5]. There is also no need

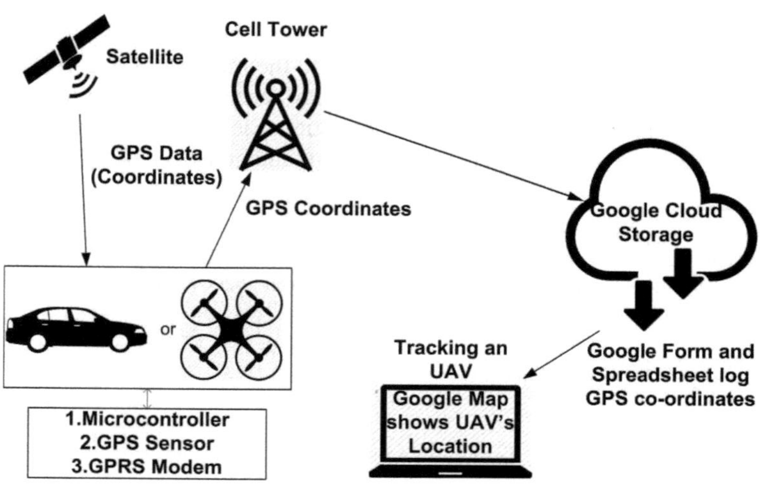

Fig. 1 The block diagram of tracking system

of static IP (Internet Protocol) for a webserver unlike in [7]. Hence, dependencies on dedicated port and port forwarding methodology get eliminated.

Figure 1 shows a block diagram of the tracking system which consists of a GPS sensor, a GPRS modem, and a microcontroller on a UAV and a PC or a laptop on the ground station.

As shown in Fig. 1, a GPS sensor receives coordinates (latitude and longitude) of a UAV from GPS satellites. Afterwards, microcontroller sends these coordinates in HTTP request using GPRS modem. On the ground station, a Google form is created which updates the responses once it receives HTTP request. The latitude and longitude values of Google form are also stored in Google spreadsheet so that the entire path of a UAV can be monitored on static Google map generated in Google spreadsheet from the ground station.

2 Hardware Specification

The tracking system consists of a microcontroller (Arduino ATmega 328), a sensing unit (GPS sensor), and a transmitting unit (GPRS modem). A microcontroller is used to process the data coming from sensors and to perform necessary actions. ATmega 328 is an 8-bit microcontroller. It has advanced RISC architecture which consists of 32 registers each of 8 bits and has a throughput of 16 MIPS at 16 MHz [8].

2.1 GPS Sensor

The GPS is a navigation system that provides location information anywhere on the earth in all weather conditions. But there must be an unobstructed line of sight from four or more GPS satellites [9]. The GPS satellites carry very stable atomic clocks which are synchronized with each other and to the ground clocks. They continuously transmit their current time and position. A GPS receiver monitors multiple satellites and solves equations to find the exact position of it. The strings which are received from a GPS sensor are called National Marine Electronics Associations (NMEA) sentences as shown in Fig. 2.

```
$GPVTG,134.75,T,,M,0.23,N,0.43,K,A*3F
$GPGGA,141932.000,2300.8578,N,07231.3166,E,1,6,1.17,91.1,M,-56.9,M,,*49
$GPGSA,A,3,13,08,11,17,01,19,,,,,,,1.52,1.17,0.96*09
$GPGSV,2,1,08,08,51,355,25,13,35,167,36,11,34,086,36,17,33,208,24*7D
$GPGSV,2,2,08,01,24,113,29,19,19,040,28,42,11,099,30,10,08,227,
$GPGLL,2300.8578,N,07231.3166,E,183503.80,A,A*64
```

Fig. 2 NMEA sentences from GPS sensor

The important string is $GPGGA, 183503.80, 2311.21644, N, 07237.69452, E, 1, 05, 1.92, 80.9, M, -54.9, M,, *4A. The parameters in $GPGGA string are as follows [10]:

1. 183504.00 → Universal Time: 18:35:04
2. 2311.21644 → Latitude: 23° and 11.21644 min
3. N → Latitude Direction
4. 07237.69452 → Longitude: 72° and 37.69452 min
5. E → Longitude Direction
6. 05 → Number of Satellites in use
7. 80.9 → Antenna Altitude above sea level (mean)
8. M → Units of altitude (m)

2.2 GPRS Module

GPRS is a packet-based mobile data service. The GPRS network enables us to use 2G, 3G, and WCDMA mobile networks to transmit IP packets to external networks [11]. A GPRS modem uses a SIM card and works like a mobile phone. When a GPRS modem is connected with arduino board using serial communication, it allows us to communicate over the mobile network.

A GPRS modem does not have a keypad and display. It receives certain commands called AT commands through serial communication and acknowledges for the same. There is a list of AT commands which are used to perform certain functions like SMS messaging, internet access etc. To test the working of GPRS modem, microcontroller sends the command AT. If it is working fine, it responds with OK, otherwise it responds with an ERROR. The AT commands used for sending HTTP request are listed below [12].

- AT + CSQ: Signal Quality Report
- AT + CGATT: To connect and disconnect from GPRS Service
- AT + SAPBR: Bearer Setting for IP applications
- AT + HTTPINIT: Initialize HTTP Service
- AT + HTTPPARA: Set HTTP Parameters Value
- AT + HTTPACTION: HTTP Action Method
- AT + HTTPREAD: Read the HTTP Server Response
- AT + HTTPSSL: Set HTTP to use SSL Function
- AT + HTTPTERM: Terminate HTTP

For testing purpose, we have used 2G GPRS service.

3 Proposed Tracking System

The GPS sensor continuously receives NMEA sentences coming from GPS satellites. These sentences are passed to a microcontroller which chooses $GPGGA sentence among them and parse it to find current latitude and longitude of a UAV. Then the microcontroller sets location information of a UAV in HTTP request as shown in Fig. 3. This HTTP request using HTTP GET method is sent by a GPRS modem which submits a response in Google form. A Google form is created with two entries named latitude and longitude on the ground station which gives a unique key. This key is used to create above HTTP request. A Google spreadsheet is also set as destination location to store these Google form responses. The screenshot of Google form responses is shown in Fig. 4.

Google apps script (GAS) is used to see the path of a UAV using latitude and longitude stored in Google spreadsheet. The GAS is a scripting language which allows us to automate the transfer of data across Google products [13]. It is linked with a Google spreadsheet to create custom functions. A custom function has been created which uses latitude and longitude column from a Google spreadsheet and maps them on static Google map using GAS. This function is configured to execute on the update trigger. This updates the location on static map on every Google sheet update. The static map and its methods which are used to see the path of a UAV are shown below.

https://docs.google.com/forms/d/18u8FUll-K40HkQBslljTH4__e5-e7W9leUCkuLZ4s/
formResponse?ifq&entry.53040747=23.125455&entry.1138846659=72.547714&submit=Submit

Latitude Longitude

Fig. 3 HTTP request

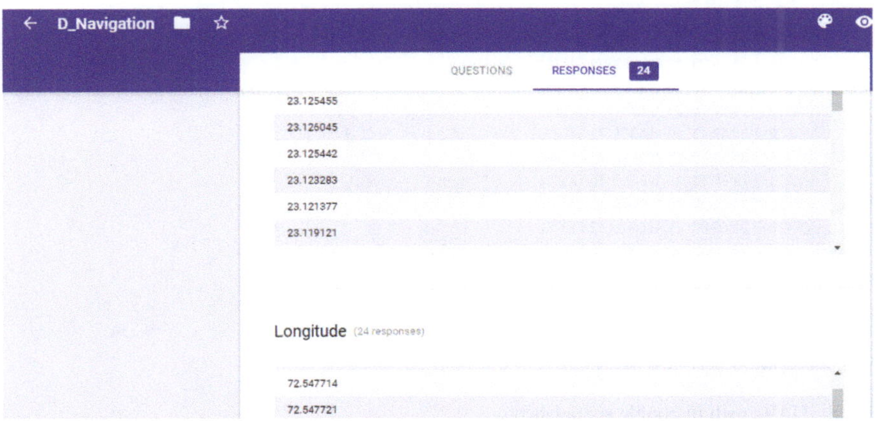

Fig. 4 Screenshot of google form responses

- To create a new StaticMap:
  var googleMap = Maps.newStaticMap()
- To create a new UI (User Interface) Application to display the map:
  var ui = UiApp.createApplication()
  Alternate to UI, image formula of sheet can be used to display static map image in sheet.
- To add a marker at the specified coordinates:
  googleMap.addMarker(latitude,longitude)
- To add that path in Google Map:
  googleMap.addPath(polyline)

4 Result

Initially for tracking system studied in this paper, the hardware components had been configured as described in Sects. 1, 2 and 3. This system is further tested to track and navigate UAV. As shown in Fig. 5. the google sheet gives the location updates in terms of latitude and longitude. Furthermore on every update of google sheet, script gets executed and static map get updated. The time required for UAVs path update is 4 s, which can be further reduced by using 3G services.

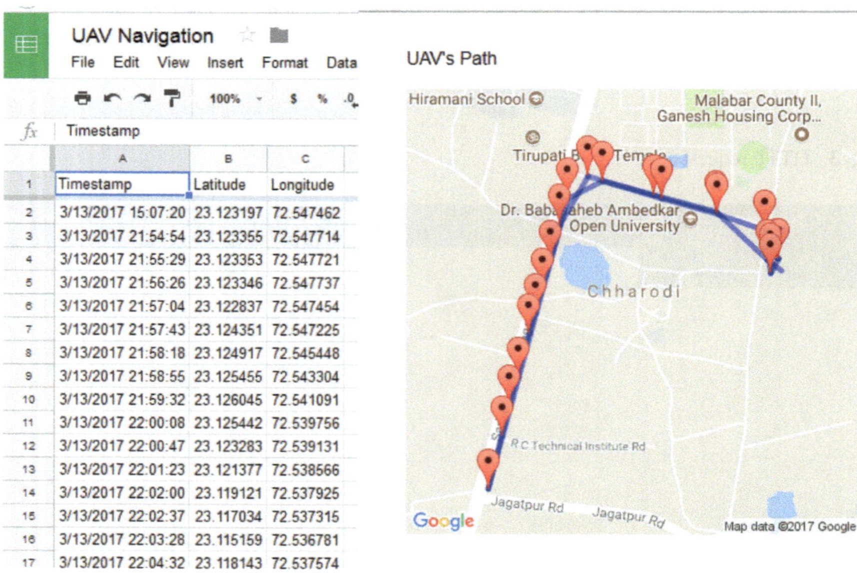

	A	B	C
1	Timestamp	Latitude	Longitude
2	3/13/2017 15:07:20	23.123197	72.547462
3	3/13/2017 21:54:54	23.123355	72.547714
4	3/13/2017 21:55:29	23.123353	72.547721
5	3/13/2017 21:56:26	23.123346	72.547737
6	3/13/2017 21:57:04	23.122837	72.547454
7	3/13/2017 21:57:43	23.124351	72.547225
8	3/13/2017 21:58:18	23.124917	72.545448
9	3/13/2017 21:58:55	23.125455	72.543304
10	3/13/2017 21:59:32	23.126045	72.541091
11	3/13/2017 22:00:08	23.125442	72.539756
12	3/13/2017 22:00:47	23.123283	72.539131
13	3/13/2017 22:01:23	23.121377	72.538566
14	3/13/2017 22:02:00	23.119121	72.537925
15	3/13/2017 22:02:37	23.117034	72.537315
16	3/13/2017 22:03:28	23.115159	72.536781
17	3/13/2017 22:04:32	23.118143	72.537574

Fig. 5 UAVs path in google spreadsheet

5 Conclusion

In this paper, the tracking system for autonomous unmanned aerial vehicle (UAV) is presented. The designed tracking system for UAV combines a GPS sensor and a GPRS modem to retrieve the current location and sends it to the ground station. On the ground station, a Google form updates the responses once it receives HTTP request using Internet connectivity and GAS processes latitude and longitude column in a Google spreadsheet to monitor the path of an UAV on Google map.

References

1. Jin S. L., Yu-Wei S., and Chung-Chou S.: A Comparative Study of Wireless Protocols: Bluetooth, UWB, ZigBee, and Wi-Fi. In: The 33rd Annual Conference of the IEEE Industrial Electronics Society (IECON), Taipei, Taiwan pp. 46–51 (Nov. 5–8, 2007)
2. Guifen G. and Guili P.: The survey of GSM wireless communication system. In: International Conference on Computer and Information Application (ICCIA), vol., no., pp. 121–124 (Dec. 3–5, 2010)
3. J. Rendon, F. Casadevall, L. Garcia, and R. Jimenez: Characterization of the GPRS Radio Interface by means of a Statistical Model. IEEE VTC, Greece, pp. 2392–2396 (2001)
4. Montaser N. Ramadan, Mohammad A. Al-Khedher, and Sharaf A. Al-Kheder: Intelligent Anti-Theft and Tracking System for Automobiles. International Journal of Machine Learning and Computing, Vol. 2, No. 1, pp. 88–92 (2012)
5. B. P. S. Sahoo and R. Satyajit: Integrating GPS, GSM and Cellular Phone for Location Tracking and Monitoring. In: Proceedings of the International Conference on Geospatial Technologies and Applications, Geomatrix' 12, IIT Bombay, Mumbai, India (Feb. 26–29, 2012)
6. L. Kong, J. Jin, and J. Cheng: Introducing GPRS technology into remote monitoring system for prefabricated substations in China. In: Proc. Int. Conf. Mobile Technol., Guangzhou, China (15–17 Nov. 2005)
7. SeokJu L., Girma T., and Jaerock K.: Design and Implementation of Vehicle Tracking System Using GPS/GSM/GPRS Technology and Smartphone Application. In: IEEE World Forum on Internet of Things (WF-IoT), Seoul, South Korea, pp 353–358 (2014)
8. Atmel, 8-bit AVR Microcontroller with 32 KBytes In-System Programmable Flash, ATmega32 datasheet, http://www.atmel.com/images/doc2503.pdf
9. T. Moore: An introduction to the global positioning system and its applications. In: Developments in the Use of Global Positioning Systems, pp. 111–116 (Feb. 1994)
10. NMEA Data, http://aprs.gids.nl/nmea/#gga
11. Roger K., Ingo M., and Michael Meyer: Wireless Internet Access Based on GPRS. IEEE Personal Communications, Vol.: 7, Issue: 2, pp. 8–18 (April 2000)
12. SIMCom, SIM900 AT Commands Manual V1.07, http://www.jechavarria.com/wp-content/uploads/2015/05/SIM900-AT-Commands-Manual_V1.07.pdf
13. Google Apps Script (GAS), Class StaticMap, https://developers.google.com/apps-script/reference/maps/static-map#methods

Mobile Robot Navigation in Unknown Dynamic Environment Inspired by Human Pedestrian Behavior

Nayan M. Kakoty, Mridusmita Mazumdar and Durlav Sonowal

Abstract Navigation in an unknown dynamic environment is one of the key challenges in mobile robotics. This paper proposes a scheme, inspired by human pedestrian behavior, for navigation of a mobile robot in an a priori unknown dynamic environment. An occupancy grid map has been built using onboard sonar sensors through successive sensor information. Inspired by human pedestrian behavior to maintain a safe direction and distance to avoid collisions with obstacles, the proposed navigation scheme trail a path for the robot following a forbidden region map concept with a velocity proportional to the distance and rate at which the obstacles are approaching or receding the robot. The reachable region of robot navigation horizon is based on the motion model predictability of the obstacles. The navigation scheme is deployed on a Fire Bird V mobile robot. The experimental result shows that the robot is able to follow a smooth and time-efficient path avoiding collisions with the mobile and stationary obstacles.

Keywords Pedestrian behavior · Velocity constraint · Reachable region

1 Introduction

The need of mobile robots sharing a common work cell with human is continuously increasing in many practical contexts. This demands for a smooth and time-efficient navigation by the robots in an unknown dynamic environment. There have been various approaches for environment mapping and navigation in the field of mobile robotics. However, most of the methods are subjected to path irregularity

---

N. M. Kakoty (✉) · M. Mazumdar · D. Sonowal
Embedded Systems and Robotics Lab, Tezpur University, Tezpur 784028, India
e-mail: nkakoty@tezu.ernet.in
URL: http://www.tezu.ernet.in/erl

M. Mazumdar
e-mail: mridusmitamazumdar24@gmail.com

D. Sonowal
e-mail: dsn@tezu.ernet.in

C. R. Panigrahi et al. (eds.), *Progress in Advanced Computing and Intelligent Engineering*, Advances in Intelligent Systems and Computing 714,
https://doi.org/10.1007/978-981-13-0224-4_40

and lower throughput leading to collisions in unknown dynamic environments. In societal and industrial settings, the implementation of a mobile robot is accepted to be enabled of navigation capabilities that meet both engineering and societal objectives [1]. Therefore, mapping and navigation in an unknown dynamic environment are largely an open research problem [2].

Based on the neuro-fuzzy concepts aiming toward collision free and minimum trajectory error, number of motion control methodologies like adaptive neural network method, sliding mode control, and back-stepping method [3] have been proposed. Simultaneous localization and mapping (SLAM) technique have been in use for building a metric map of an unknown environment [4]. Following SLAM for modeling uncertainties, unexpected disturbances, and actuator failures, an adaptive fault tracking control method have been proposed by Song et al. [5]. In the area of robot localization and navigational research, number of technologies have been utilized from mapping [4] to navigation [6]. Although aforementioned methods can realize map-building-based navigation, human efforts or exteroceptive sensor information have to be integrated for occupancy grid map building leading to limitation in smooth and time-efficient navigation of the mobile robots. Obstacles detection through infrared images [7] have been studied to facilitate the deployment of autonomous robots. Prediction of trajectories which comprises of discrete decisions for interacting agents [8], use of visual and embodied data association to build a local map [9] are explained for autonomous mobile robots. A fully distributed algorithm for robots navigation has been implemented for mutual avoidance as adopted by human have been proposed by Guzzi et al. [1]. But to address both engineering and societal aspects of the navigation in unknown dynamic environment, a robot should be equipped with the similar locomotion capability as a pedestrian.

We present a navigation scheme for a mobile robot in an unknown dynamic environment. Using successive onboard sensor information, a real-time local map has been built. Inspired by the human pedestrian behavior, the navigation scheme maintains a safe distance and direction form the obstacles by controlling its speed and heading direction. A *velocity constraint* multiplier; proportional to the velocity of the surrounding obstacles have been introduced to maintain a safe distance between the robot and obstacles. Based on motion model predictability of obstacles, the navigation scheme plan for a shorter or longer reachable region for the robot; which helps to avoid the obstacles following a simple navigation scheme. The proposed scheme has been deployed on a Fire Bird V mobile robot which ensures smooth and time-efficient navigation in terms of path irregularity and relative throughput under no obstacle, static and dynamic environments.

2 Real-Time Local Map Building

In this work, we used the Fire Bird V mobile robot customized with ultrasonic range sensors (shown in Fig. 1a) as an experimental testbed. Successive sensors information were superimposed for real-time map building of a local area; scanning a view of 360° surrounding the robot with a radius of 180 cm.

A typical ultrasonic sensor returns a radial measure of distance to the nearest obstacle within its conical field of view which lies between 10° and 30°. To avoid crosstalk possibility, we used four Ultrasonic Ranging Module HC-SR04 equally spaced 90° apart in a circular ring on the robot at a height of 15 cm from the ground. The circular ring is mounted on a servo mechanism in order to have a full 360° scan around the robot. The ultrasonic sensors were set to detect an obstacle upto 180 cm with an accuracy of ±1 cm.

Figure 1c shows schematic of four sensors as $S1$, $S2$, $S3$, and $S4$; wherein each sensor's beam of acoustic energy spreads in a cone of 30°. The typical scan time of a sensor ranges from 60 to 500 ms. The servo mechanism rotates the circular ring twice by an angle of 30° to complete a scan of a 90° cone by each sensor. The cone of the acoustic energy beam during successive rotation of sensor $S3$ for completing a scan of 90° is shown as $S3$, $S3'$ and $S3''$ in Fig. 1c. If the sensor returns a value of distance between the specified minimum and maximum range (10–180 cm), then the returned distance measurement is proportional to the distance of the nearest obstacle within the range of the sensor.

Map building requires obstacles' localization in the area with reference to the robot. For the ease of it, sensing range in front of each sensor is categorized into three zones (viz., Z_1, Z_2, and Z_3) and shown for sensor $S1$ in Fig. 1c. The zones

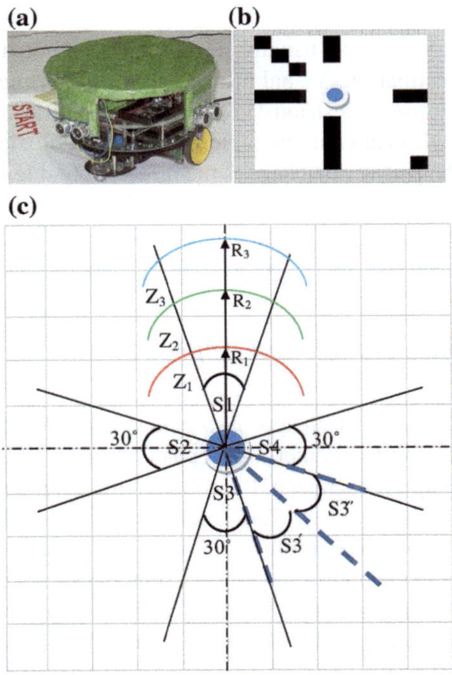

Fig. 1 **a** Experimental test bed: Fire Bird V customized with ultrasonic range sensors. **b** Typical map build with the robot as a circle in blue color and obstacles as squares in black color. **c** Schematic of the ultrasonic sensors with the cone of acoustic energy spread and sensing range

Z_1, Z_2 and Z_3 ranges upto a distance of 30 cm, 90 cm and 180 cm respectively. The incoming four sonar sensors' (S_1, S_2, S_3, and S_4) readings for three successive trials (like S_3, S_3', S_3'' for sensor S_3) are interpreted and converted to local occupancy values for map building. The grids allow the efficient accumulation of small amounts of information from individual sensor readings for increasingly accurate maps of the robot's surroundings. An occupancy grid map has been built based on the sensors information wherein a sequence of continuously changing information indicates an obstacle; a new continuous sequence after a discontinuity indicates a new obstacle; and a single measurement not related to its neighbors considered as noise. Figure 1b shows a typical map built with the robot shown as a circle in blue and the obstacles as squares in black colors. The map building through sensor data interpretation is the first phase to support the robot navigation in an unknown dynamic environment.

3 Navigation Scheme

The proposed scheme aims at human pedestrian behavior and is based on a novel cognitive science approach to determine human pedestrian behavior [10].

3.1 Human Pedestrian Behavior

A pedestrian usually selects the most convenient and efficient path for reaching the destination. Visual information is the prime source for deciding the motion strategy by the pedestrian [11]. Using the neural interface between the retina and brain, a pedestrian can estimate the time to collision with the obstacles [12]. Accordingly, the pedestrian chooses the direction that leads to the destination through the shortest path while maintaining a safe distance from the obstacles along the line of heading [10].

3.2 Navigation Approach

The navigation approach plan to scan an area of 180 cm radius with the robot's position as the center and repeats the plan after traversing the area moving toward its goal. At each step, the robot estimates the relative obstacle position and motion for choosing a way to avoid collisions. We model the velocity vector of the robot (V_{ROB}) by a superposition of the velocity due to its own actuation (V_{Act}) as a factor of a multiplier proportional to the distance between the robot and the obstacle. We introduced this multiplier as *velocity constraint* inspired by the human pedestrian behavior to maintain a safe distance with the obstacles.

$$V_{ROB} = V_{Act} + k \cdot V_{Act} \quad (1)$$

where k = *Velocity constraint* proportional to the distance between the robot and the obstacle; and is quantized as follows:

$$\begin{aligned} k &= 1 \text{ if } \Delta D_{Rob-Obs} \geq 91\,\text{cm} \\ &= 0 \text{ if } \Delta D_{Rob-Obs} = 31\text{–}90\,\text{cm} \\ &= -1 \text{ if } \Delta D_{Rob-Obs} = 10\text{–}30\,\text{cm} \end{aligned}$$

where $\Delta D_{Rob-Obs}$ = Distance between the robot and the obstacle.

The navigation scheme generates the actuation command to the robot for modifying its speed in order to avoid collision with the obstacles in its sensing range. The actuation to the robot is kept at V_{Act} = 12 cm/s with $k = 0$ initially and can have a maximum speed of 24 cm/s. To avoid collision, $k = -1$ at $\Delta D_{Rob-Obs}$ = 10–30 cm (i.e., when the obstacle is in the region Z_1) makes the robot to stop and allow change in its heading direction. The robot moves with 12 cm/s toward its goal point with $k = 0$ at $\Delta D_{Rob-Obs}$ = 31–90 cm (i.e., when the obstacle is in the region Z_2). The robot doubles its speed toward the goal point with $k = 1$ at $\Delta D_{Rob-Obs}$ = 91–180 cm (i.e., when the obstacle is in the region Z_3). The robot modifies its motion strategy successively every 4 s; out of which 3 s is required by the robot to scan the 360° around it and another 1 s for commanding the actuation including the computation for map building. The determination of the directional heading and velocity of the robot is illustrated for static and dynamic environment in the following sections.

3.2.1 Static Environment

We plan the robot navigation in position space. Position space for a short duration ΔT becomes the reachable region $R\Delta T$; which is the set of all positions that the robot can reach in time ΔT. Each obstacle corresponds to a set of directions, termed as forbidden headings that need to be avoided. We denote the forbidden heading for a given obstacle as H_{obs}.

In Fig. 2a, $R\Delta T$ shows the reachable region with the robot velocity vector V_{ROB} towards the goal position. An obstacle in between the start and goal point is represented by the occlusion points O_s and O_e in Fig. 2b. To accommodate the size of the robot of radius R_{ROB}, we extend the obstacle by this measure at the occlusion points. The heading to be avoided by the robot to prevent collision is shown as H_{Obs}; marked in the reachable region with an arc in red color and is determined as the forbidden region. The directional heading decision for passing by O_s or O_e is made through choosing the shortest path to the goal. The velocity vector V_{ROB} is determined following the Eq. 1. For multiple obstacles, multiple forbidden regions are introduced following the same approach.

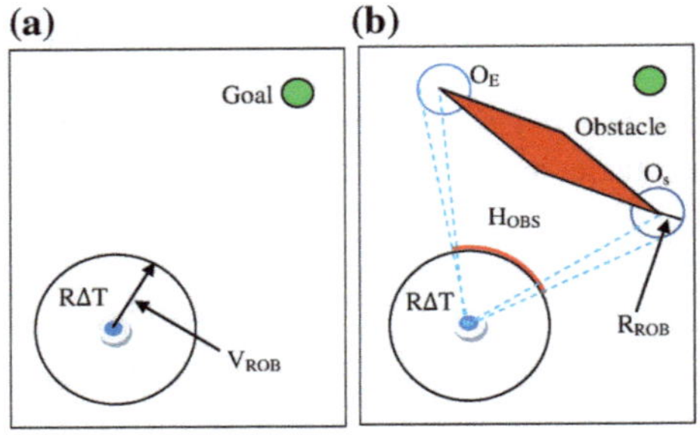

Fig. 2 **a** Reachable region and velocity vector of the robot. **b** An obstacle in red color with occlusion point O_s and O_e in between the robot and the goal and the forbidden region H_{OBS}. The robot is shown as a circle in blue color at the start point with the goal as a circle in green color

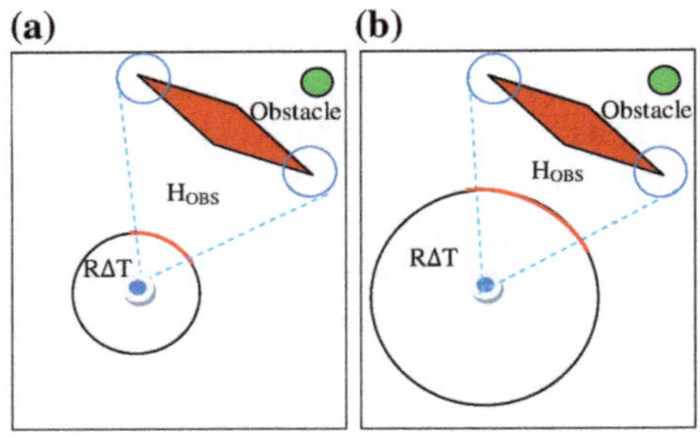

Fig. 3 **a** Choice of smaller reachable region with the obstacle in the Z_1 or Z_2 zone. **b** Choice of larger reachable region with the obstacle in the Z_3 zone

3.2.2 Dynamic Environment

The velocities of the dynamic obstacles are unknown a priori and have to be superimposed with the robot velocity in order to maintain a safe distance with the obstacles. Accordingly the Eq. 1 for determining the robot velocity is modified as follows:

$$V_{ROB} = V_{Act} + k \cdot V_{Act} \pm V_{OBS} \tag{2}$$

where V_{OBS} = Obstacle velocity measured using onboard sensors and is negative if the obstacle is proceeding towards the robot and is positive if receding from the robot.

The choice of the reachable region $R\Delta T$ depends on the distance to the nearest obstacle. If the predicted motion is considered reliable, the robot can plan for a longer reachable horizon. On the other hand, if the obstacle's motion model is highly unpredictable, the robot plans for a shorter reachable horizon. If the predicted distance to the obstacle lies in the Z_1 or Z_2 zone, the reachable region is planned with shorter time step as shown in Fig. 3a. On the other hand, if the predicted distance to the obstacle lies in the Z_3 zone, the reachable region is planned with longer time step as in Fig. 3b. With this reachable region, the directional heading is decided as for static environment.

4 Experiments and Results

4.1 Experimental Setup

Experiments are performed with the Fire Bird V mobile robot on a rectangular arena of 5.76 m^2 with 36 squares of 0.16 m^2 each in it. At first, the robot is entrusted to navigate from the start (i.e., robot origin position) to goal point (i.e., robot final position) on the arena. The data from the sonar sensors were fed to MATLAB through an UNO 328P controller. The path planned in MATLAB following the navigation approach illustrated in Sect. 3 is fed to the robot controller ATMEGA 2560 for its navigation accordingly. The experiments have been performed in the following three environments:

No Obstacles Environment: The arena is free of any obstacles and the robot travel toward the goal point from the start point.

Static Environment: The obstacles initially placed at regular intervals along the vertices of the squares in the arena and the robot was entrusted to navigate from the start to goal point.

Dynamic Environment: The remotely controlled dynamic obstacles travel obstructing the path of the robot from the start to goal point. This creates a crossroad and the robot frequently need to adjust their trajectories in order to avoid collisions.

4.2 Performance Metrics

Following performance metrics have been computed for each environments:

Throughput: It indicates the robot's time efficiency in navigating toward the goal. This measure is defined as the minimal time that the robot would take to reach the goal without any obstacles while traveling in a straight line divided by the actual time it takes while traveling from the start to goal point avoiding any collisions in the presence of obstacles.

Path Irregularity: It is defined as the amount of unnecessary turning per unit path length performed by a robot, where unnecessary turning corresponds to the total amount of robot rotation minus the minimum amount of rotation which would be needed to reach the same goal point with the most direct path. Path irregularity is measured in radian per meter and indicates the smoothness of the navigating path.

Number of Collisions: It indicates the performance of the navigation algorithm in terms of safety measuring the number of collisions occurring under the three experimental environments. It is measured as collisions per minute.

4.3 Results and Discussions

The results exploring the characteristics of the proposed navigation scheme in terms of throughput and path irregularity in three different environments: no obstacles, static and dynamic environments are reported in this section. Figure 4 through Fig. 6 shows the navigation path of the robot in no obstacle, static and dynamic environments.

Initially, the robot was kept stationary at the start point and entrusted to travel to the goal point located at a distance of ≈250 cm along the diagonal of the arena as shown in Fig. 4. Under no obstacle environment, Fig. 4 shows the robot navigation path following the shortest path from the start to the goal point. Initially, the robot starts with a speed of 12 cm/s. After a period of 4 s, the robot updates its speed to 24 cm/s following the Eq. 1. This is because the sensors could not detect any obstacles along its path towards the goal point (i.e., $k = 1$). The robot continues its navigation path at the maximum speed of 24 cm/s and completes in ≈11 s. In static environment, initially the robot follows the direction to go straight toward the goal at a speed of 12 cm/s. On the detection of an obstacle at a distance of about ≈56 cm, the robot motion is modified. Accordingly the velocity constraint value is updated as $k = -1$ following Eq. 1. Following the navigation approach illustrated in Sect. 3, the robot avoids the forbidden region and moves to one of the edges of the obstacle leading to the shortest path to the goal point as shown in Fig. 5a. Likewise, the collision avoidance of the robot with the second obstacle is shown in Fig. 5b. The second obstacle is located at a distance of ≈65 cm from that of the first obstacle's edge and the robot avoids the forbidden region with the velocity constraint value as $k = -1$ following Eq. 1. Figure 5c shows the navigation path of the robot from the start to the goal point in static environment. In dynamic environment, initially the robot follows the direction to go straight toward the goal at a speed of 12 cm/s as in static environment. The first dynamic obstacles obstructs from left side of the robot. The collision with the first obstacle is avoided as in Fig. 6a with a velocity constraint value updated to $k = -1$. The robot avoids the forbidden region following the navigation approach as illustrated in Sect. 3 with a velocity according to the Eq. 2; wherein the

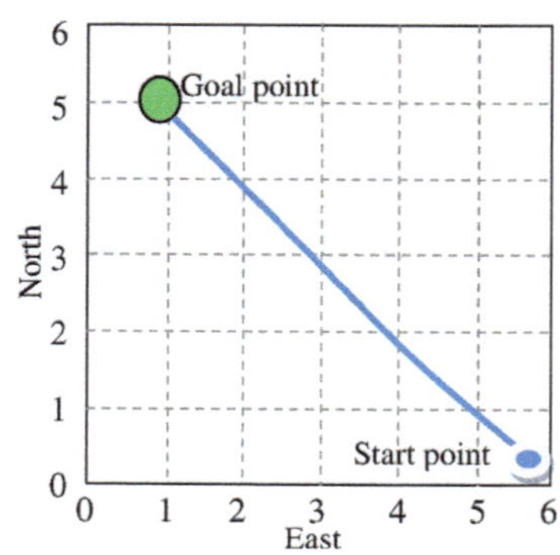

Fig. 4 Navigation of the robot in no obstacle environment

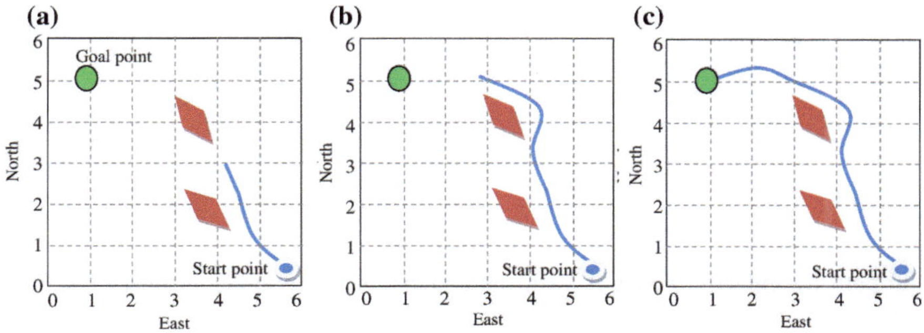

Fig. 5 Robot navigation in static environment avoiding collision with **a** first obstacle **b** second obstacle and **c** from start to the goal point

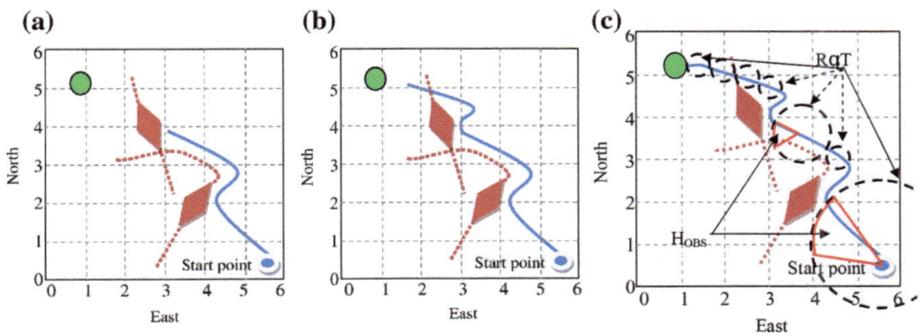

Fig. 6 Robot navigation in dynamic environment avoiding collision with **a** first obstacle **b** second obstacle and **c** from start to the goal point

Fig. 7 Throughput and path irregularity in static environment

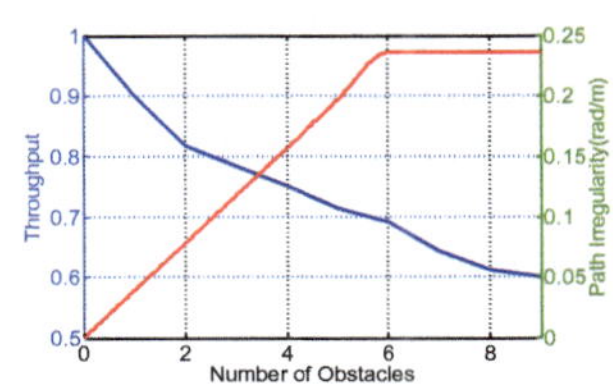

robot's velocity increases by a magnitude equal to the obstacle's velocity. Similarly, collision avoidance with the second obstacle approaching the path of the robot from right to left is shown in Fig. 6b. In this case, the robot's velocity decreases by a magnitude equal to the velocity of the obstacle. Figure 6c shows the navigation path of the robot from start to the goal point in dynamic environment with the reachable and forbidden regions as planned by the navigation scheme. Such dynamic obstacle avoidance requires a simple and fast online navigation scheme as proposed.

The throughput and path irregularity of the robot in static and dynamic environment are shown in Figs. 7 and 8. It can be observed that the throughput decreases

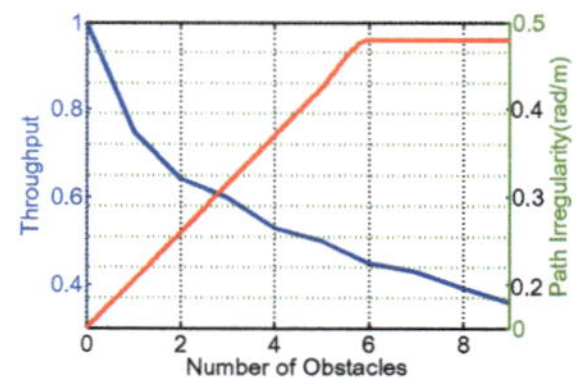

Fig. 8 Throughput and path irregularity dynamic environment

and path irregularity increases with the increase in the number of obstacles. This is because robots must follow longer and more curved path with the increase in the number of obstacles. The throughput of the robot is higher in static environment compared to the dynamic one as the robot follows more longer path to avoid dynamic obstacles compared to static obstacles. Further, path irregularity of the robot is higher in dynamic environment compared to static environment as the robot needs to follow more number of changes in heading direction for avoiding collisions with dynamic obstacles.

5 Conclusions

A navigation scheme inspired by human pedestrian behavior for a mobile robot in an a priori unknown dynamic environment is proposed. The proposed scheme is evaluated under no obstacles, static and dynamic environments; and the robot was able to avoid both stationary as well as dynamic obstacles. The experimental results show that the introduction of the *velocity constraint* and reachable regions holds promise for navigation of mobile robot. It has been observed that the throughput and path irregularity of the robot decreases and increases, respectively, with the increase in the number of obstacles. One of the opportunities to minimize it is the detection of the obstacle's orientation in the sensing cone; which is the part of ongoing research.

Acknowledgment Centre of Excellence in Machine Learning and Big Data Analysis, Tezpur University funded by Ministry of HRD, Govt. of India.

References

1. J. Guzzi, A. Giusti, L. M. Gambardella, G. Theraulaz, G. A. D. Caro, Human-friendly robot navigation in dynamic environments, in: IEEE International Conference on Robotics and Automation, Karlshure, Germany, 2013, pp. 423–430.
2. D. Tuvshinjargal, B. A. Dorj, D. J. Lee, Hybrid motion planning method for autonomous robots using kinect based sensor fusion and virtual plane approach in dynamic environments, Journal of Sensors 2015 (Article ID 471052) (2015) 1–13.
3. T. Fukao, H. Nakagawa, N. Adachi, Adaptive tracking control of a nonholonomic mobile robot, IEEE Trans Robotics and Automation 16 (5) (2000) 609–615.

4. P. Henry, M. Krainin, E. Herbst, X. Ren, D. Fox, Rgb-D mapping: Using kinect-style depth cameras for dense 3d modeling of indoor environments, International Journal of Robotics Research 31 (5) (2012) 647–663.
5. Y. D. Song, H. N. Chen, D. Y. Li, Virtual-point-based fault-tolerant lateral and longitudinal control of 4 w-steering vehicles, IEEE Trans Intel Transp Syst. 12 (4) (2011) 1343–1351.
6. A. Vieira, P. Drews, M. Campos, Spatial density patterns for efficient change detection in 3d environment for autonomous surveillance robots, IEEE Trans. Autom. Sci. Eng. 11 (3) (2014) 766–774.
7. M. Yasuno, S. Ryousuke, N. Yasuda, M. Aoki, Pedestrian detection in far infrared images, Integrated Computer-Aided Engineering 20 (4) (2013) 347–360.
8. H. Kretzschmar, M. Kuderer, W. Burgard, Learning to predict trajectories of cooperatively navigating agents, in: IEEE International Conference on Robotics and Automation, Hong Kong, 2014, pp. 4015–4020.
9. Schwendner, Jakob, S. Joyeux, F. Kirchner, Using embodied data for localization and mapping, Journal of Field Robotics 31 (2) (2014) 163–295.
10. M. Moussaï, D. Helbing, G. Theraulaz, How simple rules determine pedestrian behavior and crowd disasters, Proceedings of the National Academy of Sciences of United States 108 (17) (2011) 6884–6888.
11. M. Batty, Predicting where we walk, Nature 338 (6637) (1997) 19–20.
12. P. Schrater, D. Knill, E. Simoncelli, Mechanisms of visual motion detection, Nature Neuroscience 3 (1) (2000) 64–68.

Mobility Prediction for Dynamic Location Area in Cellular Network Using Super Vector Regression

Nilesh B. Prajapati and Dhaval R. Kathiriya

Abstract Mobility Prediction of Mobile Users in a cellular network is one of the burning issues. Once the Mobile Users location is properly predicted using mobility prediction methods in the cellular network then service-related problems can be resolved. Super Vector Regression (SVR) method is one of the methods using which mobility prediction of mobile users is possible. SVR method predicts the mobility of mobile device in cellular network better than other mobility prediction methods. SVR gives a better result for reducing location management cost by creating dynamic location area for Mobile Users. This dynamic location area is increasing prediction accuracy of Mobile Users using SVR method.

Keywords Super vector regression (SVR) · Dynamic location area (DLA) Location update · Paging · Cellular network

1 Introduction

From the past few decades, Mobile Users (MUs) increased in cellular networks drastically. To provide proper calling and mobility services with limited and pre-defined radio bandwidth is a challenging task for cellular companies. For that, the structure of the cellular network is changed by dividing the existing cell into smaller cells. Cellular companies also made so many changes in the technologies for providing better services to MUs. One of the parameters, essential requirement, for providing good QoS is location determination or prediction of MUs in the network.

N. B. Prajapati (✉)
Computer IT Engineering, B.V.M. Engineering College,
Gujarat Technological University, V. V. Nagar, India
e-mail: nbp_it53@yahoo.com

D. R. Kathiriya
Information Technology Center,
Anand Agriculture University, Anand 388110, Gujarat, India
e-mail: drkathiriya@gmail.com

C. R. Panigrahi et al. (eds.), *Progress in Advanced Computing and Intelligent Engineering*, Advances in Intelligent Systems and Computing 714,
https://doi.org/10.1007/978-981-13-0224-4_41

MUs are always attached to one of the Base Stations (BS) of the cellular network. Location of any MUs is predicted from the previous history of MUs in a cellular network. In GSM network, the network is divided into Location Area (LA), which is a group of contiguous cells, for users and bandwidth management. LA is also useful for finding MU's presence in the cellular network.

Call dropping rate can be decreased and optimal resource allocation to all MUs are possible using mobility prediction of MUs, in the cellular network. LA is two types: Static and Dynamic. Static LA has its own limitations while Dynamic Location Area (DLA) is emerging concept in the recent era. DLA creation is possible by grouping only those cells which are regularly visited by MUs. For finding such kinds of visited cells are possible by observing mobility pattern and history of the MUs. DLA is useful for reducing Location Management cost and increasing mobility prediction of MUs.

In a cellular network, numbers of MUs use same movement pattern and movement behavior in their daily lives. These kinds of users always cross the same numbers of cells with the same amount of times to reach their workplaces. According to movement behavior in the cellular network, MUs can be classified into Predictable, Semi-predictable and Random users. We can easily predict the movement of Predictable and Semi-predictable users in the cellular network while random users' movement prediction is complex. We can make a prediction of these kinds of users' very easily by first finding DLA of them using movement pattern and behavior. Mobility prediction is beneficial for resource allocations and bandwidth management. There are various methods and techniques used for mobility prediction. In this paper, SVR method is used for finding DLA and mobility prediction of MUs in a cellular network. Comparison of SVR method with the Static and Dynamic methods are also presented in the result section.

2 Location Area and Its Importance

In GSM, the system coverage area is divided into a geographical area called LA. This geographical area contains one or more cells called Base Station (BS). Each of this LA is identified by its unique Location Area Identifier (LAI). LAI is broadcasted by each BS which includes all BSs forming that particular LA. LA is useful for Location Update (LU) and Paging of the MUs. When MUs changes their place in network and go to other LA from their current LA then LU for MUs are generated. Paging is call delivery to MUs in a cellular network by finding location of MUs in respective LA. LA plays a crucial role in GSM network. Size and Shape of LA is a key factor for deciding Location Management (LM) cost which is the summation of LU cost and Paging cost. So, LA planning is crucial in GSM network. There are two extreme LA planning approaches: Always Update [1]; make LA size equal to service area of MSC, in this scenario location update cost is minimized but in turn paging cost gets increased and Never Update [1]; numbers of LA are equal to number of BS then the location update cost get increased while paging cost gets minimized. Good LA planning is achievable using dynamic as well as static methods.

Static LA consists of a contiguous number of cells which are fixed for all MUs residing in that. Static LA is formed by considering Call-to-Mobility ratio, number of users, Busy Hours call rate, Call rate, and other important parameters of the cellular network. There are so many techniques for Static LA planning that use static geographic strategies to reduce or obtain optimal signaling traffic of particular LA. Following are the main techniques used for LA Planning (LAP) for minimizing total Location Management cost of static LA. LA planning is a 0–1 linear programming problem, in which searching techniques, such as Simulated Annealing [2], Taboo search [3], Genetic Algorithm [4], Ant colony optimization [5], River Formation [6], Linear Programming [7], Spanning Tree [8], Greedy algorithm [4], and clustering algorithm [9] were employed to derive a proper planning for LAs to minimize the total number of LUs. Using these methods Location Management cost can be reduced for specific LA but these methods do not give optimal results for all MUs. These static methods are not able to give optimized LA design for each MUs as well as not able to give optimal results in zigzag movement of MUs and overlapping LAs problem. These show the requirements of such methods which give minimum LM cost for every MUs. So, DLA planning concept is introduced to fulfill the mentioned requirement.

3 Related Work

Dynamic LA Planning (DLA) methods are also known as Selective Update [1] methods; the LU occurs only when certain predefined condition or constraints are violated. DLA methods are further classified into State-based and Profile-based methods. Time-based [1, 10, 11], Distance-Based [1, 10, 11], and Movement-based [1, 10, 11] methods are main examples of state-based DLA methods. Profile-based methods use User Movement History and User Mobility Records for creating DLA. Such kinds of DLA planning methods are Cartesian product based [10], User Profile-based methods [11, 12], Artificial Neural Network [12, 13], Simulated Annealing [14], Directed Graph [1], Data mining [15], Heuristic function [14], clustering [16], etc. All these methods are useful for creation of DLA for MU with some pros and cons. Some methods are updating DLA when MUs changes its regular mobility behavior while others use concepts of overlapping DLA.

DLA is useful for reducing LU cost and paging cost. As DLA contains only those cells which are regularly visited by MUs so no need for LU. When a call comes for any MUs then only a few cells are required to page for call forwarding which can be further reduced by considering the residential probability of MUs. Mobility prediction is possible when DLA is created for MUs. In DLA, every MU most probably follows the same path with the same unit of time to reach its destination from home location. It is easy to predict next cell movement of any MU in DLA using mobility pattern and behavior. Using mobility prediction, resource allocation during ongoing call or handoff procedure of MUs can be properly possible which reduces call dropping rate. There are many methods for mobility prediction in a cellular network like Clustering [16], Ant Colony Optimization [17], Data mining [15], Hidden Markov Model [18], etc. These methods use direction, time, and behavior of MUs for mobility prediction.

4 Super Vector Regression (SVR) Method

SVM can be used for regression method which is known as Super Vector Regression (SVR). The SVR uses the same principles of SVM with minor differences. In SVR method, training data is given for train the model which is used for applying regression for future prediction. In SVR several kernel functions are used for prediction. Linear and nonlinear SVR are the main types of the SVR. The kernel functions transform the data into a higher dimensional feature space to make it possible to perform the linear separation. Here in this research, RBF kernel function is used for prediction which gives higher prediction probability then polynomial kernel function.

5 Implementation

In the cellular network, SVR method is useful for finding DLA and mobility prediction of MUs based on their mobility pattern and behavior. SVR method is implemented using Java programming. Campus wireless trace dataset of Dartmouth University [19] is used for finding mobility pattern and behavior of MUs. This dataset contains wireless user histories such as user associations with APs, duration of association, etc. This dataset is very useful to get information such as the number of unique users, their association with different APs, the probability of transitions from one AP to another, etc.

There are several steps of SVR methods which are useful for finding DLA and mobility predicting of MUs in a cellular network. The pseudo code of SVR methods is presented in below Fig. 1.

Code described in Fig. 1 is useful for DLA creation of Predictable, Semi-predictable and Random MUs. As the mobility of Random MUs is random in nature at any time so using above code mobility prediction probabilities of Predictable and Semi-predictable MUs are calculated. The results of SVR method is also compared with dynamic methods (Apriori, HMM) and static method.

Algorithm: SVR for DLA and Mobility Prediction
Input: Dataset (Mobility Traces from Dartmouth), Epsilon
Output: DLA formulation and Mobility Prediction Probability
Steps:

1. Initialize SVR model
2. Load the dataset to train SVR model.
3. Calculate regression probability of input data which is useful for formulation of DLA.
4. Use SVR with kernel function for predication.
5. To improve the performance of the SVR need to select the best parameters for the model.
6. To train a lot of models for the different couples of ϵ (Epsilon) and cost, and choose the best one.
7. Print the prediction probabilities.

Fig. 1 Pseudo code of SVR method

6 Results and Discussion

Using SVR method DLA for MUs are formulated, based on the regression probability of MUs in the cellular network, which is reducing the location management cost (location update and paging cost). In this section, results of SVR method for DLA creation and their impact on LM cost are compared with other methods. Mobility prediction accuracy of the proposed method is also compared.

Once individual MU's DLA is created, which contain only a few numbers of cells. These cells are a collection of regularly visited cells of MU whose regression probabilities are higher than other cells. So LU cost is decreased because MU freely moves within that DLA without LU and paging cost is also decreased as paging is required to done within cells' of DLA. Below Table 1 and Fig. 2 show a comparison of LU cost of different datasets. SVR method gives minimum LU cost in comparison of apriori, HMM and always update methods. Datasets contains Predicted, Semi-predictable, Random, combination of Predicted and Semi-predictable users as well as combination of Predicted and Random users.

Table 1 Comparison of LU cost

Dataset	Apriori	HMM	SVR	Always update
Dataset_P1	42.36	34.83	32.86	69.38
Dataset_E1	52.25	49.82	48.92	72.57
Dataset_R1	95.24	93.71	92.63	98.99
Dataset_PE	41.19	40.30	39.13	62.08
Dataset_PR	62.92	57.67	52.86	69.54

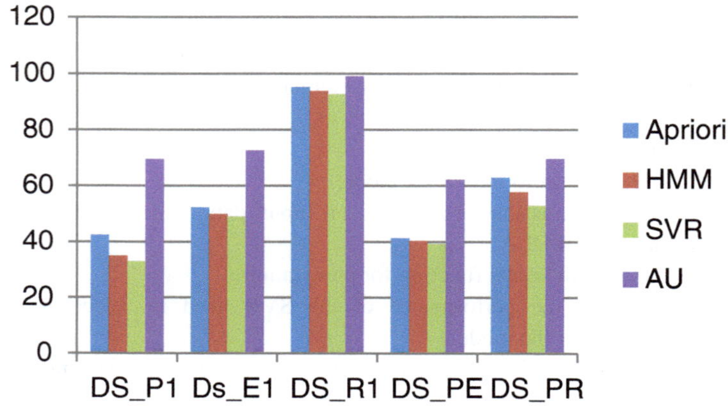

Fig. 2 Comparison of LU cost

Table 2 Comparison of paging cost

Dataset	Apriori	HMM	SVR	Never update
Dataset_P1	16.64	16.47	15.35	38.36
Dataset_E1	15.56	14.67	13.86	56.74
Dataset_R1	38.45	35.97	34.82	75.11
Dataset_PE	32.13	28.27	25.58	44.58
Dataset_PR	26.45	25.02	21.73	46.21

Comparisons of Paging Cost of each dataset are given in Table 2 and Fig. 3. SVR method gives minimum paging cost with a comparison of apriori, HMM, and Never Update method.

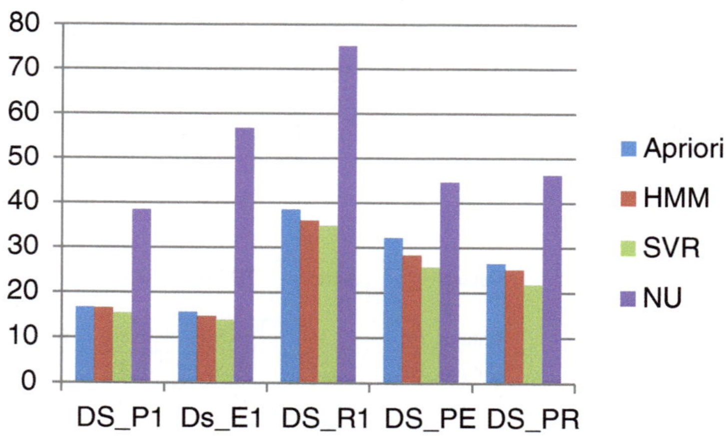

Fig. 3 Comparison of paging cost

Table 3 Comparison of mobility prediction accuracy

Dataset	Apriori	HMM	SVR
Dataset_P1	16.68	14.59	25.89
Dataset_E1	8.16	9.65	12.24
Dataset_PE	16.29	17.47	19.83

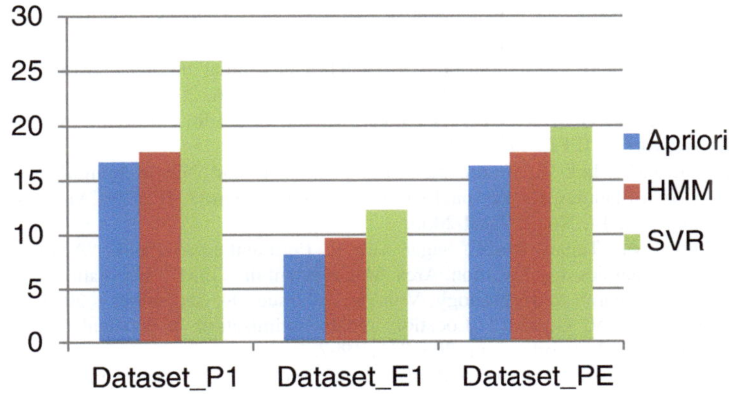

Fig. 4 Comparison of mobility prediction accuracy

Mobility Prediction of MUs in the cellular network is also more important. SVR method uses RBF kernel function for predicting mobility of MUs in the cellular network. SVR gives better mobility prediction then apriori and HMM which is shown in Table 3 and Fig. 4. Mobility prediction of Predicted and Semi-predictable MUs are easy in comparison of random MUs so in the table only these two types of MUs' prediction accuracy are given.

7 Conclusion

Location Management and providing good QoS are important tasks of a cellular network which is possible by proper planning of LA. In this work, a new regression-based dynamic method SVR is proposed for the formulation of DLA, which reduces signaling cost by minimizing LU and paging signals, and increasing mobility prediction of MUs in the cellular network. This work is implemented without making any changing in the existing architecture of the cellular network. SVR method gives minimum location management cost in comparison to apriori, HMM, and static methods. Mobility prediction for resource allocation to MUs in the cellular network is possible by SVR which give highest mobility prediction accuracy rate than other methods.

References

1. Guanling Lee, Arbee L.P. Chen, "The Design of Location Regions Using User Movement Behaviors in PCS Systems", Multimedia Tools and Applications, Volume 15, Issue 2, pp 187–202, November 2001.
2. Ilker Dermirkol, Cem Ersoy, M. Ufuk Caglayan, Hakan Delic, "Location Area Planning in Cellular Networks Using Simulated Annealing", NETLAB, Department of Computer Engineering, BUSIM Lab., Department of Electrical and Electronics Engineering, Bogazici University, Bebek 80815 Istanbul, Turkey.
3. S. Pierre, F. Houeto, "Assigning cells to switches in cellular mobile network using taboo search", IEEE trans. on system. Vol. 32, No. 3, pp. 351–356, 2002.
4. Laidi Foughali, El-Ghazali Talbi, "A Parallel Insular Model for Location Area Planning in Mobile Networks", IEEE, 978-1-4244-1694-3, 2008.
5. Yigal Bejerano, Mark A. Smith, Joseph (Seffi) Naor, and Nicole Immorlica, "Efficient Location Area Planning for Personal Communication Systems", IEEE/ACM transaction on networking, Vol. 14, No. 2, April 2006.
6. Dixa Dholakiya, Tapan Doshi, Sagar Ghiya, Prashantkumar Patel, "Advanced River Formation Dynamics for Location Area Management in GSM", International Journal of Engineering Research & Technology, Volume. 4 - Issue. 09, September – 2015.
7. A. Abutaleb and V. O. K. Li, Location update optimization in personal communication systems, Wireless Networks, 3 pp 205–216, 1997.
8. M. Munguia-Marcario, D. Munoz-Rodriguez, C. Molina, "Optimal adaptive location area design and inactive location area", in Proc. 47th IEEE Vehicular Tech. Conf., Vol. 1, pp. 510–514, 1997.
9. Jahangir khan, "Handover management in GSM cellular system", International Journal of Computer Applications (0975 – 8887) Volume 8 – No.12, October 2010.
10. S. D. Markande, S. K. Bodhe, "Cartesian Coordinate System based Dynamic Location Management Scheme", International Journal of Electronic Engineering Research, Vol-2 2009.
11. M.S. Sricharan, V. Vaidehi, "A Dynamic Distance Based Location Management Strategy Utilizing User Profiles for Next Generation Wireless Networks", First International Conference on Industrial and Information Systems, ICIIS2006, 8–11 August 2006, SriLanka.
12. B. P. Vijaykumar, P. Venkataram, "Prediction-based location management using multilayer neural networks", Indian Inst. Sci., 2002, 82, 7–21 © Indian Institute of Science.
13. Amar Pratap Singh J. and Karnan. M., "A Dynamic location management Scheme for Wireless Networks Using Cascaded Correlation Neural Network", International Journal of Computer Theory and Engineering, Vol. 2, 2010.
14. Jun Zheng, Emma Regentova, Pradip K. Srimani, "Dynamic Location Management with Personalized Location Area of Future PCS Network", Distributed Computing IWDC 2004, 6th International workshop, India, December 27–30, pages 495–501, 2004.
15. Rachida Aoudjit, Malika Belkadi, Mehammed Daoui, Lynda Chamek, "Mobility Prediction Based on Data mining", International Journal of Database Theory and Application Vol. 6, No. 2, April, 2013.
16. Javid Taheri, Albert Y. Zomaya, "Clustering techniques for dynamic location management in mobile computing", Journal of Parallel Distributed Computing, pp 430–447, 2007.
17. Ahmed Elwhishi, Issmail Ellabib, and Idris. El-Feghi, "Ant Colony Optimization For Location Area Planning In Cellular Networks", The International Arab Conference on Information Technology, University of Balamand, al Kurah, Lebanon, 2008.
18. Samir Bellahsne & Leila Kloul, "A New Markov-Based Mobility Prediction Algorithm for Mobile Networks", Computer Performance Engineering Lecture Notes in Computer Science Volume 6342, pp 37–50, 2010.
19. "Crawdad: Wireless Traces from Dartmouth" in http://crawdad.cs.dratmouth.edu/.

Object Surveillance Through Real-Time Tracking

Mayank Yadav and Shailendra Narayan Singh

Abstract In this paper, object surveillance through real-time tracking is examined where there is an analysis of the various techniques which help in the analysis of the methods that are employed in the modern society for the delivery of the right goals. The paper uses a method to help in the tracking and recognizing object in the surveillance area, focus on pixel approach which helps in arriving at the solution for the problem. The camera system is used as a sensor for the purposes of tracking the object used for the study in the surveillance area. Use of edge detection in the analysis, image segmentation process, background separation algorithm provides a clear knowledge of the foreground and the background. Finally, use of contourlet transform is used to extract features and for recognition of objects under surveillance area, pattern matching is also used for recognizing different objects in a video.

Keywords Object tracking · Features selection · Surveillance and tracking Object recognition

1 Introduction

This tracking of objects in video sequences has been of great importance in various ways in the modern society and the same skill is employed in various fields and areas among them being surveillance, human–computer interaction, smart vehicles, interactive TV, and augmented reality applications. The process of tracking objects involves two main steps, which are detection and tracking.

In the first step of detection of the object, there is the use of a common approach, and detection is used in the first frame, tracking then proceeds in the rest of the

M. Yadav (✉) · S. N. Singh
Computer Science and Engineering Department, Amity University, Sector-125, Expressway, Noida, Uttar Pradesh, India
e-mail: mail4mayankyadav@gmail.com

S. N. Singh
e-mail: snsingh36@amity.edu

C. R. Panigrahi et al. (eds.), *Progress in Advanced Computing and Intelligent Engineering*, Advances in Intelligent Systems and Computing 714,
https://doi.org/10.1007/978-981-13-0224-4_42

video. However, such an approach overlooks spatial information. There is a better approach available and it is the pursuance of a continuous integration of spatial and temporal information which combines the methods of detection and the tracking approach. The other method that can be used for tracking objects of taxonomy and it employs the use of three classes; Point tracking, Kernel tracking, and Silhouette tracking. Point tracking is where the objects that have been detected in consecutive frames are represented by points, Kernel tracking deals with the representation of objects by points such as a rectangular or elliptical shape.

In the last method, Silhouette tracking, the objects are represented through the use of contour or the region which is inside a contour, in this method, there is a detection of objects in every frame. The second category most closely fits rigid objects and utilized for an intensive period of time applications, the third category, on the opposite hand, most closely fits nonrigid and complicated objects. The Kernel chase ways use the employment of 2 models; the primary one {is used|is utilized} to collect data from the foremost recent observation and, therefore, the second model is where the various views of the object will be learned offline and employed for the chase processes.

2 Problem Identification

There is an absence of novel techniques for question following, and which meets the computational execution required for the intuitive television programs. Moreover, there is the test of following a few protests as problem areas for client cooperations among them being character determination or as locales for the constant compositing. A case of the last is either enlightening a face or appending a content to a character. There is the problem of achieving time period detection and chase objects and there are 2 main reasons for a similar. The primary reason is as a result of there's quite one object that is to be detected in every video frame and this compromises the performance of algorithms. Secondly, the high resolution videos that are used affects the process of your time as a result of a better resolution needs that a bigger space be hunted for each object. The project seeks to supply solutions to the 2 main issues in video chase.

3 Related Work

The study of object tracking in real time has attracted various forms of surveys which have proven to be of great importance to the modern society. Various authors have carried studies on identical topic with the aim of developing a strong system for time period chase. There have conjointly been known numerous strategies of object chase that are supported particle filters. In such strategies, target model of the particle filters is outlined by the employment of color data of the objects that are

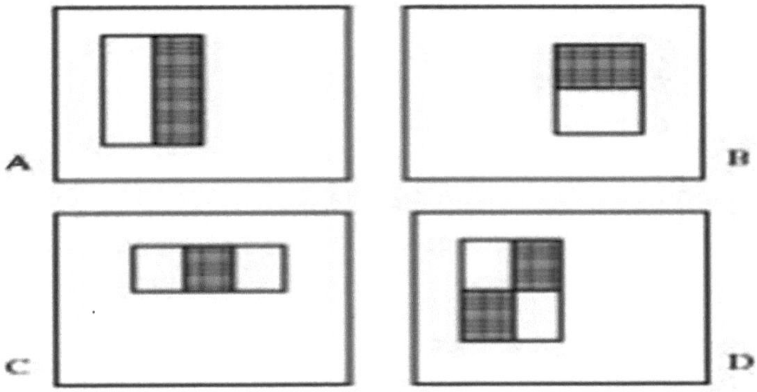

Fig. 1 Johnsen and Tews method

being half tracked. The processes of tracking objects are expensive to compute, and there have been developed methods that lead to cheaper computational costs among them are parametric models motion, optical flow, and matching blocks.

Moreover, there has been projected a reliable algorithmic program which might find objects in pictures in real time. The algorithmic program developed by Johnsen and Tews is 15 times quicker than the antecedently projected techniques (Fig. 1).

Another proposal by the author could be a technique of associate in nursing formula that is predicated on a collection of revolved Haar-like options that enriches the previous works on object following in period.

Moreover, there is a proposal by Ray, Dutta, and Chakraborty that could be a period facial feature detection on mobile devices-supported integral pictures. The advantage of the projected technique of Johnsen and Tews is that the algorithm is based on features and not the pixels thereby leading to higher performance. The study by Roth is an extension of the study of the algorithm by Johnsen and Tews to the motion and domain. The only distinction is that the study focusses a lot on the low-resolution videos of human figures underneath troublesome things. In addition, the frame rate is simply too slow.

4 Proposed Method

4.1 Search Area Reduction

In the projected technique by Johnsen and Tews, finding associate in the nursing object in a picture needs a brand new search over the image to be started anew. Such a research goes everywhere in the image by moving a window with a varied size. The window to be used begins with the littlest doable size of associate in nursing object within the image like 25 * 25 or 35 * 40. On every occasion the

```
Given J, the search window (initialized with the minimum
size of the object);
Given I, the original image;
Given λ, the scale factor;
Given Δ, the displacement of the window;
while size(J) < size(I) do
  x = 0; y = 0;
  while y + Δ < height(I) do
    while x + Δ < width(I) do
      if J contains the wanted object then
        store the actual location (x, y) of the search win-
        dow;
      end if
      increment x;
    end while
    increment y;
  end while
  scale the size of J by λ;
end while
Mark in the image the detected locations;
```

Fig. 2 Algorithm

window completes the method of looking out the entire image, its size must be inflated by an element λ, then a new search is started. The method ascertains that associate in nursing object within the image is detected notwithstanding its size. The formula may be an outline of the work by Johnsen and Tews (Figs. 2, 3).

In this case, the entire image is tested to see whether or not it contains the required object. However, the changes within the consecutive frames occur solely within the tiny regions, the search space will then be decreased through checking solely the regions that have been modified. Accomplishing this task needs the employment of a nonstatistical approach for the background segmentation by the

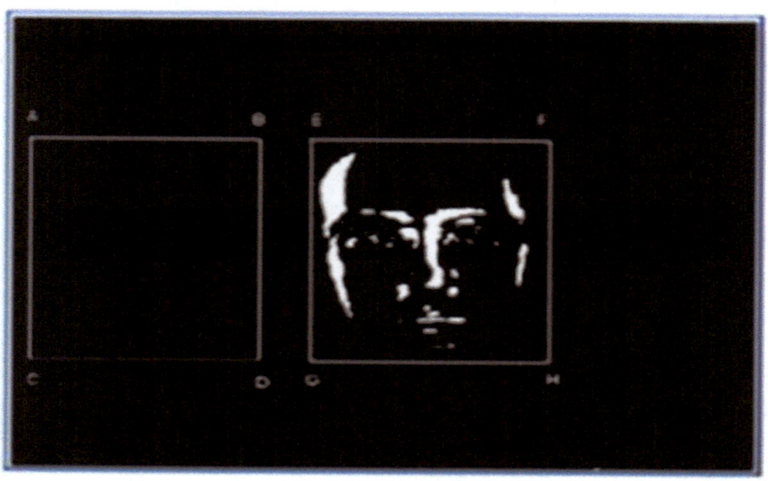

Fig. 3 Face detection

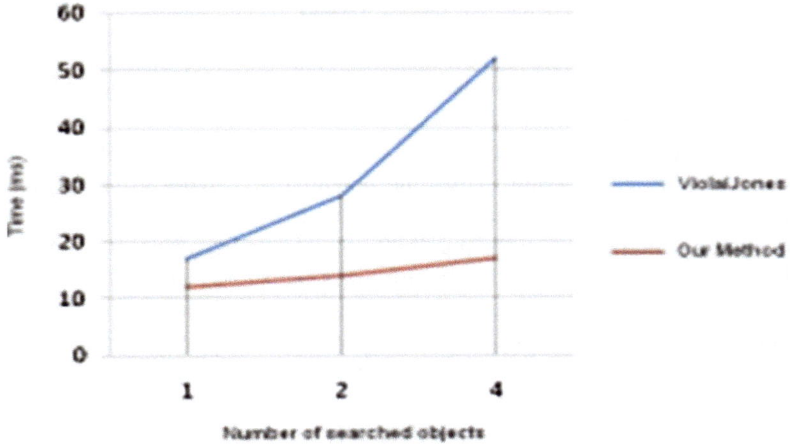

Fig. 4 Average performance of the algorithmic rule for detective work objects with and with out optimization at the side of a video resolution of 320 * 240

associate adaptive mean. The benefits of a subtraction technique through adaptive mean area unit algorithmic thereby proving not maintain a buffer memory for the storage of a background model. Finally, the technique conforms to the changes in lighting and physics and features a low procedure price (Fig. 4).

5 Object Real-Time Tracking Techniques

In the recent decades, the computers have shown improved power, functionality, ubiquity which then coupled with progress in the internet discovery and use have transformed human lives [1]. The focus of scientists has been directed toward using the unique features of technology to facilitate overreaching changes in critical societal aspects such as security. Cameras with high resolution have become essential requirements for a successful system used for real-time object tracking. Multiple or single cameras can be fitted on moving platforms such as robots, motorbikes facing any direction; forward, rear or sideways [5]. The choice of the number of cameras to use in any case depends on the cost imperatives. Multiple single camera systems installed on moving platforms such as robots or automobiles and inclined at various angles and distance that capture every detail about the object of focus [11].

The automatic reconfiguration of an object's shape requires the system to overcome three initial problems [8]. These challenges include: the movement of the robotic platforms on which the sensors and camera used to monitor an object is mounted, a wireless system of communication that provides a means through which the gathered information from the sensors is remitted to a stationary computer surveillance and; altering of the real-time projection of an object on the desired surface.

Object real-time tracking techniques currently form an integral part of the practicing security engineering. It has the potential of solving a whole range of security problems that would otherwise become intractable. Due to the difficulty in comprehending and predicting the possibility of a crime occurring, it is impossible to rely solely on human intelligence for tracking safety alerts [8]. Therefore, technology progressively undergoes harnessing to aid information gathering and dissemination, which has been critical in dismantling criminal networks and dangerous weapons. Object and individual tracking applications have been integrated with robotics technology to facilitate safety risk detection [10]. Nonetheless, one first necessity for the operations of tracking technology is that it is a must to know the elements of a scene and its changes over time before being able to interact efficiently with the environment.

The tracking of objects using artificial mechanism takes different forms including video monitoring, surveillance systems, and robotic platforms. These three fields have been researched proactively in the past [10]. In most cases, tracking devices are installed stationary at strategic points to gather diverse information that is critical to ensuring societal security [15]. However, remarkable progress continues to be made toward different forms of mobile intelligence gathering technologies such as drone fitted with highly sensitive sensors that detect and transmit real-time information from the field to computer systems located in safe places.

In stationary video tracking systems, various requirements must be met to ensure effectiveness. One important aspect of monitoring using immobile technologies is that the desired object must remain within the range of surveillance within the period of information collection [12]. Therefore, if the device or object goes outside the range, it becomes difficult to track. From a technological perspective, once an object extends beyond the surveillance range covered by the intelligence gathering equipment then it becomes intractable [8]. One way of solving this problem and ensuring sustainable information is the designing of a mobile system that uses laser range sensor and a visual spectrum to trail the moving objects and avoid obstructions by different obstacles.

Tracking of devices takes two important steps. The first phase involves training the system to learn the object of concern. The object can be a weapon such as improvised explosive devices, bomb, missile or any other object that the system launcher intends to track [14]. Making the device tracking system to learn the specific objects that should be tracked is an important milestone in generating object models. At this formative stage of monitoring system development, a time of flight range camera is integrated with the color image at the pixel. This integration results in the production of a dense range and color image for each pixel [15]. During the tracking process, elements of the field can be tracked in segments or as a whole. One of the things that the monitoring system developer must know is that no pre-recorded data is necessary for object modeling.

The absence of a need for prior information in object tracking is because object modeling grossly relies on the accurate data gathered by sensors mounted on a mobile robot. The second requirement is that a particular object framed for surveillance must first be detected and then tracked. Different methods can be used to ensure accurate object detection including a detection system that scans the environment for objects using the tracking system itself [4]. In most cases, object monitoring involves using only luminance images occasioned by removing the dependency on range images while tracking. Alternatively, robots outfitted with only luminance-based cameras can be programmed to use object models during tracking [5]. This method facilitates tracing at larger distances where range data is less accurate than camera data and facilitates the repeated use of data by robots that lack range sensors.

The establishment of a working tracking system necessitates the selection of the right features. This process of selecting appropriate system features plays a critical role in ensuring a tracking framework that overcomes some of the most perceivable detractors [4]. One of the most desirable properties of a visual tracking feature is its uniqueness that enables ease of object distinction in the future space. The selection of features has a close relationship with the object representation. For instance, in the histogram-based appearance descriptions, color is used as a function. On the other hand, contour-based image uses object edges as features. Many algorithms used in object tracking use a combination of the selected features to facilitate continued tracking of pertinent information [9].

In the cases where the color of an object being tracked forms the central point of focus, two physical factors have an influence. These factors include the spectral power distribution of the source of illumination and the properties of the objects reflection surface [9]. Edges of an object generate substantial changes in the intensity of images. In tracking, edge detections are used to identify the changes in image intensities. One distinctive feature of edges is that they are less sensitive to light than color features. In the tracking technology for objects, algorithms for determining the boundary between any closely related objects use edges as a representative aspect [15]. Optical flow is another important factor in real-time tracking of objects is optical flow. Optical flow is a dense field of displacement vectors that define the translation of each pixel in a region. As an important feature in real-time tracking of objects, optical flow facilitates motion-based segmentation and tracking applications.

6 Phases of Object Tracking in Real Time Using Artificial Intelligence

6.1 First Phase: Image Input

The first stage of developing an object monitoring system is called the object input stage. This step involves the deployment of a robot fitted with an extremely sensitive camera that takes colored images of the objects of concern for tracking. The robot platform on which the camera is installed must be adequately programmed to follow the device for a long time to learn its unique features and critical details [9]. With a recorded information, it becomes easier for the camera to detect any changes in the object over time. There are different types of sensors used in identifying and learning unique object features and the changes that occur over time. Some of these sensors include a lidar, radar, ultrasound, and stereo-based depth sensors. These technologies are available for incorporation into the system of object tracking.

Image input is a learning process of the computer vision, which is necessary for gaining new information and monitoring the fluctuations occurring around the object of focus [6]. In the tracking system, learning input of the object requires the use of sensitive cameras to capture the vision of a scene or object. The vision is unguided in nature hence can also be referred to as unsupervised.

6.2 Second Phase: Object Tracking

Object tracking phase involves the use of various approaches including the iterative method to compute the optical flow. The method should be less affected by illumination changes thus making it appropriate for tracking real-time objects. The method used in compute optical flow can also be used to calculate the motion vectors for the tracked object. The tracked object is then taken through a series of movements including moving forward, moving backward, toward the right and then to the left.

During the course of a video scene, overlaps between any two object areas may occur due to occlusion thus making the related features between them ambiguous. This ambiguity has the potential of making it difficult to monitor patterns of change within the object itself or its surrounding [3]. Cases of lost clarity between objects that form critical subjects of surveillance cause a complication that requires designing of multiple-species-based algorithms (MSBA). The reason behind an MSBA is to segregate the ground truth particles of the focus object into multiple species depending on the specified object number, which is then used in successfully modeling the relations and partial ability to see among different species.

6.3 Third Phase: Robot Control Phase

The robot control stage becomes active once the article pursuit gets a fait of watching the direction of the moving object tracked. The information collected at the follow-up period continues to provide current mass motion vector information for use in the robot control step. The robot then uses the present mass motion vector to cipher the distinction that exists between itself and also the origin.

6.4 Fourth Phase: Barrier Detection Phase

After the robot control step, it sends a command to the surveillance center. Nonetheless, the robot uses a laser scanner to detect in case of any real obstacles in its path before transmitting any information [2]. If the detector does not find any obstacle, the automaton quality part is activated mechanically. After that, the management of the article following system returns to the image input stage. On the other hand, a detected obstacle initiates another robot control command that facilitates the avoidance of the obstacle.

6.5 Fifth Phase: Obstacle Avoidance

The aim of an object tracking system is to overcome any existing barriers and provide every detail about the object being tracked including periodic changes. The achievement of this objective requires techniques that navigate through the obstructions to the clarity of information gathering. One approach for overcoming tracking system is the use of modified Potential Fields methodology. An advantage derivable from this method is their simplicity and high speed [6]. However, it also has some disadvantages including local minima problems. Therefore, the choice of the approach to avoid constraints of barriers is necessary for ensuring an efficient artificial intelligence gathering.

Any tracking technology needs to trace the focal object continuously while at the same time avoiding the impediments posed by the obstacles. Therefore, the obstacle avoidance function must be able to perform the two tasks simultaneously without losing track of the object. In the case wherever a system is unable to follow the object whereas escaping the barriers, the object may move to a brand new location outside the view of the camera [14]. Therefore, methods such as the traditional Potential Fields methods cannot work well if directly used since they work on the assumption that the goal position of the object is static. Dealing with the dynamic target area problem requires that the obstacle avoidance algorithm adjusts its path about the change of its destination during avoidance of the obstacle.

7 Conclusion

Real-time tracking of objects and humans is an important process in gathering and deamination of intelligence that contributes to crime control. In real life, tracking of objects in a video sequence using cameras and sensors is the most common approach used in real-time surveillance. A proactive real-time monitoring of objects takes two initial steps including detection of the object at the object input phase followed by a series of surveillance activities. The effective procedure used by completely different chase systems is sleuthing the object within the first frame so chase it through the remainder of the video.

References

1. Bodor, R., Jackson, B., & Papanikolopoulos, N. (2003). Vision-Based Human Tracking And Activity Recognition. In Proc. of the 11th Mediterranean Conf. on Control and Automation (Vol. 1).
2. Chen, C. H., Cheng, C., Page, D., Koschan, A., & Abidi, M. (2006). Tracking A Moving Object With Real-Time Obstacle Avoidance. Industrial Robot: An International Journal, 33 (6), 460–468.
3. Comaniciu, D., Ramesh, V., & Meer, P. (2000). Real-Time Tracking Of Non-Rigid Objects Using Mean Shift. In Computer Vision and Pattern Recognition, 2000. Proceedings, IEEE Conference on (Vol. 2, pp. 142–149). IEEE.
4. Comport, A. I., Marchand, E., Pressigout, M., & Chaumette, F. (2006). Real-Time Markerless Tracking for Augmented Reality: The Virtual Visual Servoing Framework. IEEE Transactions on visualization and computer graphics, 12(4), 615–628.
5. Fransen, B. R., Herbst, E. V., Harrison, A., Adams, W., & Trafton, J. G. (2009, October). Real-time face and object tracking. In Intelligent Robots and Systems, 2009. IROS 2009. IEEE/RSJ International Conference on (pp. 2483–2488). IEEE.
6. Hsia, C. H., Chang, W. H., & Chiang, J. S. (2012). A Real-Time Object Recognition System Using Adaptive Resolution Method for Humanoid Robot Vision Development. 15(2), 187–196.
7. Johnsen, S., & Tews, A. (2009, May). Real-time object tracking and classification using a static camera. In Proceedings of IEEE International Conference on Robotics and Automation, Workshop on People Detection and Tracking.
8. Kragic, D., Miller, A. T., & Allen, P. K. (2001). Real-Time Tracking Meets Online Grasp Planning. In Robotics and Automation, 2001. Proceedings 2001 ICRA. IEEE International Conference on (Vol. 3, pp. 2460–2465). IEEE.
9. Patricio, M. A., Dotu, I., García, J., Berlanga, A., & Molina, J. M. (2009). Discrete optimization algorithms in real-time visual tracking. Applied Artificial Intelligence, 23(9), 805–827.
10. Ray, K. S., Dutta, S., & Chakraborty, A. (2017). Detection, Recognition, and Tracking of Moving Objects from Real-time Video via SP Theory of Intelligence and Species Inspired PSO. arXiv preprint arXiv:1704.07312.
11. Roth, D. E. (2010). Real-time multi-object tracking. Konstanz: Hartung-Gorre.
12. Sobh, T. M., & International Conference on Systems, Computing Sciences and Software Engineering. (2008). Advances In Computer And Information Sciences And Engineering. New York? Springer.

13. Yilmaz, A., Javed, O., & Shah, M. (2006). Object tracking: A survey. Acm Computing Surveys (CSUR), 38(4), 13.
14. Yang, C., Duraiswami, R, Elgammal, A. & Davis, L. (2004). Real-Time Kernel-Based Tracking in Joint Feature-Spatial Spaces. Technical Report CS-TR-4567, Dept. of Computer Science, University of Maryland, College Park.
15. Zhao, P., Zhang, R., & Shibata, T. (2012). Real-Time Object Tracking Algorithm Employing On-Line Support Vector Machine And Multiple Candidate Regenerations. In Artificial Intelligence and Soft Computing (pp. 617–625). Springer Berlin/Heidelberg.

Handover Between Wi-Fi and WiMAX Technologies Using GRE Tunnel

Aroof Aimen, Saalim Hamid, Suhail Ahmad, Mohammad Ahsan Chisti, Surinder Singh Khurana and Amandeep Kaur

Abstract The next era of wireless grid inclines to be heterogeneous in the composition, i.e., wireless technologies like Wi-Fi and WiMAX networks desire co-breathe, so there is a demand for the best utilization of the accessible mixed chains. This paper considers the issue of handover between Wi-Fi and WiMAX grids with seamless connectivity. For this, first, a mobile terminal that abuts both IEEE 802.11 and IEEE 802.16 technologies was designed in the simulator. The developed mobile node was then introduced in the simulation scenario to study the various metrics. Second, we present the incorporation of GRE tunnel between the home agent and base stations for doing away with latency and packet drop and thereby improving the MOS value of the interest of consumers, giving impetus to efficiency day instant and day forth.

Keywords Wi-Fi · WiMAX · Heterogeneous network · GRE Handover · Packet loss · Packet

---

A. Aimen (✉) · S. S. Khurana · A. Kaur
Central University of Punjab, Bathinda, India
e-mail: aimenaroof@gmail.com

S. S. Khurana
e-mail: surinder.seeker@gmail.com

A. Kaur
e-mail: aman_k2007@hotmail.com

S. Hamid · S. Ahmad
University of Kashmir, Srinagar, Jammu and Kashmir, India
e-mail: saalimhamid343@gmail.com

S. Ahmad
e-mail: mir.suhail@uok.edu.in

M. A. Chisti
National Institute of Technology, Srinagar, Jammu and Kashmir, India
e-mail: ahsan@nitsri.net

C. R. Panigrahi et al. (eds.), *Progress in Advanced Computing and Intelligent Engineering*, Advances in Intelligent Systems and Computing 714,
https://doi.org/10.1007/978-981-13-0224-4_43

1 Introduction

Along the inroads in wireless technology, Wi-Fi [1–4] and WiMAX [5, 6] will coexist. The Wi-Fi has unlicensed array [7] for its order plus WiMAX shares licensed range. If a user supports both technologies, i.e., the user is served by unrestrained of expense Wi-Fi and afterward, due to the drastic decline of the incoming signal intensity [2] the ambulant station resolution change against the WiMAX, and hence a trade-off between SOS (strength of the signal) and expense. The handover (intra- and inter-technology [7, 8] handover) is required to pedestal both Wi-Fi and WiMAX which results in the continuity of service. Handover is a procedure of changing current call/session from one cell to other [9]. Horizontal Handover (HHO) [10] is used in homogeneous networks and is a comparatively simple mechanism. It is better from the user's viewpoint because it tends to be imperceptible. However, for Hybrid Networks complexity tends to be higher due to the implementation of Vertical Handover (VHO) [11] mechanisms. A serious consideration for this is that moving from one technology to the other is moving from one standard to another with certainly a different set of protocols. So this handover seems to be difficult and time consuming, causing the active sessions to break down and as a consequence user likes to continue with the current network, resulting in restricted mobility of the user. Hence, Vertical Handover is indispensable for integrated grids. The mobile terminal (MT) cannot admit packets amid the handover process until it entirely associates with the new network [12]. This handover process leads to a disruption in the latest meeting and is irritating to users [13]. To ensure a seamless handoff [14], the interruption duration (i.e., time required by total transitions of handover to be integral) should be less than the time taken by the movable depot to license its latest Access point's (AP'S) neighborhood [2].

This paper is arranged through a series of systematic investigations given as Compilation of seamless handover is presented in Sect. 1. Related work, double-layered node model, and GRE (Generic Routing Encapsulation) Tunnel are discussed in Sects. 2, 3, and 4 respectively. Section 5 describes the simulation scenarios used to assess the performance of the proposed model; with and without GRE tunnel. After running the simulation, we collect several statistics over the designed model which generates the voice traffic. Section 6 discusses the tracing of the graphs and their comparative analysis. Section 7 concludes the successful mechanism and basis for future work on this investigation.

2 Related Work

Pontes et al. [8] illustrated the wear of MIH (Media-Independent Handover) anatomy for integrating Wi-Fi and WiMAX structures to give seamless handover, through Backhaul and Dual-Mode Client Scenarios. However, it lacks the concept of multi-hop MIH. The authors in [15] have improved MIH solution by introducing a fuzzy inference engine wherein various inputs like signal intensity, distance from AP, etc., are assigned to sets. These sets are subject to the fuzzy rule base later. Gawk tools are used to analyze data, and the simulations show reduced handover packet loss and delay. However, it lacks the implementation of multi-network environment parameters. In [16], a fuzzy-based network selection model has been proposed wherein the multi-threshold mechanism makes the handover choice. This multi-criteria procedure has proven successful, but the multiple interfaced hardware node model is absent. The vertical handover algorithm proposed in [17] declares that it flaunts lower converging than established mechanisms to meet behaviors like MIH as the proposed algorithm processes on an adaption layer over the MAC layer. Also, the authors have converted the WiMAX signal into Wi-Fi signal to increase the range of the device. The end device is only Wi-Fi compatible, thereby lacking heterogeneity. Edward and Sumathy [18] analyzed various protocols for achieving seamless handover between Wi-Fi and WiMAX at different layers concluding that SIP Bicast (Session Initiation Protocol) plus MIH confirm to be the elite answers to give fast, seamless upright handoff. The performance evaluated is based on the assumption that mobile terminal supports both Wi-Fi and WiMAX but the design of such end terminal is absent. The authors in [10] reviewed various answers for handover based on parameters like handover delay, dropped packets, repetition of events, scalability, etc. To reduce the re-setup delay in WiMAX/LTE during IMS session, the author in [19] has proposed a scheme called SIP prior handover with a cross-layer design. The proposed system improved the exchange of SIP messages to 1% in contrast to the MIP. Shi et al. [7] provided a less latency handover arrangement. As per the authors, in homogeneous networks, the MT can discover its movement and estimate the handoff that diminishes the direct expense amid the MN (Mobile Node) besides its AP. In heterogeneous systems, they have used velocity as an essential metric to cause handoff. Naeem and Nyamapfene [20] proposed a decision-making algorithm for seamless handover amid Wi-Fi hot spots besides a spread WiMAX web. According to this algorithm, the MT switches from Wi-Fi to WiMAX chain when signal strength sinks inferior to the predefined satisfactory point. But for an optimal solution, seamless handovers need to be augmented with network layer information. The authors in [21] have carried out VHO between Wi-Fi and WiMAX using the NS2 simulator. It is, however, observed that the ratio of packet loss increases with the increase in the speed of MN. Similarly, the authors in [22] have summarized on how radio interfaces are selected in heterogeneous wireless networks. Wang et al. [23] used dual property that permits to receive data at the handover time. From their projected formula, the output decreased as the latency increased. Besides, the way to expeditiously operate with twin connectivity opens a variety of analysis queries. In [24], the authors have extended routing protocols for lower networks (RPL) in IoT (Internet of Things) architecture wherein handover mechanism is incorporated.

This paper aims at performing seamless VHO between Wi-Fi and WiMAX using double interfaced node. We used OPNET (Optimized Network Engineering Tool) [25] to design such node. Also, GRE tunnel [26] is used to reduce factors that decrease QoS (Quality of Service) during handover.

3 Double-Layered Node Model

Wi-Fi besides to WiMAX appear to be splendid partners to liberate favorable, inexpensive portable broadband internet utility's [1]. So there is a need for a wireless scheme that desires multiple interfaces [8]. In this paper, a double-interfaced movable station supporting Wi-Fi plus WiMAX broadcast interfaces has been designed and used in OPNET. Figure 1 shows the design of the mobile node (Double-layered [1] node). The Double-layered module composes of physical plus MAC layers for both WiMAX and Wi-Fi. The data link layer (DLL) together with the physical layer (PHY) of the OSI stack for the MN's of the twin standards can be combined keeping the upper layers alike. Figure 2 outlines the steps carried out by a double-layered node model while it comes back and forth to Wi-Fi and WiMAX base stations (BS). The MAC layers of Wi-Fi and WiMAX node take the HHO (Layer 2 handover) decision, i.e., in the module denoted as wlan_mac and wimax_mac (Fig. 1). While as the VHO (Layer 3 handover) is taken from the IP layer. The decision undertaking layer verifies the conditions described in the above algorithm and sends an interrupt to the upper layers. If the handover is

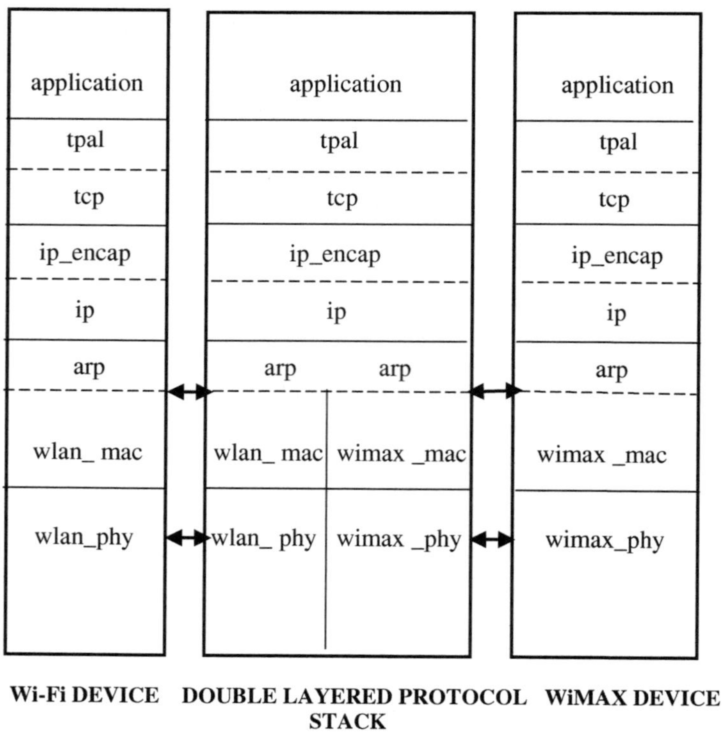

Fig. 1 Double-layered protocol stack with the Wi-Fi and WiMAX components

toward the Wi-Fi network (i.e., if SOS1 > Threshold) the voice application on the Wi-Fi node starts receiving data and voice application on WiMAX port ceases data generation. The operations are performed in reverse order if the algorithm is toward WiMAX network.

The primary requirement for simulating the vertical handover in OPNET is the design of a double-layered node. We achieve this aim by combining protocol stacks of these two different technologies in OPNET.

4 GRE Tunnel

Tunneling is a mechanism which involves encapsulation and decapsulation of the packets such that the path traversed by the packet does not depend on the address of the destination, but is prespecified [27]. The GRE tunnel is such a virtual point-to-point data passage that allows encapsulation of packets of one protocol in the body of another protocol [28]. Thus, an entirely new packet forms after stamping IP packets with GRE header. The GRE is used when packets, demand to be sent from one network to alternative, without being processed like IP packets by intervening routers, hence desirable for handover. We also choose the least hop path while drawing the GRE tunnel due to which processing occurs at the lesser number of intermediate routers. GRE also increases the efficiency and security of topology. The efficiency of handover is increased by tunneling the data from the home agent to foreign agents due to which minimum packet loss occurs when the end devices switch the network. Unlike IP-in-IP tunnel used in MIP solution, [29] in which data travels in unencrypted form, the data traverses in encrypted form in GRE thereby making eavesdropping difficult and enhancing security. GRE configuration does not include static access lists to traffic data rather the dynamic protocols like RIP, OSPF, etc., are used for network management. GRE also supports multicast. It uses the keep-alive mechanism in which the router keeps its port up even if the other side's router is unreachable. Further, GRE tunnel has not degraded the throughput due to the addition of new header because the maximum MTU (Maximum Transfer Unit) is greater than the fragment size, which has been manually set [28]. Figure 3 depicts the Signaling Scheme for vertical handover using GRE Tunnel.

1. WiFi_AP (Foreign Agent: FA _current) sends the beacon to MN which includes various QoS parameters provided by it.
2. MN compares the SOS of WiMAX_BS2 (FA_previous) and WiFi_AP to decide handover and sends an Association Request to WiFi_AP which in turn establishes a connection by sending Association Response.
3. In the meantime, the server sends data destined for the MN to HA (Home Agent). The MN is doubly connected while performing handover so it can receive the packets from HA through WiMAX_BS1 although it is associated with WiFi_AP too.

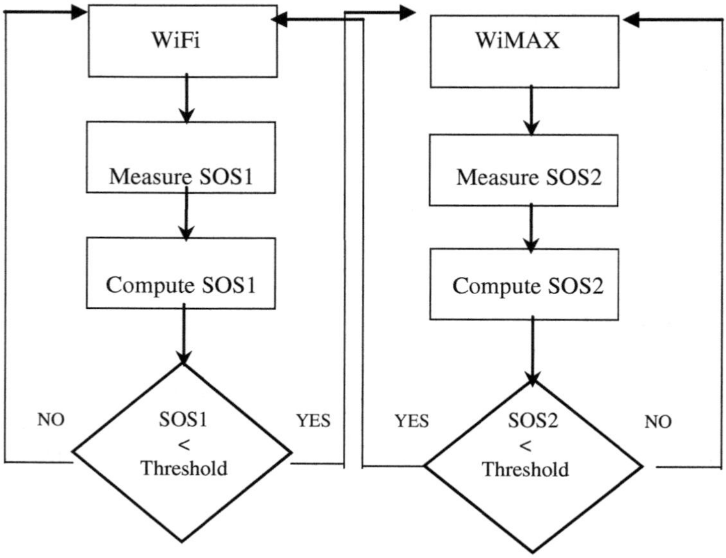

Fig. 2 Proposed flowchart for seamless vertical handover

4. It then informs the HA about its new COA (Care of Address) by sending it a Registration Message through WiFi_AP. HA replies WiFi_AP back with Registration Reply and forwards it to the MN. In the meantime, a tunnel establishes between WiFi_AP and HA. Further, the connection between WiMAX_BS2 and HA terminates.
5. Now all the data destined to MN are forwarded to WiFi_AP by HA which in turn sends it to the MN.

5 Simulation Scenario Setup

The OPNET simulation tool was used [25] to examine the performances of the proposed node model in a network during the handover. We evaluated performance parameters under the following scenarios:

5.1 Scenario 1

In Scenario 1 as shown in Fig. 4a, we deployed two WiMAX base stations (WIMAX_BS_1 as Home Agent) and one Wi-Fi Access point. A Point-to-Point

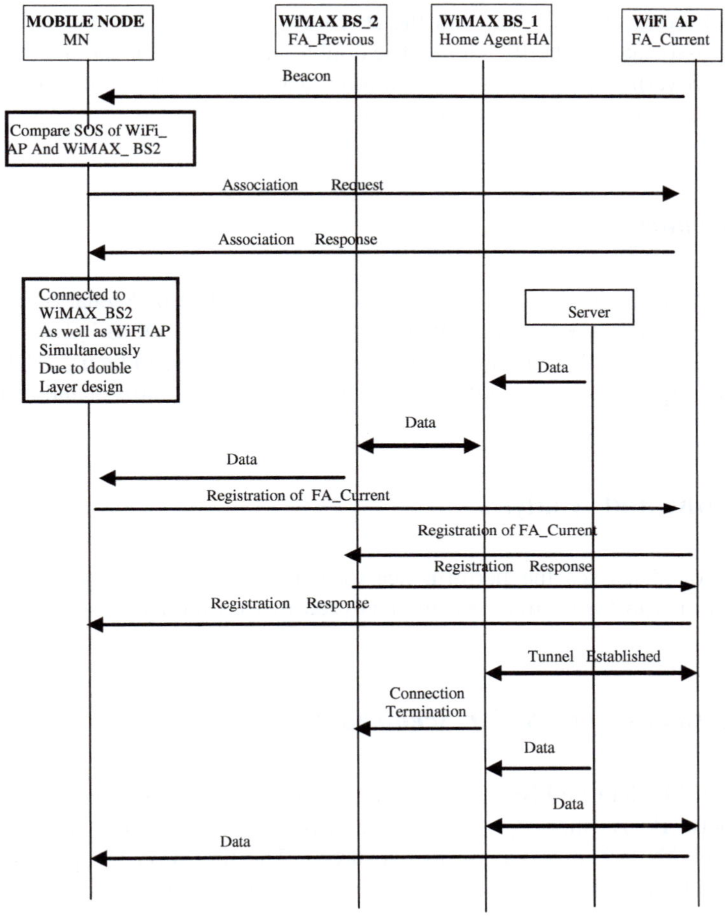

Fig. 3 Signaling scheme

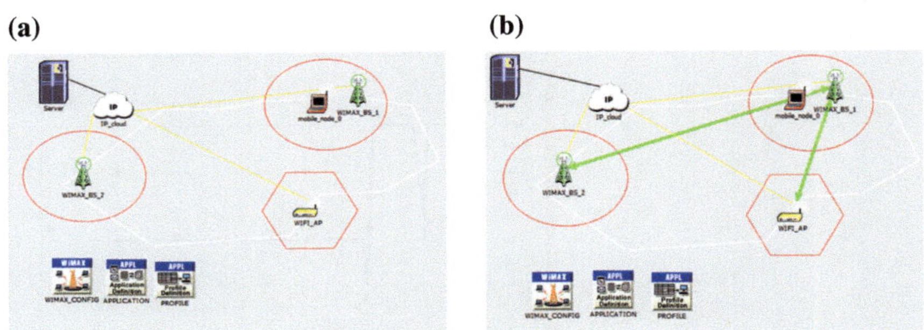

Fig. 4 **a** Scenario 1 for handover process, **b** Scenario 2 for handover process

(PPP) link connects each BS with the IP cloud. The server (corresponding node CN) also associates to the IP_ cloud. The double-layered node is moving at varying pace and covering different distances throughout the scenario giving rise to horizontal and vertical handovers respectively.

5.2 *Scenario 2*

Scenario 2 as shown in Fig. 4b, is a homogeneous replica of Scenario 1 but with GRE tunnels established between WiMAX_BS_1 and WiMAX_BS_2 and between WiMAX_BS_1 and Wi-Fi AP. The fundamental purpose of inducing GRE tunnel in the topology is to decrease the loss of user information during handover.

6 Results and Analysis

This section discusses the performance evaluation of deployed topologies. The Simulation is run for 60 min, and the analyzed metrics are addressed below.

6.1 *Results Obtained for Scenario 1*

Global Traffic Sent and Received: Figure 5 shows global traffic sent and received on the network. As mobile_node_0 moves from WiMAX_BS1 to WiMAX_BS2 then to WLAN AP, it gives rise to two drops. The first drop (50 packets/s lost) occurs due to the HHO while the second one (20 packets/s lost) takes place due to the VHO. More packets are lost during HHO than VHO because simple MAC layer takes HHO decision and the complex IP layer takes the VHO decision.

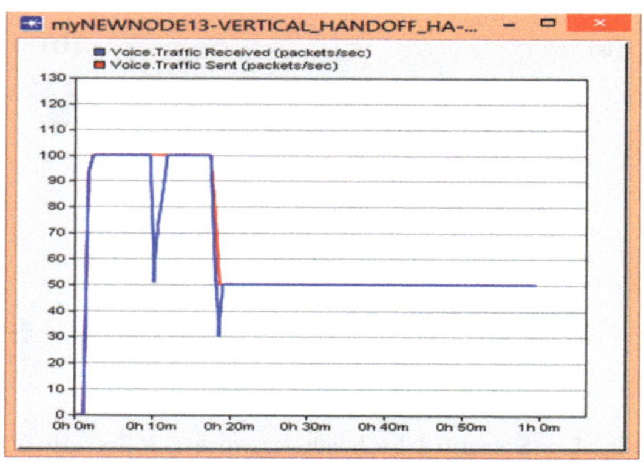

Fig. 5 Scenario 1: global voice traffic sent and received

The predefined trajectory compels the mobile node to progress through the different points and then returns to WiMAX_BS1.

6.2 *Overlapped Results Obtained for Scenarios 1 and 2*

In Scenario 2, the analysis of the same performance parameters is carried out again.

Voice traffic received by mobile_node_0: The voice traffic received by the mobile_node_0 (Fig. 6) is the average number of packets per second forwarded to all voice applications at the transport layer in the network. This figure shows that the packets dropped during the HHO were nearly 45(50-5) Packets/s in Scenario 1. However, by using GRE tunnel, the packet drop reduces to 23(50-27) packets/s. Similarly, for the VHO, the number of packets dropped were initially 22 Packets/s which reduces to 14 packets/s in Scenario 2 indicating that packet drop was impressively decreased using GRE tunnels.

MOS value for mobile_node_0: MOS (Mean Opinion Score) is a mathematical formula of expressing the perceived class of the media received and is expressed in names of digits, ranging from 1 to 5, 1 living in the inferior and 5 is the ultimate [30]. The MOS value of mobile_node_0 in both scenarios is compared in Fig. 7. Initially, in Scenario 1, the topology started with the MOS value of 3.5 but at 10 m 0 s due to HHO the MOS value reduced to 3.3; which further degraded to 2.9 at 19 m 0 s owing to VHO. It then started to escalate and acquired a permanent value of almost 3. However, in the case of Scenario 2, the very appraise of 3.2 is obtained which tends to be superior. Since the number of packets dropped is inversely proportional to MOS value, so fewer packets dropped implies higher MOS value and vice versa.

Packet End-to-End delay: Packet End-to-End Delay refers to the duration taken to deliver a packet from source to sink across the network [31]. Figure 8 compares the

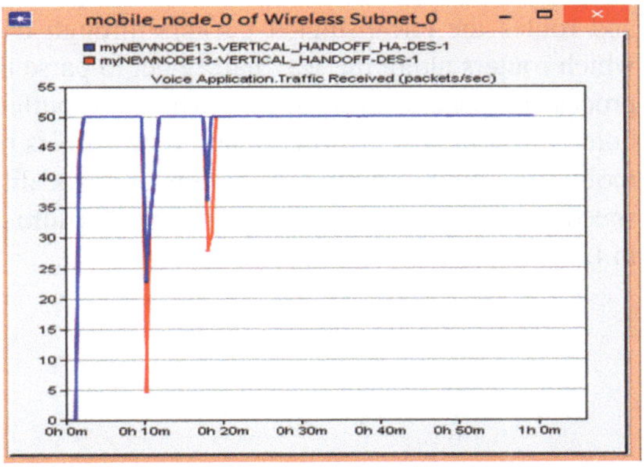

Fig. 6 Overlapped voice traffic sent and received by mobile_node_0

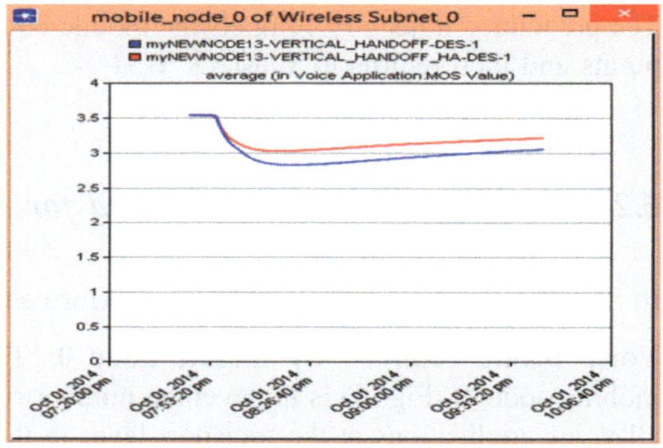

Fig. 7 Overlapped MOS value of mobile_node_0

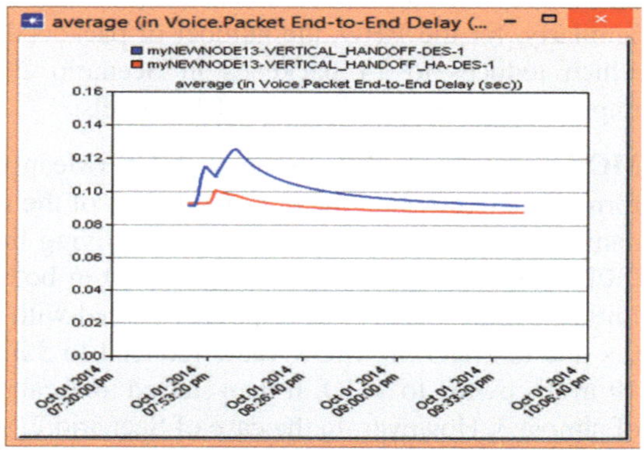

Fig. 8 Overlapped global packet end-to-end delay value in both scenarios

Global Packet End-to-End values of both the scenarios. It is evident from the graph that due to the GRE tunnel Packet End-to-End delay value decreases as GRE routes packets quickly. In Scenario 1, the delay finally acquires the value of 0.95 s, but in the second one, it reduced to approximately 0.94 s.

We may conclude from the results that GRE-implemented scenario resulted in fewer packet drops and the end-to-end delay because GRE encapsulates the payload and routes the encapsulated packets through desired IP networks. As a result of which routers along the way hardly get to parse the whole payload leading to lower processing pace at every router. Thus, the buffered packets swiftly route from the home agent to new BS such that the new BS is ready to transmit the packets to the mobile station as readily as it befalls in its locality. Introducing GRE tunnel simply speeds up the conventional handover procedure and thereby minimizing delay and in turn increasing the MOS value.

7 Conclusion and Future Work

This paper examined seamless handover between two different well liked technologies, viz., Wi-Fi and WiMAX. We incorporated and implemented Wi-Fi and WiMAX technologies in the double-layered node model to perform seamless vertical handover. We also examined the impact of GRE Tunnels to obtain a low latency handover arrangement as a result of which the user can move from a network of one standard to another standard while continuing the session. The simulation results depict that the implementation of GRE tunnels in such heterogeneous environment improves the performance and thereby reducing packet drops. The maximum delay obtained was 0.13 s which was outperformed by the GRE-implemented scenario to 0.10 s. The simulations carried also revealed that the packet drop considerably reduced to 23 packets/s against 45 packets in Scenario 1.

As in this paper path loss was selected to be "Free Space", consequently, transmitter beside recipient own a clear line of sight among them, i.e., no additional source of impairment. This path loss is hardly practical. Therefore, in future more realistic path loss models need to be assessed. The future work further involves proposing a better and an efficient handover scheme between Wi-Fi, WiMAX, and other technologies such as LTE, UMTS, and ZigBee, etc. that can significantly decrease the two most important parameters in handover, i.e., packet loss and delay.

References

1. Paul, T., Ogunfrunmiri, T.: Wireless LAN comes of age: Understanding the IEEE 802.11 n amendment. IEEE Circuits and Systems Magazine, 8(1) (2008) 28–54
2. 802.11n Next-Generation Wireless LAN Technology. White Paper, Broadcom Corporation (2006)
3. Watson, R., Huang, D.: Understanding the IEEE 802.11 ac Wi-Fi standard. Preparing for the next gen of WLAN (2012)
4. Schelstraete, S.: An Introduction to 802.11 ac. Quantenna Communications (2011)
5. Ohrtman, F.: WiMAX handbook: Building 802.16 networks. McGraw Hill Professional (2005)
6. Andrews, J. G., Ghosh, A., Muhamed, R. Fundamentals of WIMAX. Prentice Hall publication (2007)
7. Shi, F., Li, K., Shen, Y.: Seamless handoff in Wi-Fi and WiMAX heterogeneous networks. Future Generation Computer Systems, 26(8) (2010) 1403–1408
8. Pontes, A. B., dos Passos Silva, D., Jailton, J., Rodrigues, O., Dias, K. L.: Handover management in integrated WLAN and mobile WiMAX networks. IEEE Wireless Communications, 15(5) (2008)
9. Stojmenovic, I.: Handbook of wireless networks and mobile computing (Vol. 27). Wiley (2003)
10. Damhuis, J. R.: Seamless handover within WiMAX and between WiMAX and WLAN. In Proceedings of 8th Twenty Student Conference on IT, Enschede, Netherlands (2008)
11. Saini, A., Bhalla, P.: A Review: Vertical Handover between Wi-Fi and WiMAX. International Journal of Advanced Research in Computer Science and Software Engineering, 3(6) (2013)

12. Singh, N. P., Singh, B.: Proxy Mobile IPv6-Based Handovers for VoIP Services in Wireless Heterogeneous Networks. International Journal of Engineering and Technology, 4(5) (2012) 527–531
13. Kadhim, D. J., Abed, S. S.: Performance and handoff evaluation of heterogeneous wireless networks (HWNs) using opnet simulator. International Journal of Electronics and Communication Engineering & Technology (IJECET), 4(2), (2013) 477–496
14. Fathi, H., Chakraborty, S. S., Prasad, R.: Voice over IP in wireless heterogeneous networks: Signaling, mobility and security. Springer Science & Business Media (2008)
15. Tu, J., Zhang, Y. J., Zhang, Z., Ye, Z. W., Chen, Z. L.: Performance analysis of vertical handoff in Wi-Fi and wimax heterogeneous networks. In Computer Network and Multimedia Technology, 2009. CNMT 2009. International Symposium on IEEE (2009) 1–5
16. Khan, M., Ahmad, A., Khalid, S., Ahmed, S. H., Jabbar, S., Ahmad, J.: Fuzzy based multi-criteria vertical handover decision modeling in heterogeneous wireless networks. Multimedia Tools and Applications (2017) 1–26
17. Saeed, R. A., Mohamad, H., Ali, B. M., Abbas, M.: WiFi/WiMAX heterogeneous seamless handover. In Broadband Communications, Information Technology & Biomedical Applications, 2008 Third International Conference on IEEE (2008) 169–174
18. Edward, E. P., Sumathy, V.: A survey of seamless vertical handoff schemes for Wi-Fi/ WiMAX heterogeneous networks. In Signal Processing and Communications (SPCOM), 2010 International Conference on IEEE (2010) 1–5
19. Edward, E. P.: A novel seamless handover scheme for WiMAX/LTE heterogeneous networks. Arabian Journal for Science and Engineering, 41(3) (2016) 1129–1143
20. Naeem, B., Nyamapfene, A.: Seamless vertical handover in WiFi and WiMAX networks using RSS and motion detection: An investigation. The Pacific Journal of Science and Technology, 12(1) (2011) 298–304
21. Bhosale, S. K., Daruwala, R. D.: Simulation of vertical handover between WiFi and WiMax and its performance analysis—An installation perspective. In India Conference (INDICON), 2011 Annual IEEE (2011) 1–4
22. Miyim, A. M., Ismail, M., Nordin, R.: Vertical handover solutions over LTE-advanced wireless networks: An overview. Wireless personal communications, 77(4) (2014) 3051–3079
23. Wang, H., Rosa, C., Pedersen, K. I.: Dual connectivity for LTE-advanced heterogeneous networks. Wireless Networks, 22(4) (2016) 1315–1328
24. Fotouhi, H., Moreira, D., Alves, M.: mRPL: Boosting mobility in the Internet of Things. Ad Hoc Networks, 26 (2015) 17–35
25. Riverbed Modeler, https://www.riverbed.com/in/products/steelcentral/opnet.html?redirect=opnet
26. Hanks, S., Meyer, D., Farinacci, D., Traina, P.: Generic routing encapsulation (GRE) (2000)
27. Perkins, C.: IP encapsulation within IP. (1996)
28. Point-to-Point, G. R. E. over IPSec Design Guide. Cisco System, San Jose, USA (2006)
29. Schiller, J. H.: Mobile communications. Pearson Education (2003)
30. Mean Opinion Score, https://en.wikipedia.org/wiki/Mean_opinion_score
31. Demichelis, C., Chimento, P.: RFC 3393. IP packet delay variation metric for IP performance metrics (IPPM) (2002)

Coplanar Waveguide UWB Bandpass Filter Using Defected Ground Structure and Interdigital Capacitor

Pratibha Verma, Tamasi Moyra, Dwipjoy Sarkar, Priyansha Bhowmik, Sarbani Sen and Dharmvir Kumar

Abstract In this present work, a coplanar waveguide (CPW) ultra-wideband (UWB) bandpass filter (BPF) is proposed using defected ground structure (DGS) and interdigital capacitor having the exact passband frequency range of (3.1–10.6) GHz, minimum passband insertion loss of 0.14 dB and stopband rejection level below 20 dB from 11.8 to 16 GHz. Lumped equivalent circuit model of CPW UWB BPF is also extracted. Mathematical modelling to calculate values of generated inductances, capacitances and resistance of proposed BPF is indicated. Circuit size is miniaturized to an area of 7 mm × 9.3 mm and BPF is simulated on Fr4 substrate with dielectric constant of 4.4 and thickness of 1.6 mm. The proposed BPF is applicable for Bluetooth and wireless devices.

Keywords CPW · UWB · BPF · DGS · Interdigital capacitor Stopband rejection · Insertion loss · Lumped equivalent circuit

---

P. Verma (✉) · T. Moyra · D. Sarkar · P. Bhowmik · S. Sen · D. Kumar
Department of Electronics and Communication Engineering, NIT Agartala, Jirania, India
e-mail: vermapratibha1007@gmail.com

T. Moyra
e-mail: tamasi_moyra@yahoo.co.in

D. Sarkar
e-mail: dwipjoysarkar@gmail.com

P. Bhowmik
e-mail: priyansha.bhowmik@gmail.com

S. Sen
e-mail: Sensarbani77@gmail.com

D. Kumar
e-mail: dharmvir151@gmail.com

C. R. Panigrahi et al. (eds.), *Progress in Advanced Computing and Intelligent Engineering*, Advances in Intelligent Systems and Computing 714,
https://doi.org/10.1007/978-981-13-0224-4_44

1 Introduction

Ultra-wideband (UWB) filter plays a very important role for hand-held systems that allows passing signal of frequency range (3.1–10.6) GHz and rejects the other frequency components [1]. CPW is used because of its various advantages over microstrip lines, strip lines, conductor-backed coplanar waveguide (CB-CPW), etc., such as the presence of only planar ground that provides very low insertion loss because of the minimum effect of parasitic capacitance [2]. In both CPW and CB-CPW, characteristic impedance and dielectric constant can be easily adjusted just by adjusting the ratio of the width of the signal plane and the gap between the signal plane and the ground plane, i.e. changing ratio of the thickness of the signal plane and the thickness of the dielectric substrate is not needed such as in microstrip line. Ease of fabrication is more in CPW because of its planar structure. In [3], a simple designed short-circuited CPW multiple mode resonators for UWB BPF is depicted but having a passband insertion loss of 1.5 dB and not having exact passband range as defined by FCC for UWB BPF. CPW-feeder BPF is proposed in [4] with exact passband range (3.1–10.6) GHz having an improved insertion loss of 0.5 dB. Still, stopband rejection level below 20 dB is only up to 13 GHz. Thereafter, using DGS, spurious rejection up to 16 GHz below 20 dB is shown in [5], but it has a disadvantage of the high value of insertion loss of 0.9 dB and structure on both sides of the substrate having via which can create complexity during fabrication. In [6], UWB BPF with a combination of single-ring resonator(SRR) and DGS having good selectivity and insertion loss less than 0.16 dB are shown. But the effect of DGS to suppress harmonic is not effective and also complexity in fabrication is not removed. In [7], stopband rejection level is below 25 dB up to 20 GHz using hybrid microstrip DGS. Still, selectivity and insertion loss are not satisfactory. Selectivity and insertion loss are highly improved in [8] based on microstrip to CPW transition. However, the complexity of fabrication is not eliminated. The proposed UWB BPF is made from the concepts of cascading of low-pass filter (LPF) and high-pass filter (HPF). Interdigital capacitor for HPF and DGS section for LPF as well as miniaturization of the circuit due to slow-wave effect are adopted from [9] and [10] respectively. Moreover, to eliminate spurious harmonics and to improve the sharpness of the proposed UWB BPF of present work, the concept of defected ground structure (DGS) is also taken from [11–14]. The respective mathematical equation to extract the equivalent capacitive value of the interdigital capacitor, inductance value of any transmission line, the parasitic capacitance value of CPW line and the equivalent circuit of DGS section are taken from [2, 15, 16, and 17].

In this present work, UWB BPF having passband range (3.1–10.6) GHz, least insertion loss of only 0.14 dB due to open stub section, stopband rejection level below 20 dB from 11.8 to 16 GHz and least circuit area of only 9.3 mm * 7 mm compared to the work stated above is composed of cascading interdigital capacitor and DGS. The problem of complexity in fabrication is also removed because of the presence of whole structure only in the same plane. Two unit DGS cells are incorporated to get stopband rejection level lower than 20 dB. Lumped equivalent circuit model is extracted from mathematical equations studied from above-stated literature. Input and output ports have an impedance of 50 Ω and the calculated value of a characteristic impedance is 46.91 Ω. Thus, there is less chance of distortion of signal due approximation in the matching of input impedance with a characteristic impedance.

2 Modelling of the Proposed CPW UWB BPF

The proposed BPF is simulated using IE3D EM simulator in FR4 substrate ($\varepsilon_r =$ 4.4, $h =$ 1.6 mm & $\tan\theta =$ 0.002).

2.1 *Two-Unit DGS Cell*

The phase velocity, $\upsilon_p = \frac{1}{\sqrt{L_0 C_0}}$ of a lossless transmission line, is dependent upon the inductance (L_0) and the capacitance (C_0) per unit length of the transmission line, where ω is the operating angular frequency. As L_0 and C_0 are increased, the phase velocity is minimized and slow-wave effects are achieved, which can be realized by etching symmetrical patterns on both sides of the ground plane along the direction of propagation. Transmission line wavelength is reduced compared to normal wavelength according to the higher value of slow-wave factor which causes compactness in size of the proposed filter. In this work, the concept of a DGS cell is taken from [10]. The current distribution is changed by the etched patterns leading to variation in distributed capacitance and inductance and low-pass performance is obtained. The structure of cascaded two unit DGS cell of different dimensions and its response are displayed in Fig. 1. More stopband rejection compared to single DGS cell is obtained through two unit cell structures. Open stub section is created to reduce the passband insertion loss. The optimized LPF (Fig. 1b) has 3 dB cut-off frequency of 10.9 GHz, the insertion loss in the passband region is from 0.09 to 0.7 dB and the rejection is better than 20 dB from 12.3 to 16.6 GHz.

2.2 *Interdigital Capacitor Used in the Proposed UWB BPF*

Desired capacitance at the design frequency in a reasonable area is easily obtained by IDC with the response of high-pass characteristic (Fig. 2). Capacitance is raised

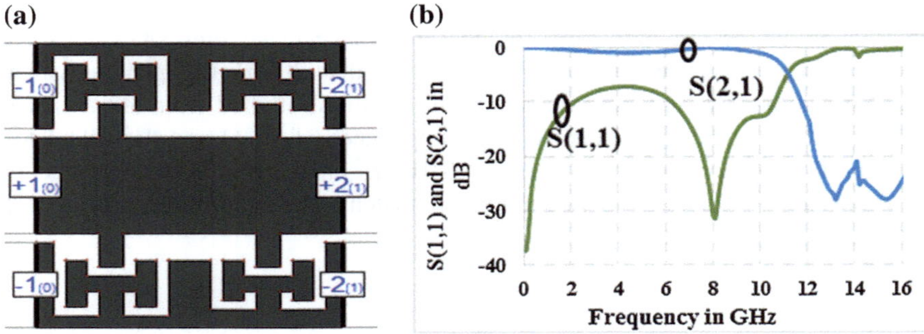

Fig. 1 **a** CPW two unit DGS cell. **b** S-parameter response of two unit DGS cell

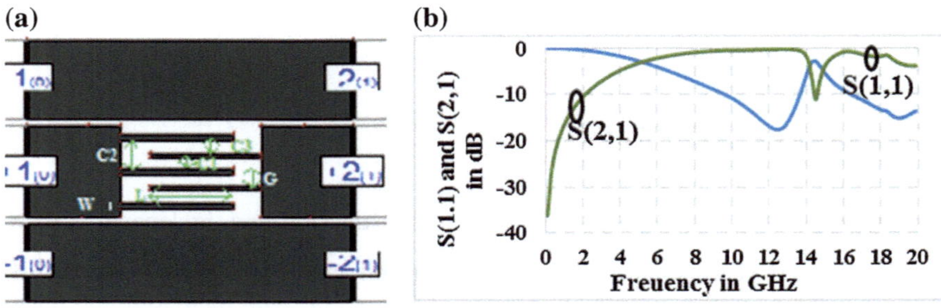

Fig. 2 **a** Interdigital capacitor. **b** S-parameter response of interdigital capacitor

with enhancement of width or length of fingers as because more area to store charge and also decrease in characteristic impedance leads to higher effective capacitance. The capacitance increases as the gaps are decreased because of more coupling between the input and output port across the gap in the IDC structure. High-pass characteristic is noticed up to 12.9 GHz in Fig. 2b.

2.3 Filter Design

In Fig. 3, the layout of the proposed CPW UWB BPF using the DGS and IDC is depicted. The dimensions are tabulated in Table 1.

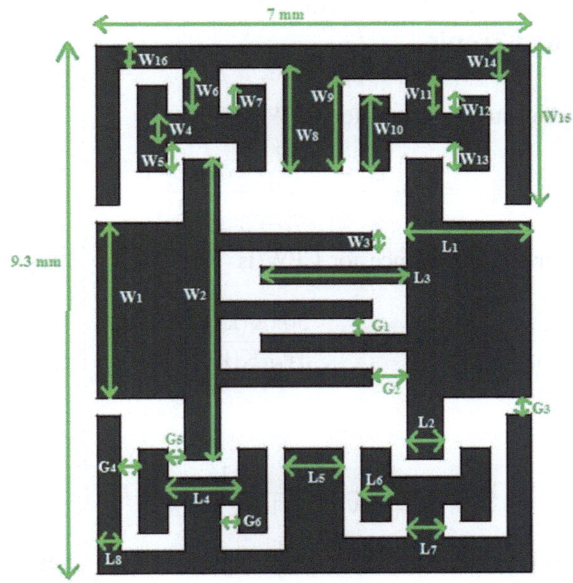

Fig. 3 Layout of the proposed CPW UWB BPF

Table 1 Dimension of the proposed UWB BPF

Length and gap	Value of length and gap in mm	Width	Value of width in mm
L1	2	W1	3.1
L2	0.6	W2	5.3
L3	2.6	W3, W12	0.3
L4	1.2	W4, W5, W7, W13	0.5
L5	1	W6	0.8
L6	0.5	W8	1.8
L7	0.6	W9	1.6
L8	0.4	W10	1.3
G1, G3, G4, G5, G6	0.3	W11, W14	0.6
G2	0.6	W15	2.8
		W16	0.4

3 Mathematical Modelling for Extraction of Lumped Equivalent Circuit

Since effective dielectric constant for CPW is

$$\varepsilon_{eff} = 1 + \frac{\varepsilon_r - 1}{2} = 1 + \frac{4.4 - 1}{2} = 2.7 \tag{3.1}$$

Since characteristic impedance for CPW is

$$Z_0 = \frac{30\pi K(k_0')}{\sqrt{\varepsilon_{eff}} K(k_0)} \tag{3.2}$$

where

$$k_0 = \frac{S}{S + 2W} = \frac{3.1}{3.1 + 2 \times 0.3} = 0.84 \tag{3.3}$$

Equation (3.3) is the dimension of filter design, the width of the 50 Ω input port, $S = W_1 = 3.1$ mm and gap between the input port and upper ground plane, $W = G_3 = 0.3$ mm.

K(k) denotes the complete integral of the first kind.

$$K(k_0) = \frac{\frac{\pi}{2}}{\mathrm{agm}\left(1, \sqrt{1 - k_0^2}\right)} = 2.085 \tag{3.4}$$

and

$$K(k_0') = \frac{\frac{\pi}{2}}{\mathrm{agm}\left(1, \sqrt{1 - k_0'^2}\right)} = 1.706 \tag{3.5}$$

where agm is the arithmetic geometric mean.

Hence, the characteristic impedance is

$$Z_0 = \frac{30\pi K(k_0')}{\sqrt{\varepsilon_{eff}} K(k_0)} = \frac{30\pi \times 1.706}{\sqrt{2.7} \times 2.085} = 46.91 \tag{3.6}$$

To calculate the capacitance of interdigital capacitor

The effective capacitance can be contributed by a combination of three capacitive values as shown in Fig. 2a.

The capacitance of the structure (Fig. 2a) can be estimated using the equations stated below.

The capacitance between inside consecutive fingers of the IDC,

$$C_1 = \varepsilon_0 \varepsilon_{eff} \frac{K(k_{01}^{'})}{K(k_{01})} L \tag{3.7}$$

where

$$k_{01} = \sqrt{1 - \left(\frac{W}{W + G}\right)^2} = 0.866 \tag{3.8}$$

and

$$k_{01}^{'} = \sqrt{1 - 0.866^2} = 0.5 \tag{3.9}$$

Above Eqs. (3.7) and (3.8) are for W = 0.15 mm and G = 0.15 mm

$$K\left(k_{01}^{'}\right) = \frac{\frac{\pi}{2}}{\text{agm}(1,\ 0.866)} = 1.688 \tag{3.10}$$

and

$$K(k_{01}) = \frac{\frac{\pi}{2}}{\text{agm}(1,\ 0.5)} = 2.16 \tag{3.11}$$

$$\therefore C_1 = 8.825 \times 10^{-12} \times 2.7 \times \frac{1.688}{2.16} \times 1.9 \times 10^{-3} = 3.538 \times 10^{-14}\,\text{F} = 0.0354\,\text{pF} \tag{3.12}$$

The capacitance between alternative fingers of IDC

$$C_2 = 2\varepsilon_0 \varepsilon_{eff} \frac{K(k_{01})}{K(k_{01}^{'})} L_{ext} = 0.04298\ \text{pF} \tag{3.13}$$

Here, the length of the finger except L is $L_{ext} = 0.7\,\text{mm}$
The capacitance between consecutive inside and outside fingers of IDC,

$$C_3 = 4\varepsilon_0 \varepsilon_{eff} \frac{K(k_{02}^{'})}{K(k_{02})} L, \tag{3.14}$$

where

$$k_{02} = \sqrt{\frac{G}{2(2W + G)}} = 0.408 \tag{3.15}$$

Above Eq. (3.15) is for W = 0.15 mm and G = 0.15 mm

$$K(k_{02}) = \frac{\frac{\pi}{2}}{\text{agm}\left(1, \sqrt{1 - k_{02}^2}\right)} = 1.642 \quad (3.16)$$

and

$$K(k_{02}^{'}) = \frac{\frac{\pi}{2}}{\text{agm}(1, k_{02})} \simeq 2.34 \quad (3.17)$$

$$\therefore C_3 = 4 \times 8.825 \times 10^{-12} \times 2.7 \times \frac{2.34}{1.62} \times 1.9 \times 10^{-3} = 0.258\,\text{pF} \quad (3.18)$$

Therefore, effective value of capacitance of IDC structure is

$$C_{IDC} = (n - 3)C_1 + (n - 1)C_2 + 2C_3 = (5 - 3) \times 0.034 + (5 - 1) \times 0.04298 + 2 \times 0.258 \simeq 0.756\,\text{pF} \quad (3.19)$$

The response of BPF has a resonance frequency at $f_0 = 4.3$ GHz in the passband region.

Inductance created due to coupling between fingers of IDC is

$$L_{IDC} = \frac{1}{4 \times 3.14^2 \times f_0^2 \times C_{IDC}} = \frac{1}{4 \times 3.14^2 \times 4.3^2 \times 10^{18} \times 0.756 \times 10^{-12}} = 1.85\,\text{nH} \quad (3.20)$$

To calculate the inductance, capacitance and resistance of LPF part

The LPF is produced due to the presence of DGS section.

The equivalent frequency of the parallel circuit present in DGS is

For the LPF part, cut-off frequency, $f_c = 10.6$ GHz

For the LPF part, resonant frequency, $f_0 = 12.5$ GHz

$$C_{DGS} = \frac{f_c}{4\pi Z_0(f_0^2 - f_c^2)} = 0.4099\,\text{pF} \quad \text{and} \quad L_{DGS} = \frac{1}{4\pi^2 f_0^2 C_{DGS}} = 0.395\,\text{nH} \quad (3.21)$$

$$R_{DGS} = \frac{2Z_0}{\sqrt{\frac{1}{|S_{11}(\omega_0)|^2} - \left(2Z_0\left(\omega_0 C_{DGS} - \frac{1}{\omega_0 L_{DGS}}\right)\right)^2 - 1}} = 350.468\,\Omega \quad (3.22)$$

To calculate the parasitic capacitance of CPW

The parasitic capacitance due to lower dielectric layer is

$$C_{pd} = 2\varepsilon_0(\varepsilon_{r1} - 1)\frac{K(k_1)}{K(k_1')} \times L = 0.06172\ \text{pF}, \quad \text{For}\, L = 1\ \text{mm} = 10^{-3}\text{m} \quad (3.23)$$

Parasitic capacitance due to air

$$C_{air} = 4\varepsilon_0 \frac{K(k_0)}{K(k_0')} \times L\ \text{Farad} = 0.04314\,\text{pF} \quad (3.24)$$

Total parasitic capacitance

$$C_P = C_{pd} + C_{air} = 0.06172 + 0.04314\,\text{pF} = 0.10486\,\text{pF} \quad (3.25)$$

To calculate the capacitance and inductance due to open stub section

Since the area of open stub section is $A = 0.6 \times 10^{-3} \times 1.1 \times 10^{-3}\,\text{m}^2$, the distance between the signal plane and the ground plane is $d = 0.3 \times 10^{-3}$ m and f_0 due to the open stub section is 15.9 GHz.

The capacitance and the inductance due to open stub section is

$$C_{stub} = \frac{\varepsilon_0 \varepsilon_{eff} A}{d} = 0.0524\,\text{pF} \quad (3.26)$$

and

$$L_{stub} = \frac{1}{4\pi^2 f_0^2 C_{stub}} = 1.9\ \text{nH} \quad (3.27)$$

To calculate the inductance of the 50 Ω transmission line, i.e. i/o port

$\ell = 3.1$ mm, $W = 2$ mm and $t = 0.004$ mm.

$$L_T = 5.08 \times 10^{-3}\ell\left[\ln\left(\frac{\ell}{W + t}\right) + 1.190 + 0.022\left(\frac{W + t}{\ell}\right)\right]\text{nH/mil} = 0.93\,\text{nH} \quad (3.28)$$

3.1 Lumped Equivalent Circuit of the Proposed BPF with its Response

In Fig. 4, the lumped equivalent circuit of the proposed UWB BPF is pictured according to above mathematical modelling and simulated in 'RF sim 99' simulator.

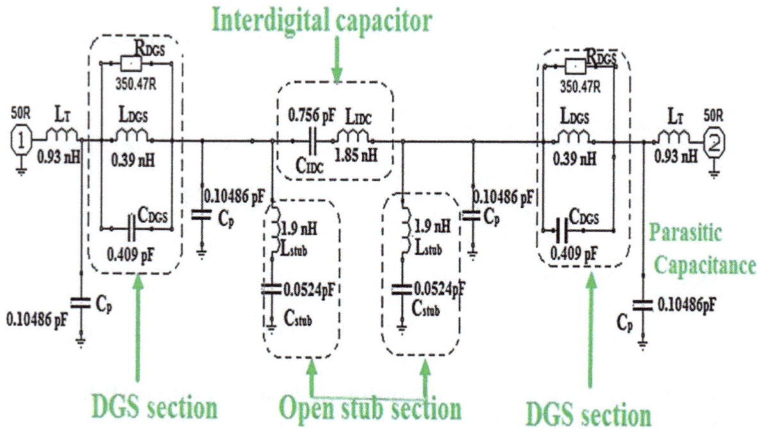

Fig. 4 Lumped equivalent circuit of the proposed UWB BPF

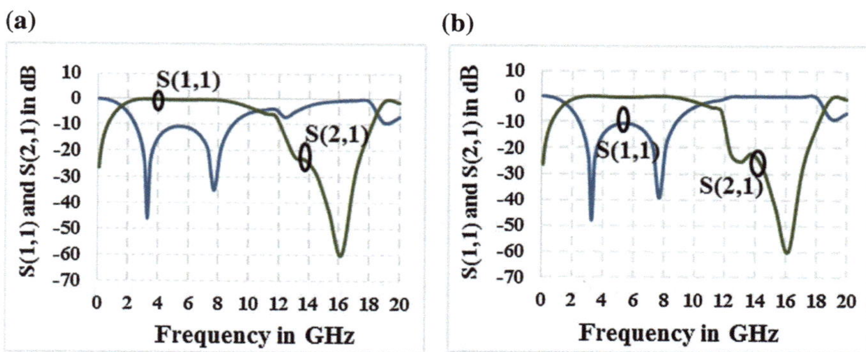

Fig. 5 **a** S-parameter response of lumped equivalent circuit considering resistance. **b** S-parameter response of lumped equivalent circuit without considering resistance

Its S-parameter response considering resistance and without considering resistance is depicted in Fig. 5. The insertion loss of Fig. 5a is 0.47 dB and of Fig. 5b is 0.3 dB with their respective passband range of (2–10.4) GHz and (1.7–10.6) GHz.

4 Results and Discussion

In Fig. 6, the S-parameter response of the simulated IE3D and the S-parameter response of the lumped equivalent circuit without considering the resistance is compared having their passband frequency ranges are (3.1–10.6) GHz and (1.7–10.6) GHz, respectively, with their corresponding resonance frequencies

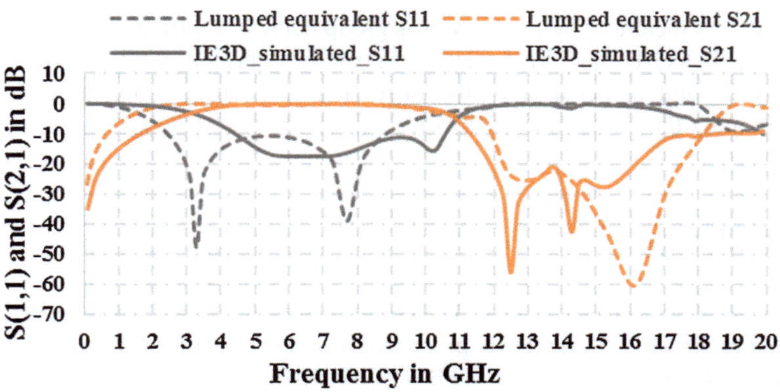

Fig. 6 Comparison between S-parameter of simulated IE3D and S-parameter response of lumped equivalent circuit without considering resistance

Table 2 Detailed performance properties of the proposed UWB BPF

Properties	Values	Properties	Values
Fractional bandwidth	130.89%	Group delay	0.15 ns
Passband insertion loss	0.14–0.7 dB	3 dB bandwidth	7.5 GHz
Return loss	17.4 dB	Roll-off factor	27.48 dB/GHz
Average rejection level	20 dB up to 16 GHz	30 dB Shape factor	1.61

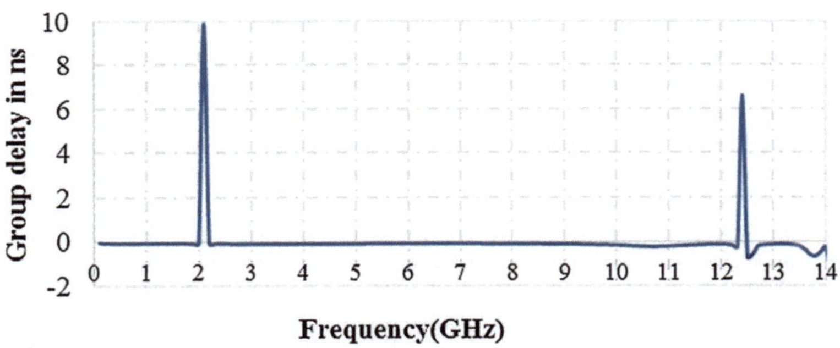

Fig. 7 Frequency versus group delay

(fc1 = 5.5 GHz, fc2 = 10.2 GHz) and (fc1 = 3.2 GHz, fc2 = 7.8 GHz). The wide harmonic suppression is achieved up to 16 and 17.4 GHz below 20 dB in the simulated IE3D and the lumped equivalent circuit respectively.

Their corresponding minimum passband insertion losses are 0.14 and 0.3 dB which is very little using CPW waveguide as compared to the other waveguides such as microstrip lines, stripline, CB-CPW, etc., for the same fr4 substrate. The performance properties of the proposed BPF are detailed in Table 2.

In Fig. 7, a picture of frequency versus group delay is depicted, where spikes are noticed at 2.1 GHz and 12.1 GHz with group delay of 9.89 ns and 6.6 ns, respectively, and in between of the mentioned frequencies, the group delay is approximately flat with value of 0.15 ns.

5 Conclusion

In this work, a compact 5.73 GHz centre frequency of the CPW UWB BPF passband range (3.1–10.6) GHz is proposed, which is based on IDC and DGS. The simulated BPF provides 130.89% FBW, 0.14 dB passband insertion loss, group delay 0.15 ns for the entire passband and the average 20 dB spurious rejection level up to 16 GHz. The insertion loss of the UWB BPF is reduced by creating the open stub section and using the CPW. The ease of fabrication is achieved by designing the total structure in the same plane. The calculated value of characteristic impedance of 46.91 Ω is approximately matched with the impedance of 50 Ω in the i/o ports that indicates the low insertion loss of the signal. The lumped equivalent circuit is also extracted from the mentioned mathematical equations above. The proposed BPF is suitable for the Bluetooth and wireless application devices.

References

1. J.-S. Hong, Microstrip Filters for RF/Microwave Applications, 2nd Edition, Wiley, 2011.
2. C. P. Wen, "Coplanar waveguide: a surface strip transmission line suitable for nonreciprocal gyromagnetic device applications," IEEE Trans. Microwave Theory Tech., Vol. MTT-17, No. 12, pp. 1087–1090, Dec. 1969.
3. Gao, Jing, et al. "Short-circuited CPW multiple-mode resonator for ultra-wideband (UWB) bandpass filter." IEEE Microwave and Wireless Components Letters 16.3 (2006): 104–106.
4. Yao, Chunhui, et al. "A New Type of Miniature Ultra-Wideband Band-Pass Filter with Coplanar Waveguide Fed." International Journal of Infrared and Millimeter Waves 28.10 (2007): 859–863.
5. Sekar, Vikram, and Kamran Entesari. "Miniaturized UWB bandpass filters with notch using slow-wave CPW multiple-mode resonators." IEEE Microwave and Wireless Components Letters 21.2 (2011): 80–82.
6. Fang, Jinyong, et al. "Compact printed ultra-wide band filter employing SRR-DGS." Ubiquitous Wireless Broadband (ICUWB), 2016 IEEE International Conference on. IEEE, 2016.
7. Kumar, Mukesh, and Suresh Kumar. "Different Methods of Designing Ultra Wideband Filters in Various Applications-A Review." International Journal of Wired and Wireless Communications Vol. 3, Issue 1, October, 2014.
8. Oh, Sangyeol, et al. "UWB bandpass filter with dual notched bands based on microstrip to CPW transition." Wireless and Microwave Technology Conference (WAMICON), 2015 IEEE 16th Annual. IEEE, 2015.

9. Aryanfar, Farshid, and Kamal Sarabandi. "Characterization of semilumped CPW elements for millimeter-wave filter design." IEEE transactions on microwave theory and techniques 53.4 (2005): 1288–1293.
10. Xu, Mingming, et al. "Design of a novel V-band coplanar waveguide low-pass filter based on defected ground structures." Microwave Technology & Computational Electromagnetics (ICMTCE), 2011 IEEE International Conference on. IEEE, 2011.
11. Ehab K. L. Hamad, Amr M. E. Safwat, and Abbas S. Omar, "L-Shaped Defected Ground Structure for Coplanar Waveguide", IEEE Antennas and Propagation Society International Symposium, 0-7803-8883-6/05/$20.00 ©2005 IEEE, vol. 2B, pp. 663–666, 2005.
12. Hu jiang and Liu gang, "Triangle-shaped defected ground structure for coplanar waveguide", IEEE 2008.
13. L. H. Weng, Y. C. Guo, X. W. Shi, and X. Q. Chen, "An overview on Defected Ground Structure", Progress in Electromagnetics Research B, Vol. 7, pp. 173–189, 2008.
14. M. F. Karim, A. Q. Liu, A. Alphones, X. J. Zhang and A. B. Yu, "CPW bandstop filter using unloaded and loaded EBG structure", IEE Proceeding - Microwaves, Antennas and Propagation, IET Journals & Magazines, Vol. 152, pp. 434–440, 2005.
15. Ajayan, K. R., and K. J. Vinoy. "Planar Inter Digital Capacitors on Printed Circuit Board." IEEE Trans. Microw. Theory Tech 41.9 (1993): 191–194.
16. Peddireddy, Prathibha, and Chhavi Kush. "Micromachined wide bandpass filter." Communications and Signal Processing (ICCSP), 2015 International Conference on. IEEE, 2015.
17. Khandelwal, Mukesh Kumar, Binod Kumar Kanaujia, and Sachin Kumar. "Defected Ground Structure: Fundamentals, Analysis, and Applications in Modern Wireless Trends." International Journal of Antennas and Propagation 2017 (2017).

Improved SLReduct Framework for Stress Detection Using Mobile Phone-Sensing Mechanism in Wireless Sensor Network

Prabhjot Kaur and Sheenam Malhotra

Abstract Stress is a major issue for every person. There are various machine learning methods and sensor systems that are widely used to detect the stress. Mobile phone-sensing mechanism is a cheaper technique to detect the stress, as mobile phones are easily available and every single person is using it. The work here deals with the detection of stress by measuring the physiological parameters of the human body. The results show good performance of the proposed system. Hybrid approach that involves the combination of heuristic algorithm and Bayesian classifier with the neural network used here provides a good accuracy of 92.86% with the involvement of Blood Pressure Measurement (BPM) as one physiological parameter and 85.71% with the Heart Rate (HR) as another physiological parameter of human body to detect the stress of a person.

Keywords Hybrid approach · Stress · Physiological parameter

1 Introduction

Detection of stress is the most common problem a human is suffering from. Stress is a major problem that causes "flight or fight" response, that makes a person flight in the situation or run away from the situation [1]. Stress can lead to various cardiovascular diseases that may be hypotension, heart disease, hypertension, etc. The technology has been developed rapidly to monitor human physiological changes for the detection of human stress. Stress can be physical stress, emotional stress, survival stress, internal stress, mental stress, etc. When a person is under stress then there is a hormone secretion that makes a person to fight the situation. The physical stress, internal stress or external stress can be overcome by following some exercises and

P. Kaur · S. Malhotra (✉)
Sri Guru Granth Sahib World University, Fatehgarh Sahib 140406, Punjab, India
e-mail: sheenam.malhotra@gmail.com

P. Kaur
e-mail: prabh.caur16@gmail.com

C. R. Panigrahi et al. (eds.), *Progress in Advanced Computing and Intelligent Engineering*, Advances in Intelligent Systems and Computing 714,
https://doi.org/10.1007/978-981-13-0224-4_45

yoga that could relax the person. The emotional stress is caused due to an imbalance in emotions the person face. Survival stress is caused when a person is in fear of his/ her survival problems. The study surveyed depicts that the adult's age group between 18 and 35 are more in a stressed state. The women are more stressed than then men due to the loneliness, poor eating habits, home–work routine, etc. [2]. Using the human physiological parameters like blood pressure, heart rate, skin temperature, sweating, EEG signals the stress can be detected and analysed [3]. The technology is growing very faster and the market is adapting the new changes. For the growing technology, there is a need for such a system that helps the people to detect stress easily, accurately and also in a cheaper way. This goal is achieved in our study that uses the mobile phone mechanism to detect stress of a person [4, 5].

Here, in this paper an improved SLReduct framework is introduced to detect the stress more accurately using physiological parameters of the human body. The analysation is performed using the hybrid approach which is composed of Bayesian classifier and heuristic technique in the neural network. The paper is summarized as follows. Section 1 consists of Introduction, Sect. 2 comprises of Related Work, Sect. 3 comprises of Improved SLReduct Framework, Sect. 4 comprises of Results and Implementation, Sect. 5 comprises of Comparison of Existing technique and Proposed technique, Sect. 6 comprises of Conclusion and Future work.

2 Related Work

Stress is evaluated using various machine learning techniques by involving controlled experiments to investigate the effect of stress on physiological parameters. The outcome depicts the best classification results [6]. Supervised machine learning algorithms are used to diagnose stress that involves finger temperature and heart rate variability [7]. Various types of physiological sensors are discussed in existing studies for the detection of stress that involves ECG, PPG, EDA, BVP, EMG, EEG and TEPR [8, 9]. But the physiological signals SCL, ECG and EMG perform better to detect the stress of a person and are considered more successful signals in the detection scenario. But every signal has its own limitations [4]. The existing study for stress detection is composed of SLReduct framework that refers to the stress level reduction framework [10]. The process started with a survey that includes the people's interest in electronic gadgets. The survey concludes the results that people under stress can overcome their situation using the relaxation technique that is depicted in the mobile phone after knowing their stress level and then SLReduct framework was proposed. The framework was designed to detect the stress using the mobile phone-sensing mechanism by measuring the BPM values as shown in Fig. 1. The normal human blood pressure value is 120/80 mm Hg. The upper value is the systolic pressure value and lower is the diastolic pressure value. The parameter values are feed into the mobile phone and database. The analysation is performed which involves the heuristic algorithm to detect the stress and database is used for the online analysis. The database includes the attribute values that involve

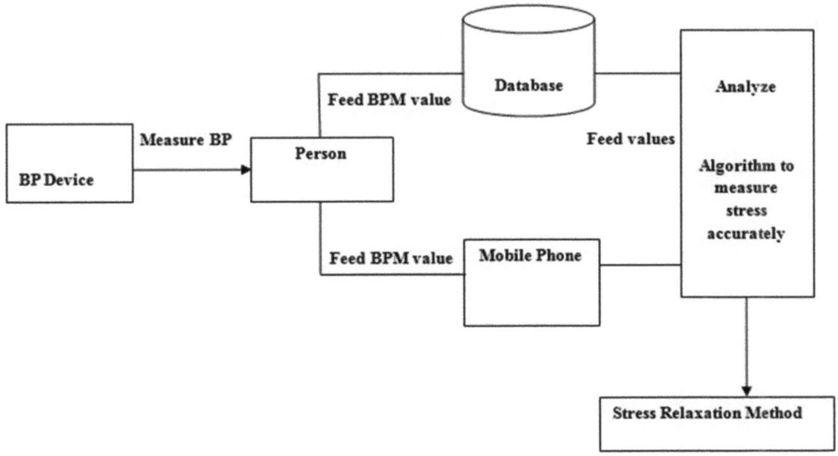

Fig. 1 Architecture of the existing SLReduct framework

age of the person, BP value, name, email id, phone no., sex, stress relaxation method. After the analysation, a relevant relaxation technique was presented that involves listening to music, yoga and exercise to reduce the stress level of the person from which he/she is suffering in daily routine due to the pressure of work and other personal reasons. The algorithm involves the determination and comparison of the BP values with the normal BP value. Stress levels are designed based on the BPM values which display high stress, low stress and medium stress [11]. The approach provides good accuracy to the system.

3 Improved SLReduct Framework

Here in this proposed work, the SLReduct framework is improved for detecting the stress of the person. The improved SLReduct framework is composed of two physiological parameters (BPM, HR) and a hybrid approach to detect the stress of a person as shown in Fig. 2. The hybrid approach involves bayesian classifier and heuristic technique in the neural network to improve the existing SLReduct framework and to enhance the accuracy of the system. The normal HR value is 60–100 bpm and normal BPM value is 120/80 mm Hg. As depicted in the framework a person is equipped with the sensors to detect HR rate, BP measurements. A sphygmomanometer is used to detect BPM value of the person which involves systolic pressure and diastolic pressure rate. With the help of ECG and heart rate, device heart rate value can be determined. The data of the person and stress parameters are feed into the database and mobile phone. The database includes the

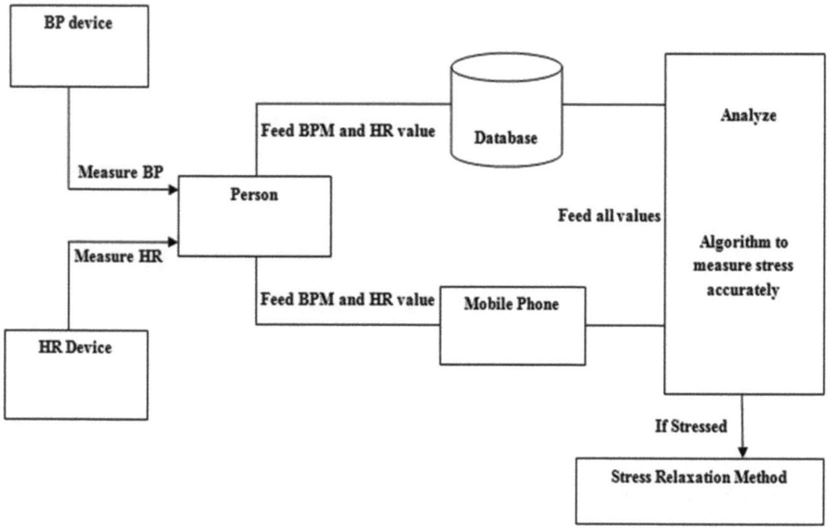

Fig. 2 Architecture of the improved SLReduct framework

attribute values that involve age of person, BP value, HR value, name, email id, phone no., sex, stress relaxation method.

The database is used for the online help so that the caretaker and the person get to know the health condition regarding the stress. In case of emergency situations, the caretaker or doctor can inform the situation of the patient to their relatives. Mobile phone gathers the raw data from the sensors embedded in the phone and after collecting the information various machine learning algorithms and data mining techniques are applied. When the processing gets completed, the results are informed to the medical centre and hospitals. The mobile phone is the main tool which involves the analyzation process to detect the stress and after analyzation the relaxation method is introduced.

The processing of algorithm is referred to as analysation mechanism, which occurs through the mobile phone system. The hybrid approach is used in the analyzation part that involves Bayesian classifier and heuristic algorithm in the neural network. Neural network works as our human brain, they are inspired by the biological nervous system. Neural network is composed of interconnected elements called neurons that work in unity to solve a specific problem. Bayesian classifier is a special class of Bayesian network. It is based on the idea that the role of a class is to predict the values of features for members of that class. If an agent knows the class, it can predict the values of other features. It is a probabilistic model where the classification is a latent variable that is probabilistically related to the observed variables. The heuristic algorithm has different ways of picking the points. The simplest way is to pick points randomly [12]. It is a method for solving the computationally hard optimization problems. The algorithms move from solution to solution in the space of candidate solution by local changes, until a solution deemed optimal is found or a time bound is elapsed. The hybrid approach represents the combination mechanism of two or more techniques or approaches which are discussed as follows:

```
Step 1- Load the dataset for the input.
Step 2- Execute the while loop.
Step 2.1- To apply the systematic analysis of the input data.
Step 2.2- Highlight the major features of the input data and
while.
Step 3- Execute while loop until highlight dataset
classified.
Step 3.1- Form the Bayesian rules on the basis of input data-
Bayesian (rule) =
```

$$\mathrm{P(O)/P(D)} * P(O) \tag{1}$$

```
Step 3.2- Classify the data
Data classification-
```

$$\mathrm{P}(O/E) = P(E) * P(O)/P(E) \tag{2}$$

```
Step 4- Display classified results and show accuracy.
Results calculated-
Stress detection using BPM values-
```

$$\begin{gathered}\text{stress} = \sqrt{(sum(Dx1 - Dy1).\hat{\ }2)}\ ./\ n \\ \text{Final stress} = \text{stress} * 100\end{gathered} \tag{3}$$

```
Stress detection using HR values-
```

$$\begin{gathered}\text{stress} = \sqrt{(sum(Dx1 + Dy1).\hat{\ }2)}\ ./\ n \\ \text{Final stress} = \text{stress} * 100\end{gathered} \tag{4}$$

4 Results and Implementation

The work is implemented in the MATLAB simulation tool. The system used is the Intel(R) Celeron(R) CPU N3060 @ 1.60 GHz with 4 GB RAM, 64-bit Operating system, x64-based processor. The dataset used for the implementing the proposed work is BPM values and HR rate values taken randomly. The inputs taken for detecting the stress are weight and node number. Neural network input sets (pattern and desired output values) are also used that provide the classification criteria using Bayesian classifier. The pattern represents the three input values that are 0, 1 and −1. '0' represents the 'not stressed' state, '1' represents the 'stressed' state and '−1' represents the 'happy' state that helps for the classification which is represented in Fig. 3 and Fig. 4 using BPM values and HR values respectively. The learning rate used is 0.08 that defines learning speed to sense the previous datasets and depicts newly featured datasets. The data is gathered randomly and executed. For the transmission of data, gateway nodes are added which aggregates the data from the nodes and then the data is transmitted from the gateway nodes to the base station that helps in consumption of energy of sensor nodes. The execution process involves heuristic technique for the systematic analysis of the input data and highlights major features that can lead to the stress and not stressed states. After the execution process of the heuristic technique, the Bayesian classifier is used that executes the loop until highlighted dataset gets classified. The Bayesian classifier forms bayesian rules that are based on the input data and classifies data according to specified rules. After the data gets classified results are depicted in the form of graphs discussed below in Sects. 4.1 and 4.2. The stress is calculated from the data classified using the Eqs. 3 and 4 as described in Sect. 3.

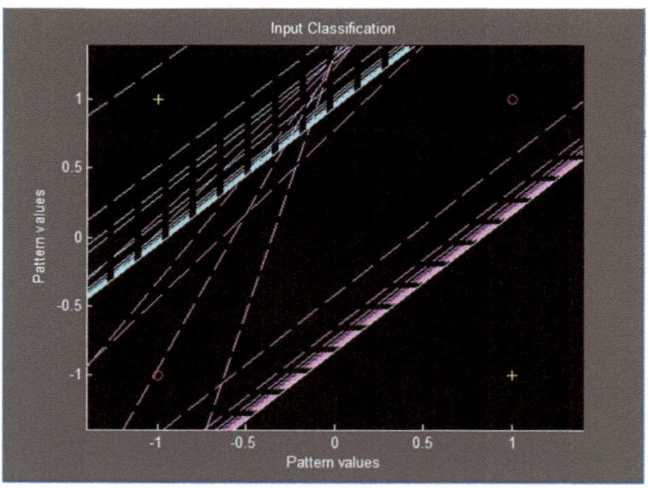

Fig. 3 Classification of stressed and non-stressed regions using BPM values

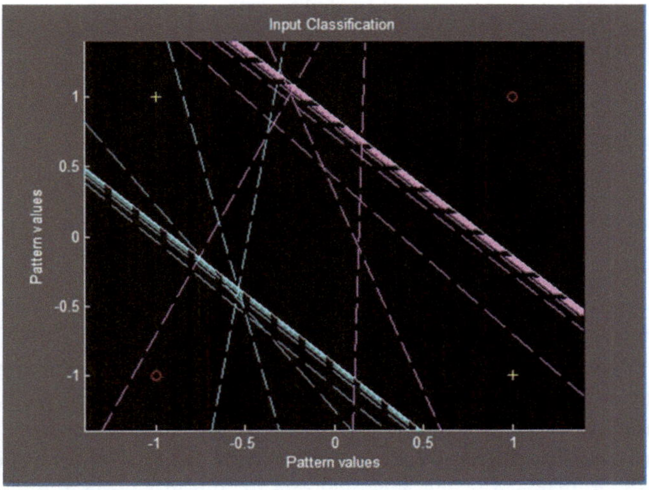

Fig. 4 Classification of stressed and non-stressed regions using heart rate values

4.1 Stress Detection Using the Blood Pressure Measurement (BPM) Values

Figure 3 represents the stressed and not stressed states using the BPM values. The values which are under stress are represented in the area which is under positive state and the not stressed values are depicted in the zero regions of graph. Stress is calculated using Eq. 3.

4.2 Stress Detection Using Heart Rate (HR) Values

Figure 4 represents the stressed and not stressed states using heart rate values. The values which are under stress are represented in the area which is under positive state and the not stressed values are depicted in zero region of graph. Stress is calculated using the Eq. 4 for the heart rate values.

5 Comparison of the Existing Study and the Proposed Technique

The proposed technique is mainly based on the detection of stress which classifies that whether the person is under stressed state or not. Existing study is enhanced and improved for detecting the stress of a person by implementing the proposed

Table 1 Comparison of the existing study with the proposed technique

Compared sections	Existing technique	Proposed technique
Technique	Heuristic approach	Hybrid approach heuristic + Bayesian classifier in the neural network
Physiological parameters used	BPM	BPM, HR
Accuracy	78.57%	BPM—92.86% HR—85.71%

technique which involves proposing more physiological parameters (BPM, HR) and hybrid approach (heuristic and Bayesian classifier) that represented good results and accuracy as depicted in Table 1. The work is performed on the analyzation part of the framework for measuring the stress of a person more accurately. The hybrid approach used shows different accuracy rates with different physiological parameters. There is a decrease in the accuracy rate of hybrid approach when heart rate values are used to detect the stress of a person due to the fluctuations in the heart rate values. The approach represents high accuracy rate when BPM values are used as a physiological parameter to detect the stress of person.

6 Conclusion and Future Work

Mobile phone-sensing mechanism for the detection of stress provides better relief to the problem, as smartphones are most widely used due to their personal usage, low cost, good efficiency and easy availability. It becomes very easy for the healthcare taker as well as for the person to know about his/her stress level. Various machine learning techniques are used in the previous studies for good accuracy of the outcomes. Here, in this proposed technique hybrid approach is implemented that shows an interesting scenario which classified the stressed and not stressed states. The results of detection of stress using BPM and HR values are compared. More physiological parameters can be proposed for more accurate results. In future, the method can be further extended for the prediction and analyzation of other human problems that involves skin problems, BP problems, etc. The future work may also involve the detection of stress using the skin temperature measurements as a physiological parameter. The further work can be analysed for the detection of stress using the real-time sensors to accurately examine the physiological parameter values.

References

1. Sharma Nandita, Gedeon Tom.: Modelling Stress Recognition in Typical Virtual Environments. *In* 7th *International Conference on Pervasive Computing Technologies for Healthcare and Workshops*, pp. 17–23, 5–8 May 2013.
2. Abouelenien, M., Burzo, M., & Mihalcea, R.: Human Acute Stress Detection via Integration of Physiological Signals and Thermal Imaging. *In Proceedings of the 9th ACM International Conference on PErvasive Technologies Related to Assistive Environments*, 2016.
3. Scully, C., Lee, J., Meyer, J., Gorbach, A. M., Granquist-Fraser, D., Mendelson, Y., et al.: Physiological parameter monitoring from optical recordings with a mobile phone. *IEEE Transactions on Biomedical Engineering*, 59(2), pp. 303–306, 2012.
4. Sioni Riccardo & Chittaro Luca.: Stress Detection Using Physiological Sensors. *In The IEEE Computer Society*. pp. 26–36, 2015.
5. Akane Sano, Rosalind W. Picard.: Stress Recognition using Wearable Sensors and Mobile Phones. *In IEEE Humanie Association Conference on Affective Computing and Intelligent Interaction*, pp. 671–676, 2013.
6. Smets, E., Casale, P., Großekathöfer, U., Lamichhane, B., De Raedt, W., Bogaerts, K., & Van Hoof, C.: Comparison of machine learning techniques for psychophysiological stress detection. *In International Symposium on Pervasive Computing Paradigms for Mental Health*, pp. 13–22, September 2015.
7. Barua S., Begum, S., & Ahmed, M. U.: Supervised machine learning algorithms to diagnose stress for vehicle drivers based on physiological sensor signals. *In pHealth*, pp. 241–248, 2015.
8. Carbonaro, N., Anania, G., Mura, G. D., Tesconi, M., Tognetti, A., Zupone, G. et al.: Wearable biomonitoring system for stress management: A preliminary study on robust ECG signal processing. *In 2011 IEEE international symposium on a World of Wireless, Mobile and Multimedia Networks* (*WoWMoM*), pp. 1– 6, 20–24 June 2011.
9. Vanitha, V., & Krishnan, P.: Real time stress detection system based on EEG signals. *Biomedical Research*, 2016.
10. Dhulipala, V. S., Devadas, P., & Murthy, P. T.: Mobile Phone Sensing Mechanism for Stress Relaxation using Sensor Networks: A Survey. *Wireless Personal Communications*, 86(2), pp. 1013–1022, 2015.
11. Jung, Y., & Yoon, Y. I.: Multi-level Assessment model for wellness service based on Human Mental Stress level. *In Springer publication*, pp. 1–13, 14 March 2016.
12. Sharma Nandita, Gedeon Tom.: Hybrid Genetic Algorithms for Stress Recognition in Reading. *In European Conference on Evolutionary Computation, Machine Learning and Data Mining in Bioinformatics* (*EvoBio 2013*), pp. 117–128, 2013.

Part VI
Algorithms, Emerging Computing and Intelligent Engineering

NavIC—An Indigenous Development for Self-reliant Navigation Services

Dinesh Kumar Misra and Mohammad Zaheer Mirza

Abstract Navigation with the aid of satellite has got realistic importance in the aviation industry, ship movement and strategic purpose throughout the world for all countries. A GPS system of America is providing navigation support worldwide. India has taken an initiative to develop an independent navigation system under Indian control having a constellation of seven GSO satellites and has established navigation facilities for this purpose. The Indian NavIC system is broadcasting navigation data for Indian continent as well as 1500 km beyond its geopolitical boundary. This paper brings ideas about India's efforts to develop its own indigenous navigation system and its application in India and the surrounding region.

Keywords Navigation · Satellites · Geopolitical · MEO · GEO orbit satellites · Accuracy

1 Introduction

Indian Regional Navigation Satellite system, abbreviated as IRNSS and now functionally renamed by the Government as "NavIC—Navigation with Indian Constellation" is an indigenous development and effort of India within very short span of time and limited budget. NavIC functional area covers whole India as well as the geopolitical periphery of 1500 km of Indian Ocean [1–4].

This NavIC [5] system of India was conceived in 1999. Indian strategic operations were totally dependent on American-controlled Global Positioning System (GPS). Experiencing total dependency on GPS navigation data, Indian Government has taken an initiative and given clearance to develop its own indigenous system of

D. K. Misra (✉) · M. Z. Mirza
Department of Space, ISTRAC/ISRO, Lucknow, India
e-mail: misradkm@gmail.com

M. Z. Mirza
e-mail: mirza@istrac.org

C. R. Panigrahi et al. (eds.), *Progress in Advanced Computing and Intelligent Engineering*, Advances in Intelligent Systems and Computing 714,
https://doi.org/10.1007/978-981-13-0224-4_46

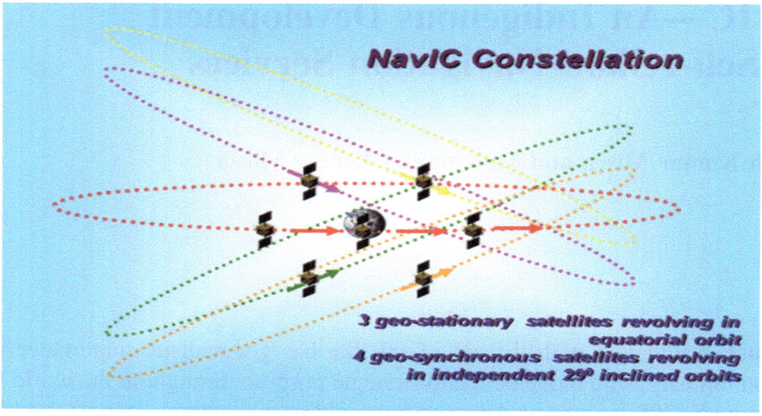

Fig. 1 India's NavIC system [6, 7]

navigation in 2005. The Indian NavIC system provides two level services: the standard positioning services for civilian applications and encrypted services for strategic users. Indian NavIC system is perused in Fig. 1.

2 Navigation Services of Other Space Agencies

2.1 GPS-SPS System of Navigation

Established in 1995 by United State Airforce, US, Global Positioning System-Standard Positioning Service provides three-dimensional positions, velocity and time information. The system provides precise positioning services with full system accuracy to designated users and standard positioning service provides an accurate position to all users. GPS has a constellation of 24 satellites in six orbital planes in a circular orbit of 20,200 km. These satellites have an inclination angle of 55° with orbital period 12 h. GPS-SPS services provide a general navigation and harbor approach with a horizontal accuracy of 9 m. American GPS satellite constellation is shown in Fig. 2 [2, 3, 8].

There are 31 satellites in a constellation at present in the GPS system. A modernization program of system aims to improve the accuracy and availability to all users by involving some more satellites. It has proposed four new additional signals, three for civilian L2C, L1C and L5 and one for strategic purpose coded as M-code. The next generation of GPS-III satellites is expected in 2021 [8].

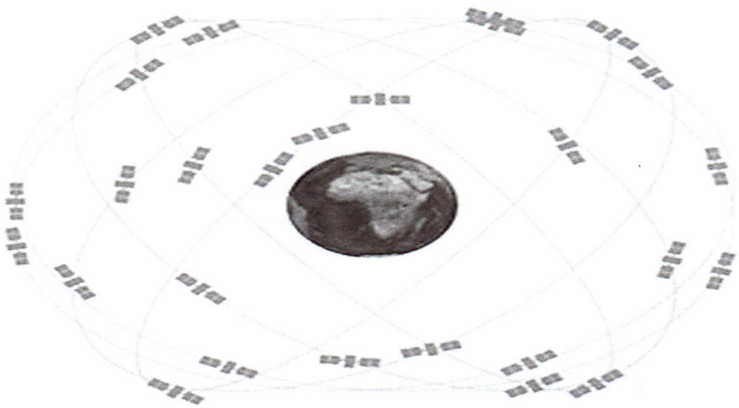

Fig. 2 Constellation of GPS 24 satellites in 6 orbital planes [3]

2.2 GLONASS Navigation System

Global Navigation Satellite System (GLONASS) [3] controlled and managed by the Russian Space Agency provides three-dimensional information in terms of position, velocity and time for Russian Federation. It has a constellation of 24 satellites in MEO orbit with angular spacing between orbits 120° at the altitude of 19,100 km. In the constellation, 8 satellites are positioned in each three- orbital plane. The orbital period of satellites is 11 h and 15 min. GLONASS services would satisfy general navigation achieving horizontal position accuracy of 45 m. The recent launch of GLONASS-M with civil signal L2 and GLONASS-K with civil signal L3 has improved the positional accuracy. GLONASS-K was having a differential correction, integrity information, search and rescue function also. GLONASS-K was using FDMA and in the later satellites CDMA access also. GLONASS constellation of the satellite is perused in Fig. 3.

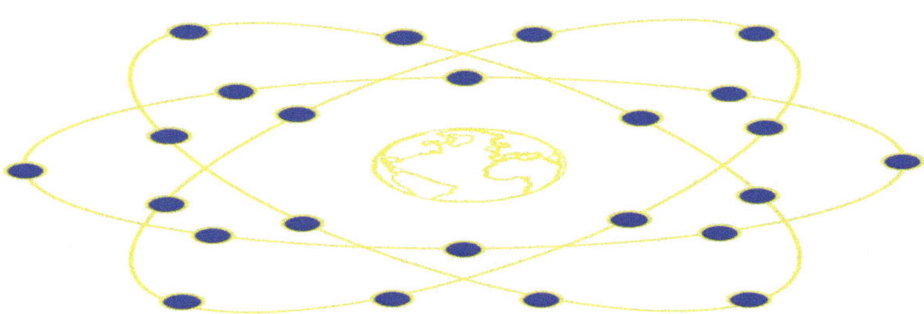

Fig. 3 GLONASS satellite constellation [3]

Fig. 4 Constellation of Galileo satellites [3]

2.3 GALILEO Navigation System

Galileo [3] is the European Space Agency navigation system under civilian control attaining constellation of 30 satellites in orbit, 27 satellites in operation and 3 satellites as an active spare in each three plane. There are 9 satellites in each of three planes re-known as the Walker constellation (27/3/1). Satellites are positioned in MEO orbit inclined to 56° from the equator at the altitude of 23,222 km and zero eccentricity. The orbital period is 14 h 21.6 min with ground track repetitivity of 3 days. Constellation diagram is shown in Fig. 4.

Galileo is providing 5 levels of services: Open services (OS), Safety of life services (SoL), Commercial services (CS), Public-regulated services (PRS) and Search and Rescue service (SAR). Galileo is expected to provide timely warnings of integrity failure, within a very few seconds. Among all of above its novel service providing distress message with the aid of COSPAS-SARSAT satellites has gained much fame and saved many lives.

2.4 BeiDou/Compass Navigation System

The space agency of China is also engaged in developing its own navigation system. Its first five Compass [3] satellites were placed in Geostationary Orbit. It has also positioned some satellites in MEO orbit to cater navigation services. As of 2016, China has 21 operational satellites: 6 in GSO, 8 satellites inclined at 55° IGSO orbit and 7 in Medium earth orbit. A constellation of satellites is shown in Fig. 5.

Compass is currently using two GEO satellites to cater 100 m horizontal position accuracy. BeiDou 2 is planned to have constellation of 30 MEO satellites or 4GEO and 12 MEO with goal of position accuracy less than 20 m.

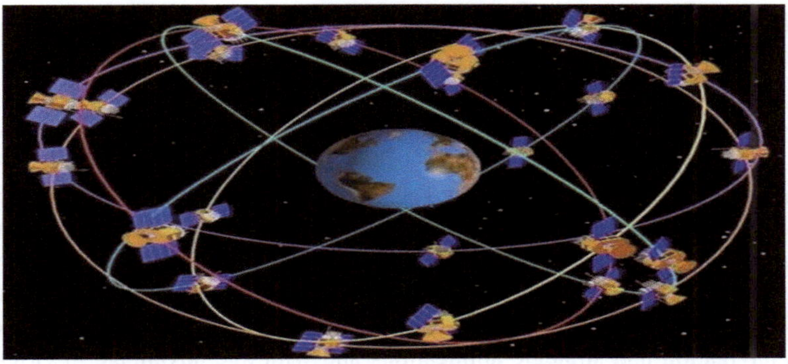

Fig. 5 Constellation of satellites of BeiDou China [3]

3 NavIC System Description

With the aid of Satellite, either it is global or regional constellation of satellite: position, navigation data and time can be determined at any part of the globe. NavIC [1–3] has a constellation of 3 satellites in Geosynchronous orbit with the inclination 5° crossing the equator at 32.5°, 83°, and 129.5° respectively, whereas 4 satellites in Geosynchronous orbit with the inclination 29° crossing the equator at 55° and 111.75° respectively. NavIC system services, e.g. Standard positioning and Encrypted services are available in L5 (1176.5 MHz) and S-band (2492.028 MHz) frequency. Modulation module of SPS is 1 MHz BPSK and for encrypted signal, it is the BOC (5, 2). Navigation signals are transmitted over the S-band frequency with the application of Phased Array Antenna. This antenna will provide required coverage and signal strength. NavIC satellite system constellation is intended to provide positional accuracy better than 10 m in the Indian continent and better than 20 m in the Indian Ocean approximately 1500 km around India. NavIC system is designed to provide timing accuracy less than 10 ns. It is intended to provide navigation services with high service availability of 99.97% on 24 × 7 bases. Indian NavIC constellation is perused in Fig. 6.

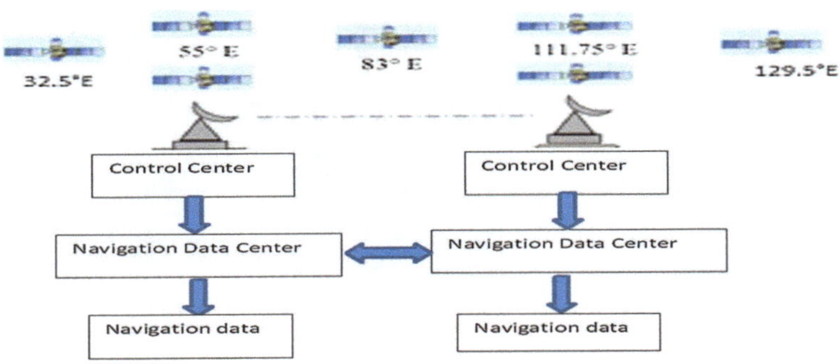

Fig. 6 NavIC control and data centre [6]

4 NavIC Constellation Requirement

In NavIC constellation, four satellites are in inclined orbit and trace a Fig. 8 pattern on earth. When 2 satellites are at equator other 2 satellites are at an extreme position as shown in Fig. 7. In any navigation system requirement of satellites does arise as per unknown parameters to find out. The General basic thumb rule says that four unknown can be solved by four equations, same is applied to space segment, i.e., we may require four satellites at least in space to solve four unknown as shown in Fig. 8. Thus, the Indian NavIC system was thought to cover Indian continent and its geopolitical boundaries [5, 9]. The Indian NavIC system is a self-reliant service for Indian navigation and strategic applications and with due advantage in GSO orbit and satellite control from Indian continent, a constellation of seven satellites was launched and controlled to provide navigation data [5].

4.1 Augmentation Process

As the accuracy and integrity attained from core system is not sufficient for some strategic applications. GEO overlay augmentation systems are meant to improve the accuracy and integrity as shown in Fig. 9 below. In a GEO overlay augmentation system bent-pipe type payloads are used for such application [10]. GPS signals accuracy and integrity is augmented by WAAS, EGNOS, MSAS and GAGAN systems as shown in Fig. 10 below: [10]. Same way NavIC system is augmented.

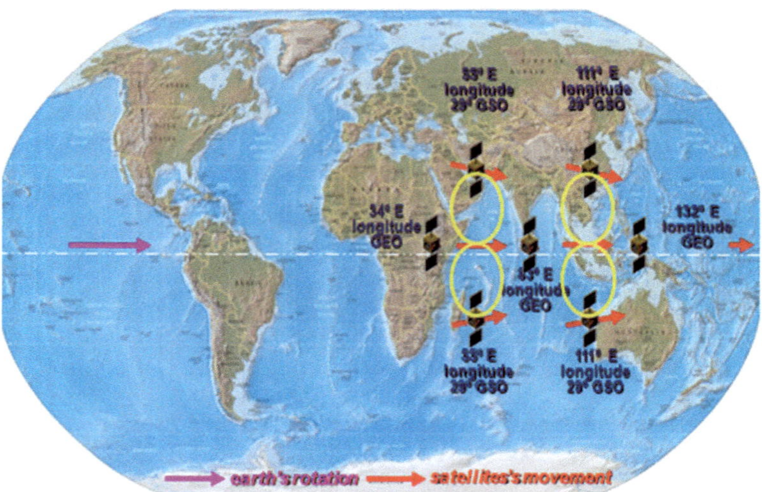

Fig. 7 NavIC constellation and satellite position [3, 9]

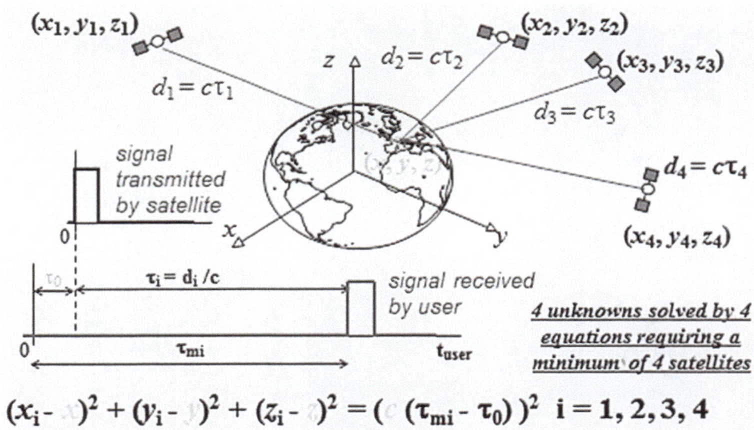

Fig. 8 Four satellite requirement [1, 9]

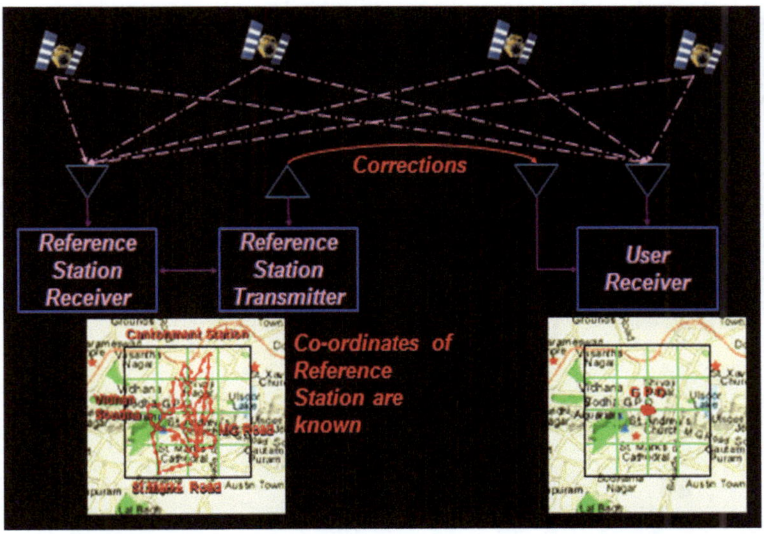

Fig. 9 Augmentation process [3]

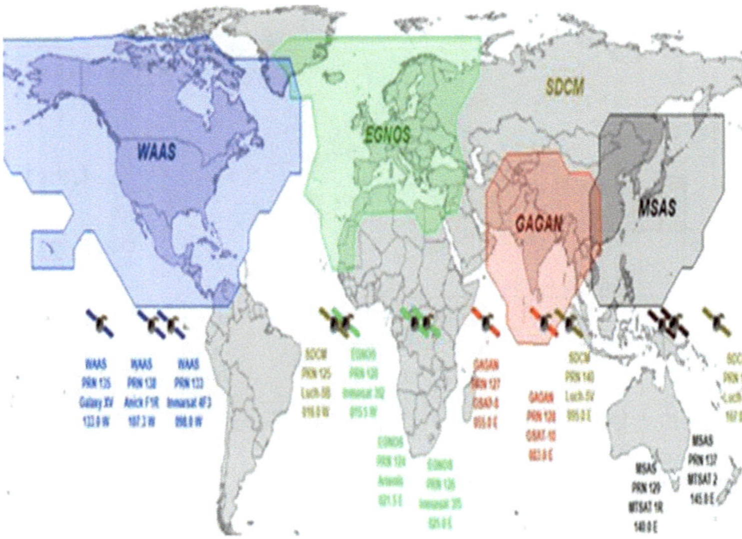

Fig. 10 Augmentation system coverage [3]

5 NavIC Position Accuracy

On above subsection, it was perused that how with the application of augmentation accuracy and integrity can be is enhanced. Dilution of Precision and accuracy measurement has a large impact on position accuracy Fig. 11.

DOP can be found out as

$$DOP = \sqrt{Trace}\left[\mathrm{A}^{\mathrm{T}}\mathrm{A}\right]^{-1}$$

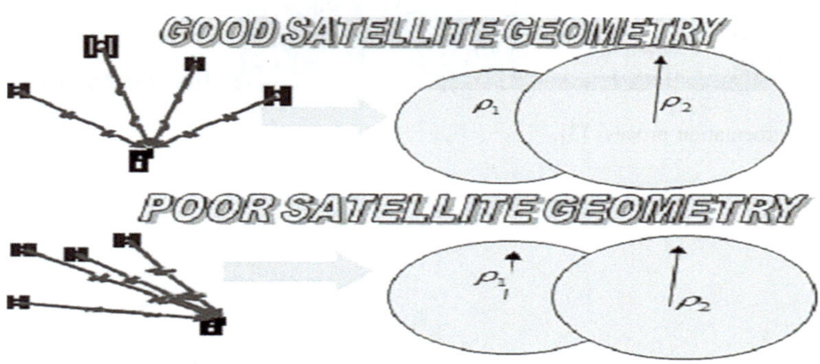

Fig. 11 Satellite geometry for position accuracy measurement [9]

where

$$\dot{A} - \left[\frac{x_i - x_u}{\rho_i}\ \frac{y_i - y_u}{\rho_i}\ \frac{z_i - z_u}{\rho_i}\right]_{i-1.4}$$

And ρ = Range

There are many errors in position determination and can be determined by the equation given below [9].

Measurements, Range, Receiver Clock Bias, Satellite Ephemeris & Clock Errors, Ionospheric Delay, Troposhepheric Delay, Multipath and Receiver Noise, Integer Ambiguity

$$\rho_i^k = r_i^k + \delta t_i - \delta t^k - I^k + T^k + \varepsilon_i^k \quad \text{(m)}$$

Pseudorange

$$\phi_j^k = (r_i^k + \delta t_i - \delta t^k - I^k + T^k + \tilde{\varepsilon}_i^k)/\lambda + N_i^k \quad \text{(cycles)}$$

6 NavIC Coverage Area

NavIC constellation of satellite was meant primarily for strategic and navigation requirement and its fulfillment to Indian continent and its geopolitical boundary. But it is broadcasting a signal to some extended service area as shown in Fig. 12.

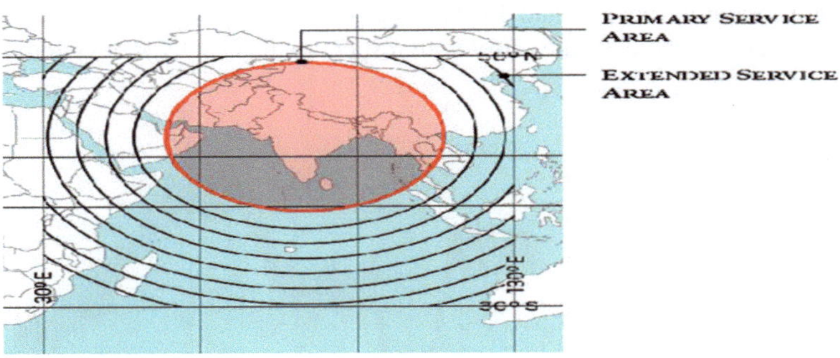

Fig. 12 Service area of NavIC [6]

7 NavIC Services and Application Area

The Indian NavIC system will provide two types of services: [3].

- Standard Positioning services with the position accuracy less than 20 m for all users.
- Restricted services are encrypted service and provided to authorized users only.

Some of the applications are summarized below:

1. Terrestrial, Aerial and Marine Navigation
2. Disaster Management
3. Vehicle tracking and Fleet Management
4. Integration with Mobile Phones
5. Precise Timing
6. Mapping and Geodetic data capture
7. Terrestrial navigation aids to hikers and travelers
8. Visual and Voice communication for drivers, etc.…

8 Results and Discussions

The NavIC system with the constellation of seven GSO satellite has achieved its identified goal by providing position accuracy less than 10 m in Indian continent and less than 20 m in its geopolitical boundary of 1500 km. Timing accuracy maintained and achieved by NavIC system is less than 20 ns. Figures 13, 14, 15 and 16 justify the achieved goal.

NavIC system time offsets achieved with respect to GPS and GLONASS time system are shown in Figs. 13 and 14. With respect to GPS system, it is well within the range of 20 ns whereas time offset with respect to the GLONASS system shown

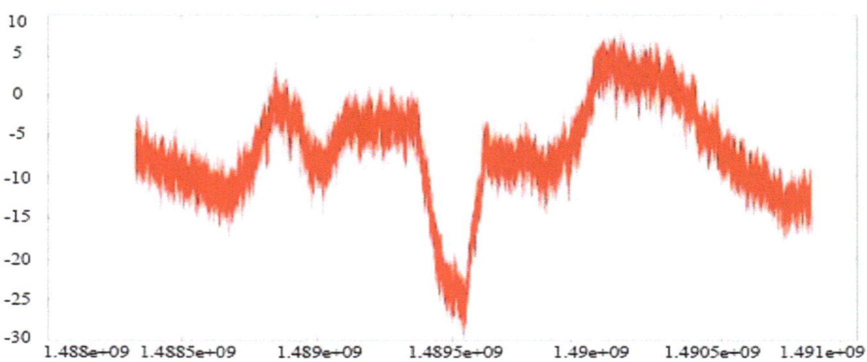

Fig. 13 Offset between GPS and NavIC time [6, 7]

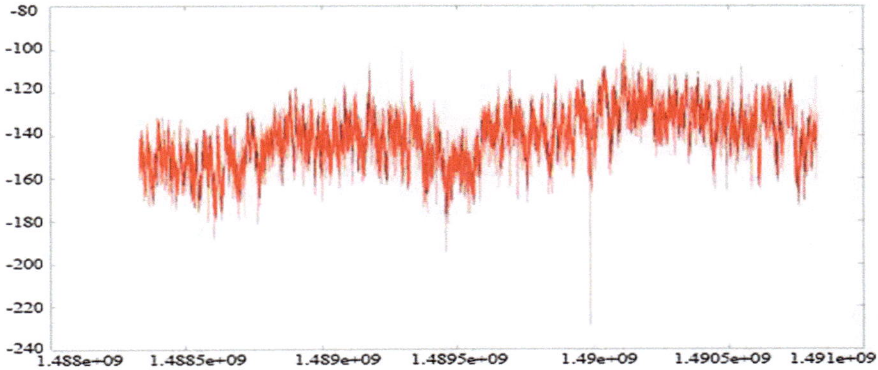

Fig. 14 Offset between GLONASS and NavIC time [6, 7]

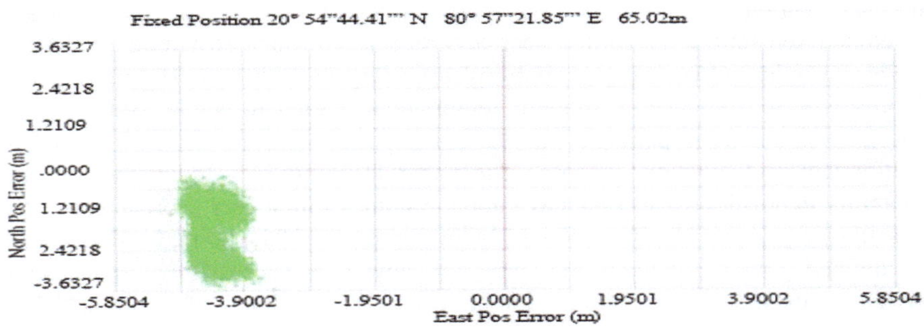

Fig. 15 Position accuracy of NavIC system's primary [6, 7]

in Fig. 15 is also on average 40 ns IS achieved. NavIC position accuracy is targeted as 10 m in the primary area and the accuracy achieved is less than 10 m. GDOP found in the range of 4–5 m is well ahead than targeted. It is shown in Figs. 15 and 16 respectively. Integrity information received is in encrypted form, and only authorized user can receive it [3]. It is also perused that all above Figs. 13, 14, 15 and 16 are plotted by the data collected at operation control centre of ISRO/ ISTRAC [6, 7]. Timing and position accuracy achieved by Indian operation center are very much optimum than specified.

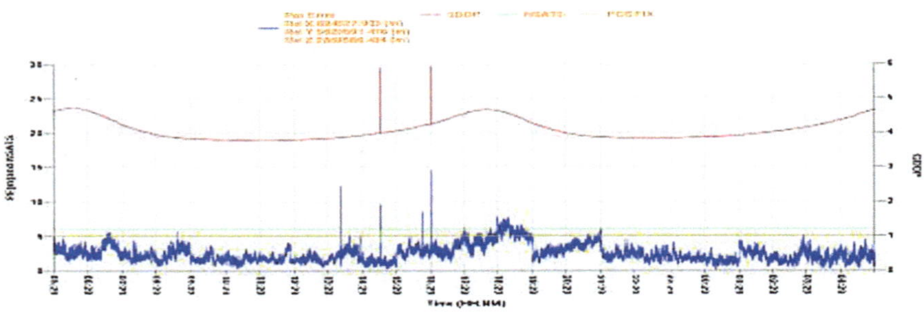

Fig. 16 GDOP and position accuracy of NavIC system [6, 7]

9 Conclusion

In conclusion remark and scope of the paper, it is to bring to perusal that India has developed self-reliant navigation service which is not only providing navigation service to India but covering geopolitical boundary of 1500 km. The service will also provide self-reliance on strategic requirement. Earlier to NavIC system, India was having a total dependency on GPS navigation data for its civil, strategic requirement and aviation purpose. The paper also provides India's vision and capability in developing its own navigation system within a very short span of time by setting a goal and NavIC system of India is a result of this. Started in 2013, with the defined goal of accuracy and integrity, now it is successfully broadcasting navigation data. With the constellation of seven GSO satellites, India has achieved its civilian as well strategic need by providing very good position, time and integrity information. Table 1 shows a comparative study of all countries having navigation facility [3, 5, 11].

Table 1 NavIC accuracy comparison

System/types	Country	Satellites	Coverage	Launched year	Precisions
GPS	US	31	Global	1978	<10 m
GLONASS	Russia	24	Global	1982	<45 m
Galileo	EU	30	Global	2011	10 m-public 10 cm-stratgic
BeiDou	China	21	Global	2000	10 m-public 10 cm-stratgic
NavIC	India	7	1500 km Geo-pol. boundary	2013	20 m-public 10 m-stratgic

Acknowledgments The Authors have heartfelt gratitude to the department for providing assistance and opportunity to write research papers. The authors have the open idea to include any suggestion referred by honorable reviewers. We will be highly thankful for all honorable peer reviewers/researchers to provide comments and suggestion to improve the strength of paper.

References

1. W.H. Guir, "Genesis of Satellite Navigation", APL Technical digest, Vol.9, no.1, pp. 178–181, 1997.
2. H.J. Christopher,"Evolution of Global Navigation System", IEEE proceedings, Vol.96, pp. 1902–1917, 2008.
3. IALA Aids to Navigation Manual-Naveguide, 2010.
4. Nel Samama, "Global Positioning: Technologies and Performances", Wiley, pp. 65–72, 2008.
5. Ishan Srivastava, "How Kargil spurred India to design its own GPS", TOI, 2014.
6. www.istrac.org.
7. www.isro.org.
8. S.M. Grewal, R.W.Lawrence, "Global Positioning System, Inertial Navigation", 2nd edit, Wiley, pp. 92–93, 2007.
9. Pratap Misra and Per Enge, "Global Positioning System-Signals, Measurement and Performance", 2nd edition, GPS handbook, 2011.
10. S.K. Chauhan, SK Gupta and ULNV Subramanian, "Mining GPS Data to Determine Interesting Location", IWII, 2008.
11. E. Howell, "Navstar: GPS Satellite Network", Space Communication, 2013.

Modelling of Force and Torque Due to Solar Radiation Pressure Acting on Interplanetary Spacecraft

Aman Kumar Sinha and Mirza Mohd Zaheer

Abstract In this paper Planck's quantum theory is considered to explain the effect of SRP (Solar Radiation Pressure) on interplanetary spacecraft, which not only disturbs the orbit of spacecraft but also exerts torque (absorbed by reaction wheel), thereby creating the requirement for momentum desaturation using thrusters. The force acting on the spacecraft and hence the torque due to SRP can be computed with the knowledge of the optical properties of the material used, i.e. absorption, reflection and transmission. Further, using Newtonian mechanics and momentum conservation principle a mathematical relation is derived between the force and coefficient of the material properties. The model, thus, derived can be used to predetermine the disturbance torque acting on the spacecraft for various orientations and thus optimum orientation of the spacecraft can be chosen where disturbance on the platform is minimum. The model is already verified for Mars Orbiter Mission during cruise phase (spacecraft in heliocentric orbit).

Keywords Planck's quantum theory · Solar radiation pressure
Torque

1 Introduction

An interplanetary spacecraft during its journey from Earth to the desired planet has to travel through various phases, which include Earth-bound orbit (orbit around Earth), cruise phase (orbit with Sun as the parent body) and planet capture (orbit around the desired planet). In cruise phase, the only significant perturbation acting on the spacecraft is the force and torque due to solar radiation pressure. The torque exerted on spacecraft due to SRP is absorbed by reaction wheel. However, reaction

A. K. Sinha (✉) · M. M. Zaheer
Department of Space, Istrac-Isro, Sector-G, Jankipuram, Lucknow 226021, India
e-mail: aman.itbhu@gmail.com

M. M. Zaheer
e-mail: mirza@istrac.org

C. R. Panigrahi et al. (eds.), *Progress in Advanced Computing and Intelligent Engineering*, Advances in Intelligent Systems and Computing 714,
https://doi.org/10.1007/978-981-13-0224-4_47

wheel gets saturated rapidly if disturbance torque acting on spacecraft is high and so the wheel has to be desaturated by momentum dumping using thrusters. When solar radiation interacts with the spacecraft surfaces, partly they are reflected (specular and diffused), transmitted and absorbed. Depending on the type of material, solar radiation interacts differently with different parts of the spacecraft. This paper derives a mathematical model for solar radiation force and also the torque created due to this force on the spacecraft.

2 Mathematical Model

2.1 Solar Radiation

Electromagnetic radiation is quantized in particles called photons, particle aspect of light. Photons are best explained by quantum mechanics [1]. They have the properties of energy and momentum and thus exhibit the property of mass as they travel at light's speed. The momentum of a photon with energy E is given by

$$P = \frac{E}{c} \tag{1}$$

If I_1 is solar radiation energy falling per unit area for a given time duration, then it can be clearly written as

$$E = I_1 \times A$$

Substituting this value of E in Eq. (1), we get

$$P = I_1 \times \frac{A}{c}, \quad \frac{P}{A} = \frac{I_1}{c}$$

Further if I is solar radiation energy falling per unit area per sec, then

$$I_1 = I \times t, \frac{P}{A} = \frac{I \times t}{c}, \frac{P}{A \times t} = \frac{I}{c}$$

The rate of change of momentum of photons on a unit surface area can be called as force with which photons are striking on per unit area of the surface.

Thus,

$$F = \frac{I}{c} \tag{2}$$

Sun is considered to produce a constant amount of energy [2]. At the surface of the Sun, intensity of solar radiation is about 6.33×10^7 W/m^2 (Fig. 1). Without loss of generality, it can be assumed that radiating source is concentrated at the centre of Sun and radiates isotropically as per inverse square law is given by

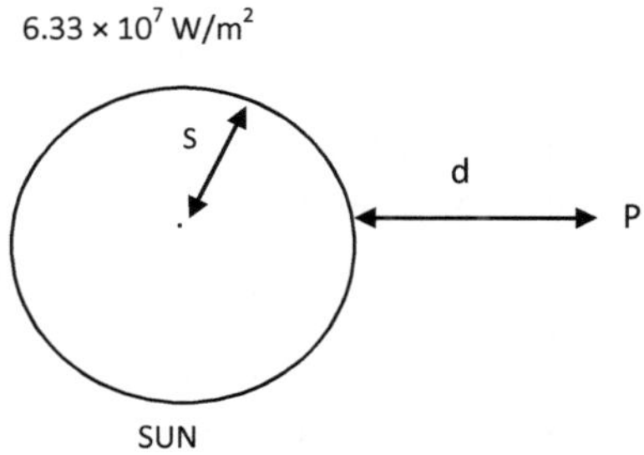

Fig. 1 Solar intensity calculation at any arbitrary point outside the Sun

$$P_s = \frac{P_c}{4\pi s^2}$$

P_s solar intensity at surface of the Sun
P_c solar intensity at centre of the Sun (hypothesis)
s radius of Sun

Thus,

$$P_c = P_s \times 4\pi s^2 = 6.33 \times 10^7 \times 4 \times 3.14 \times 4.83 \times 10^{17}\,\mathrm{W/m^2}$$
$$= 3.85 \times 10^{26}\,\mathrm{W/m^2}$$

Thus, solar intensity at a distance d away from Sun is given by

$$I = \frac{P_c}{4\pi(d+s)^2}$$

2.2 *Force*

Let F_N be the force due to solar radiation per unit area for normal incidence. If radiation is inclined to angle θ with normal, we get

$$F_1 = F_N \cos\theta \,(\text{represented in magnitude only}) \tag{3}$$

where F_1: incident force per unit area (Fig. 2)

Component of F_1 normal to surface is given by

$$F_{1y} = -F_1 \cos\theta\, y = -F_N \cos^2\theta\, y \tag{4}$$

Component of F_1 tangential to surface is

$$F_{1x} = F_1 \sin\theta\, x = F_N \sin\theta \cos\theta\, x \tag{5}$$

x unit vector tangential to surface
y unit vector normal to surface

Let R be the fraction of solar radiation which is reflected, then 1-R will be the fraction of radiation absorbed by the surface (transmission can be practically assumed to be 0 for components of spacecraft). The part of solar radiation absorbed by the surface (1-R) is re-radiated isotropically in all directions and so does not create any net impact on the platform. Now assume out of this R fraction, S part is getting specularly reflected and remaining (1-S)R is having a diffuse reflection. For specular reflection (Fig. 3).

$$F_r = RSF_1 \text{ (represented in magnitude only)} \tag{6}$$

where F_r is reflected solar radiation force per unit area. The component of F_r normal to surface is expressed as

$$F_{ry} = F_r \cos\theta\, y = RSF_1 \cos\theta\, y = RSF_N \cos^2\theta\, y \tag{7}$$

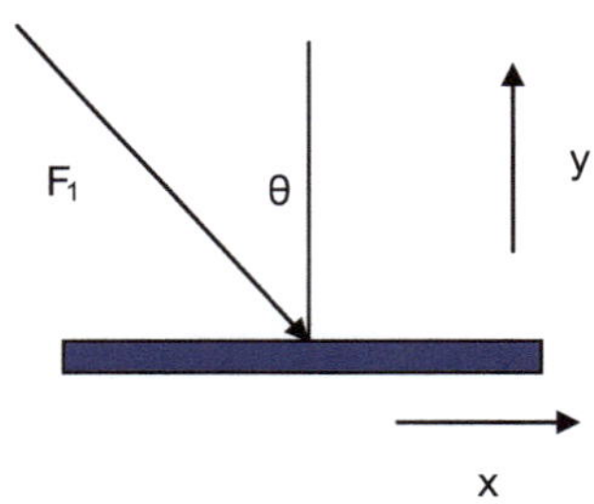

Fig. 2 Sun vector inclined at θ with respect to surface normal

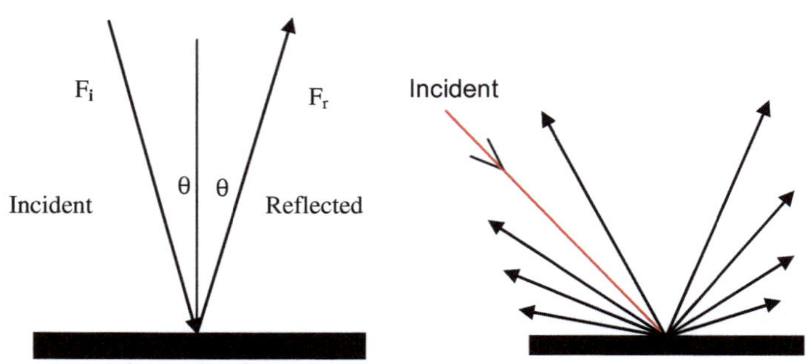

Fig. 3 Specular (*left*) and diffuse (*right*) reflection of solar radiation

And the component of F_r tangential to surface is given as

$$F_{rx} = F_r \sin\theta\, x = RSF_1 \sin\theta\, x = RSF_N \sin\theta \cos\theta\, x \quad (8)$$

Diffuse reflection is the reflection of light from a surface such that an incident ray is reflected at many angles rather than at just one angle as in the case of specular reflection [3]. An ideal diffuse reflecting surface will have equal luminance from all directions, which lie in the half-space adjacent to the surface (Fig. 3). Force with which photons are coming out of the surface due to diffused reflection part (1-S)R can be computed as follows (Fig. 4).

Let e be the solar intensity of diffused reflected radiation in the normal direction. Then as per Lambert's law intensity at an angle θ from the horizontal plane is e sinθ (e cosθ if θ is measured from vertical axis) [4]. Now, as per law of conservation of energy, total solar power due to diffused reflected radiation can be computed by integrating diffused radiation intensity over entire hemispherical area. Let I' be the part of solar radiation which gets diffuse reflected from the surface. Then

$$I' = \int_0^{\pi/2} e \sin\theta (2\pi R \cos\theta) R d\theta = \pi R^2 e \quad (9)$$

If we project the force due to solar radiation in the horizontal plane, net force is 0 as it is symmetrical in all directions. However, component of force in vertical direction will be summed up and is proportional to

$$\int_0^{\pi/2} e \sin\theta (2\pi R \cos\theta) \sin\theta R d\theta = 2\pi R^2 e/3 = 2I'/3 \text{ (Eq. 9)} \quad (10)$$

As force due to solar radiation is directly proportional to intensity of solar radiation (Eq. 2) and so

$$F_d = \frac{2}{3}(1 - S) R F_1 y$$

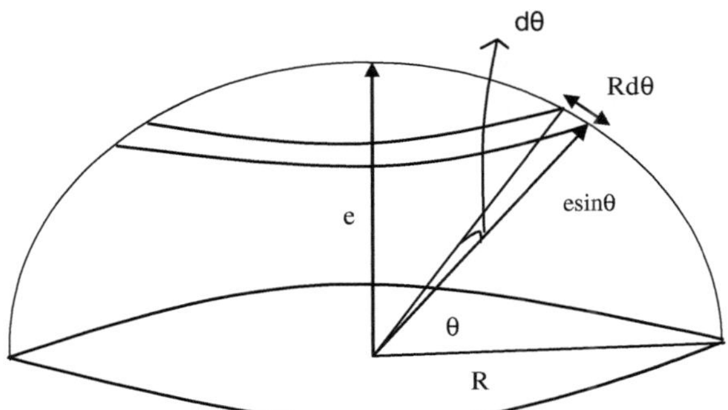

Fig. 4 Diffuse reflected radiation distributed over the hemispherical area in upper half of the reflecting surface

Thus, the normal part of force due to diffusion per unit area is given as

$$F_d = F_{dy} = \frac{2}{3}(1 - S)RF_N \cos\theta\, y \tag{11}$$

and the tangential part is given by

$$F_T = 0 \tag{12}$$

As per momentum conservation principle, momentum imparted to surface is equal to difference of incident momentum and momentum of photons reflected from the surface. Thus, the net tangential force per unit area on the surface is difference between the tangential part of incident and reflected radiations and is expressed as difference of Eqs. 5 and (8 plus 12).

$$\begin{aligned} F_{Tangential} &= F_N \sin\theta \cos\theta\, x - RSF_N \sin\theta \cos\theta\, x \\ &= (1 - RS)\, F_N \sin\theta \cos\theta\, x \end{aligned} \tag{13}$$

Similarly, the net normal force per unit area on the surface is difference between the normal part of incident and reflected radiation and can be seen as difference of Eqs. 4 and (7 plus 11), which can be written as

$$\begin{aligned} F_{normal} &= F_{1y} - \left(F_{ry} + F_{dy}\right) \\ &= -F_N \cos^2\theta\, y - \left[RSF_N \cos^2\theta + \frac{2}{3}(1 - S)RF_N \cos\theta\right] y \\ &= -F_N \left[\cos^2\theta(1 + RS) + \frac{2}{3}(1 - S)R \cos\theta\right] y \end{aligned}$$

Hence, the net force per unit area is expressed as

$$\begin{aligned} F_{net} &= F_{tangential} + F_{normal} \\ &= (1 - RS)F_N \sin\theta \cos\theta\, x - F_N \left[\cos^2\theta\,(1 + RS) + \frac{2}{3}(1 - S)R \cos\theta\right] y \end{aligned} \tag{14}$$

where F_N is the force per unit area due to the solar radiation on the surface if rays are impinging on the surface normally. Let I be the energy of the solar radiation impinging on the surface per unit area per second and c be speed of the radiation, then

$$F_N = \frac{I}{c}$$

The product RS denotes the fraction of specularly reflected radiation, say S_c (coefficient of specular reflection), i.e. $S_c = RS$. Now the product (1-S)R indicates the fraction of diffused radiation, say D_c (coefficient of diffuse reflection).

Substituting F_N, RS and (1-S)R in Eq. 14, we have

$$F_{net} = \frac{I}{c}\left[(1 - S_c)\cos\theta\sin\theta x - \left\{(1 + S_c)\cos^2\theta + \frac{2}{3}D_c\cos\theta\right\}y\right]$$

Hence, the force acting on the elementary surface area dA is given by

$$dF = \frac{I}{c}\left[(1 - S_c)\cos\theta\sin\theta x - \left\{(1 + S_c)\cos^2\theta + \frac{2}{3}D_c\cos\theta\right\}y\right]dA$$

Therefore, the total force over the entire surface area is given as

$$F = \frac{I}{c}\iint\left[(1 - S_c)\cos\theta\sin\theta x - \left\{(1 + S_c)\cos^2\theta + \frac{2}{3}D_c\cos\theta\right\}y\right]dA \quad (15)$$

2.3 *Torque*

Let

r_{dA} Coordinate of an elemental surface component of spacecraft
dF Force on elementary surface area of spacecraft
r_g Centre of gravity coordinate of S/C

Then, moment arm for the elemental component is $r = r_{dA} - r_g$

Torque acting on spacecraft is $\tau = \int(r_{dA} - r_g)\ dF$

$$\tau = \frac{I}{c}\iint(r_{dA} - r_g)\left[(1 - S_c)\cos\theta\sin\theta x - \left\{(1 + S_c)\cos^2\theta + \frac{2}{3}D_c\cos\theta\right\}y\right]dA \quad (16)$$

3 Implementation on MOM (Mars Orbiter Mission)

On 13 February 2014, MOM attitude was as shown in Fig. 5.

Sun to spacecraft distance $= 1.585 \times 10^8$ kms

Solar panel rectangular in shape and cross-sectional area = 7.56 m^2

S_c (coefficient of specular reflection for panel) = 0.08

D_c (coefficient of diffuse reflection for panel) = 0.03

Sun incident angle = −25° (Angle between panel normal and sun vector)

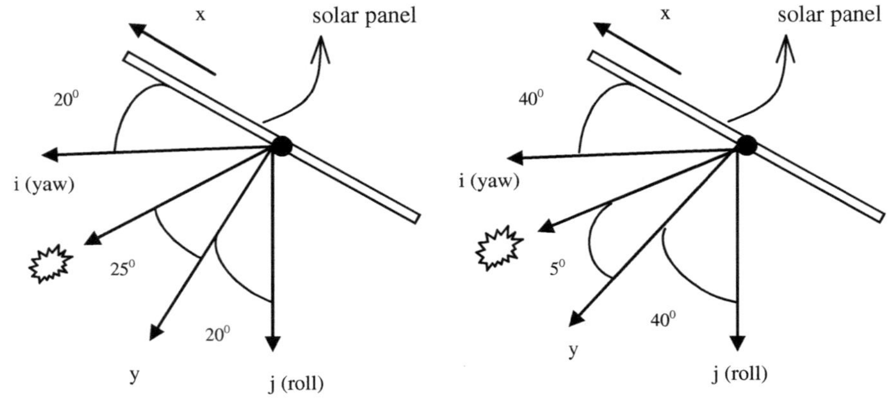

Fig. 5 Solar panel orientation for SPDM offset −20° and −40° respectively

x unit vector along the panel
y unit vector normal to the panel

Body axes are represented by 3 mutually perpendicular axes defined as yaw, roll and pitch where yaw and roll are as shown in the figure and pitch axis is given by the cross product of yaw and roll.

i unit vector along yaw axis
j unit vector along roll axis
k unit vector along pitch axis

Sun lies in yaw–roll plane making an angle of 45° with roll axis and 25° with solar panel normal. Solar panel was having SPDM (solar panel drive motor) offset of −20°. A zero SPDM offset corresponds to panel normal along roll axis. Torque acting on spacecraft as seen from telemetry was

$$\tau(\mathrm{s/c}, -20^\circ) = [22.2\mathrm{i} - 33.33\mathrm{j} + 0\,\mathrm{k}] \times 10^{-6}\,\mathrm{NM} \quad (17)$$

where τ (s/c, −20°): Torque acting on spacecraft for SPDM offset −20°.

Now, let us compute torque on the spacecraft due to SRP effect on solar panel. Representing x and y in terms of body axes (i, j and k), we have

$$\mathrm{x} = \cos(20^\circ)\,\mathrm{i} - \sin(20^\circ)\,\mathrm{j} = 0.9396\mathrm{i} - 0.3420\mathrm{j}$$
$$\mathrm{y} = \sin(20^\circ)\,\mathrm{i} + \cos(20^\circ)\,\mathrm{j} = 0.3420\mathrm{i} + 0.9396\mathrm{j}$$

Now, $\frac{I}{c} = 4 \times 10^{-6}$ J/m^3 (for above mentioned Sun–spacecraft distance)

θ(Sun incident angle) = −25° (Figs. 2 and 5)

$$F = \frac{I}{c}\iint [(1 - S_c)\cos\theta\sin\theta\, x - \left\{(1 + S_c)\cos^2\theta + \frac{2}{3}D_c\cos\theta\right\} y]dA \text{ (Eq.) 15}$$

Putting the values

$$\begin{aligned} F = {} & 4 \times 10^{(-6)}[(1 - 0.08)\cos(-25°) \times \sin(-25°)(0.9396i - 0.3420j) - \{(1 + 0.08)\cos^2 25° \\ & + \frac{2}{3} \times 0.03 \times \cos 25^{\circ}\}(0.3420i + 0.9396j)] \\ & \times 7.56\ N = (-19.37i - 22.08j) \times 10^{-6} N \end{aligned} \tag{18}$$

$$\tau = \frac{I}{c}\iint \left(r_{dA} - r_g\right)\left[(1 - S_c)\cos\theta\sin\theta\, x - \left\{(1 + S_c)\cos^2\theta + \frac{2}{3}D_c\cos\theta\right\} y\right] dA$$

Since the panel is rectangular and uniform in shape and so for all practical purposes centre of pressure for the panel will be its geometrical centre. Thus, moment arm for the panel about centre of gravity of spacecraft is the difference between r_{cp} (panel geometrical centre or centre of pressure) and r_{cg} (centre of gravity) and is given by

$$r_{cp} - r_{cg} = 0.044i + 0.103j + 2.41\ k(\text{Data from MOM handbook}) \tag{19}$$

So,

$$\begin{aligned} \tau(\text{panel}, -20°) &= \left(r_{cp} - r_{cg}\right) \times F \\ &= [0.53i - 0.46j + 0.01\ k] \times 10^{-4}\ \text{NM (Eqs. 18, 19)} \end{aligned} \tag{20}$$

τ (panel, −20°): Torque acting on spacecraft due to SRP effect on solar panel for SPDM offset −20°. It was desired to check the torque experienced by spacecraft for various SPDM offsets without disturbing body attitude (orientation).

3.1 Technique Involved

Assume S/C is maintained in a particular attitude with a SPDM offset θ_1. Torque acting on S/C can be defined as

Torque on S/C for SPDM offset θ_1 = Torque due to panel for SPDM offset θ_1 + Torque due to rest of body for SPDM offset θ_1

$$\tau(s/c,\,\theta_1) = \tau(\text{panel},\,\theta_1) + \tau(\text{rest of body},\,\theta1)$$

If the offset is changed to θ2 by maintaining the same attitude

$$\tau(s/c,\,\theta_2) = \tau(\text{panel},\,\theta_2) + \tau(\text{rest of body},\,\theta_2)$$

Since attitude is the same for both offsets and so it is clear that

$$\begin{aligned}
&\tau(\text{rest of body}, \theta_2) = \tau(\text{rest of body}, \theta_1)\\
&\text{or, } \tau(s/c, \theta_1) - \tau(\text{panel}, \theta_1)\\
&= \tau(s/c\ , \theta_2) - \tau(\text{panel}\ , \theta_2)\\
&\tau(s/c, \theta_2) = \tau(s/c\ , \theta_1) + [\tau(\text{panel}\ , \theta_2) - \tau(\text{panel}\ , \theta_1)]
\end{aligned} \tag{21}$$

3.2 *Torque Prediction for SPDM Offset –40°*

$$\tau(s/c,\ -40°) = [29.9i - 39.66\ j + 0.14k] \times 10^{-6}\,\text{NM [seen from telemetry]}$$

Now, let us compute torque acting on spacecraft due to panel, τ (panel, –40°) Representing x and y in terms of body axes (i, j and k) (Fig. 5).

$$\begin{aligned}
&x = \cos(40°)\ i - \sin(40°)\ j = 0.7660i - 0.6427j\\
&y = \sin(40°)\ i + \cos(40°)\ j = 0.6427i + 0.7660j\\
&\theta\,(\text{sun inciden angle}) = -5°
\end{aligned}$$

$$\begin{aligned}
F = {} & 4 \times 10^{-6}[(1 - 0.08)\cos(-5°) \times \sin(-5°)(0.7660i - 0.6427j) - \{(1 + 0.08)\cos^2 5°\\
& + \frac{2}{3} \times 0.03 \times \cos 5°\}(0.6427i + 0.7660j)] \times 7.56\,\text{N (Eq. 15)}
\end{aligned} \tag{22}$$

$$\begin{aligned}
&\tau(\text{panel},\ -40°) = (r_{cp} - r_{cg}) \times F\\
&= [0.65i - 0.47j + 0.01k] \times 10^{-4}\text{NM (Eqs. 19, 22)}\\
&\tau(s/c,\ -40°) = \tau(s/c,\ -20°) + [\tau(\text{panel},\ -40°) - \tau(\text{panel},\ -20°)]\ \text{(Eq. 21)}\\
&= [22.2i - 33.33j + 0k] \times 10^{-6}\text{NM} + [0.65i - 0.47j\ + 0.01k] \times 10^{-4}\\
&\quad - [0.53i - 0.46j + 0.01k] \times 10^{-4}\ \text{(Eqs. 7, 20, 23)}\\
&= [34.2i - 34.3j + 0.0k] \times 10^{-6}\,\text{NM}
\end{aligned} \tag{23}$$

Table 1 Torque acting on spacecraft (Predicted and Observed)

Torque for SPDM Offset −40°	Predicted (NM)	Observed (NM)
Yaw	34.2×10^{-6}	29.9×10^{-6}
Roll	-34.3×10^{-6}	-39.6×10^{-6}
Pitch	0	0.14×10^{-6}

4 Result and Discussion

The torque acting on the spacecraft platform (predicted and observed) due to solar radiation pressure is summarized in Table 1.

The table depicts that the predicted value using SRP model is very close to the actual value as observed from the telemetry. The mathematical model of force and torque due to solar radiation pressure derived in this article can be implemented on any interplanetary spacecraft travelling in heliocentric orbit where the most significant perturbation acting on the spacecraft is solar radiation pressure.

5 Conclusion

In this paper, particle nature of wave-particle duality of electromagnetic radiation is considered and applying the Newtonian mechanics a mathematical model is derived to estimate the disturbance acting on the spacecraft platform due to solar radiation pressure. The model requires the knowledge of optical properties of the material and area distribution of components of spacecraft. The model is highly useful to estimate the best possible orientation of spacecraft where the effect of SRP is minimum. The model is already verified for Mars Orbiter Mission during its journey in heliocentric orbit (Table 1).

References

1. Planck constant (2017, April 10). Retrieved from https://en.wikipedia.org/wiki/ Planck_constant.
2. Solar Energy Reaching the Earth's Surface. (n.d.). Retrieved from http://www.itacanet.org/-sun-as-a-source-of-energy/part-2-solar-energy-reaching-the-earths-surface.
3. Diffuse Reflection. (2017, April 2). Retrieved from https://en.wikipedia.org/wiki/Diffuse_reflection.
4. Lambert's cosine law (2017, April 13). Retrieved from https://en.wikipedia.org/wiki/ Lambert %27s_cosine law.

A Tree-Based Graph Coloring Algorithm Using Independent Set

Harish Patidar and Prasun Chakrabarti

Abstract This paper introduces a tree data structure-based graph coloring algorithm. Algorithm explores vertices in the tree form to finds maximal independent set, than these independent sets are colored with minimum colors. Proposed algorithm is tested on various DIMACS standard of graph instances. Algorithm is design to solve graph coloring problem for high degree graphs, i.e. the proposed algorithm is highly efficient for those graphs which has number of edges to number of vertices ratio is very high. Worst and best case time complexity of proposed algorithm is also discussed in this paper.

Keywords Maximal independent set · Complement edge table
Graph coloring · Chromatic number

1 Introduction

Graph coloring problem has three different areas of problem. Fist is vertex coloring; second edge coloring and third one is phase coloring. This paper is focused on vertex coloring problem. In the vertex coloring for any given graph G = (V, E), where V is set of vertices E is set of edges, and C = {1, 2, 3, …, K} is set of colors, each vertex must assign a color from set of color in such a way that colors of two connected vertices must be different, i.e., f: V_C such that for each [u, v], f(v) is not equal to find f(u).

H. Patidar (✉) · P. Chakrabarti
Department of Computer Science and Engineering,
Sir Padampat Singhania University, Udaipur, India
e-mail: harish.patidar@gmail.com

P. Chakrabarti
e-mail: prasun.chakrabarti@spsu.ac.in

C. R. Panigrahi et al. (eds.), *Progress in Advanced Computing and Intelligent Engineering*, Advances in Intelligent Systems and Computing 714,
https://doi.org/10.1007/978-981-13-0224-4_48

1.1 Applications of Graph Coloring Problem

Graph coloring problem has a wide area of applications like nearest neighbor search [1], register allocation in compiler [2], social networking, puzzles like Sudoku solving, frequency allocation in cellular network, cognitive radio dynamic frequency distribution and many more.

1.2 Algorithms of Graph Coloring Problem

Vertex coloring or more formally it can be called as graph coloring algorithms are designed and implemented to solve the graph coloring problems. There are certain objectives to design graph coloring algorithm. The primary objective of most graph coloring algorithm is to find minimum number of colors for coloring vertices of graphs. Many algorithms also try to achieve some other objectives like optimization of time and space complexity, improvement in execution success rate [3], finding efficient algorithm for large graphs where number of vertices in graph is high and many more.

On the basis of execution pattern, graph coloring algorithm can be sequential and parallel. Graph coloring algorithm is broadly divided into two categories one is exact approach and another next one is approximate [4]. Exact method's execution success rate is high but they are not efficient for the large graphs. Approximate algorithm give results on optimum time for large graphs but their execution success rate is low.

There are many algorithms already proposed by researchers like ant colony optimization algorithm [5, 6], Cuckoo optimization, Parallel genetic algorithm [7], Modified cuckoo optimization GCA [3], constructive hyper heuristic algorithm [8], and many more.

2 Proposed Algorithm

This paper proposed an algorithm to solve the vertex coloring problem for higher degree graph. Proposed algorithm is based on finding maximum independent set using tree data structure.

2.1 Maximum Independent Sets

The subsets of the graph containing those vertices that are not connected, i.e. no element in the set is connected to any other element of the same set, are known as independent sets. And, as the name suggests "Maximal Independent Sets" are those independent sets that contain maximum number of vertices.

2.2 Proposed Algorithm on Petersen Graph

For the graph in Fig. 1 the independent sets are shown in Table 1.

In Table 1, there are two maximal independent sets starting from vertex 1, i.e., Set1 and Set5. Based on the rules of proposed algorithm, Set1 is selected for further exploration.

Now, as Vertex 2 is not included in Set1, Vertex 2 is to explore independent sets for with those vertices that are not included in Set1. Table 2 shows the independent set starts from vertex 2.

Again, based on the rules, according to algorithm Set11 is selected and so Vertex 5 is explored as shown in Table 3, being the first un-included vertex till now.

As, all the vertices of graph are now included, three maximal independent sets can be selected, Set1, Set11, and Set17. Now, as it is known that each element in a maximal independent set is disconnected with others. A single color can be assigned to all the elements of each maximal independent set. Thus, assigning every maximal independent set a different color, proper minimum coloring can be

Fig. 1 Petersen graph

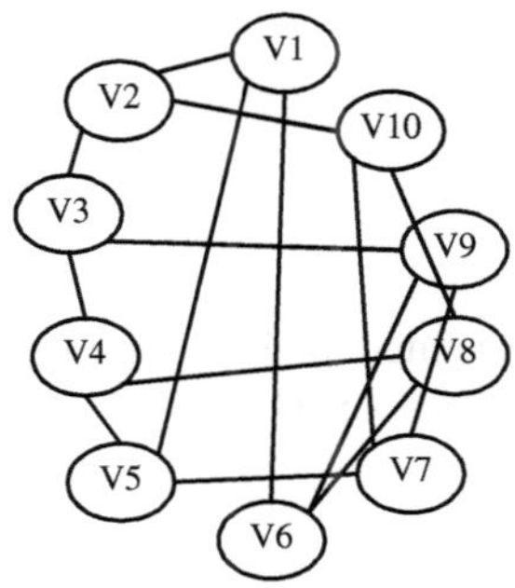

Table 1 Independent sets starting from vertex 1

S. No.	Set	Vertices
1	Set1	V1, V3, V7, V8
2	Set2	V1, V3, V8
3	Set3	V1, V3, V10
4	Set4	V1, V4, V7
5	Set5	V1, V4, V9, V10
6	Set6	V1, V4, V10
7	Set7	V1, V7, V8
8	Set8	V1, V8, V9
9	Set9	V1, V9, V10
10	Set10	V1, V10

Table 2 Independent sets starting from vertex 2

S. No.	Set	Vertices
1	Set11	V2, V4, V6
2	Set12	V2, V4, V9
3	Set13	V2, V5, V6
4	Set14	V2, V5, V9
5	Set15	V2, V6
6	Set16	V2, V9

Table 3 Independent sets starting from vertex 3

S. No.	Set	Vertices
1	Set17	V5, V9, V10
2	Set18	V5, V10

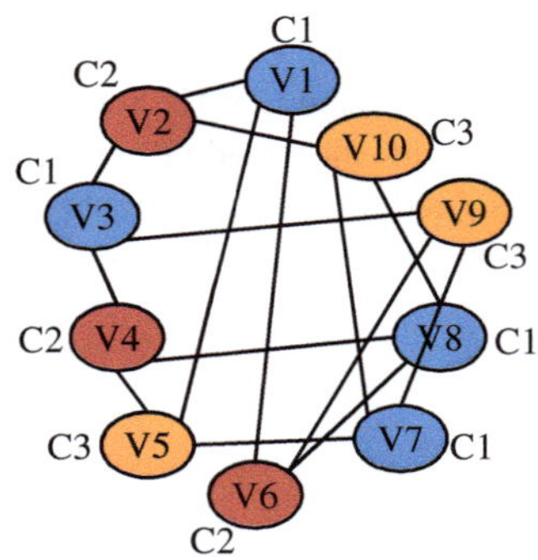

Fig. 2 Colored Petersen graph using the proposed algorithm

assigned as shown in Fig. 2. C1 color is assigned to Set1, C2 color to Set11, and C3 color to Set17.

2.3 *Algorithm*

Here, this paper proposed a vertex coloring algorithm that calculates minimum colors for graphs. Entire algorithm is divided into three steps. The first step is the development of complement edge table. Second, is finding maximum independent sets. And the third step is coloring of the maximal independent sets.

Step 1: *Complement Edge Table*

Complement edge table is the opposite of edge table. In order to find maximal independent sets, it is required to put together those vertices that are not connected by each other. And, so if complement edge table defines which vertices are not connected, it would reduce the time complexity significantly.

By scanning the edge table, make a new edge table that comprises of only those edges which were not in the original edge table. Also, include only those edges which originate from a vertex of smaller numbering than its destination.

In this step, when making the complement edge table, the algorithm also calculates the number of occurrences of every vertex, i.e., the degree of each vertex.

Step 2: *Finding Maximal Independent Sets*

This is the core of proposed algorithm. In this step, algorithm proceeds by exploring a tree for each maximal independent set. This step itself is a multistep process, which are as follows:

i. *Initiation Step*: To find the sets, algorithm needs to commence from somewhere, and so, the initiation step is defined to select the first vertex that is not yet included in any maximal independent set. If, there is no maximal independent set yet, algorithm can be start from vertex 1. Make a new empty maximal independent set.

ii. *Tree exploration*: This section has two main activities.

a. Every vertex that is greater in numbering than the selected vertex and is a connection in the complement edge table is made a child of the selected vertex given it is not included in any maximal set till now.

b. For all children, explore one by one by making the vertices that are in connection with complement edge table, are not included in any maximal set, and are siblings of the vertex being explored. This step is repeated till there are no more vertices which can be explored.

iii. *Path selection*: Then select the path with maximum length. If more than one path has maximum length, then the sum of degree for each such path is calculated and selects the path with minimum sum of degrees. The sum of degree is calculated as

$$\text{Sum} = \text{sum of degrees of all vertices in the path} - \text{L} * (\text{L} - 1)/2 \qquad (1)$$

where L is the length of the path.

If, more than one longest path has minimum sum of degrees, first traversing left to right is selected.

iv. Path to the maximal independent set is added for each vertex in the paths, degree of all its connections is decremented.
v. Step i through iv is repeated until all the vertices are included in some maximal independent set.

Step 3: *Coloring the Maximal Independent Sets*

This is the final step of the proposed minimum coloring algorithm. This step assigns a different color to each maximal independent set, i.e. all the vertices that belong to

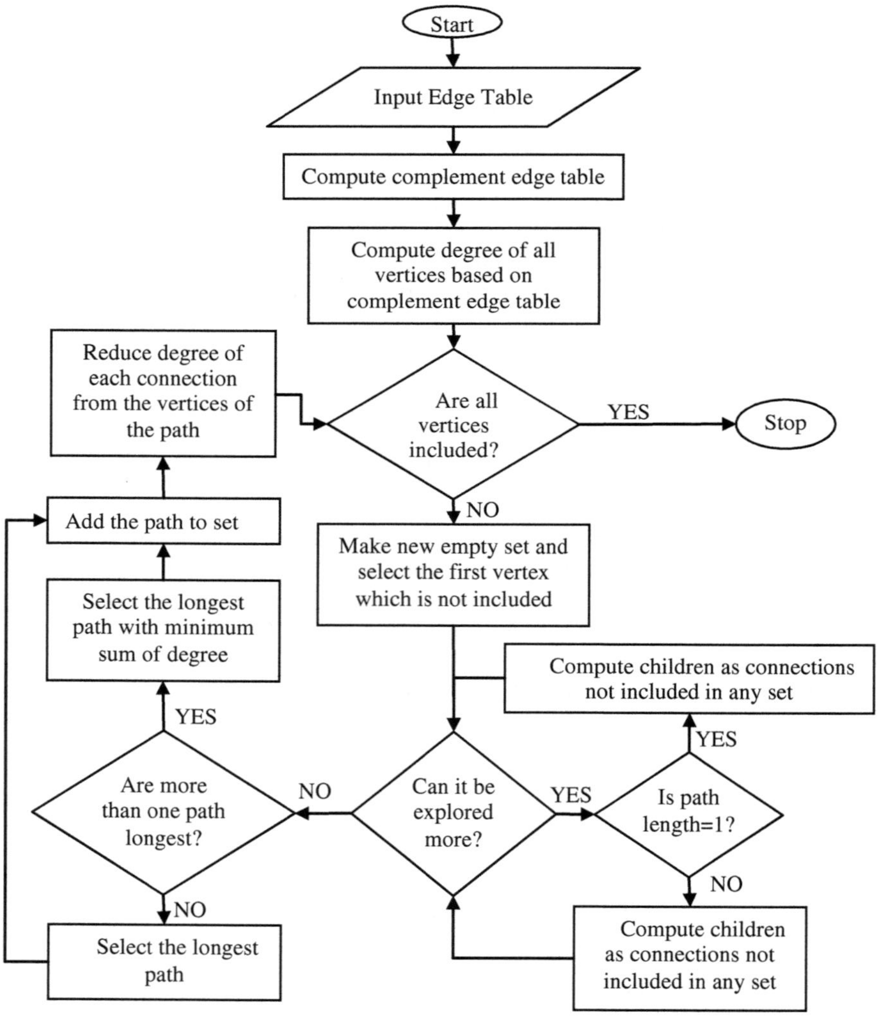

Fig. 3 Flowchart of the proposed algorithm

the same maximal independent set are assigned the same color and all the vertices that belong to different independent sets, now have different colors.

In Fig. 3 entire flow of the proposed algorithm has been shown.

3 Complexity Analysis

Worst-Case Complexity: The complexity of the proposed algorithm depends on the tree exploration. And, the maximum number of nodes would be explored when all the vertices are isolated, i.e., the edge table is empty, making $^{n}C_2/2$ edges in the complement edge table, where n is the number of vertices. So, complexity can be calculated for exploring the tree when all the vertices are isolated.

If n is the number of vertices, then V1 would have n − 1 connections. Similarly, V2 would have n − 2, V3 would have n − 3 connections, and so on, since the algorithm is accounting only the edges to vertices greater in numbering. Equation 2 is the recursion equation for the exploration.

$$T(n) = T(n-1) + T(n-2) + T(n-3) + \cdots + T(2) + T(1) \tag{2}$$

And the initial conditions would be

T(1) = 0 and T(2) = 1

From Eq. (2)

$$T(n-1) = T(n-2) + T(n-3) + \cdots + T(2) + T(1) \tag{3}$$

Subtracting (3) from (2)

$$\begin{gathered} T(n) - T(n-1) = T(n-1), \\ T(n) = 2 * T(n-1) \end{gathered} \tag{4}$$

By Eq. 4, the worst-case complexity of the algorithm is O(2n − 1).

4 Result Analysis

4.1 Test Data Sets

DIMACS graph instances are taken as data set for experiment results analysis of the proposed algorithm. DIMACS (Center for Discrete Mathematics and Theoretical Computer Science) defined a format to represent undirected graphs [9], which has been used by most of the researchers for analysis of their graph coloring algorithms. In this format graph data are stored in an input file, which contains all information about graph. In this input file, nodes are numbered from 1 to n. Edges are stored in the form of edge list like “e 1 2”.

4.2 Algorithm Implementation Platform

The proposed algorithm is implemented using Java programming language. Window 7 Ultimate 64-bit operating system platform with Intel(R) Core(TM) 2 Duo 2.10 GHz with 2 GB Installed memory (RAM) is used for algorithm execution.

4.3 Experimental Results

The proposed algorithm is tested on 26 DIMACS graph instances, includes queen graphs, DSJC series graphs, miles graphs, random series graphs, insertion and full insertion graphs. In Table 4 all experimental results are shown. Table contains the

Table 4 Experimental results of the proposed algorithm

Instance	V	E	Avg degree	K (Result)
myciel3	11	20	3.64	4
myciel4	23	71	6.17	5
queen5_5	25	320	25.60	5
1-FullIns_3	30	100	6.67	4
queen6_6	36	580	32.22	10
2-Insertions_3	37	72	3.89	4
myciel5	47	236	10.04	6
queen7_7	49	952	38.86	7
queen8_8	64	1456	45.50	10
queen9_9	81	2112	52.15	12
queen8_12	96	2736	57.00	15
queen10_10	100	2940	58.80	13
queen11_11	121	3960	65.45	15
DSJC125.9	125	6961	111.38	52
miles1500	128	10,396	162.44	74
miles1000	128	6432	100.50	45
DSJC250.9	250	27,897	223.18	82
DSJC250.5	250	15,668	125.34	35
DSJC500.9	500	224,874	899.50	150
DSJR500.1c	500	121,275	485.10	104
latin_square_10	900	307,350	683.00	119
DSJC1000.9	1000	449,449	898.90	259
R100_9 g	100	4438	88.76	42
R100_9 gb	100	4438	88.76	42
R50_9g	50	1092	43.68	25
R50_9gb	50	1092	43.68	24

Table 5 Experimental results of the proposed algorithm on geometric series graphs

Instance	V	E	Avg degree	K (Result)
GEOM20	20	40	4.00	5
GEOM20a	20	57	5.70	5
GEOM20b	20	52	5.20	4
GEOM30	30	80	5.33	6
GEOM30a	30	111	7.40	6
GEOM30b	30	111	7.40	5
GEOM40a	40	186	9.30	7
GEOM40b	40	197	9.85	7
GEOM50a	50	288	11.52	10
GEOM50b	50	299	11.96	10
GEOM60a	60	399	13.30	11
GEOM60b	60	426	14.20	11

Table 6 Comparison of the proposed algorithm with HPGAGCP

Instance	K (Proposed)	K (HPGAGCA)
myciel3	4	4
myciel4	5	5
queen5_5	5	5
myciel5	6	6
queen7_7	7	8
queen8_8	10	10

instance name, number of vertices in graph (V), number of edges in graph (E), average degree of vertices in graph (AvgDegree) and number of colors required to color the graph, which are generated by the proposed algorithm (K).

From the experimental results, it has been found that proposed algorithm gives optimum results for high degree of graphs.

The proposed algorithm implementation is also tested on 12 geometric series graphs, these are weighted with bandwidth. But proposed algorithm implementation is not for weighted graphs so in algorithm weight is ignored from the data records. Also there is no use of bandwidth in the proposed algorithm, so that bandwidth is also ignored. In Table 5 experimental results of geometric series graphs (GEOM) can be seen.

In Table 6 few experimental results are also compared with a well-known hybrid genetic algorithm for graph coloring problem (HPGAGCP) [9]. Some interesting results are found in Table 6.

5 Conclusion

Generally complexity of any graph coloring algorithm is high for high degree graphs. But the proposed algorithm is based on complementary edge table. If the degree of graph is higher then size of complement edge table is small. Exploring tree through this complement edge table does not extend up to high level, i.e., algorithm is executed in optimum time and space complexity.

Tree-based graph coloring algorithm using independent set (proposed algorithm) is an efficient algorithm to calculate the number of colors required and assign to vertices. Time complexity of this algorithm is optimum for high degree graphs. The algorithm gives number of colors precisely equal to the chromatic number of graphs.

References

1. Berchtold, S., Böhm, C., Braunmüller, B., Keim, D. A., and Kriegel, H.-P.: Fast Parallel Similarity Search in Multimedia Databases, In ACM SIGMOD Int. Conf. on Management of Data, (1997)
2. Chaitin, G. J., Auslander, M. A., Chandra, A. K., Cocke, J., Hopkins, M. E., and Markstein, P. W.: Register Allocation via Coloring. Computer Languages, Vol. 6, Issue 1, (1981) 47–57
3. Mahmoudi, S., Lotfi, S.: Modified cuckoo optimization algorithm (MCOA) to solve graph coloring problem. ELSVIER, Applied Soft Computing, (2015) 48–64
4. Gupta A., Patidar H.: A Survey on Heuristic Graph Coloring Algorithm. International Journal for Scientific Research & Development, Vol. 4, Issue 04, (2016) 297–301
5. Salari, E., and Eshghi, K.: An ACO Algorithm for the Graph Coloring Problem. Interracial Journal Contemp. Math Sciences, Vol. 3, no. 6 (2008) 293–304
6. Thang, N. Bui, Nguyen, T. H., Patel, C. M., and Kim-Anh Phan, T.: An Ant-Based Algorithm for Coloring Graphs. Discrete Applied Mathematics 156 (2008) 190–200
7. Chen, B., Chen, Bo., Liu, H., Zhang X.: A Fast Parallel Genetic Algorithm for Graph Coloring Problem Based on CUDA. IEEE International Conference on Cyber-Enabled Distributed Computing and Knowledge Discovery, (2015) 145–148
8. Sabar, N. R., Ayob M., Qu, R., Kendall, G.: A Graph Coloring Constructive Hyper-Heuristic for Examination Time Tabling Problems. Online publication, Springer Science Business Media, (2011)
9. Hindi, M., and Yampolskiy, R. V.: Genetic Algorithm Applied to the Graph Coloring Problem, Proc. 23rd Midwest Artificial Intelligence and Cognitive Science Conf, (2012) 61–66

Low-Complexity MPN Preemption Policy for Real-Time Task Scheduling

Kiran Arora, Savina Bansal and Rakesh Kumar Bansal

Abstract Preemption plays a vital role in deciding and guaranteeing schedulability of real-time tasks. Over time, many preemption policies have been suggested in the literature ranging from no-preemption, full preemption, limited preemption, and recently a new MPN (mixed preemptive/non-preemptive) preemption policy. Optimal Algorithm (OA) given by Lee et al. for the disallowance of preemption can be further improved for low-time complexity at a little cost of reducing the number of schedulable task sets found. In this work, it is conjectured that if only a certain number of higher density tasks are selected for disallowance of preemption, then time complexity of Optimal Algorithm can be reduced substantially. Accordingly, a new Low-Complexity MPN (LCMPN) is proposed, implemented, and analyzed which has a lower time complexity than Optimal Algorithm. Simulated results of the proposed LCMPN in comparison to the Optimal Algorithm on the tested constrained task sets, justifies our conjecture.

Keywords Scheduling · Preemption · EDF · MPN

K. Arora (✉)
Department of RIC, IKGPTU, Jalandhar, India
e-mail: erkiranarora@gmail.com

K. Arora
Department of CSE, BHSBIET, Lehragaga, India

S. Bansal · R. K. Bansal
Department of ECE, GZSCCET, Bathinda, India
e-mail: savina.bansal@gmail.com

S. Bansal · R. K. Bansal
MRSPTU, Bathinda, India
e-mail: drrakeshkbansal@gmail.com

C. R. Panigrahi et al. (eds.), *Progress in Advanced Computing and Intelligent Engineering*, Advances in Intelligent Systems and Computing 714,
https://doi.org/10.1007/978-981-13-0224-4_49

1 Introduction

Silicon chip advancements have paved the way for technological revolution. Due to this rapid development in processor technology, real-time applications have got a boost. The output of Real-Time System is acceptable not only when it gives logically correct results, but also on its timeliness. Task scheduling plays a vital role in improving performance of real-time applications and hence being paid much attention by researchers.

The traditional real-time scheduling algorithms like Earliest Deadline First (EDF) and Rate Monotonic Scheduling (RMS) are based on the ideal characteristics of tasks. EDF has received significant attention in real-time scheduling for uniprocessor platform due to its optimal properties and hence chosen as baseline algorithm for scheduling in this work.

Under the ideal case of negligible preemption cost, preemptive scheduling is shown to be more efficient than non-preemptive scheduling. This is because if lower priority tasks are executed non-preemptively, than they will result in blocking of higher priority tasks and produce the blocking time. Blocking sometimes results in missed deadlines. In practice, preemptions generate a nonzero run-time overhead and can result in significant variation in the task's total execution time. This reduces system predictability.

Many limited preemption techniques have been presented in the literature to overcome the preemption's limitations. Various techniques are being proposed for dealing with preemption overheads [6–8, 13]. Recently, Lee et al. proposed controlled preemption for uniprocessor and MPN (Mixed Preemptive/ Non-preemptive) for multicore platforms [21, 22]. In the policies suggested, few of the tasks are designated as non-preemptive by assigning certain parameters based on their transactional properties [21].

In this work, MPN policy for preemption has been taken into consideration. It is conjectured that allowance/disallowance of preemption should not be done randomly as it depends on the property of a task. The Optimal Algorithm for the disallowance of preemption given in [21] can be improved for its time complexity by considering high-density tasks first for disallowance of preemption. In this paper, a Low-Complexity Algorithm for allowance/disallowance of preemption has been proposed and presented. Further, the organization of paper is as follows. Sect. 2 discusses the related work done. System model has been presented in Sect. 3. Section 4 describes schedulability tests for MPN policy. Section 5 elaborates the improved LCMPN algorithm and Sect. 6 evaluates simulation results. Lastly, Sect. 7 concludes the paper.

2 Related Work

Liu and Layland [23] proposed an optimal dynamic priority scheduling algorithm for uniprocessor known as Earliest Deadline First. He showed that in EDF scheduling algorithm, active job with smallest deadline is chosen for execution at every instant of time. Dertouzos [14] proposed the EDF for sporadic tasks. Both these works considered zero preemption overhead, an assumption that was later shown unjustified as context switching overhead gets added to the execution time of a job [17–19, 25].

Other works dealing with non-preemptive and fully preemptive (FP) policies include [1, 18]. Some of them worked on improving the schedulability of FP policy [9]. Jeffay et al. [16] focused on non-preemptive EDF. Baruah [4] took up the policy of limited preemption to EDF. In this policy, the maximum duration for which the task executes non-preemptively is prespecified for each job.

A lot of work has been done on multiprocessor platform for schedulability test of fully preemptive algorithms, but only a few studies concentrate on non-preemptive algorithms [5, 15]. Lee et al. proposed a new policy where few tasks are disallowed for preemption, making them non-preemptive. So, the task set contains a mix of preemptive and non-preemptive tasks. Schedulability test has also been proposed by him for both uniprocessor platform [22] and multicore platform [21].

3 System Model

This paper focuses on sporadic real- time task model [24]. In this model task $\tau_i \epsilon \tau$, τ is a task set. Let the total number of tasks in the task set is 'n'. τ_i is modeled as (P_i, C_i, D_i) where P_i is the minimum separation between two consecutive task invocations, C_i is worst-case execution time which is the maximum amount of time required for the completion of execution of task τ_i on processor and D_i is its relative deadline. Each task τ_i invokes a series of jobs, where two consecutive jobs are separated by no less than P_i time units. Constrained deadline task set model has been considered here for analysis. If task parameter $Y_k = 1(=0)$, then the kth task is considered as preemptive (non-preemptive).

The target processor platform is multicore platform with ϖ identical cores. Global work conserving algorithm has been considered where job can be executed on any processing core. No core will remain idle if there is an incomplete job in a queue waiting for execution.

4 Schedulability Test

Bertogna et al. [11] presented the sufficient schedulability test for sporadic task set with constrained deadline based on task density which is as follows:

$$\delta_{sum} = \varpi - (\varpi - 1)\delta_{max} \tag{1}$$

where,

δ_{sum} is total density of task set
δ_{max} is the maximum density of any task in the task set
ϖ is number of processors.

Later, the schedulability test for work conserving algorithm like EDF for the same model was proposed by Bertogna et al. [10] is as follows:

Response Time Analysis (Theorem 3 in [10]): A task $\tau_k \in \tau$ is schedulable, if every job J_k^p invoked by τ_k satisfies following equation for some $C_k \leq \mathbb{Z} \leq D_k$:

$$C_k + \left\lfloor \frac{1}{m} \sum_{\tau_i \epsilon \tau - \{\tau_k\}} \min\left(\hat{I}_k^i\left(r_k^p, r_k^p + \mathbb{Z}\right), \mathbb{Z} - C_k + 1\right) \right\rfloor \leq \mathbb{Z} \quad (2)$$

In Eq. (2), RTA for EDF considers the following upper bound:

$$\hat{I}_k^i\left(r_k^p, r_k^p + \mathbb{Z}\right) \leq \min(\Psi_i\left(\mathbb{Z}, S_i\right), E_i(D_k, S_i)) \quad (3)$$

where,
$\Psi_i(\mathbb{Z}, S_i)$ is an upper bound on the workload of task τ_i for interval $\mathbb{Z}$ and
$E_i(D_k, S_i)$ is the interference of a task τ_i on a task τ_k in an interval of size equal to D_k.
$\Psi_i(\mathbb{Z}, S_i)$ and $E_i(D_k, S_i)$ are define below:

Workload

$$\Psi_i\left(\mathbb{Z}, S_i\right) = \eta_i(\mathbb{Z})\, C_i + \min(C_i, \mathbb{Z} + D_i - C_i - \eta_i(\mathbb{Z})\, P_i) \quad (4)$$

$\eta_i(\mathbb{Z})$ is maximum number of jobs task τ_i in an interval $\mathbb{Z}$ and it is defined as follows:

$$\eta_i(\mathbb{Z}) = \left\lfloor \frac{\mathbb{Z} + D_i - C_i}{P_i} \right\rfloor \quad (5)$$

Interference

$$E_i\left(D_k, S_i\right) = \left\lfloor \frac{D_k}{P_i} \right\rfloor \cdot C_i + \max(0, \min(C_i, D_k - \left\lfloor \frac{D_k}{T_i} \right\rfloor \cdot P_i - S_i)) \quad (6)$$

Lee and Shin [21] further modified the above Response Time Analysis for MPN policy, to consider the response time of non-preemptive tasks. MPN-* scheduling algorithm has also been proposed in [21], which is similar to the work conserving EDF algorithm on multiprocessor except with a difference that if a new high priority job j_k is invoked and all the processors are busy with non-preemptive jobs then j_k has to wait in a queue. The schedulability of task set for MPN-* algorithm can be found as follows.

RTA for mpn-* (Theorem 1 in [21]): For MPN-* scheduling algorithm, in case of schedulable task set, the upper bound on the response time of preemptive task $\tau_k|Y_k = 1 \in \tau$ is $\mathbb{R}_k = \mathbb{R}_k^x$ such that $\mathbb{R}_k^{x+1} \leq \mathbb{R}_k^x$ holds in the following expression, initializing from $R_k^0 = C_k$:

$$\mathbb{R}_k^{x+1} \leftarrow C_k + \left\lfloor \frac{1}{m} \sum_{\tau_i \epsilon \tau - \{\tau_k\}} \min\left(\hat{I}_k^i\left(\mathbb{R}_k^x\right), \mathbb{R}_k^x - C_k + 1\right) \right\rfloor \tag{7}$$

And for non-preemptive tasks $\tau_i|Y_k = 0 \in \tau$, an upper bound on response time is $R_k = F_k^x + C_k + 1$ such that $F_k^{x+1} \leq F_k^x$ holds in the following expression, starting from $F_k^0 = 1$:

$$F_k^{x+1} \leftarrow 1 + \left\lfloor \frac{1}{m} \sum_{\tau_i \epsilon \tau - \{\tau_k\}} \min\left(\hat{I}_k^i\left(F_k^x\right), F_k^x\right) \right\rfloor \tag{8}$$

if $\forall \tau_k \in \tau$, $\mathbb{R}_k \leq D_k$ holds, then τ is schedulable by the algorithm, otherwise the task τ_k is deemed to be unschedulable. Iterations in above given equations continue until $\mathbb{R}_k \leq D_k$ and $F_k^x + C_k + 1 \leq D_k$.

Now, $\forall \tau_k \in \tau$, following inequalities hold for all $0 \leq \mathbb{Z} \leq D_k$ (Lemma 3 in [21]). If $\tau_k|Y_k = 1$,

$$\sum_{\tau_i \epsilon \tau - \{\tau_k\}} \min\left(\hat{I}_k^i(\mathbb{Z}), \mathbb{Z} - C_k + 1\right) \text{ in Eq. (7)} \leq \sum \min\left(\Psi_i\left(\mathbb{Z}, S_i\right), E\left(D_k, S_i\right), \mathbb{Z} - C_k + 1\right) + \sum_{\tau_i|Y_i=0\epsilon\tau-\{\tau_k\}} \min\left(\Psi_i\left(\mathbb{Z}, S_i\right), \mathbb{Z} - C_k + 1\right) \tag{9}$$

And if $\tau_k|Y_k = 0$,

$$\sum_{\tau_i \epsilon \tau - \{\tau_k\}} \min\left(\hat{I}_k^i(\mathbb{Z}), \mathbb{Z}\right) \text{ in Eq. (8)} \leq \sum \min\left(\Psi_i\left(\mathbb{Z}, S_i\right), E_i\left(D_k, S_i\right), \mathbb{Z}\right) + \sum_{mlargest\ \tau_i|Y_i=0\&D_i>D_k\epsilon\tau-\{\tau_k\}} \max(0, \min\left(\Psi_i\left(\mathbb{Z}, S_i\right), C_k - 1, \mathbb{Z}\right) - \min(\Psi_i\left(\mathbb{Z}, S_i\right), E_i\left(D_k, S_i\right), \mathbb{Z})) \tag{10}$$

5 Assignment of $\{Y_i\}$ for MPN-*

The Optimal Algorithm (OA) for the assignment of $\{Y_i\}$ for MPN-* given in [21] has a time complexity $O(n^4 \cdot max_{\tau_i \epsilon \varphi} D_i^2)$, where $O(n^3 \cdot max_{\tau_i \epsilon \varphi} D_i^2)$ is the time complexity of finding the task schedulability and $O(n)$ times it checks the schedulability of task set after making the unschedulable tasks non-preemptive until all tasks are non-preemptive or the task set is schedulable. If rather than checking the schedulability for all n tasks, only ϖ higher density tasks are made non-preemptive one at a time until the task set is schedulable, then time complexity can be reduced to $O(\varpi \cdot n^3 \cdot max_{\tau_i \epsilon \varphi} D_i^2)$. ϖ higher density tasks are only selected because generally high-density tasks are unschedulable due to having greater interference and workload resulting in large response time.

Proposed LCMPN algorithm is explained in the Algorithm 1 given below:

Algorithm 1 :Y_k Assignment for $\tau_i \epsilon \tau$

Require: Task set having tasks sorted in non-increasing order based on density.
```
 1: if task set is schedulable then return.
 2: Set counter = 1
 3: while counter ≤ ϖ do
 4:     Select the highest density preemptive task τ_k and set Y_k = 0
 5:     Check if task set is schedulable with respect to MPN-*
 6:     if schedulability = true then
 7:         schedulable = true return schedulable
 8:     else
 9:         schedulable = false
10:     end if
11: end while
        return schedulable
```

If $\varpi \ll n$, then there is significant amount of reduction in the time complexity of algorithm at a little cost of making the unschedulable task sets schedulable. It saves up to $O((n - \varpi) \cdot n^3 \cdot max_{\tau_i \epsilon \varphi} D_i^2)$ time which is a very large amount if the number of tasks is significantly larger than the number of processors.

6 Evaluation

6.1 Generation of Task Set

Task sets are generated as suggested in [2], which is also used by other researchers [12, 20, 21]. Constrained task sets are generated where $D_i \leq P_i$. Number of cores for which task sets are generated are 2, 4, and 16. The task utilization ($\frac{C_i}{D_i}$) distribution used is exponential with parameter 0.1, 0.3, and 0.5 (for the parameter value given as

$(1/\lambda)$, value for utilization is selected based on exponential distribution with pdf as $\lambda \cdot \exp(-\lambda \cdot p))$. For each task, T_i is uniformly distributed in (11,000) and execution time is calculated implicitly. Deadline is uniformly distributed in $[C_i, P_i]$.

For each experiment 10,000 task sets are generated for any given ϖ, according to the following procedure:

1. Initially, $\varpi + 1$ tasks are generated.
2. Generated task set is checked for necessary feasibility condition [3] (for the exclusion of unschedulable task sets).
3. A new set is created by adding a new task to the old set only if the task set passes the test in the previous step. Then again go to step 2 for feasibility check. Otherwise, if task set fails go to step 1 for next task set generation.

6.2 Simulation Results

Simulation is done on Windows 10 operating system having Intel Core i7-6700HQ CPU @ 2.60 GHz with 16 GB RAM by closing all the applications along with antivirus and disabling internet connectivity. The platform for coding is Eclipse Neon.1 IDE for JAVA EE.

Run-time reduction comparison of Optimal Algorithm and proposed Low-Complexity MPN algorithm are shown in Figs. 1, 2 on pp. 7, 8 and Fig. 3 on p. 8 for different number of processors.

When the difference between number of tasks (n) and number of processors (ϖ) is large, there is a significant difference in execution time of the two algorithms that

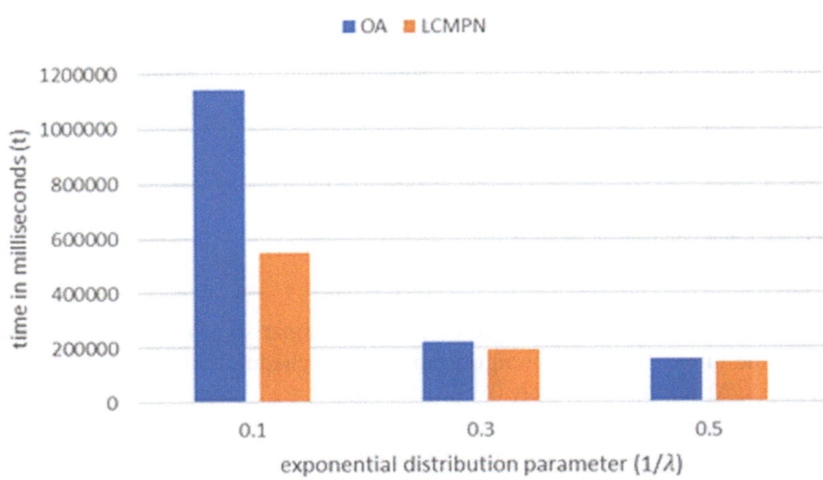

Fig. 1 Run-time comparison for $\varpi = 16$

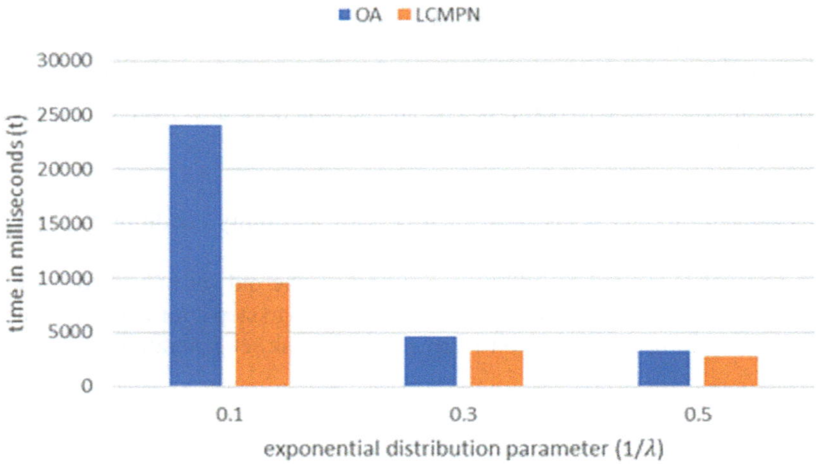

Fig. 2 Run-time comparison for $\varpi = 4$

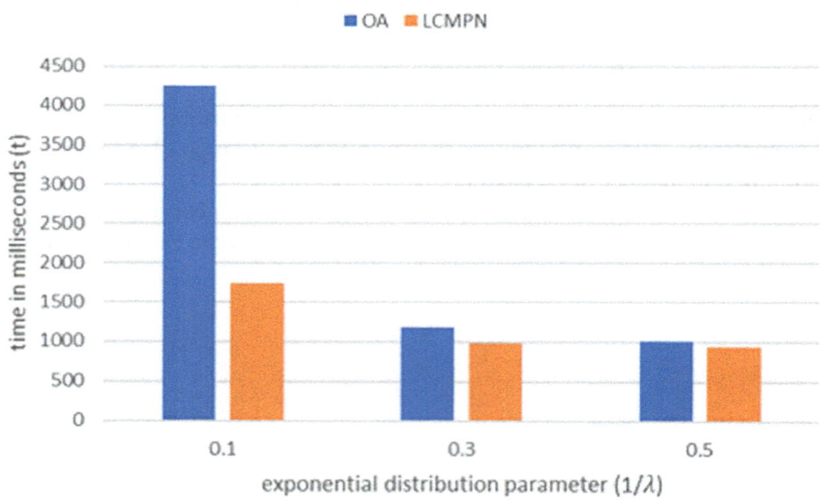

Fig. 3 Run-time comparison for $\varpi = 2$

can be seen from Fig. 1 on p. 7 where run-time comparison for $\varpi = 16$ is shown. Further, the impact of LCMPN algorithm on task schedulability can be seen from the Table 1. At a little loss of only 0.33% of finding extra number of schedulable task sets by Optimal Algorithm (OA), LCMPN is saving considerable execution time. Although the results are not substantial at small ϖ, but still there is improvement in execution time.

Table 1 Comparison of MPN and LCMPN in terms of schedulability (exponential distribution parameter $(1/\lambda) = 0.1$)

Number of processors (ϖ)	Average no of tasks in a task set (n)	Total task sets schedulable by OA	Total task sets schedulable by LCMPN	Extra task sets schedulable by OA
16	83	1797	1791	6 (0.33%)
4	20	4069	3993	76 (0.87%)
2	10	3090	3063	27 (1.86%)

7 Conclusion

In this paper, LCMPN algorithm has been suggested and implemented. It has been demonstrated that time complexity of Optimal Algorithm for the disallowance of preemption can be reduced by only considering the higher density tasks. So, whenever the time saving is more desirable (in terms of program execution time) by the application in comparison to a number of schedulable task sets, then LCMPN is a better choice.

References

1. Altmeyer, S., Davis, R.I., Maiza, C.: Cache Related Pre-emption Delay Aware Response Time Analysis for Fixed Priority Pre-emptive Systems. 2011 IEEE 32nd Real-Time Systems Symposium pp. 261–271 (2011)
2. Baker, T.P.: Comparison of Empirical Success Rates of Global vs. Partitioned Fixed-Priority and EDF Scheduling for Hard Real Time. Technical Report pp. 1–14 (2005)
3. Baker, T.P., Cirinei, M.: A necessary and sometimes sufficient condition for the feasibility of sets of sporadic hard-deadline tasks. Proceedings - Real-Time Systems Symposium (0509131), 178–187 (2006)
4. Baruah, S.: The limited-preemption uniprocessor scheduling of sporadic task systems. Proceedings - Euromicro Conference on Real-Time Systems 2005, 137–144 (2005)
5. Baruah, S.K.: The Non-preemptive Scheduling of Periodic Tasks upon Multiprocessors. Real-Time Systems 32(1–2), 9–20 (Feb 2006)
6. Baruah, S.K., Mok, A.K., Rosier, L.E.: Preemptively scheduling hard-real-time sporadic tasks on one processor. In: Proceedings - Real-Time Systems Symposium. pp. 182–190 (1990)
7. Bertogna, M., Baruah, S.: Limited preemption EDF scheduling of sporadic task systems. IEEE Transactions on Industrial Informatics 6(4), 579–591 (2010)
8. Bertogna, M., Buttazzo, G., Marinoni, M., Yao, G., Esposito, F., Caccamo, M.: Preemption points placement for sporadic task sets. In: Proceedings - Euromicro Conference on Real-Time Systems. pp. 251–260 (2010)
9. Bertogna, M., Buttazzo, G., Yao, G.: Improving feasibility of fixed priority tasks using non-preemptive regions. In: Proceedings - Real-Time Systems Symposium. pp. 251–260 (2011)
10. Bertogna, M., Cirinei, M.: Response-time analysis for globally scheduled symmetric multiprocessor platforms. Proceedings - Real-Time Systems Symposium pp. 149–158 (2007)
11. Bertogna, M., Cirinei, M., Lipari, G.: Improved schedulability analysis of EDF on multiprocessor platforms. In: Proceedings - Euromicro Conference on Real-Time Systems. vol. 2005, pp. 209–218 (2005)

12. Bertogna, M., Cirinei, M., Lipari, G.: Schedulability analysis of global scheduling algorithms on multiprocessor platforms. IEEE Transactions on Parallel and Distributed Systems 20(4), 553–566 (2009)
13. Buttazzo, G.C., Bertogna, M., Yao, G.: Limited Preemptive Scheduling for Real-Time Systems. A Survey. IEEE Transactions on Industrial Informatics 9(1), 3–15 (2013)
14. Dertouzos, M.L.: Control Robotics: The Procedural Control of Physical Processes. In: Proceedings of IFIP Congress. pp. 807–813 (1974)
15. Guan, N., Yi, W., Deng, Q., Gu, Z., Yu, G.: Schedulability analysis for non-preemptive fixed-priority multiprocessor scheduling. Journal of Systems Architecture 57(5), 536–546 (2011)
16. Jeffay, K., Stanat, D., Martel, C.: On non-preemptive scheduling of period and sporadic tasks. In: 1991 Proceedings Twelfth Real-Time Systems Symposium. pp. 129–139. No. December (1991)
17. Ju, L., Chakraborty, S., Roychoudhury, A.: Accounting for cache-related preemption delay in dynamic priority schedulability analysis. Proceedings - Design, Automation and Test in Europe, DATE pp. 1623–1628 (2007)
18. Lee, C.G., Hahn, J., Seo, Y.M., Min, S.L., Ha, R., Hong, S., Park, C.Y., Lee, M., Kim, C.S.: Analysis of cache-related preemption delay in fixed-priority preemptive scheduling. IEEE Transactions on Computers 47(6), 700–713 (1998)
19. Lee, C.G., Lee, K., Hahn, J., Seo, Y.M., Min, S.L., Ha, R., Hong, S., Park, C.Y., Lee, M., Kim, C.S.: Bounding cache-related preemption delay for real-time systems. IEEE Transactions on Software Engineering 27(9), 805–826 (2001)
20. Lee, J., Easwaran, A., Shin, I., Lee, I.: Zero-laxity based real-time multiprocessor scheduling. Journal of Systems and Software 84(12), 2324–2333 (2011)
21. Lee, J., Shin, K.: Improvement of Real-Time Multi-Core Schedulability with Forced Non-Preemption. IEEE Transactions on Parallel and Distributed Systems 25(5), 1233–1243 (2014)
22. Lee, J., Shin, K.G.: Preempt a job or not in EDF scheduling of uniprocessor systems. IEEE Transactions on Computers 63(5), 1197–1206 (2014)
23. Liu, C.L., Layland, J.W.: Scheduling Algorithms for Multiprogramming in a Hard-Real-Time Environment Scheduling Algorithms for Multiprogramming. Journal of the Association for Computing Machinery 20(1), 46–61 (1973)
24. Mok, A.K.: Fundamental design problems of distributed systems for the hard-real-time environment (1983)
25. Phavorin, G., Richard, P.: Cache-Related Preemption Delays and Real-Time Scheduling: A Survey for Uniprocessor Systems. Tech. rep. (2015)

A Non-autonomous Ecological Model with Some Applications

Jai Prakash Tripathi, Vandana Tiwari and Syed Abbas

Abstract Here, we propose a non-autonomous predator–prey system with feedback controls. Some of its possible applications in some branches of advanced computing have also been discussed. In particular, the global attractivity of the almost periodic solution (APS) is proved. The effect of control parameters has also been observed. The paper ends with some current possible applications of the work in bioinformatics, social networks, wireless sensor networks (WSNs), etc.

Keywords Finite automata mappings · Lotka–Volterra system · Almost periodic solution · Global attractivity · Social networks

1 Introduction

Almost periodicity and periodicity play an important role in several branches of sciences and engineering. Several branches of advanced computing and intelligent engineering (e.g., bioinformatics, social networks, wireless sensor networks, etc.) are not untouched with the concept of periodicity and almost periodicity. Periodicity in solenoid protein structure may be hidden at sequence level, however, evident at structure level. Periodic data prediction is one of the significant ways to reduce the number of transmissions in WSNs [1]. Similarly, spectral analysis of social network is

J. P. Tripathi
Department of Mathematics, Central University of Rajasthan, Bandar Sindri, Kishangarh, Ajmer 305817, Rajasthan, India
e-mail: jtripathi85@gmail.com

V. Tiwari (✉)
Department of Mathematical Sciences, Indian Institute of Technology (BHU) Varanasi, Varanasi 221005, Uttar Pradesh, India
e-mail: vandanpp@gmail.com

S. Abbas
School of Basic Sciences, Indian Institute of Technology Mandi, Mandi 175001, Himachal Pradesh, India
e-mail: sabbas.iitk@gmail.com

C. R. Panigrahi et al. (eds.), *Progress in Advanced Computing and Intelligent Engineering*, Advances in Intelligent Systems and Computing 714,
https://doi.org/10.1007/978-981-13-0224-4_50

done to identify periodicity [2, 3]. In brief, the generalization of periodicity (almost periodicity) becomes an interesting question as far as solenoid protein structure and WSNs are concerned. Moreover, the almost periodicity also comes in picture in the theory of finite automata mappings [4, 5]. In this respect, one can ask a question of generalization of the concept of periodicity in above fields like, bioinformatics, social networks, wireless sensor networks, etc. The main objective of the present study is the analysis of a non-autonomous model along with some of its possible applications in some branches of advanced computing and intelligent engineering.

A non-autonomous (systems with time-dependent periodic or almost periodic parameters) model system indicates the presence of temporal irregularity in the environment. In the recent years, almost periodicity in ecological modelling has been extensively studied by several authors [6–10]. Here we investigate the following predator–prey model system [11–15] incorporating feedback controls and prey refuge:

$$\begin{aligned}
\frac{du_1(t)}{dt} &= u_1(t)\left[a_1(t) - b_1(t)u_1(t) - \frac{\alpha(t)(1-m)u_1(t)u_2(t)}{\beta^2 u_2^2(t) + (1-m)^2 u_1^2(t)} - c_1(t)z_1(t)\right],\\
\frac{du_2(t)}{dt} &= u_2(t)\left[d_1(t) - \frac{e_1(t)u_2(t)}{(1-m)u_1(t)} - p_1(t)z_2(t)\right],\\
\frac{dz_i(t)}{dt} &= \mu_i(t) - v_i(t)z_i(t) + \delta_i(t)u_i(t).
\end{aligned} \tag{1}$$

Here, $z_i(t)$ denotes the control variables ($i = 1, 2$), $u_1(t)$ and $u_2(t)$ denote the densities of prey and predator species respectively. Here $m \in [0, 1)$. Except m, β, a_1 and α_1, other coefficients involved in (1) are non-negative almost periodic functions (AFSs). β^2 is a positive constant while a_1 and α_1 are AFSs. Moreover, for the ecological meanings of the coefficients involved in the system (1), and detailed study of feedback controls and prey refuge, one can refer to [16–19].

2 Boundedness and Permanence

In this section, we establish the positive invariance, boundedness, permanence and global asymptotic stability. Let $R_+^4 = \{(u_1, u_2, z_1, z_2) \in R^4 | u_1 > 0, u_2 > 0, z_1 > 0, z_2 > 0\}$.

Theorem 1 *Under some appropriate restrictions on the parameters involved in (1),*

$$\begin{aligned}
\kappa := \Big\{(u_1, u_2, z_1, z_2) \in R^4 | m_1 \le u_1 \le M_1, m_2 \le u_2 \le M_2,\\
m_3 \le z_1 \le M_3, m_4 \le z_2 \le M_4, \Big\}
\end{aligned}$$

is positively invariant. Moreover system (1) is permanent.

Proof For the proof, one can see [19].

3 Global Attractivity and Almost Periodicity

One can refer to [6], for the detailed study of AFSs.

Theorem 2 *Let there exist* $\zeta_1 > 0, \zeta_2 > 0, \chi_1 >$ *and* $\chi_2 > 0$ *such that*

$$\pi_1 = -\chi_1\delta_1(t) + \zeta_1 b_1(t) - \frac{\zeta_2 e_1(t) M_2}{(1-m)^2 m_1^2} + \frac{\zeta_1 \alpha(t)(1-m)m_2}{\beta^2 M_2^2 + (1-m)^2 M_1^2} + \frac{2\zeta_1(1-m)\alpha(t) m_1^2 m_2}{\beta^2 M_2^2 + (1-m)^2 M_1^2} > 0,$$

$$\pi_2 = -\chi_2\delta_2(t) - \frac{\zeta_1(1-m)\alpha(t)M_1}{\beta^2 m_2^2 + (1-m)^2 m_1^2} + \frac{\zeta_2 e_1(t)}{(1-m)M_1} - \frac{2\beta^2(1-m)\zeta_1\alpha(t)M_1 M_2^2}{(\beta^2 m_2^2 + (1-m)^2 m_1^2)^2} > 0,$$

$\chi_1 v_1(t) - \zeta_1 c_1(t) > 0, \chi_2 v_2(t) - \zeta_2 p_1(t) > 0$. *If conditions of Theorem 1 hold, then (1) has a globally attractive solution.*

Proof Consider $U(t) = (u_1(t), u_2(t), z_1(t), z_2(t))$, $V(t) = (v_1(t), v_2(t), x_1(t), x_2(t))$ as any two positive solutions of (1). Then, clearly we have $m_1 \le u_1 \le M_1, m_2 < u_2 < M_2, m_3 \le z_1 \le M_3, m_4 < z_2 < M_4$.

Consider $S(t) = \sum_{i=1}^{2} \zeta_i |\ln u_i(t) - \ln v_i(t)| + \chi_i |\ln z_i(t) - \ln x_i(t)|$, where $i = 1, 2$ and $t \in R$. Then, the upper Dini derivative of $S(t)$, is given by

$$\begin{aligned}
D^+S(t) &= \sum_{i=1}^{2} \zeta_i D^+ |\ln u_i(t) - \ln v_i(t)| + \chi_i D^+ |\ln z_i(t) - \ln x_i(t)| \\
&= \zeta_1 sgn[u_1(t) - v_1(t)]\Big(- b_1(t)(u_1(t) - v_1(t)) - c_1(t)(z_1(t) - x_1(t))\Big) \\
&\quad - (1-m)\alpha(t)\Big(\frac{u_1(t)u_2(2)}{\beta^2 u_2^2(t) + (1-m)^2 u_1^2(t)} - \frac{v_1(t)v_2(2)}{\beta^2 v_2^2(t) + (1-m)^2 v_1^2(t)}\Big) \\
&\quad + \chi_1 sgn[z_1(t) - x_1(t)][-v_1(t)(z_1(t) - x_1(t)) + \delta_1(t)(u_1(t) - v_1(t))] \\
&\quad - \chi_2 sgn[z_2(t) - x_2(t)][-v_2(t)(z_2(t) - x_2(t)) + \delta_2(t)(u_2(t) - v_2(t))] \\
&\quad + \zeta_2 sgn[u_2(t) - v_2(t)]\Big[- \frac{e_1(t)}{(1-m)}\Big(\frac{u_2(t)}{u_1(t)} - \frac{v_2(t)}{v_1(t)}\Big) - p_1(t)(z_2(t) - x_2(t))\Big].
\end{aligned}$$

Thus, one can easily compute that,

$$\begin{aligned}
D^+S(t) \le &-\Big[\zeta_1 b_1(t) - \chi_1\delta_1(t) - \frac{\zeta_2 e_1(t) v_2(t)}{(1-m)^2 u_1(t) v_1(t)}\Big]|u_1(t) - v_1(t)| \\
&\Big[- \frac{\zeta_2 e_1(t)}{(1-m)u_1(t)} + \chi_2\delta_2(t)\Big]|u_2(t) - v_2(t)| \\
&- [\chi_1 v_1(t) - \zeta_1 c_1(t)]|z_1(t) - x_1(t)| - [\chi_2 v_2(t) - \zeta_2 p_1(t)]|u_2(t) - v_2(t)| \\
&- (1-m)\zeta_1\alpha(t) sig\{u_1(t) - v_1(t)\}\Big[\frac{u_2(t)(u_1(t) - v_1(t))}{\beta^2 u_2^2(t) + (1-m)^2 u_1^2(t)} \\
&+ \frac{v_1(t)(u_2(t) - v_2(t))}{\beta^2 u_2^2(t) + (1-m)^2 u_1^2(t)} \\
&+ \frac{v_1(t)v_2(t)(\beta^2 v_2^2(t) - \beta^2 u_2^2(t) + (1-m)^2(v_1^2(t) - u_1^2(t)))}{(\beta^2 u_2^2(t) + (1-m)^2 u_1^2(t))(\beta^2 v_2^2(t) + (1-m)^2 v_1^2(t))}\Big].
\end{aligned}$$

Now, if we choose

$$\pi = \min[\chi_1 v_1(t) - \zeta_1 c_1(t), \chi_2 v_2(t) - \zeta_2 p_1(t), \pi_1, \pi_2],$$

$$\pi_1 = -\chi_1\delta_1(t) + \zeta_1 b_1(t) - \frac{\zeta_2 e_1(t)M_2}{(1-m)^2 m_1^2} + \frac{\zeta_1\alpha(t)(1-m)m_2}{\beta^2 M_2^2 + (1-m)^2 M_1^2} + \frac{2\zeta_1(1-m)\alpha(t)m_1^2 m_2}{\beta^2 M_2^2 + (1-m)^2 M_1^2}$$

$$\pi_2 = -\chi_2\delta_2(t) - \frac{\zeta_1(1-m)\alpha(t)M_1}{\beta^2 m_2^2 + (1-m)^2 m_1^2} + \frac{\zeta_2 e_1(t)}{(1-m)M_1} - \frac{2\beta^2(1-m)\zeta_1\alpha(t)M_1 M_2^2}{(\beta^2 m_2^2 + (1-m)^2 m_1^2)^2},$$

then we have

$$D^+S(t) \le -\pi\big[|u_1(t) - v_1(t)| + |u_2(t) - v_2(t)| + |z_1(t) - x_1(t)| + |z_2(t) - x_2(t)|\big].$$

Hence, one can find that

$$\lim_{t\to+\infty} |u_1(t) - v_1(t)| = 0, \lim_{t\to+\infty} |u_2(t) - v_2(t)| = 0$$
$$\lim_{t\to+\infty} |z_1(t) - x_1(t)| = 0, \lim_{t\to+\infty} |z_2(t) - x_2(t)| = 0.$$

Thus, the existence of a globally attractive solution of (1) is established. Moreover, existence and uniqueness of APS can be easily discussed [9].

4 Numerical Simulations

Here, we provide some numerical examples and their simulation results. We consider an important aspect of co-existence of species. We start with the following set of parametric values:

$$\begin{aligned} &a_1 = 8.5 + \sin 3t, b_1 = 3, \alpha = \frac{1}{2}, \beta = 1.2, c_1 = \frac{1}{7}(1 + \sin(\sqrt{2}t)), d_1 = 8 + \cos 2t, \\ &p_1 = \frac{1}{7}(1 + \cos(\sqrt{2}t)), \mu_1 = 2.5 + \sin 2t, e_1 = 4, v_1 = \frac{1}{5}, \delta_1 = \frac{1}{6}(1 + \cos t), \\ &\mu_2 = 2 + \cos 2t, v_2 = \frac{1}{5}, \delta_2 = \frac{1}{9}(1 + \cos 3t) \end{aligned} \tag{2}$$

First, we simulate the system (1) with the parametric values given in (2) and three different values of prey refuge parameter, i.e. $m = 0, m = 0.8$ and $m = 0.99$. In this case, Fig. 1 ensures the existence of APS while Figs. 2 and 3 show that as the value of m increases, the predator species gets close to x–axis. However, species y persist even at higher value of m due to availability of additional food.

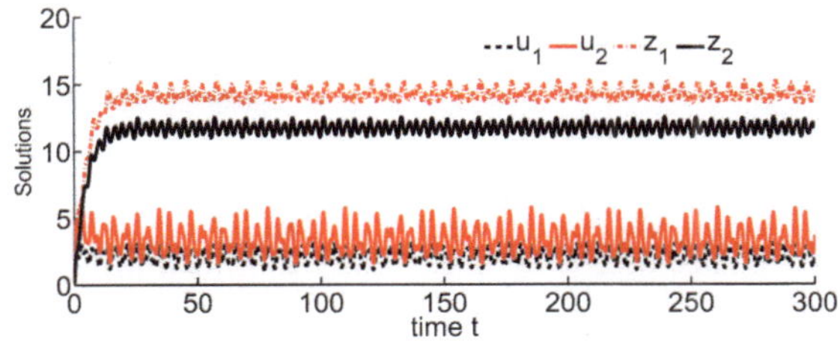

Fig. 1 Existence of a APS for $m = 0$

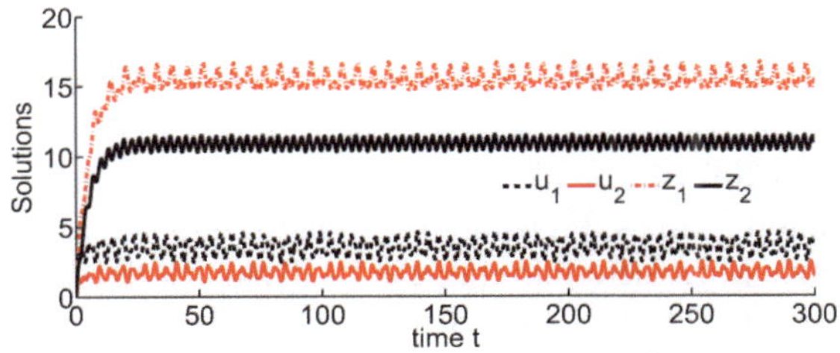

Fig. 2 Solution for $m = 0.80$

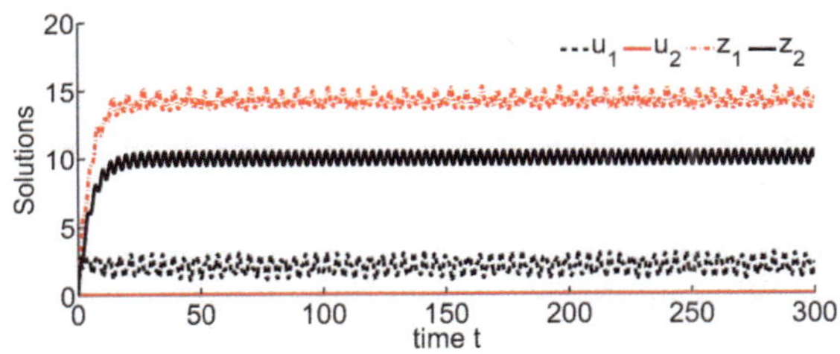

Fig. 3 Solution curves for $m = 0.99$

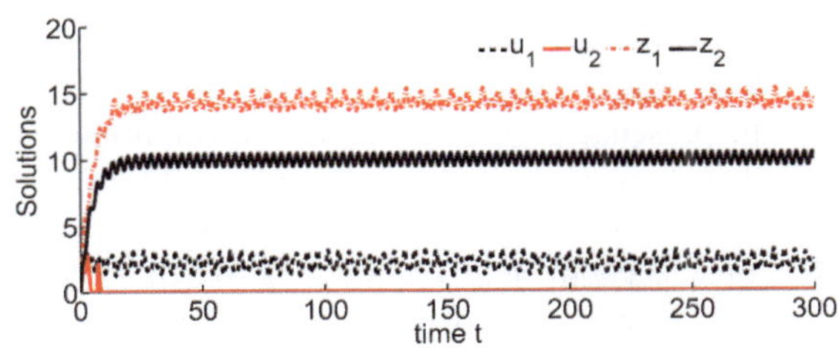

Fig. 4 For $m = 0.50$ and $p_1 = 1 + \cos(\sqrt{2}t)$

Now, we simulate the system (1), with parametric values in (2) except $m = 0.50$ and $p_1 = 1 + \cos(\sqrt{2}t)$. Figure 4 determines that as the control parameter associated to predator species y crosses a threshold value, the predator species starts oscillating very closed to $x-$axis. Similar kind of dynamics can also be observed corresponding to control parameter c_1 (refer the Fig. 5).

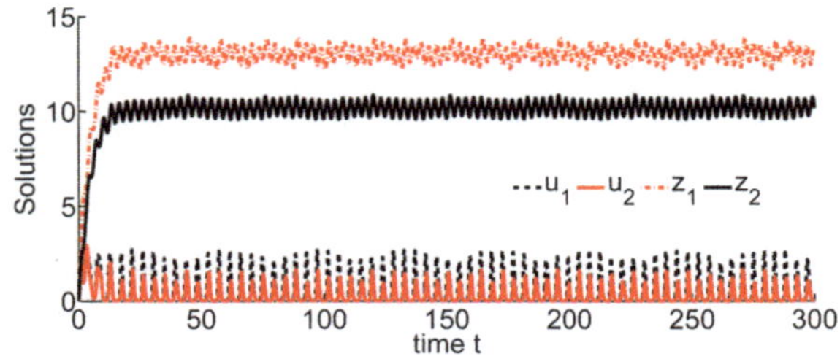

Fig. 5 For $m = 0.99$ and $c_1 = \frac{1}{2}(1 + \cos(\sqrt{2}t))$. Here, both x and y fluctuate very closed to the x−axis

5 Discussion: Applications and Future Scope

Variability in environmental factors plays a critical role in shaping intrinsic population dynamics. Predator–prey relationship is one of the basic links among populations which determine population dynamics and trophic structure. Classical predator–prey model has commonly been studied in idiosyncratic fashion, without considering variability in the surrounding environment in which the population grows and survives. In this paper, environmental variability is captured in the model parameters with time-dependent AFSs, which makes the model non-autonomous in nature. In this work, we have analysed a non-autonomous ecological system with almost periodic coefficients, feedback controls and prey refuge. The results of the present work are summarized as follows:

(i) The introduction of feedback controls signifies the presence of unpredictable forces in the environment. The incorporation of prey refuge is important in the sense of co-existence of species. We have investigated the effect of feedback control parameters as well as prey refuge on the dynamics of the system (1).

(ii) The sufficient conditions required for the permanence (see, Theorem 1) can be derived which help to prove the global stability of solution. One can also observe the presence of m, in the sufficient conditions of global stability of the solution of system (1), however, sufficient conditions of the Theorem 2 are highly complex. Hence, we have tried to investigate the effect of prey refuge and feedback controls parameters, numerically. When the feedback control parameters c_1 and p_1 take value above a threshold level, species densities get very closed to x−axis and adverse environmental conditions may cause extinction (refer the Figs. 4 and 5). Moreover, the results obtained for almost periodic model system (1) also hold for the associated periodic system.

(iii) The concept of almost periodicity generalizes the concept of periodicity. Periodicity (or periodic solutions) is a universally existing phenomenon, e.g. bioinformatics: periodicity in proteins structures at solenoid-level models, social networks: delay-induced periodicity in social networks [3], wireless sensor networks: periodicity in proactive type sensor networks [2] etc. In this perspective, the current study may be provide interesting applications and generalization of the concept of periodicity in bioinformatics, social networks, wireless sensor network, modelling of criminal and non-criminal, etc. Of course, a nice and challenging task can be validation of mathematical results with empirical data.

References

1. Zhao, J., Liu, H., Li, Z., Li, W.: Periodic data prediction algorithm in wireless sensor networks. China Conference on Wireless Sensor Networks, Springer, Berlin Heidelberg. 695–701 (2012)
2. Al-Karaki, J.N., Ahmed, E.K.: Routing techniques in wireless sensor networks: a survey. IEEE Wireless Communications. 11(6), 6–28 (2004)
3. Eubank S., et al.: Modelling disease outbreaks in realistic urban social networks, Nature. 429, 180–184 2004
4. Baianu, I.C.: Computer models and automata theory in biology and medicine. Math. Model. 7(9-12) 1513–1577 (1986)
5. Pritykin, Y.L.: Almost periodicity, finite automata mappings, and related effectiveness issues. Russian Mathematics (Iz VUZ). 54(1), 59–69 (2010)
6. Bohr, H.: Almost periodic functions, American Mathematical Society, (1947)
7. Guo, H., Chen, X.: Existence and global attractivity of positive periodic solution for a Volterra model with mutual interference and Beddington-DeAngelis functional response. Appl. Math. Comput. 217(12), 5830–5837 (2011)
8. Tripathi, J.P.: Almost periodic solution and global attractivity for a density dependent predator-prey system with mutual interference and Crowley-Martin response function. Differ Equ Dyn Syst. https://doi.org/10.1007/s12591-016-0298-6 (2016)
9. Tripathi, J.P., Abbas, S.: Almost Periodicity of a Modified LeslieGower Predator-Prey System with CrowleyMartin Functional Response. Mathematical Analysis and Its Applications, pp. 309–317. Springer, New Delhi (2015)
10. Abbas, S., Banerjee, M., Hungerbuhler, N.: Existence, uniqueness and stability analysis of allelopathic stimulatory phytoplankton model. J. Math. Anal. Appl. 367, 249–259 (2010)
11. Lotka, A.: Elements of Mathematical Biology. Dover, New york, (1956)
12. Tripathi, J.P., Meghwani, S.S., Thakur, M., Abbas, S., A modified LeslieGower predator-prey interaction model and parameter identifiability. Commun. Nonlinear Sci. Numer. Simulat. 54, 331–346 (2018)
13. Jana, D., Tripathi, J.P.: Impact of generalist type sexually reproductive top predator interference on the dynamics of a food chain model. Int. J. Dynam. Control. 80, https://doi.org/10.1007/s40435-016-0255-9 (2015)
14. Parshad, R. D., Basheer, Jana, D., Tripathi, J. P.: Do prey handling predators really matter: Subtle effects of a Crowley-Martin functional response. Chaos, Solitons and Fractals, 103, 410–421 (2017)
15. Abbas, S., Tripathi, J.P., Neha, A.A.: Dynamical analysis of a model of social behaviour: criminal versus non-criminal. Chaos Solitons and Fractals. 98, 121–129 (2017)
16. Chen, F., Cao, Y.: Existence of almost periodic solution in a ratio-dependent Leslie system with feedback controls. J. Math. Anal. Appl. 341, 1399–1412 (2008)
17. Lin, X., Chen, F.: Almost periodic solution for a Volterra model with mutual interference and Beddington-DeAngelis functional response. Appl. Math. Comput. 214, 548–556 (2009)
18. Tripathi, J.P., Abbas, S.: Global dynamics of autonomous and nonautonomous SI epidemic models with nonlinear incidence rate and feedback controls. Nonlinear Dyn. 86(1), 337–351 (2016)
19. Du, Z., Lv, Y.: Permanence and almost periodic solution of a LotkaVolterra model with mutual interference and time delays. Appl. Math. Model. 37, 1054–1068 (2013)

Algebraic Characterization of IF-Automata

Vijay K. Yadav, Swati Yadav, M. K. Dubey and S. P. Tiwari

Abstract In the present work, we have introduced the concept of layers of IF-automaton, provide the characterization of algebraic concepts of an IF-automaton from its layer point of view, and investigate relationship between IF-automata and upper semilattices. We also confer a decomposition of an IF-automaton and investigate a method of formation of an IF-automaton for a given finite poset. Interestingly, we have demonstrated that there exists an isomorphism between an upper semilattice and the poset of class of subautomata of an IF-automaton.

Keywords Layers of IF-automata · IF-subautomaton · Upper semilattice

1 Introduction

Automata as general computational systems over discrete space play a key role within computing science due to its importance in both theoretical as well as application point of view. It is well known that many aspects of classical theoretical computer science is advanced by implementation of several concepts from algebra, e.g., the algebraic study of automata in completely different approaches have been done in (Bavel [3], Holcombe [7] and Ito [9]), while in case of fuzzy atomata study of

V. K. Yadav (✉)
Department of Mathematics, School of Mathematics, Statistics and Computational Sciences, Central University of Rajasthan, NH-8, Bandar Sindari, Ajmer 305817, Rajasthan, India
e-mail: vkymaths@gmail.com

S. Yadav · M. K. Dubey · S. P. Tiwari
Department of Applied Mathematics, Indian Institute of Technology (Indian School of Mines), Dhanbad 826004, India
e-mail: yswatimaths@gmail.com

M. K. Dubey
e-mail: maheshdubey6@gmail.com

S. P. Tiwari
e-mail: sptiwarimaths@gmail.com

C. R. Panigrahi et al. (eds.), *Progress in Advanced Computing and Intelligent Engineering*, Advances in Intelligent Systems and Computing 714,
https://doi.org/10.1007/978-981-13-0224-4_51

algebraic aspects was initiated by Malik, Mordeson, and Sen [15, 16] and has been further improved by several researchers (cf., e.g., Jin [10], Jun [11–13] and Kumbhojkar and Chaudhri [14]). Holcombe [7], pointed out that the concepts of decompositions (cf., [22]) and products of automata come into picture due to curiosity of perceiving the behavior pattern of a system in certain circumstances and have a fundamental place in the advancement of computer science. The emerging practical applications of fuzzy finite automata have been reported in numerous branches of science and technology, some of the most significant examples are clinical monitoring, communication networks, pattern recognition, and fuzzy discrete-event systems (cf., [6, 17, 18] and references therein).

After the introduction of intuitionistic fuzzy set (abbreviated as IF-set in [19–21, 25, 26]) by Atanassov [1, 2], Jun [11, 12] introduce the concept of an intuitionistic fuzzy finite state machine as a generalized version of the concept of fuzzy automata. Interestingly, Chen [4, 5] successfully implemented the concept of IF-automaton for a detailed investigation and analysis of problems related to social sciences, management, and industry.

The concept of layers of automata is coined by Ito [9], while layers of fuzzy automata is introduced and studied by Tiwari et al. [23]. Through this paper, we introduced similar concept of layers of an IF-automaton to characterize existing algebraic concepts of an IF-automaton from its layers point of view. Interestingly, we associate upper semilattices with an IF-automaton and establish an isomorphism between an upper semilattice and the poset of class of subautomata of an IF-automaton, meanwhile we confer a $\oplus$-composition of IF-automata and a decomposition of IF-automaton from its layer point of view.

2 Preliminaries

The basic definitions and notations from IF-sets, IF-automata, which will be required in the subsequent sections is recalled here. Throughout, I stand for interval $[0, 1]$, for a finite set S, $|S|$ denotes its cardinality, unless stated Σ^* denotes free monoid generated by a nonempty set Σ. We denote the identity element of this monoid by e.

Definition 1 [1] A pair (μ_B, ν_B) of fuzzy sets in Σ defines an **IF-set** B in Σ if $\mu_B(a) + \nu_B(a) \leq 1$; $\forall a \in \Sigma$, where the maps $\mu_B : \Sigma \to I$ and $\nu_B : \Sigma \to I$, respectively, represent the membership and nonmembership degree of an element $a \in \Sigma$ to the set B.

Remark 1 A map $\eta : \Sigma \to I$ defines a **fuzzy set** η on a nonempty set Σ.

Remark 2 An IF-set $B = (\mu_B, \nu_B)$ in Σ can be viewed as a map $B : \Sigma \to I \times I$, defined by $B(a) = (\mu_B(a), \nu_B(a)), a \in \Sigma$, with $\mu_B(a) + \nu_B(a) \leq 1$.

Definition 2 [19] Let M be a nonempty set, Σ be a monoid having identity element e, and λ be an IF-subset of $M \times \Sigma \times M$, then $\mathscr{M} = (M, \Sigma, \lambda)$ define an **IF-automaton** (*IFA*), where M, Σ and λ are, respectively, known as set of states, input monoid, and transition function of $\mathcal{M}$. The transition function $\lambda = (\lambda_1, \lambda_2)$: $M \times \Sigma \times M \to I \times I$ has property that $\forall \alpha, \beta \in M$ and $\forall a, b \in \Sigma$,

$$\lambda_1(\alpha, e, \beta) = \begin{cases} 1 \text{ if } \alpha = \beta \\ 0 \text{ if } \alpha \neq \beta, \end{cases} \quad \lambda_2(\alpha, e, \beta) = \begin{cases} 1 \text{ if } \alpha \neq \beta \\ 0 \text{ if } \alpha = \beta, \end{cases}$$

with $\lambda_1(\alpha, ab, \beta) = \vee\{\lambda_1(\alpha, a, \gamma) \wedge \lambda_1(\gamma, b, \beta) : \gamma \in M\}$, and $\lambda_2(\alpha, ab, \beta) = \wedge\{\lambda_2(\alpha, a, \gamma) \vee \lambda_2(\gamma, b, \beta) : \gamma \in M\}$.

Remark 3 In Definition 2, Jun [11] assumed Σ to be a set instead of monoid.

Definition 3 Let $\mathcal{N} = (N, \Sigma, \eta)$ be an *IFA*. An *IFA* $\mathcal{M} = (M, \Sigma, \lambda)$ is called an **IF-subautomaton** (*IFSA*) of $\mathcal{N}$ if $M \subseteq N$, $s_N(M) = M$, $\eta_{1/M\times\Sigma\times M} = \lambda_1$ and $\eta_{2/M\times\Sigma\times M} = \lambda_2$. The *IFSA* $\mathcal{M}$ is called **separated** *IFSA* if $s_N(N - M) \cap M = \phi$.

Definition 4 ([11]) Let $\mathcal{M} = (M, \Sigma, \lambda)$ be an *IFA* and $N \subseteq M$, then **intuitionistic source** and **intuitionistic successor** of N, are, respectively, defined as the sets

$\sigma_M(N) = \{\alpha \in M : \lambda_1(\alpha, w, \beta) > 0$ and $\lambda_2(\alpha, w, \beta) < 1$, for some $(w, \beta) \in \Sigma \times N\}$, and

$s_M(N) = \{\beta \in M : \lambda_1(\alpha, w, \beta) > 0$ and $\lambda_2(\alpha, w, \beta) < 1$, for some $(w, \alpha) \in \Sigma \times N\}$.

Proposition 1 *Let* $\mathcal{M} = (M, \Sigma, \lambda)$ *be an IFA and* $T \subseteq M$. *Then* $s(M - T) = M - T$ *iff* $\sigma(T) = T$.

Definition 5 An *IFA* $\mathcal{M} = (M, \Sigma, \lambda)$ is called

(i) **strongly connected** if $\forall \beta, \alpha \in M, \beta \in s(\alpha)$, and
(ii) **retrievable** if $\lambda_1(\alpha, w, \beta) > 0, \lambda_2(\alpha, w, \beta) < 1$, for some $(\alpha, w, \beta) \in M \times \Sigma^* \times M \Rightarrow \lambda_1(\beta, w', \alpha) > 0, \lambda_2(\beta, w', \alpha) < 1$, for some $w' \in \Sigma^*$.

Definition 6 An *IFA* $\mathcal{M} = (M, \Sigma, \lambda)$ is called **cyclic** if $\forall\ \beta \in M\ \exists\ \alpha_0 \in M$ and $a \in \Sigma^*$ such that $\lambda_1(\alpha_0, a, \beta) > 0$, and $\lambda_2(\alpha_0, a, \beta) < 1$.

Definition 7 ([11]) A pair (h, k) of maps defines a **homomorphism** from an *IFA* $\mathcal{M} = (M, \Sigma_1, \lambda)$ to an *IFA* $\mathcal{N} = (N, \Sigma_2, \eta)$, where $h : M \to N$ and $k : \Sigma_1 \to \Sigma_2$, such that $\forall(\alpha, w, \beta) \in M \times \Sigma_1 \times M$, $\eta_1(h(\alpha), k(w), h(\beta)) \geq \lambda_1(\alpha, a, \beta)$ and $\eta_2(h(\alpha), k(w), h(\beta)) \leq \lambda_2(\alpha, w, \beta)$.

Remark 4 If $\Sigma_1 = \Sigma_2$ & $k = Id_{\Sigma_1}$, then h is called homomorphism from $\mathcal{M}$ to $\mathcal{N}$.

The concepts associated with lattices used in this paper are standard and we refer to Vickers [24] for these concepts (cf., [9, 23] also).

3 Layers of IF-Automata

Consider an *IFA* $\mathcal{M} = (M, \Sigma, \lambda)$, a relation S on state set M of $\mathcal{M}$ is defined as $(\beta, \alpha) \in S$ iff $\{\lambda_1(\beta, t, \alpha) > 0, \lambda_2(\beta, t, \alpha) < 1\}$ and $\{\lambda_1(\alpha, w, \beta) > 0, \lambda_2(\alpha, w, \beta) < 1\}$, for some $t, w \in \Sigma^*$. Obviously, S define an equivalence relation on M. Let $\beta \in M$ be fixed, the set $L_\beta = \{\alpha \in M : (\beta, \alpha) \in S\}$ is called a **layer** of $\mathcal{M}$. Let L_β and L_α be two layers of $\mathcal{M}$, define $L_\beta \preceq_{\mathcal{M}} L_\alpha$ if $\lambda_1(\alpha, t, \beta) > 0$, and $\lambda_2(\alpha, t, \beta) < 1$, for some $t \in \Sigma^*$. Obviously $(\{L_\beta : \beta \in M\}, \preceq_{\mathcal{M}})$ is a poset, we denote this poset by $(T_{\mathcal{M}}, \preceq_{\mathcal{M}})$. For an *IFA* $\mathcal{M} = (M, \Sigma, \lambda)$, throughout $T_{\mathcal{M}}$ and $T^m_{\mathcal{M}}$ denote the set of all its layers and minimal layers, respectively.

Proposition 2 *An IFA* $\mathcal{N} = (N, \Sigma, \eta)$ *is a subautomaton of IFA* $\mathcal{M} = (M, \Sigma, \lambda)$ *iff*

(i) $\exists\ L_{\beta_1}, L_{\beta_2}, \ldots, L_{\beta_r} \in T_{\mathcal{M}}$ *with* $N = \{\alpha \in M : L_\alpha \preceq_{\mathcal{M}} L_{\beta_i}$, *for some* $i \in \{1, 2, \ldots, r\}\}$, *and*
(ii) *For all* $\beta, \alpha \in N$ *and* $\forall w \in \Sigma, \eta_i(\alpha, w, \beta) = \lambda_i(\alpha, w, \beta), i = 1, 2.$

Proof Suppose $\mathcal{N} = (N, \Sigma, \eta)$ be an *IFSA* of the given *IFA* $\mathcal{M} = (M, \Sigma, \lambda)$, so by definition we have $N \subseteq M, s(N) = N$, $\eta_1 = \lambda_{1/N\times\Sigma\times N}$ and $\eta_2 = \lambda_{2/N\times\Sigma\times N}$. But, $s(N) = N \Rightarrow N = \{\alpha \in M : \lambda_1(\beta, w, \alpha) > 0$, and $\lambda_2(\beta, w, \alpha) < 1$, for some $(w, \beta) \in \Sigma^* \times N\}$, or that $\exists\, L_{\beta_i} \in T_{\mathcal{N}} = \{L_\beta : \beta \in N\}$ such that $N = \{\alpha \in M : L_\alpha \preceq_{\mathcal{M}} L_{\beta_i}\}$, i.e., $\exists\ L_{\beta_1}, L_{\beta_2}, \ldots, L_{\beta_r} \in T_{\mathcal{M}}$ satisfying $N = \{\alpha \in M : L_\alpha \preceq_{\mathcal{M}} L_{\beta_i}$, for some $i \in \{1, 2, \ldots, r\}\}$. The part (ii) of proposition holds clearly as $\eta_1 = \lambda_1|_{N\times\Sigma\times N}$ and $\eta_2 = \lambda_2|_{N\times\Sigma\times N}$.

For converse part, suppose $\alpha \in s(N)$, then $\exists\ \beta \in N$ and $w \in \Sigma^*$ such that $\lambda_1(\beta, w, \alpha) > 0$, and $\lambda_2(\beta, w, \alpha) < 1$. Since, $\beta \in N$ it follows that $L_\beta \preceq_{\mathcal{M}} L_{\beta_i}$, for some $i \in \{1, 2, \ldots, r\}$, i.e., $\exists\ v \in \Sigma^*$ with $\lambda_1(\beta_i, v, \beta) > 0$, and$\lambda_2(\beta_i, v, \beta) < 1$. Since $\lambda_1(\beta_i, vw, \alpha) \geq \lambda_1(\beta_i, v, \beta) \wedge \lambda_1(\beta, w, \alpha) > 0$ and $\lambda_2(\beta_i, vw, \alpha) \leq \alpha_2(\beta_i, v, \beta) \vee \lambda_2 (\beta, w, \alpha) < 1$ it follow that $L_\beta \preceq_{\mathcal{M}} L_{\beta_i}$, whereby $\alpha \in N$. Hence $s(N) \subseteq N$.

Proposition 3 *Consider an IFA,* $\mathcal{M} = (M, \Sigma, \lambda)$, *then*

(i) *retrievablity of* $\mathcal{M}$, *implies* $\forall \alpha \in M$, $s(\alpha) \in T_{\mathcal{M}}$, *and*
(ii) *strongly connectedness of* $\mathcal{M}$, *implies* $M \in T_{\mathcal{M}}$.

Proposition 4 *Consider an IFA* $\mathcal{M} = (M, \Sigma, \lambda)$, *then* $\mathcal{N} = (N, \Sigma, \eta)$ *is a separated IFSA of* $\mathcal{M}$ *iff*

(i) $\exists\ L_{\beta_1}, L_{\beta_2}, \ldots, L_{\beta_k} \in T_{\mathcal{M}}$ *with* $N = \{\alpha \in M : L_\alpha \preceq_{\mathcal{M}} L_{\beta_i}$ *and* $L_{\beta_j} \preceq_{\mathcal{M}} L_\alpha$, *for some* $i, j \in \{1, 2, \ldots, k\}\}$, *and*
(ii) $\eta_1(\alpha, w, \beta) = \lambda_1(\alpha, w, \beta)$ and $\eta_2(\alpha, w, \beta) = \lambda_2(\alpha, w, \beta), \forall \alpha, \beta \in N$ *and* $\forall w \in \Sigma$.

Proof Following Definition 3 and Propositions 1 and 2, the proof of $\sigma(N) = N$ iff $\alpha \in N$ and $L_{\beta_j} \preceq_{\mathcal{M}} L_\alpha$, for some $j \in \{1, 2, \ldots, k\}$ is obvious.

Proposition 5 *For any IFA there exists at least one strongly connected IFSA.*

Proof Consider an *IFA* $\mathcal{M} = (M, \Sigma, \lambda)$. Let $\beta \in M$ be such that $L_\beta \in T^m_{\mathcal{M}}$. If $\alpha \in s(L_\beta)$, then $\exists\ a \in \Sigma^*$ and $\gamma \in L_\beta$ satisfying $\lambda_1(\gamma, a, \alpha) > 0$ and $\lambda_2(\gamma, a, \alpha) < 1$. But, $\gamma \in L_\beta \Rightarrow \exists\ b \in \Sigma^*$ satisfying $\lambda_1(\beta, b, \gamma) > 0$ and $\gamma_2(\beta, b, \gamma) < 1$, whereby $\lambda_1(\beta, ba, \alpha) \geq \gamma_1(\beta, b, \gamma) \wedge \lambda_1(\gamma, a, \alpha) > 0$ and $\lambda_2(\beta, ba, \alpha) \leq \gamma_2(\beta, b, \gamma) \vee \lambda_2(\gamma, a, \alpha) < 1$. Now, as $L_\beta \in T^m_{\mathcal{M}}$ and $L_\beta \preceq_{\mathcal{M}} L_\alpha$, we have $\lambda_1(\alpha, w, \beta) > 0$ and $\lambda_2(\alpha, w, \beta) < 1$, for some $w \in \Sigma^*$. Hence $\forall\ \alpha \in s(L_\beta), \alpha \in L_\beta$, or that $(L_\beta, \Sigma, \lambda_{/L_\beta \times \Sigma \times L_\alpha})$ is a *IFSA* of $\mathcal{M}$. Next, suppose $\alpha, \gamma \in L_\beta$, so $\exists\ a, b \in \Sigma^*$ satisfying $\lambda_1(\beta, a, \alpha) > 0, \lambda_1(\gamma, b, \beta) > 0$ and $\lambda_2(\beta, a, \alpha) < 1, \lambda_2(\gamma, b, \beta) < 1$, or that $\lambda_1(\gamma, ba, \alpha) > 0$ and $\lambda_2(\gamma, ba, \alpha) < 1$, i.e., $\alpha \in s(\gamma)$, hence *IFSA* $(L_\beta, \Sigma, \lambda_{/L_\beta \times \Sigma \times L_\beta})$ is strongly connected. This completes the proof.

Proposition 6 *For any cyclic IFA $\mathcal{M}$, there exists a maximal layer which is unique and maximum in $T_{\mathcal{M}}$.*

The directable automaton has been studied in [8], we present such study here in case of *IFA*.

Definition 8 An *IFA* $\mathcal{M} = (M, \Sigma, \lambda)$ is called **directable** if $\forall\ \beta, \alpha \in M, \exists, \gamma \in M$ and $a \in \Sigma^*$ such that $\lambda_1(\beta, w, \gamma) > 0$ and $\lambda_2(\beta, w, \gamma) < 1$ as well as $\lambda_1(\alpha, w, \gamma) > 0$ and $\lambda_2(\alpha, w, \gamma) < 1$.

Now, we have following result.

Proposition 7 *For any directable IFA there exists a unique member in $T^m_{\mathcal{M}}$.*

For an *IFA* having a unique member in $T^m_{\mathcal{M}}$, we now demonstrate a method to construct an *IFA* which has singleton as a unique member in $T^m_{\mathcal{M}}$. It is notable here that the resulting *IFA* is a homomorphic image of the given *IFA*.

Consider an *IFA*, $\mathcal{M} = (M, \Sigma, \lambda)$ with L_β as unique member of $T^m_{\mathcal{M}}$. For a new state γ design an *IFA* $\mathcal{M}' = ((M \backslash L_\beta) \cup \{\gamma\}), \Sigma, \eta)$, where the map $\eta : ((M \backslash L_\beta) \cup \{\gamma\}) \times \Sigma \times ((M \backslash L_\beta) \cup \{\gamma\}) \to I$ has property that $\forall (\alpha, a, \gamma) \in ((M \backslash L_\beta) \cup \{\gamma\}) \times \Sigma \times ((M \backslash L_\beta) \cup \{\gamma\})$,

$$\eta_1(\alpha, a, \xi) = \begin{cases} \lambda_1(\alpha, a, \xi), & \text{if } \beta, \alpha \in M \backslash L_\beta \\ 1, & \text{otherwise;} \end{cases}$$

$$\eta_2(\alpha, a, \xi) = \begin{cases} \gamma_2(\alpha, a, \xi), & \text{if } \beta, \alpha \in M \backslash L_\beta \\ 0, & \text{otherwise.} \end{cases}$$

Hence, $\{\xi\}$ is a unique member of $T^m_{\mathcal{M}'}$, in view of definition of $\mathcal{M}'$.

Proposition 8 *The IFA $\mathcal{M}'$ is a homomorphic image of $\mathcal{M}$.*

Proof For all $\alpha \in M$, define a map $h : \mathcal{M} \to \mathcal{M}'$ as

$$h(\alpha) = \begin{cases} \alpha, & \text{if } \alpha \in M \backslash L_\beta \\ \gamma, & \text{otherwise.} \end{cases}$$

Then, the cases which may arise are as follows.
Case 1: If $\alpha, \xi \in M \backslash L_\beta$, then $\eta_i(h(\alpha), a, h(\xi)) = \lambda_i(\alpha, a, \xi), i = 1, 2$.

Case 2: If $\alpha, \xi \in L_\beta$, then $\eta_1(h(\alpha), a, h(\xi)) = \eta_1(\xi, a, \xi) = 1 \geq \lambda_1(\alpha, a, \xi)$ and $\eta_2(h(\alpha), a, h(\xi)) = \eta_2(\gamma, a, \gamma) = 0 \leq \lambda_2(\alpha, a, \xi)$.

Case 3: If $\alpha \in M \backslash L_\beta$, $\gamma \in L_\beta$, then $\eta_1(h(\alpha), a, h(\gamma)) = \eta_1(\alpha, a, \gamma) = 1 \geq \lambda_1(\alpha, a, \gamma)$ and $\eta_2(h(\alpha), a, h(\gamma)) = \eta_2(\alpha, a, \gamma) = 0 \leq \lambda_2(\alpha, a, \gamma)$.

Case 4: If $\gamma \in M \backslash L_\beta, \alpha \in L_\beta$, then $\eta_1(h(\alpha), a, h(\gamma)) = \eta_1(\gamma, a, \gamma) = 1 \geq \lambda_1(\alpha, a, \gamma)$ and $\eta_2(h(\alpha), a, h(\gamma)) = \eta_2(\gamma, a, \gamma) = 0 \leq \lambda_2(\alpha, a, \gamma)$.

Hence $\forall(\alpha, a, \gamma) \in M \times \Sigma \times M, \eta_1(h(\alpha), a, h(\gamma)) \geq \lambda_1(\alpha, a, \gamma)$ and $\eta_2(h(\alpha), a, h(\gamma)) \leq \lambda_2(\alpha, a, \gamma)$. The surjectivity of h follows from its definition, this complete the proof.

Definition 9 Suppose $\mathscr{M} = (M, \Sigma, \lambda)$ be an *IFA* with L_β as a unique member of $T^m_{\mathscr{M}}$. A pair of IF-automata $\{\mathscr{M}_1, \mathscr{M}_2\}$ is called a **decomposition** of $\mathscr{M}$, where $\mathscr{M}_1 = (L_\beta, \Sigma, \lambda_{/L_\beta \times \Sigma \times L_\beta})$ and $\mathscr{M}_2 = ((M \backslash L_\beta) \cup \{\gamma\}), \Sigma, \eta)$, the γ appeared in state set of $\mathscr{M}_2$ is a new state and $\eta : ((M \backslash L_\beta) \cup \{\gamma\}) \times \Sigma \times ((M \backslash L_\beta) \cup \{\gamma\}) \to I$ is a map such that $\forall(\alpha, a, \xi) \in ((M \backslash L_\beta) \cup \{\gamma\}) \times \Sigma \times ((M \backslash L_\beta) \cup \{\gamma\})$,

$$\eta_1(\alpha, a, \xi) = \begin{cases} \lambda_1(\alpha, a, \xi), & \text{if } \beta, \alpha \in M \backslash L_\beta \\ 1, & \text{otherwise;} \end{cases}$$

$$\eta_2(\alpha, a, \xi) = \begin{cases} \lambda_2(\alpha, a, \xi), & \text{if } \beta, \alpha \in M \backslash L_\beta \\ 0, & \text{otherwise.} \end{cases}$$

Proposition 9 *Let $\{\mathscr{M}_1, \mathscr{M}_2\}$ be decomposition of an IFA $\mathscr{M} = (M, \Sigma, \lambda)$ and L_β as a unique member of $T^m_{\mathscr{M}}$, then $\mathscr{M}$ is directable iff $\mathscr{M}_1$ is directable.*

Proof Let L_β be a unique member of $T^m_{\mathscr{M}}$ and $\mathscr{M}$ be directable. Let $\alpha, \gamma \in L_\beta$, then $\exists\ a \in \Sigma^*$ and $\gamma' \in M$ such that $\lambda_1(\alpha, a, \gamma') > 0$ and $\lambda_2(\alpha, a, \gamma') < 1$ as well as $\lambda_1(\gamma, a, \gamma') > 0$ and $\lambda_2(\gamma, a, \gamma') < 1$, or that $\lambda_{1/L_\beta \times \Sigma \times L_\beta}(\alpha, a, \gamma') > 0$ and $\lambda_{2/L_\beta \times \Sigma \times L_\beta}(\alpha, a, \gamma') < 1$ as well as $\lambda_{1/L_\beta \times \Sigma \times L_\beta}(\gamma, a, \gamma') > 0$ and $\lambda_{2/L_\beta \times \Sigma \times L_\beta}(\gamma, a, \gamma') < 1$, whereby $\mathscr{M}_1$ is directable.

For Converse part, let us assume that $\mathscr{M}_1$ be a directable *IFA* and L_β be a unique member of $T^m_{\mathscr{M}}$. Let us consider $\alpha, \xi \in M$, then $\exists\ \alpha', \xi' \in M$ with $\alpha \in L_{\alpha'}$ and $\xi \in L_{\xi'}$, i.e., $\exists\ a, b \in \Sigma^*$ satisfying $\lambda_1(\alpha, a, \alpha') > 0$ and $\lambda_2(\alpha, a, \alpha') < 1$ as well as $\lambda_1(\xi, b, \xi') > 0$ and $\lambda_2(\xi, b, \xi') < 1$. Now, L_β being a unique member of $T^m_{\mathscr{M}}$, $\exists\ a', b' \in \Sigma^*$ and $\alpha'', \xi'' \in L_\beta$ satisfying $\lambda_1(\alpha', a', \alpha'') > 0$ and $\lambda_2(\alpha', a', \alpha'') < 1$ as well as $\lambda_1(\xi, b', \xi'') > 0$ and $\lambda_2(\xi, b', \xi'') < 1$. Hence we have, $\lambda_1(\alpha, aa', \alpha'') = \vee\{\lambda_1(\alpha, a, \alpha') \wedge \lambda_1(\alpha', a', \alpha'')\} > 0$ and $\lambda_2(\alpha, aa', \alpha'') = \wedge\{\lambda_2(\alpha, a, \alpha') \vee \lambda_2(\alpha', a', \alpha'')\} < 1$ as well as $\lambda_1(\xi, bb', \xi'') = \vee\{\lambda_1(\xi, b, \xi') \wedge \lambda_1(\xi', b', \xi'')\} > 0$ and $\lambda_2(\xi, bb', \xi'') = \wedge\{\lambda_2(\xi, b, \xi') \vee \lambda_2(\xi', b', \xi'')\} < 1$. Further, $\mathscr{M}_1$ being directable and $\alpha'', \xi'' \in L_\beta$, $\exists\ \beta' \in L_\beta$ and $w \in \Sigma^*$ satisfying $\lambda_1(\alpha'', w, \beta') > 0$ and $\lambda_2(\alpha'',$

$w, \beta') < 1$ as well as $\lambda_1(\xi'', w, \beta') > 0$ and $\lambda_2(\xi'', w, \beta') < 1$. Thus $\lambda_1(\alpha, aa'w, \beta') = \vee\{\lambda_1(\alpha, aa', \alpha'') \wedge \lambda_1(\alpha'', w, \beta')\} > 0$ and $\lambda_2(\alpha, aa'w, \beta') = \wedge\{\lambda_2(\alpha, aa', \alpha'') \vee \lambda_2 (\alpha'', w, \beta')\} < 1$ as well as $\lambda_1(\xi, bb'w, \beta') = \vee\{\lambda_1(\xi, bb', \xi'') \wedge \lambda_1(\xi'', w, \beta')\} > 0$ and $\lambda_2(\xi, bb'w, \beta') = \wedge\{\lambda_2(\xi, bb', \xi'') \vee \lambda_2(\xi'', w, \beta')\} < 1$, i.e., $\forall\ \alpha, \xi \in M\ \exists\ \beta' \in M$ and $aa'w, bb'w \in \Sigma^*$ satisfying $\lambda_1(\alpha, aa'w, \beta') > 0$ and $\lambda_2(\alpha, aa'w, \beta') < 1$ as well as $\lambda_1(\xi, bb'w, \beta') > 0$ and $\lambda_2(\xi, bb'w, \beta') < 1$, whereby $\mathscr{M}$ is directable.

4 Semilattices Associated with IF-Automata

We have established here the connection between upper semilattices and *IF*-automata and tried to fill-up gap between these two concepts. We have introduced a method to construct an *IFA* which is associated with a given finite poset. We have also shown here that there exists an isomorphism between an upper semilattice and the poset of class of *IF*-subautomata of an *IFA*.

Proposition 10 *Corresponding to a finite poset* $(S, \preceq)$, $\exists$ *an IFA* $\mathscr{M}$ *such that* $T_{\mathscr{M}} \cong (S, \preceq)$.

Proof Suppose $(S, \preceq)$ be a finite poset, and $\beta \in S$. Let $\alpha_1, \alpha_2, \ldots, \alpha_r$ be the predecessors of β. An *IFA* $\mathscr{M} = (M, \Sigma, \lambda)$ is now defined by taking $M = S$ and $\Sigma = \{a_1, a_2, \ldots, a_n\}$, where $n = (|S|)$ and $\forall \beta, \alpha \in M$ and $\forall a_j \in \Sigma$, the transition function $\lambda : M \times \Sigma \times M \to I \times I$ is defined as

$$\lambda_1(\beta, a_j, \alpha) = \begin{cases} \xi \in (0, 1], & \text{if}\ \alpha = \alpha_j, 1 \le j \le r \\ 0, & \text{if}\ \alpha = \alpha_j, r+1 \le j \le n \\ 1, & \text{if}\ \alpha = \beta, r+1 \le j \le n \\ 0, & \text{if}\ \alpha = \beta, 1 \le j \le r. \end{cases}$$

$$\lambda_2(\beta, a_j, \alpha) = \begin{cases} (1-\xi) \in (0, 1], & \text{if}\ \alpha = \alpha_j, 1 \le j \le r \\ 1, & \text{if}\ \alpha = \alpha_j, r+1 \le j \le n \\ 0, & \text{if}\ \alpha = \beta, r+1 \le j \le n \\ 1, & \text{if}\ \alpha = \beta, 1 \le j \le r. \end{cases}$$

Clearly, $L_\beta = \{\beta\}, \forall \beta \in M$. Now, define a map $h : (S, \preceq) \to T_{\mathscr{M}}$ by $h(\beta) = L_\beta, \forall \beta \in S$, this map h is bijective, and $\forall\ i = 1, 2, \ldots, r$, $\alpha_j \preceq \beta$ iff $L_{\alpha_j} \preceq_{\mathscr{M}} L_\beta$, i.e., $h(\alpha_j) \preceq h(\beta)$, whereby $T_{\mathscr{M}} \cong (S, \preceq)$.

Now, let $\mathscr{M}$ be an *IFA*, and $\mathscr{S}(\mathscr{M})$ denote the class of all its *IF*-subautomata. For $\mathscr{N}, \mathscr{N}' \in \mathscr{S}(\mathscr{M})$, the notion $\mathscr{N} \sqsubseteq \mathscr{N}'$ represent $\mathscr{N}$ is an *IFSA* of $\mathscr{N}'$. Obviously $(\mathscr{S}(\mathscr{M}), \sqsubseteq)$ define a poset, and a finite upper semilattice as well.

Proposition 11 *For an IFA* $\mathscr{M}$, $(\mathscr{S}(\mathscr{M}), \sqsubseteq)$ *is a finite upper semilattice.*

Now, we have a nice result which ensure the existence of an *IFA* $\mathscr{M}$ such that $(\mathscr{S}(\mathscr{M}), \sqsubseteq)$ is isomorphic to given tree.

Proposition 12 *If a tree $\mathscr{L}$ has more than two minimal elements, then no IFA $\mathscr{M}$ exists such that $(\mathscr{S}(\mathscr{M}), \sqsubseteq) \cong \mathscr{L}$.*

In the literature several compositions of automata and fuzzy automata were studied (cf., e.g., [16]), for IF-automata, $\oplus$-composition is introduced here.

Definition 10 Let $\mathscr{M}' = (M', \Sigma, \lambda')$ and $\mathscr{M}'' = (M'', \Sigma, \lambda'')$ be two IF-automata such that $M' \cap M'' = \phi$. Let the collection of all minimal and maximal layers of $\mathscr{M}'$ and $\mathscr{M}''$, respectively, be denoted by S and T, and are such that $\forall\, R \in S$, $\exists$ a maximal layer $S_R \in T$ satisfying $\{S_R : R \in S\} = T$. Then, a **$\oplus$-composition** of $\mathscr{M}'$ and $\mathscr{M}''$ is an *IFA* $\mathscr{M}' \oplus \mathscr{M}'' = (M' \cup M'', \Sigma, \eta)$, where $\forall \beta, \alpha \in M' \cup M''$ and $\forall a \in \Sigma$ the transition function $\eta : (M' \cup M'') \times \Sigma \times (M' \cup M'') \rightarrow I \times I$ is defined as

$$\eta_1(\beta, w, \alpha) = \begin{cases} \lambda'_1(\beta, w, \alpha), & \text{if } \beta, \alpha \in M', \beta \,\&\, \alpha \notin \text{ a member of } T^m_{\mathscr{M}'} \\ \lambda''_1(\beta, a, \alpha), & \text{if } \beta, \alpha \in M'' \\ 1, & \text{if } \beta \in \text{ a member } R \text{ of } T^m_{\mathscr{M}'}, \,\&\, \alpha = \alpha_w \\ & \text{for unique } \alpha_w \in S_R \\ 0, & \text{otherwise.} \end{cases}$$

$$\eta_2(\beta, w, \alpha) = \begin{cases} \lambda'_2(\beta, w, \alpha), & \text{if } \beta, \alpha \in M', \beta \,\&\, \alpha \notin \text{ a member of } T^m_{\mathscr{M}'} \\ \lambda''_2(\beta, w, \alpha), & \text{if } \beta, \alpha \in M'' \\ 0, & \text{if } \beta \in \text{ a member } R \text{ of } T^m_{\mathscr{M}'}, \,\&\, \alpha = \alpha_w \\ & \text{for unique } \alpha_w \in S_R \\ 1, & \text{otherwise.} \end{cases}$$

Proposition 13 *Let* $\mathscr{M} = \mathscr{M}' \oplus \mathscr{M}''$. *Then* $(\mathscr{S}(\mathscr{M}), \sqsubseteq) \cong (\mathscr{S}(\mathscr{M}'), \sqsubseteq) \oplus (\mathscr{S}(\mathscr{M}''), \sqsubseteq)$.

Proof Let IF-automata $\mathscr{M} = (M, \Sigma, \lambda)$, $\mathscr{M}' = (M', \Sigma, \lambda')$ and $\mathscr{M}'' = (M'', \Sigma, \lambda'')$ be such that $\mathscr{M} = \mathscr{M}' \oplus \mathscr{M}''$. Proposition 10, implies that the layers of each of $\mathscr{M}'$ and $\mathscr{M}''$ is a singleton set. Let $\{\alpha\} \in T^m_{\mathscr{M}'}$ and $S_\alpha = \cup\{\beta \in M : \lambda_1(\alpha, t, \beta) > 0 \text{ and } \lambda_2(\alpha, t, \beta) < 1, t \in \Sigma^*\}$, so we have $\mathscr{N}_\alpha = (S_\alpha, \Sigma, \lambda_{/S_\alpha \times \Sigma \times S_\alpha}) \in \mathscr{S}(\mathscr{M})$ but $\mathscr{N}_\alpha \notin \mathscr{S}(\mathscr{M}'')$ as $\alpha \notin M''$. Next, assume that $\mathscr{N} = (T, \Sigma, \lambda''_{/T \times \Sigma \times T}) \in \mathscr{S}(\mathscr{M}'')$, so for some α, $T \subset \mathscr{N}_\alpha$, where $\{\alpha\} \in T^m_{\mathscr{M}'}$. Further, for all $\mathscr{N} \in \mathscr{S}(\mathscr{M})$, a map $h : \mathscr{S}(\mathscr{M}) \rightarrow \mathscr{S}(\mathscr{M}') \oplus \mathscr{S}(\mathscr{M}'')$ be defined as

$$h(\mathscr{N}) = \begin{cases} \mathscr{N}, & \text{if } \mathscr{N} \in \mathscr{S}(\mathscr{N}_\alpha), \text{ where } \{\alpha\} \in T^m_{\mathscr{M}'} \\ \mathscr{N}', & \text{if } \mathscr{N} \in \mathscr{S}(\mathscr{M}) \backslash \mathscr{S}(\mathscr{M}''), \text{ where } \mathscr{N}' = (T \cap M'', \Sigma, \\ & \quad \lambda^{\mathscr{N}'}/_{(T \cap M'') \times \Sigma \times (T \cap M'')}) \text{ and } \mathscr{N} = (T, \Sigma, \lambda_{T \times \Sigma \times T}). \end{cases}$$

Clearly, h is an isomorphism, whereby $(\mathscr{S}(\mathscr{M}), \sqsubseteq) \cong (\mathscr{S}(\mathscr{M}'), \sqsubseteq) \oplus (\mathscr{S}(\mathscr{M}''), \sqsubseteq)$.

Proposition 14 *For an IFA $\mathscr{M}$, $\exists$ positive integers $j_1, j_2, \ldots, j_r$ such that $(\mathscr{S}(\mathscr{M}), \sqsubseteq) \cong \mathscr{L}(j_1) \oplus \mathscr{L}(j_2) \oplus \cdots \oplus \mathscr{L}(j_r)$.*

Proof Let $\{M_1, M_2, \ldots, M_{j_1}\} = T^m_{\mathscr{M}}$, $\mathscr{M} = (M, \Sigma, \lambda)$, and $\mathscr{M}_{j_1} = (M_1 \cup M_2 \cup \cdots \cup M_{j_1}, \Sigma, \lambda_{/(M_1 \cup M_2 \cup \cdots \cup M_{j_1}) \times \Sigma \times (M_1 \cup M_2 \cup \cdots \cup M_{j_1})})$. By Definition 2.10 of [23], $\mathscr{S}(\mathscr{M}_{j_1})$ is a finite upper semilattice, i.e., $(\mathscr{S}(\mathscr{M}_{j_1}), \sqsubseteq) \cong \mathscr{L}(j_1)$. Now, construct an *IFA* $\mathscr{M}' = (M \backslash M_1 \cup M_2 \cup \cdots \cup M_{j_1}, \Sigma, \eta)$, where $\eta_1(\beta, t, \alpha) = \lambda_1(\beta, t, \alpha)$ and $\eta_2(\beta, t, \alpha) = \lambda_2(\beta, t, \alpha)$, $\forall \beta, \alpha \in M \backslash M_1 \cup M_2 \cup \cdots \cup M_{j_1}$ and $\forall t \in \Sigma$. Then $\mathscr{M} = \mathscr{M}' \oplus \mathscr{M}_{j_1}$, hence Proposition 13 implies $(\mathscr{S}(\mathscr{M}), \sqsubseteq) \cong (\mathscr{S}(\mathscr{M}_{j_1}), \sqsubseteq) \oplus (\mathscr{S}(\mathscr{M}'), \sqsubseteq) \cong \mathscr{L}(j_1) \oplus (\mathscr{S}(\mathscr{M}'), \sqsubseteq)$. Similarly for $\mathscr{M}'$ we have $(\mathscr{S}(\mathscr{M}), \sqsubseteq) \cong \mathscr{L}(j_1) \oplus \mathscr{L}(j_2) \oplus \mathscr{S}(\mathscr{M}''), \sqsubseteq)$, for some *IFA* $\mathscr{M}''$. Continuing in the same way we have $(\mathscr{S}(\mathscr{M}), \sqsubseteq) \cong \mathscr{L}(j_1) \oplus \mathscr{L}(j_2) \oplus \cdots \oplus \mathscr{L}(j_r)$.

Proposition 15 *If for some positive integers $m_1, m_2, \ldots, m_r$, $\mathscr{L} \cong \mathscr{L}(m_1) \oplus \mathscr{L}(m_2) \oplus \cdots \oplus \mathscr{L}(m_r)$ holds for an upper semilattice $\mathscr{L}$, then $\exists$ an IFA $\mathscr{M}$ such that $(\mathscr{S}(\mathscr{M}), \sqsubseteq) \cong \mathscr{L}(m_1) \oplus \mathscr{L}(m_2) \oplus \cdots \oplus \mathscr{L}(m_r)$.*

Proof Suppose $\mathscr{L} \cong \mathscr{L}(m_1) \oplus \mathscr{L}(m_2) \oplus \cdots \oplus \mathscr{L}(m_r)$ is true for an upper semilattice $\mathscr{L}$. Now, define an *IFA* $\mathscr{M} = (M, \Sigma, \lambda)$, where $M = \{1, 2, \ldots, m_r\}$ $|\Sigma| = \max\{m_1, m_2, \ldots, m_r\}$ and for all $\beta, \alpha \in M$ and for all $a \in \Sigma$, $\lambda : M \times \Sigma \times M \to I \times I$ be such that:

$$\lambda_1(\beta, a, \alpha) = \begin{cases} \xi \in (0, 1], & \text{if } \beta = \alpha \\ 0, & \text{otherwise;} \end{cases} \quad \lambda_2(\beta, a, \alpha) = \begin{cases} 1 - \xi \in (0, 1], & \text{if } \beta = \alpha \\ 1, & \text{otherwise.} \end{cases}$$

Hence $(\mathscr{S}(\mathscr{M}), \sqsubseteq) \cong \mathscr{L}(m_r)$. Next, assume that for minimal value of positive integer k, *IFA* $\mathscr{M}' = (M', \Sigma, \lambda')$ satisfies, $(\mathscr{S}(\mathscr{M}), \sqsubseteq) \cong \mathscr{L}(m_k) \oplus \mathscr{L}(m_{k+1}) \oplus \cdots \oplus \mathscr{L}(m_r)$. The result holds for $k = 1$. If $k > 1$, let *IFA* M'' satisfies $(\mathscr{S}(\mathscr{M}''), \sqsubseteq) \cong \mathscr{L}(m_{k-1})$, then $(\mathscr{S}(\mathscr{M}' \oplus \mathscr{M}''), \sqsubseteq) \cong (\mathscr{S}(\mathscr{M}''), \sqsubseteq) \oplus (\mathscr{S}(\mathscr{M}'), \sqsubseteq) \cong \mathscr{L}(m_{k-1}) \oplus \mathscr{L}(m_k) \oplus \cdots \oplus \mathscr{L}(m_r)$, contradicting the fact that k is minimal. Hence $k = 1$, so $\exists$ an *IFA* $\mathscr{M}'$ satisfying $(\mathscr{S}(\mathscr{M}'), \sqsubseteq) \cong \mathscr{L}(m_1) \oplus \mathscr{L}(m_2) \oplus \cdots \oplus \mathscr{L}(m_r)$.

Finally, by Propositions 14 and 15 we have following result.

Proposition 16 *For a finite upper semilattice $\mathscr{L}$, $\exists$ an IFA $\mathscr{M}$ satisfying $(\mathscr{S}(\mathscr{M}), \sqsubseteq) \cong \mathscr{L}$ iff for some positive integers $m_1, m_2, \ldots, m_r$, $\mathscr{L} \cong \mathscr{L}(m_1) \oplus \mathscr{L}(m_2) \oplus \cdots \oplus \mathscr{L}(m_r)$.*

5 Conclusion

A novel concept of layers of IF-automata is introduced and algebraic concepts of IF-automata is characterized from layer point of view. For a given *IFA*, $\mathscr{M}$ having unique member in $T^m_{\mathscr{M}}$, we have provided a method to construct an *IFA*, N which has singleton as a unique member in $T^m_{\mathscr{N}}$. Interestingly, it is notable that the resulting *IFA* is a homomorphic image of the given *IFA*, meanwhile we have shown that every cyclic (or directable) *IFA* has a unique minimal layer. We have also provide a decomposition and $\oplus$-composition of *IFA*. Finally, we have associated upper semilattices with IF-automata and establish the isomorphism between an upper semilattice and the poset of class of subautomata of an *IFA*.

References

1. Atanassov, K.T., Intuitionistic fuzzy sets, *Fuzzy Sets and Systems*, **20**, 87–96 (1986).
2. Atanassov, K.T., More on Intuitionistic fuzzy sets, *Fuzzy Sets and Systems*, **33**, 37–45 (1989); Transition preserving functions of finite automata, *Journal of Association for Computing Machinery*, **15**, 135–158 (1968).
3. Bavel, Z., Structure and transition preserving functions of finite automata, *Journal of Association for Computing Machinery*, **15**, 135–158 (1968).
4. Chen, T.Y., Chou, C.C., Fuzzy automata with Atanassov's intuitionstic fuzzy sets and their applications to product involvement, *Journal of the Chinese Institute of Industrial Engineers*, **26**, 245–254 (2009).
5. Chen, T.Y., Wang, H.P., Wang, J.C., Fuzzy automata based on Atanassov fuzzy sets and applications on consumers, advertising involvement, *African Journal of Business Management*, **6**, 865–880 (2012).
6. Deng, W.L., Qiu, D.W., Supervisory Control of Fuzzy Discrete Event Systems for Simulation Equivalence, *IEEE Transactions on Fuzzy Systems*, **23(1)**, 178–192 (2015).
7. Holcombe, W.M.L., Algebraic Automata Theory, *Cambridge University Press, Cambridge*, (1982).
8. Ito M., Duske, J., On cofinal and definite automata, *Acta Cybernetica*, **6**, 181–189 (1983).
9. M. Ito, Algebraic structures of automata, *Theoretical Computer Science*, **429**, 164–168 (2012).
10. Jin, J., Li, Q., Li, Y., Algebraic properties of *L*-fuzzy finite automata, *Information Sciences*, **234**, 182–202 (2013).
11. Jun, Y.B., Intuitionistic fuzzy finite state machines, *Journal of Applied Mathematics and Computing*, **17**, 109–120 (2005).
12. Jun, Y.B., Intuitionistic fuzzy finite switchboard state machines, *Journal of Applied Mathematics and Computing*, **20**, 315–325 (2006).
13. Jun, Y.B., Quotient structures of intuitionistic fuzzy finite state machines, *Information Sciences*, **177**, 4977–4986 (2007).
14. Kumbhojkar, H.V., Chaudhri, S.R., On proper fuzzification of fuzzy finite state machines, *International Journal of Fuzzy Mathematics*, **4**, 1019–1027 (2008).
15. Malik, D.S., Mordeson, J.N., M.K. Sen, Submachines of fuzzy finite state machine, *Journal of Fuzzy Mathematics*, **2**, 781–792 (1994).
16. Mordeson, J.N., Malik, D.S., Fuzzy Automata and Languages: Theory and Applications, *Chapman and Hall/CRC. London/Boca Raton*, (2002).
17. Qiu, D. W., Supervisory Control of Fuzzy Discrete Event Systems: A Formal Approach, *IEEE Transactions on Systems, Man and Cybernetics-Part B*, **35(1)**, 72–88 (2005).
18. D. W. Qiu, F.C. Liu, Fuzzy Discrete Event Systems under Fuzzy Observability and a Test-Algorithm, *IEEE Transactions on Fuzzy Systems*, **17(3)**, 578–589 (2009).
19. Srivastava, A.K., Tiwari, S.P., IF-topologies and IF-automata, *Soft Computing*, **14**, 571–578 (2010).
20. Tiwari, S.P., Singh, Anupam K., On bijective correspondence between IF-preorders and saturated IF-topologies, *International Journal of Machine Learning and Cybernetics*, **4**, 733–737 (2013).
21. Tiwari, S.P., Singh, Anupam K., IF-preorder, IF-topology and IF-automata, *International Journal of Machine Learning and Cybernetics*, **6**, 205–211 (2015).

22. Tiwari, S.P., Srivastava, A.K., On a decomposition of fuzzy automata, *Fuzzy Sets and Systems*, **151**, 503–511 (2005).
23. Tiwari, S.P., Yadav, Vijay K., Singh, A.K., On algebraic study of fuzzy automata, *International Journal of Machine Learning and Cybernetics* **6**, 479–485 (2015).
24. Vickers, S., Topology via Logic, *Cambridge University Press*, (1989).
25. Yadav, Vijay K., Gautam, Vinay, Tiwari, S.P., On minimal realization of IF-languages: A categorical approach, *Iranian Journal of Fuzzy Systems*, **13(3)**, 19–34 (2016).
26. Zhang, X., Li, Y., Intuitionistic fuzzy recognizers and intuitionistic fuzzy finite automata, *Soft Computing*, **13**, 611–616 (2009).

Efficient Traffic Management on Road Network Using Edmonds–Karp Algorithm

V. Rajalakshmi and S. Ganesh Vaidyanathan

Abstract Consequential magnification in urbanization over the past decades has brought in an immense burden on city transportation system. Traffic congestion is a vital problem that increments queuing of conveyances on roads and brings in delay in our daily routine. The motivation of this research work is to address this problem by developing a Traffic Management System utilizing Edmonds–Karp algorithm. Edmonds–Karp algorithm is a network flow algorithm that is used to find the various routes to divert and regulate traffic flow during traffic congestion. The Traffic Management System is efficient by executing the Edmonds–Karp algorithm on small road networks in parallel that reduces the computation cost of finding the maximum flow in the network.

Keywords Augmenting path · Parallelization · Breadth first search
Edmonds–Karp algorithm and traffic management system

1 Introduction

Traffic Management aims at monitoring and controlling traffic flow on roads in order to eschew traffic congestion. In spite of rapid magnification in technology for traffic management, traffic congestions remains to subsist. Several algorithms endeavor to predict traffic congestion [1] in advance, but it is quite a challenging one. Traffic congestions are quite common, managing and redirecting traffic flow on

V. Rajalakshmi (✉)
Department of Computer Science and Engineering,
Sri Venkateswara College of Engineering, Chennai, India
e-mail: vraji@svce.ac.in

S. Ganesh Vaidyanathan
Department of Electronics and Communication Engineering,
Sri Venkateswara College of Engineering, Chennai, India
e-mail: gvaidyas@svce.ac.in

C. R. Panigrahi et al. (eds.), *Progress in Advanced Computing and Intelligent Engineering*, Advances in Intelligent Systems and Computing 714,
https://doi.org/10.1007/978-981-13-0224-4_52

congestions should additionally be devised. This research paper makes use of Edmonds–Karp algorithm to handle situations of traffic congestions.

Edmonds–Karp algorithm [2] is a variant of Ford Fulkerson network [3] flow algorithm that finds the shortest augmenting paths and maximum flow along the path. This algorithm states that if there exists a path from the source (start node) to the sink (end node), with available capacity on all edges in the path, flow along path can be sent. This process is reiterated until there is no path from the source to the sink.

Traffic Management using Edmonds–Karp algorithm is a new approach for redirecting traffic flow during traffic congestion. The running time of the algorithm is $O(nm^2)$, where '*n*' is the number of junctions on the road network considered and '*m*' is the number of roads connecting the junctions. The Traffic Management System is efficient by reducing the running time [4] to $O(nm^2/k)$, where k is the number of threads created.

2 Proposed Traffic Management System Formulation

Edmonds–Karp algorithm provides a solution to find the maximum flow in a flow network. This research work models a traffic management system to represent traffic flow on the road network. The maximum traffic flow represents the maximum number of vehicles that can pass through the given network.

The proposed traffic management system is represented as a directed road network graph, where the edges represent road capacity and the vertices represent road junctions in the road network. The road capacity is the maximum traffic flow on the given road using all available lanes. The traffic management system considers that road network contains one source node and one sink node. The system additionally considers no traffic flow enters the source node and leaves the sink node. The amount of traffic flow entering a junction must leave the junction.

2.1 Formal Definition of Traffic Management System

The transportation network system $S = (G, V, E, s, t, C, F, MF)$ is formally defined as follows:

1. *G:* (V, E) is the graph generated for the given transport network.
2. *V*: Set of vertices representing road junctions.
3. *E:* Set of edges representing the roads.
4. *s*: Source node.
5. *t*: Sink node.
6. *C:* Set of road capacities, where C_{AB} is maximum traffic flow between the junctions *A* and *B*.
7. *F*: Set of traffic flows with their path, where F_{st} is the traffic flow along the path from source *s* to sink *t*.

8. *MF*: Maximum Traffic Flow, which is the sum of all traffic flows, where $MF = \sum F_{st}$.

3 Algorithm to Find Maximum Flow

Algorithm 1 is an implementation of Edmonds–Karp algorithm to find the maximum traffic flow on the given transportation road network. This algorithm makes a function call to the Breadth First Search strategy that returns the flow and the augmenting path along the flow. The augmenting path is a path from *s* to *t*, such that the capacity of each edge along the path is greater than zero. The residual matrix is updated with the left out capacity after the flow. The procedure is reiterated until there is no augmenting path along the network.

```
program Elmonds - Karp
  Input :  Road Network Graph G = (V,E)
           Capacity Matrix C
           Residual Matrix R [Initially R = C]
           Source Node s
           Sink Node t

  Output:  Maximum Traffic Flow MF
  var      flow = 0;                //FLOW FOR EACH PATH
           Path[];                  //EACH PATH FROM s to t

  begin
    repeat
      flow, Path = BreadthFirstSearch(C, R, s, t);
      if flow = 0
          break;
      MF := MF + flow;
      //BACK SEARCH TO UPDATE RESIDUAL MATRIX
      y := t;
      repeat
        x = Path[y];
        R[x, y] := R[x, y] - flow;
        R[y, x] := R[y, x] + flow;
        y := x;
      until y != s
    until no PATH exist
  end.
```

Algorithm 1. Elmonds Karp Algorithm

Algorithm 2 is an implementation of Breadth First Search to find the flow and the augmenting path along the flow for the given transportation road network. The algorithm takes residual matrix, source node and sink node as its input and traverses by visiting adjacent nodes starting from the source node till it reaches the sink node. The path traversed is the augmenting path and the minimum capacity value along the path is the flow value of the augmenting path. If no such augmenting path exists, the algorithm returns the flow value to be 0.

```
program BreadthFirstSearch
  Input :  Residual Matrix R
           Source Node s
           Sink Node t

  Output:  Flow for each path flow
           Path from s to t   Path[]
  var      n                 //NUMBER OF NODES ON THE PATH

  begin
    for u in 1…n
          Path[u] := -1;
    Path[s] := -2;
    flow := MaxValue;
    Create a queue Q;
    Enqueue s onto Q;
    repeat
      Dequeue x from Q;
      For each adjacent node y of x in G
        if R[x, y] > 0 and Path[y] == -1
          Path[y] = x;
          flow = minimum(flow, R[x, y]);
          if y != t
             Enqueue y onto Q;
          else
             return flow, Path[]
    until Q.size() == 0
  return 0, Path[]
  end.
```

Algorithm 2. Breadth First Search

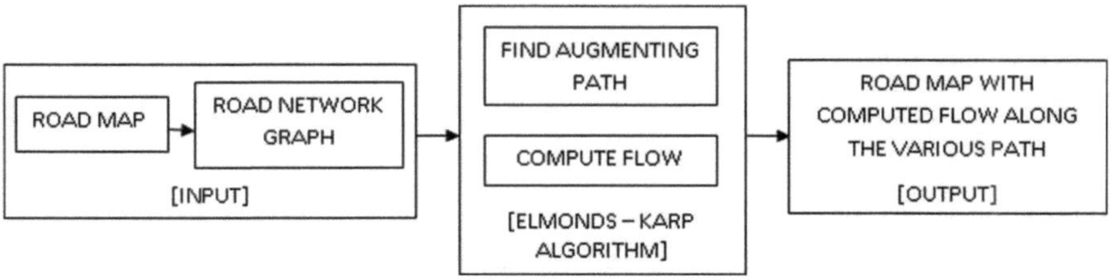

Fig. 1 Architecture of transportation management system

4 Architecture of Transport Management System

Consider there is traffic congestion at a particular junction on the road. On selecting that junction as source to any destination on the roadmap, the road network graph G can be generated. The graph G will be represented as adjacency matrix containing road capacities between the various junctions on the selected area. The road junctions are the points where the traffic signals are located. But the road capacity between two junctions will not be the same. Hence, consider the system generates the road junctions by using the road capacity. For example, if roads say from junction A to C contains road capacity 5 initially and if the road shrinks after and the capacity reduces to 3, then create a junction B, such that capacity from A to B is 5 and B to C is 3.

The capacity matrix, with the source node and destination node are given as input to the Edmonds–Karp algorithm. The algorithm computes the maximum traffic flow of the network. The architecture of the traffic management system is shown in Fig. 1.

5 Complexity Analysis

Let 'n' be the number of junctions on the road network and 'm' be the number of roads connecting the junctions. Breadth First Search strategy executes the loop 'nm' times in total to find each augmenting path. Edmonds–Karp algorithm calls Breadth First Search, 'm' times to find the different augmenting paths. Hence, the total running time of the Edmonds–Karp algorithm is $O(nm^2)$.

As the number of junctions 'n' on the road network and the number of roads 'm' connecting the junctions increases, time taken for computation of maximum flow increases in polynomial time. Hence, when there is a huge road network, divide the network into small networks. Edmonds–Karp algorithm is executed on each individual network to compute the traffic flow and paths on the network. Finally, the maximum traffic flow among traffic flow in each network is found as the traffic flow of the entire network.

Table 1 Serial execution on graph with 11 nodes and 17 edges

Execution time $O(nm^2)$ (s)	Maximum flow	Number of augmenting paths
0.015	15	7

Table 2 Serial execution on divided graph

Network number	Execution time $O(nm^2)$ (s)	Maximum flow	Number of augmenting paths
Network I with 5 nodes and 6 edges	0.0024	8	3
Network II with 5 nodes and 6 edges	0.0021	11	3
Total	0.0043	11 (maximum among the 2 flows)	6

Table 3 Parallel execution on divided graph

Network number	Execution time $O(nm^2)$ (s)	Maximum flow	Number of augmenting paths
Network I with 5 nodes and 6 edges	0.0024	8	3
Network II with 5 nodes and 6 edges	0.0021	11	3
Total	0.0026	11 (maximum among the 2 flows)	6

Better results are achieved by executing dividing small networks *in parallel.* This is implemented by creating multiple threads using OpenMP programming. Hence, *Total Execution Time* = $O(nm^2/k)$, where k is the number of threads created.

Table 1 shows the result of execution on a sample road network graph with 11 nodes and 17 edges. The given sample road network graph is divided into two small networks. Table 2 shows the result of serial execution of the divided network graphs. Table 3 shows the result of parallel execution of the divided network graphs with 2 threads.

6 Conclusion and Future Work

This paper focuses on implementation of Traffic Management System for road networks during traffic congestions. The Traffic Management System proposed in this paper models a road network graph. The traffic congestions are solved using Edmonds–Karp algorithm by identifying different paths on the road network graph to divert the traffic flow. The complexity analysis and the results presented in the paper prove that the proposed Traffic Management System using Edmonds–Karp algorithm is efficient. This work can be extended to automate the partitioning of road network into subnetworks with one source and sink node each.

Acknowledgments We would like to thank our colleagues from our institution for providing deeper insights and expertise that helped our research to great extent.

References

1. Khaled Rabieh, Mohamed M. E. A. Mahmoud, Mohamed Younis: Privacy-Preserving Route Reporting Schemes for Traffic Management Systems. IEEE Transaction on Vehicular Technology. Vol. 66 (3). (2017) 2703–2713.
2. Otsuki. K, Kobayashi. Y, Murota. K:Improved max-flow min-cut algorithms in a circular disk failure model with application to a road network. Eur. J. Oper. Res. 248(2), 396–403 (2016).
3. Elisa Valentina Moisi, Benedek Nagy, Vladimir loan Cretu: Maximum flow minimum cost algorithm for reconstruction of images represented on triangular grids. IEEE 8th International Symposium on Applied Computational Intelligence and Infomatics. (2013) 35–40.
4. Dancoisne. B, Dupont. E, Zhang. W: Distributed max-flow in spark (2015).

Quadrature Synchronization of Two Van der Pol Oscillators Coupled by Fractional-Order Derivatives

Aman K. Singh and R. D. S. Yadava

Abstract The paper presents a theoretical analysis of the synchronization behavior of two coupled Van der Pol oscillators, where the coupling is defined by fractional-order derivatives. The condition for frequency synchronization is obtained for the two oscillators being in-phase quadrature. It is found that the synchronization frequency oscillates rapidly with respect to the deviations from phase quadrature and the order of fractional derivative. The linear stability analysis is carried out by analyzing the roots of Jacobian on phase error.

Keywords Coupled Van der Pol oscillators · Synchronization
Fractional-order coupling · Quadrature oscillators

1 Introduction

Fractional calculus provides an alternate approach for efficient modeling of some nonlinear dynamical systems in physics and engineering that involve many time scales [1]. An overview of the applications of fractional derivatives in electromagnetic theory and in automation and control can be seen in [2]. The fractional derivatives have been used in modeling the frequency-dependent damping in viscoelastic systems and in real materials [2–9], in controlling chaos in MEMS oscillators [10], and in frequency synchronization of the coupled Van der Pol oscillators [11]. In the last mentioned study [11], the two Van der Pol oscillators coupled by fractional-order derivatives were analyzed for the stable in-phase and out-of-phase synchronization modes. In electronic communication systems, the multiphase harmonic oscillators are needed for a variety of signal processing

A. K. Singh (✉) · R. D. S. Yadava
Sensors & Signal Processing Laboratory, Department of Physics, Institute of Science, Banaras Hindu University, Varanasi 221005, India
e-mail: aman.strgtr@gmail.com

R. D. S. Yadava
e-mail: ardius@gmail.com; ardius@bhu.ac.in

C. R. Panigrahi et al. (eds.), *Progress in Advanced Computing and Intelligent Engineering*, Advances in Intelligent Systems and Computing 714,
https://doi.org/10.1007/978-981-13-0224-4_53

operations like mixing, modulation/ demodulation, and noise suppression filtering [12]. In this work, we extend the analysis of fractionally coupled Van der Pol oscillators for stable quadrature phase (0 and $\pi/2$ phase shifted) synchronization. The quadrature phase oscillators are of great importance for the modern digital communication [13].

The oscillator system in this study has been defined by a pair of Van der Pol oscillators, whose outputs are coupled by the fractional-order model of a capacitor. The current flow through the capacitor C is modeled as $i = CD^{\beta}x$ with $D^{\beta} = \frac{d^{\beta}}{dt^{\beta}}$ being the fractional-order time derivative; x denotes the time dependent voltage across the capacitor, $0 < \beta < 1$, and $C = \frac{\Gamma(1-\beta)}{h}$ with $\Gamma(.)$ denoting the Gamma function and h being a parameter. The parameters h and β are related to the dielectric properties and losses in the capacitor respectively [14].

The fractional-order derivative of a function $z(t)$ is defined in terms of Riemann–Liouville operator as [1]

$$D^{\gamma}z(t) = \frac{1}{\Gamma(1-\gamma)}\frac{d}{dt}\int_{0}^{t}\frac{z(\tau)}{(t-\tau)^{\gamma}}d\tau, \quad 0<\gamma<1. \tag{1}$$

2 Equations of Motion

The Van der Pol oscillators consist of a nonlinear current source and LCR components. The dimensionless form of Van der Pol equation is written as

$$\ddot{u} - \varepsilon(1-u^2)\dot{u} + u = 0. \tag{2}$$

where $\varepsilon > 0$ is the strength of nonlinear damping. When $\varepsilon \gg 1$, the oscillations are relaxation type and for $\varepsilon \ll 1$ the system undergoes limit cycle oscillations [15]. We will consider the identical current sources in the coupled systems. The current of a nonlinear source can be written as [16]

$$i_{d1} = -g_1x + g_2x^3 \quad \text{and} \quad i_{d2} = -g_1y + g_2y^3. \tag{3}$$

The equation of motion of coupled Van der Pol system can be obtained using Kirchoff's law from the circuit shown in Fig. 1

$$\frac{1}{L_1}\int xdt + \frac{x}{R_1} + C_1\frac{dx}{dt} - i_{d1} = CD^{\beta}(y-x) \tag{4}$$

$$\frac{1}{L_2}\int ydt + \frac{y}{R_2} + C_2\frac{dy}{dt} - i_{d2} = CD^{\beta}(x-y). \tag{5}$$

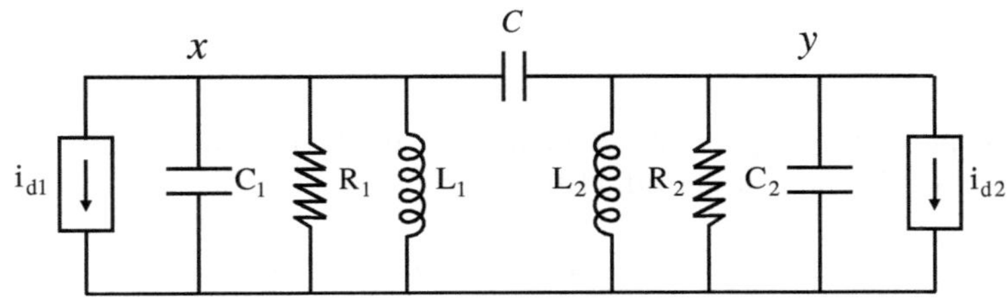

Fig. 1 Schematic of two Van der Pol oscillators coupled by fractional capacitor

Differentiating one more time and substituting the values of nonlinear currents from Eq. (1), we obtain

$$\ddot{x}-2(\delta_0-\delta_2x^2)\dot{x}+\sigma_1^2x=\lambda_1D^\alpha(y-x) \quad (6)$$

$$\ddot{y}-2(\gamma_0-\gamma_2y^2)\dot{y}+\sigma_2^2y=\lambda_2D^\alpha(x-y). \quad (7)$$

where the constants are defined below

$$\begin{aligned}&\delta_0=\frac{g_1-\frac{1}{R_1}}{2C_1},\gamma_0=\frac{g_1-\frac{1}{R_2}}{2C_2},\delta_2=\frac{3g_1}{2C_1},\gamma_2=\frac{3g_1}{2C_2},\\&\sigma_1^2=\frac{1}{L_1C_1},\sigma_2^2=\frac{1}{L_2C_2},\lambda_1=\frac{C}{C_1},\lambda_2=\frac{C}{C_2},\alpha=\beta+1\text{ and }C=\frac{\Gamma(2-\alpha)}{h}.\end{aligned} \quad (8)$$

3 Synchronization

Equations (6) and (7) describe the system of a pair of Van der Pol oscillators coupled by fractional-order derivative $(1<\alpha<2)$. We choose the parameters defined in Eq. (8) in such a way that the coupled oscillators can be assumed to be weakly nonlinear oscillators with weak coupling. There are various analytical approaches in the domain of perturbation techniques [17, 18] to deal with nonlinear systems with weak nonlinearities. We apply averaging method [18] which suggests the solutions of harmonic form with a small perturbation from limit cycles of individual free running oscillators. Let us assume the solution of Eqs. (6) and (7) of the form

$$x=P\cos\omega t-Q\sin\omega t,\quad y=R\sin\omega t-S\cos\omega t. \quad (9)$$

where the coefficients of harmonics are slowly varying time dependent functions and ω is the frequency of synchronization. We can also define the velocities as

$$\dot{x} = -\omega(P \sin \omega t + Q \cos \omega t), \quad \dot{y} = \omega(R \cos \omega t + S \sin \omega t). \tag{10}$$

with assumption that P, Q, R, and S are the slowly varying functions of time, we can get the accelerations using Eq. (10) as

$$\begin{aligned} \ddot{x} &= -\omega^2(P \cos \omega t - Q \sin \omega t) - \omega(\dot{P} \sin \omega t + \dot{Q} \cos \omega t), \\ \ddot{y} &= -\omega^2(R \sin \omega t - S \cos \omega t) + \omega(\dot{R} \cos \omega t + \dot{S} \sin \omega t). \end{aligned} \tag{11}$$

Substitution of the transformations from Eqs. (9)–(11) into Eqs. (6) and (7) and evaluating the fractional derivative [4] from Eq. (1) followed by averaging over quadrature components [18] we obtain a system four couple first order ODE asODE as

$$\begin{aligned} \omega \frac{dP}{dt} &= (\omega^2 - \sigma_1^2)Q + 2\omega P\left[\delta_0 - \frac{\delta_2(P^2+Q^2)}{4}\right] + \lambda_1\omega^\alpha\left[(P+S)\cos\alpha\frac{\pi}{2} - (Q+R)\sin\alpha\frac{\pi}{2}\right] \\ \omega \frac{dQ}{dt} &= -(\omega^2 - \sigma_1^2)P + 2\omega Q\left[\delta_0 - \frac{\delta_2(P^2+Q^2)}{4}\right] - \lambda_1\omega^\alpha\left[(P+S)\sin\alpha\frac{\pi}{2} + (Q+R)\cos\alpha\frac{\pi}{2}\right] \\ \omega \frac{dR}{dt} &= -(\omega^2 - \sigma_2^2)S + 2\omega R\left[\gamma_0 - \frac{\gamma_2(R^2+S^2)}{4}\right] - \lambda_2\omega^\alpha\left[(P+S)\sin\alpha\frac{\pi}{2} + (Q+R)\cos\alpha\frac{\pi}{2}\right] \\ \omega \frac{dS}{dt} &= (\omega^2 - \sigma_2^2)R + 2\omega S\left[\gamma_0 - \frac{\gamma_2(R^2+S^2)}{4}\right] + \lambda_2\omega^\alpha\left[(P+S)\cos\alpha\frac{\pi}{2} - (Q+R)\sin\alpha\frac{\pi}{2}\right] \end{aligned} \tag{12}$$

For the steady-state solutions, we substitute $\dot{P} = \dot{Q} = \dot{R} = \dot{S} = 0$ in Eq. (12)

$$\begin{aligned} &(\omega^2 - \sigma_1^2)Q + 2\omega P\left[\delta_0 - \frac{\delta_2(P^2+Q^2)}{4}\right] + \lambda_1\omega^\alpha\left[(P+S)\cos\alpha\frac{\pi}{2} - (Q+R)\sin\alpha\frac{\pi}{2}\right] = 0 \\ &-(\omega^2 - \sigma_1^2)P + 2\omega Q\left[\delta_0 - \frac{\delta_2(P^2+Q^2)}{4}\right] - \lambda_1\omega^\alpha\left[(P+S)\sin\alpha\frac{\pi}{2} + (Q+R)\cos\alpha\frac{\pi}{2}\right] = 0 \\ &-(\omega^2 - \sigma_2^2)S + 2\omega R\left[\gamma_0 - \frac{\gamma_2(R^2+S^2)}{4}\right] - \lambda_2\omega^\alpha\left[(P+S)\sin\alpha\frac{\pi}{2} - (Q+R)\cos\alpha\frac{\pi}{2}\right] = 0 \\ &(\omega^2 - \sigma_2^2)R + 2\omega S\left[\gamma_0 - \frac{\gamma_2(R^2+S^2)}{4}\right] + \lambda_2\omega^\alpha\left[(P+S)\cos\alpha\frac{\pi}{2} - (Q+R)\sin\alpha\frac{\pi}{2}\right] = 0. \end{aligned} \tag{13}$$

For determining the condition of phase quadrature synchronization, we substitute $Q = 0$, $R = M \cos\phi$, $S = M \sin\phi$ in Eq. (13)

$$\omega P\left(\delta_0 - \frac{\delta_2 P^2}{4}\right) + \lambda_1\omega^\alpha\left[P \cos\alpha\frac{\pi}{2} - M \sin\left(\alpha\frac{\pi}{2} - \phi\right)\right] = 0. \tag{14}$$

$$-(\omega^2-\sigma_1^2)P-\lambda_1\omega^\alpha\left[P\sin\alpha\frac{\pi}{2}+M\cos\left(\alpha\frac{\pi}{2}-\phi\right)\right]=0. \tag{15}$$

$$-(\omega^2-\sigma_2^2)M\sin\phi+\omega M\cos\phi\left(\gamma_0-\frac{\gamma_2 M^2}{4}\right)-\lambda_2\omega^\alpha\left[P\sin\alpha\frac{\pi}{2}-M\cos\left(\alpha\frac{\pi}{2}+\phi\right)\right]=0 \tag{16}$$

$$(\omega^2-\sigma_2^2)M\cos\phi+\omega M\sin\phi\left(\gamma_0-\frac{\gamma_2 M^2}{4}\right)-\lambda_2\omega^\alpha\left[P\cos\alpha\frac{\pi}{2}-M\sin\left(\alpha\frac{\pi}{2}+\phi\right)\right]=0 \tag{17}$$

From Eq. (15), we find

$$\left((\omega^2-\sigma_1^2)+\lambda_1\omega^\alpha\sin\alpha\frac{\pi}{2}\right)P=-\left(\lambda_1\omega^\alpha\cos\left(\alpha\frac{\pi}{2}-\phi\right)\right)M. \tag{18}$$

and using Eqs. (16) and (17), we have

$$\left((\omega^2-\sigma_2^2)-\lambda_2\omega^\alpha\sin\alpha\frac{\pi}{2}\right)M=-\left(\lambda_2\omega^\alpha\cos\left(\alpha\frac{\pi}{2}-\phi\right)\right)P. \tag{19}$$

For obtaining the condition for synchronization, we eliminate P and M from Eqs. (18) and (19)

$$\begin{aligned}&(\omega^2-\sigma_1^2)(\omega^2-\sigma_2^2)+(\lambda_1-\lambda_2)\omega^{2+\alpha}\sin\alpha\frac{\pi}{2}+(\lambda_2\sigma_1^2-\lambda_1\sigma_2^2)\omega^{\alpha+2}\sin\alpha\frac{\pi}{2}\\&\quad=\lambda_1\lambda_2\omega^{2\alpha}\cos^2\phi\end{aligned} \tag{20}$$

Equation (20) is the condition of mutual synchronization in-phase quadrature and it has more than one root (real as well as complex). We have chosen only one real root of Eq. (20) to proceed further. Other real roots are equally important but they will not make much qualitative difference in the analyses.

4 Results and Discussion

The parameters of the systems are taken as [16]:
$g_1=8.5\times10^{-3}\,\Omega^{-1}$, $g_2=7.3\times10^{-3}\,\Omega^{-1}/\text{V}$, $L_1=1.601$ nH, $C_1=7.5$ pF, $R_1=200\ \Omega$, $L_2=1.60$ nH, $C_2=8$ pF, $R_2=200\ \Omega$, $\sigma_1=9.126$ GHz, $\sigma_2=8.840$ GHz.

FigFigure 2 shows the variation of synchronization frequency with the order of derivative α for the condition of perfect phase quadrature synchronization ($\phi=0°$) for different values of h. It can be seen that the coupled system synchronizes to

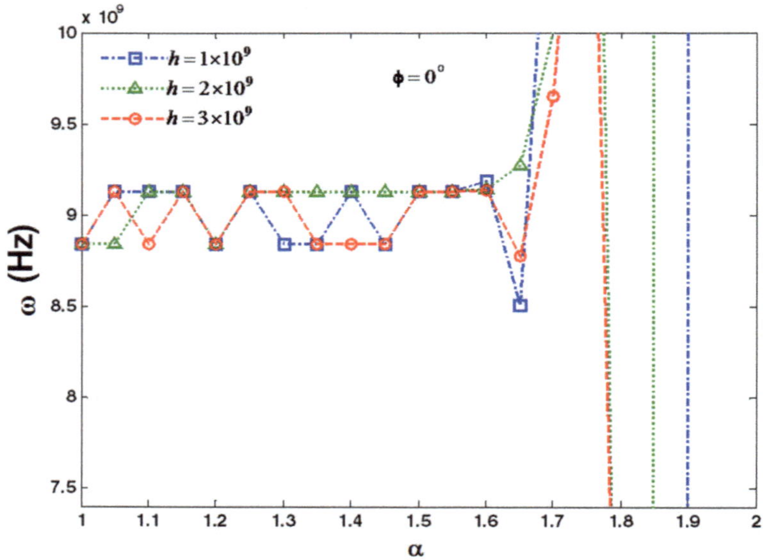

Fig. 2 Variation of frequency with the order of fractional derivative (α) for a phase error of 0°

either σ_1 or σ_2 depending on the values of h and α. It can be tuned to be in any of these two quadrature states by varying α for a given h. In Figs. 2 and 3 only those points for which a quadrature solutions exist are shown. The system can be switched from one quadrature to the other by varying α. However, note that this switching occurs only with discrete changes in the values of α. There is no quadrature solution for the values of α in between.

Figure 3a, b shows the variation of synchronization frequency with the order of fractional derivative (α) for the phase error of $-1°$ and $1°$ respectively. Each curve is generated for a fixed value of coupling strength h (in units of $\mathrm{V/Asec}^{\alpha}$). Qualitatively the results are similar to that shown in Fig. 2. At higher values of α the synchronization frequency diverges indicating instabilities of the system. There are certain ranges of α where no synchronization state exists. In certain intervals, however, the synchronization frequency is less sensitive to the variations in α, for example, $\alpha = 1 - 1.2$ in Fig. 3a.

Figure 4 shows the variation of synchronization frequency with the phase error ϕ (in degree) at a fixed value of order of derivative $(\alpha = 1.5)$ and coupling strength $(h = 1 \times 10^9\ \mathrm{V/ASec}^{\alpha}, \alpha = 1.5)$. We observe that the synchronization frequency oscillates rapidly between the frequencies σ_1 and σ_2 of free running oscillators as a function of ϕ. Thus, there is the occurrence of phase error-induced sharp transition of synchronization frequency from the frequency of one oscillator to another

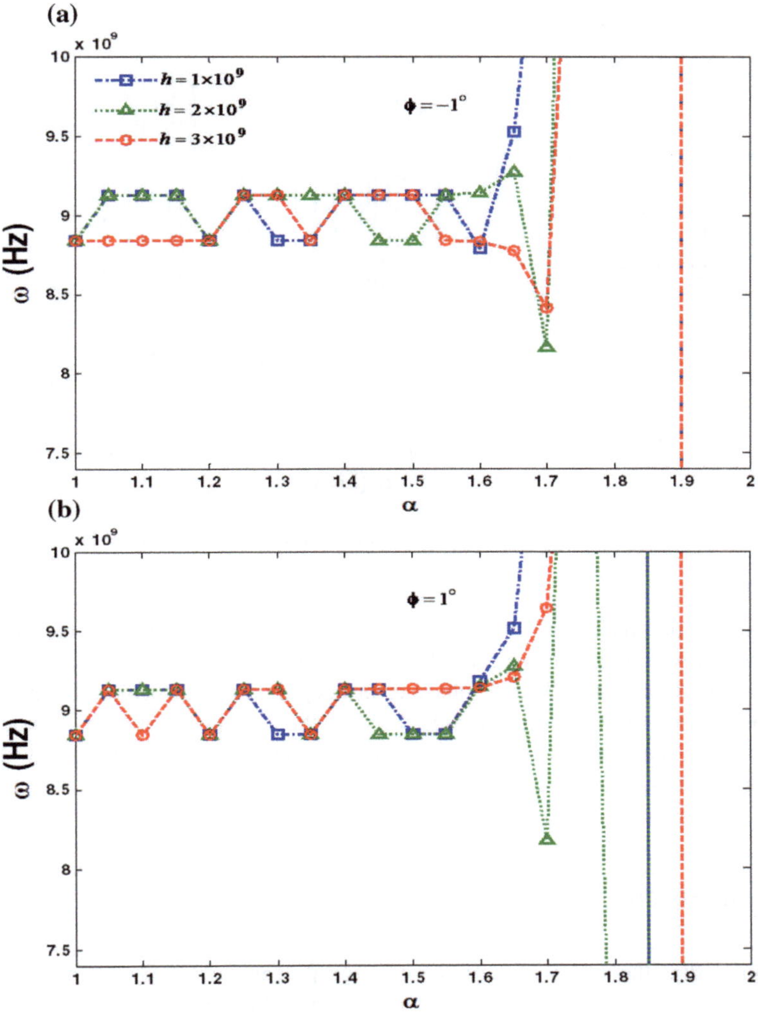

Fig. 3 Variation of frequency with the order of fractional derivative (α) for phase error **a** $\phi = -1°$ **b** $\phi = 1°$

oscillator. There are only two frequencies of synchronization in-phase quadrature state. We have not discussed in and out-of-phase synchronization that may happen at intermediate frequencies also.

The linear stability analysis [19] has been presented in Fig. 5 that shows the variation of roots of the Jacobian matrix of Eq. 12 (taking $Q = 0$ from the

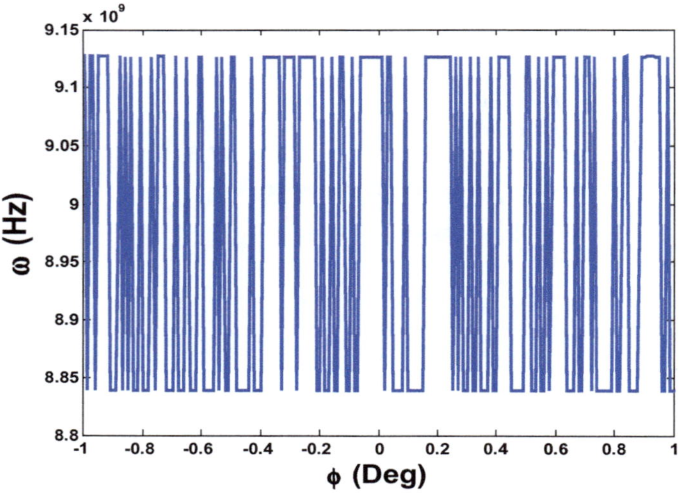

Fig. 4 Variation of frequency with the phase error for $\alpha = 1.5$ and $h = 1 \times 10^9$ V/ASec$^\alpha$

beginning) with phase error. If all the roots have the negative real part the system is asymptotically stable. We see that all the three roots (since Q is zero) of Jacobian are negative for certain values of phase error keeping all other parameters fixed. We can easily find these values of ϕ for which the synchronization is stable. The synchronized state also oscillates in between the stable and unstable states when the phase error is varied.

5 Conclusion

We found that the order of derivative α (or β) can be tuned to find a desired synchronization frequency for a given phase error. For higher α, the synchronization frequency diverges which indicate that the system becomes unstable and is not of interest to study (Figs. 2 and 3). For a fixed value of derivative order ($\alpha = 1.5$) the synchronization frequency makes rapid transitions from one oscillator's frequency to the other when the phase error ϕ is increased on either side of the perfect quadrature (Fig. 4). The linear stability analysis ensures that the system is asymptotically stable in the assumed range ($-1°$ to $1°$) of phase error with $\alpha = 1.5$ and $h = 1 \times 10^9$ V/Asec$^\alpha$ (other values may also provide stability).

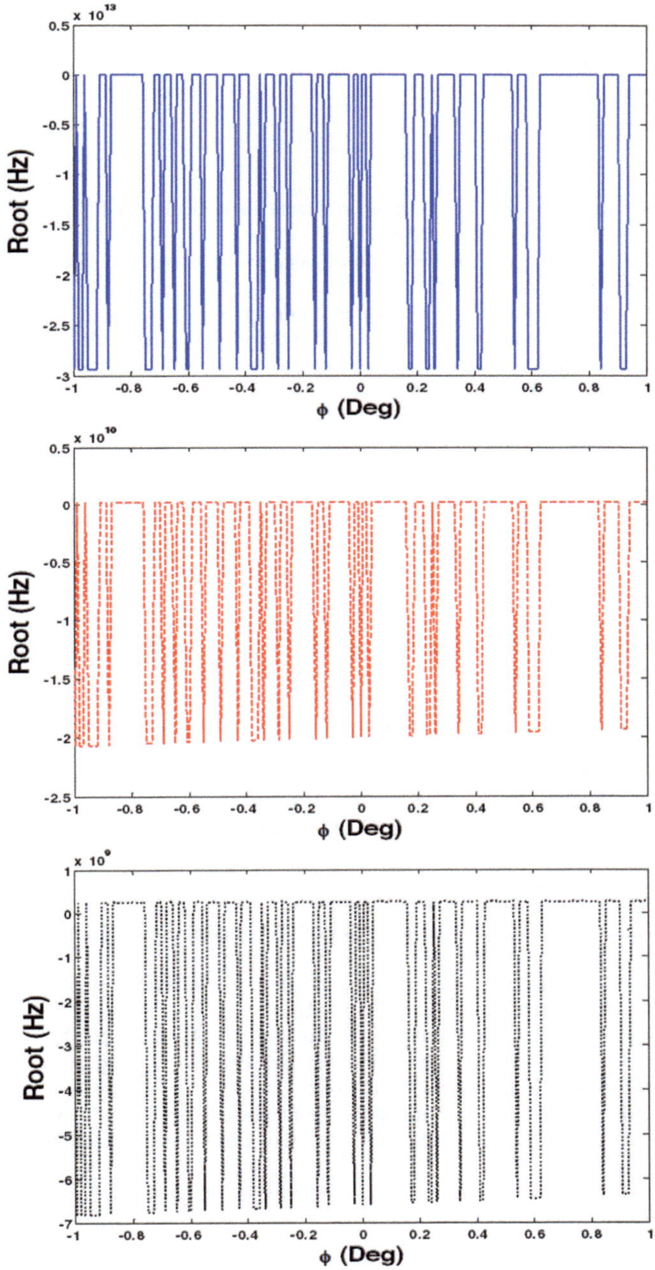

Fig. 5 Variation of roots of Jacobian matrix with the phase error ϕ for $\alpha = 1.5$, $h = 1 \times 10^9$

Acknowledgments The author AKS is thankful to University Grants Commission (UGC, Govt. of India, Delhi) for financial assistance.

References

1. Miller, K., Ross, B.: An Introduction to Fractional Calculus and Fractional Differential Equations, Wiley, New York (1993)
2. Bagley, R.L., Torvik, P.J.: On the fractional calculus model of viscoelastic behavior, Journal of Rheology 30 (1984) 133–155
3. Torvik, P.J., Bagley, R.L.: On the appearance of fractional derivative in the behaviour of real materials, Transaction of the ASME, Journal of Applied Mechanics 51 (1984) 294–298
4. Mainardi, F.: Fractional relaxation-oscillation and fractional diffusion-wave phenomena, Chaos, Solitons and Fractals 7 (1996) 1461–1477
5. Wahi, P., Chatterjee, A.: Averaging oscillations with small fractional damping and delayed term, Nonlinear Dynamics 38 (2004) 3–22
6. Galucio, A.C., Deu, J.F., Ohayon, R.: Finite element formulation of viscoelastic sandwich beams using fractional derivative operators, Computational Mechanics 33 (2004) 282–291
7. Jesus, I.S., Machado, J.A.T.: Implementation of fractional order electromagnetic potential through a genetic algorithm, Commun Nonlinear Sci Numer Simulat 14 (2008) 1838–1843
8. Rand, R.H., Sah, S.M., Suchorsky, M.K.: Fractional Mathieu equation, Commun Nonlinear Sci Numer Simulat 15 (2010) 3254–3262
9. Machado, J.A.T., Silva, M.F., Barbosa, R.S., Jesus, I.S., Reis, C.M., Marcos, M.G. Galhano, A.F.: Some applications of fractional calculus in engineering, Mathematical Problems in Engineering 2010 (2010) 1–34
10. Tusset, A.M., Balthazar, J.M., Bassinello, D.G., Pontes Jr., B.R., Felix, J.L.P.: Statements on chaos control designs, including a fractional order dynamical system applied to a MEMS comb-drive actuator, Nonlinear Dynamics 69 (2012) 1837–1857
11. Suchorsky, M.K., Rand, R.H.: A pair of Van der Pol oscillators coupled by fractional derivatives, Nonlinear Dynamics 69 (2012) 313–324
12. Rappaport, T.S.: Wireless Communication: Principles and Practice. 2nd edn. Prentice Hall, (2002)
13. Elbadry, M., Harjani, R.: Quadrature Frequency Generation for Wideband Wireless Application, Springer, New York (2015)
14. Wasterlund, S., Ekstam, L.: Capacitor theory, IEEE Transaction on Dielectric and Electrical Insulation 1 (1994) 826–839
15. Ginoux, J.M., Letellier, C.: Van der Pol and the history of relaxation oscillations: Toward the emergence of a concept, Chaos 22 (2012) 1–15
16. Dumitrescu, I., Bachir, S., Cordeau, D., Paillot, J. –M., Iordache, M.: Modelling and characterization of oscillator circuits by Van Der Pol model using parameter estimation, J. Circuits Systems Computers 21 (2012) 1–15
17. Nayfeh, A.H.: Introduction to Perturbation Techniques, Wiley, New York (1993)
18. Adronov, A.A, Vitt, S.E., Khaikin, S.E.: Theory of Oscillators, 2nd edn. Pergamon, Oxford (1966)
19. Török, J.S.: Analytical Mechanics with an Introduction to Dynamical Systems, Wiley, New York (2000)

On the Category of Quantale-Semimodules

M. K. Dubey, Vijay K. Yadav and S. P. Tiwari

Abstract The concepts from quantale theory is applied in this work to bring forward and study an abstract idea of quantale-semimodule (Q-semimodule) and Q-sets. We have introduced the category $\boldsymbol{Q}$**-Mod** of Q-semimodule and confer an adjunction between the category $\boldsymbol{Q}$**-Mod** and the well-known category **Set**, and another between the category $\boldsymbol{L}$**-Set** of L-sets and the category $\boldsymbol{Q}$**-Set** of Q-sets. Finally, we have shown that the category $\boldsymbol{Q}$**-Mod** forms a monoidal category.

Keywords Quantale · Q-semimodule · Morphism · Category

1 Introduction

The theoretical computer science is developed significantly by using the concepts from fuzzy set theory and category theory (cf., e.g., [25–29]). The concept of L-fuzzy set (or L-set) was first coined by Goguen [6] as a generalized version of Zadeh's (cf., [30]) fuzzy sets. This work of Goguen provides a direction to study fuzzy sets having membership values in different lattice structures (cf., [3, 4, 7, 9, 14, 15, 20, 21, 24, 28]), which also leads to widespread application of fuzzy sets in different branches of science, like theoretical computer science(cf., e.g., [26, 27]), physics, biology, mathematical sciences [17, 23, 24]. In particular, fuzzy sets in [3] known as the L-subsets of a set, have membership value in complete distributive lattice, while in

M. K. Dubey (✉) · S. P. Tiwari
Department of Applied Mathematics, Indian Institute of Technology
(Indian School of Mines), Dhanbad 826004, India
e-mail: maheshdubey6@gmail.com

S. P. Tiwari
e-mail: sptiwarimaths@gmail.com

V. K. Yadav
Department of Mathematics, School of Mathematics, Statistics
and Computational Sciences, Central University of Rajasthan, NH-8, Bandar Sindari,
Ajmer 305817, Rajasthan, India
e-mail: vkymaths@gmail.com

C. R. Panigrahi et al. (eds.), *Progress in Advanced Computing and Intelligent Engineering*, Advances in Intelligent Systems and Computing 714,
https://doi.org/10.1007/978-981-13-0224-4_54

[28] poset, in [27] complete residuated lattice and in [24] lattice-ordered monoid were underlying structures for membership values of fuzzy sets. In recent years, the structure of distributive lattices, compete lattices, complete residuated lattices attract the attention of researchers and therefore studied in different directions (cf., e.g. [2, 5, 18, 19, 29].

With the help of complete lattices, the notion of quantale was introduced by Mulvey in [12]. The quantales and quantale modules were found successfully applicable in several areas of science and technology, some of the most relevant examples are logic, image processing, and computer science [15, 16]. Rosenthal in [15], provided a systematic introduction of quantale theory and its application to ideal theory of rings and linear logic, whereas some aspects of algebraic and categorical properties of quantales and quantale modules were studied in (cf., [1, 10, 11, 16, 22]), among these works, Solovyov in [22], specifically presents the collection of results regarding the category ***Q*-Mod** of quantale module which is monadic.

In this paper, using category theoretic approach and the concept of quantale we make further some contribution on theory of quantale module. We have introduced the concept of quantale-semimodule (Q-semimodule), free Q-semimodule, and quantale set (Q-set), and define the categories ***Q*-Mod** and ***Q*-Set** of Q-semimodules and Q-sets, respectively. The existence of an adjuction between ***Q*-Mod** and well-known category **Set**, as well as between ***Q*-Set** and L-Set is also investigated.

2 Preliminaries

The basic concept and notions from fuzzy set theory, lattice theory, and quantale theory, which will be needed in subsequent sections for completion of this paper is recalled here.

Definition 1 [13] A partially ordered set $L = (L, \leq)$ is called **lattice**, if for any $\alpha, \beta \in L$ have both an infimum and a supremum denoted, respectively, by $\alpha \wedge \beta$ and $\alpha \vee \beta$. This lattice is often denoted as $L = (L, \vee, \wedge)$.

An algebra $L = (L, \vee, \wedge, 0, 1)$, having 0 and 1, respectively, as the least and the greatest element of L, is called a **complete lattice**.

Definition 2 [6] A pair (A, μ) is said to be an ***L*-Set**, where A is a nonempty set and $\mu : A \to L$ is a map having membership value in a complete lattice L, called the ***L*-valued map**.

Eklund et al. (cf., [3]), demonstrated that L-sets and their morphisms form category ***L*-Set**. The object class of this category is the class $\{L^A | \ A \in$ **Set**$\}$ and for its two objects L^A and L^B the class of morphisms is defined as $\{\widetilde{f} : L^A \to L^B | \ f : A \to B \ in$ **Set**$\}$ such that $\forall \ \mu \in L^A$,

$$\widetilde{f}(\mu)(b) = \begin{cases} \bigvee \mu(a) & \text{if } a \in f^{-1}(b) \quad \text{and} \quad \mu(a) > 0, \\ 0 & \text{otherwise.} \end{cases}$$

From the category **Set** to category ***L*-Set**, the existence of a covariant L-valued power-set functor $\mathcal{L}$ was also demonstrated, this functor sends a member A in **Set** to the member $\mathcal{L}(A) = L^A$ in ***L*-Set** and morphism $f : A \to B$ in **Set** to the morphism $\mathcal{L}(f) : L^A \to L^B$ in ***L*-Set** such that $\mathcal{L}(f)(\mu)(b) = \bigvee_{\{a|f(a)=b\}} \mu(a)$, for $\mu \in L^A$.

Definition 3 [15] A complete lattice Q together with a binary operation $\star$ defines a **quantale** if $\star$ is associative and $\forall \alpha, \beta_i \in Q$, $i \in I$ (an index set) it satisfies the following conditions:

(i) $\alpha \star (\vee_i \beta_i) = \vee_i (\alpha \star \beta_i)$, and
(ii) $(\vee_i \beta_i) \star \alpha = \vee_i (\beta_i \star \alpha)$

Also, a quantale Q is called

(i) **unital**, if $\exists\ e_Q \in Q$ (called an identity or a unit element) such that $e_Q \star \alpha = \alpha \star e_Q = \alpha, \forall\ \alpha \in Q$.
(ii) **commutative**, if $\forall\ \alpha, \beta \in Q, \alpha \star \beta = \beta \star \alpha$.

It can be easily observed that every commutative unital quantale forms a semiring. In the rest of the paper, a quantale is assumed to be a semiring.

Definition 4 A map $\phi : M \to N$ define a **morphism** between two quantales M and N, if $\forall\ \alpha, \beta \in M$ and $\forall S \subseteq M$,

(i) $\phi(\bigvee S) = \bigvee\{\phi(\alpha') : \alpha' \in S\}$,
(ii) $\phi(\alpha \star \beta) = \phi(\alpha) \star \phi(\beta)$, and
(iii) $\phi(e_M) = e_N$.

Definition 5 Let M and A be a quantale and a set, respectively, then M is called **free** over A, if for given a quantale N and a map $h : A \to N$, $\exists$ a map $\psi : A \to M$ and a unique quantale morphism $\phi : M \to R$ such that $\phi \circ \psi = h$.

3 Category of Quantale Semimodule (Q-Mod)

The concept of semimodule, free semimodule over a commutative semiring have been studied in [8]. We introduced here the notion of a quantale semimodule (Q-semimodule), a free Q-semimodule, and Q-set and study them in detail. The existence of functor and an adjunction between ***L*-Set** and ***Q*-Set** is also demonstrated.

Definition 6 A **left *Q*-semimodule** over Q is a commutative monoid $(S, \odot, e_S)$, along with a map $\cdot : Q \times S \to S$ such that $\alpha \cdot s \mapsto \alpha s$ known as scalar multiplication, if $\forall\ \beta, \alpha \in Q$, $Q_1 \subseteq Q$ and $s, s' \in S$, the following conditions hold:

(i) $(\alpha \vee \beta) \cdot s = \alpha \cdot s \odot \beta \cdot s$,
(ii) $\alpha \cdot (s \odot s') = \alpha \cdot s \odot \alpha \cdot s'$,
(iii) $(\alpha \star \beta) \cdot s = \alpha \cdot (\beta s)$,
(iv) $1 \cdot s = s, \quad 1 \in Q$,
(v) $\alpha \cdot e_S = e_S = 0 \cdot s, \quad 0 \in Q$.

Throughout the paper, Q be a commutative quantale, whereby, any left Q-semimodule be also a right Q-semimodule (which can be defined in a similar fashion) and vice-versa.

Example 1 The structure $(L, \vee, \cdot, 0)$ defined a Q-semimodule over Q, if $(L, \vee, 0)$ is a join-semilattice having least element for which $\forall\, \beta \in Q$ and $n \in L, \beta \cdot n := \beta \wedge n$ holds.

Example 2 A Q-semimodule over Q is four tuples $(Q^X, \vee, \cdot, 0)$, where

(i) Q^X is a set of Q-valued functions over X,
(ii) 0 is a zero function, and
(iii) the multiplication '$\cdot$' is defined as $(\alpha \cdot \mu)(x) := \alpha \star \mu(x)$, $\forall\, \alpha \in Q$ and $\mu \in Q^X$.

Definition 7 A ***Q*-semimodule morphism** between Q-semimodules $(\mathfrak{M}, \odot_{\mathfrak{M}}, e_{\mathfrak{M}})$ and $(\mathfrak{R}, \odot_{\mathfrak{R}}, e_{\mathfrak{R}})$ is a map $\varphi : \mathfrak{M} \to \mathfrak{R}$ such that

(i) $\varphi(\gamma \odot_{\mathfrak{M}} \gamma_1) = \varphi(\gamma) \odot_{\mathfrak{R}} \varphi(\gamma_1)$,
(ii) $\varphi(\alpha \cdot \gamma) = \alpha \cdot \varphi(\gamma)$,
(iii) $\varphi(e_{\mathfrak{M}}) = e_{\mathfrak{R}}$,

$\forall \gamma, \gamma_1 \in \mathfrak{M}$ and $\forall \alpha \in Q$.

Proposition 1 *The Q-semimodules and Q-semimodule morphisms define a category.*

We have denoted this category by ***Q*-Mod**.

Definition 8 Let $\mathfrak{M}$ be a Q-semimodule and B be a set, then $\mathfrak{M}$ called **free** over B, if for a given Q-semimodule $\mathfrak{N}$ and a map $k : B \to \mathfrak{N}$, $\exists$ a map $\psi : B \to \mathfrak{M}$ and a unique $\varphi : \mathfrak{M} \to \mathfrak{N}$, a Q-semimodule morphism with $\varphi \circ \psi = k$.

We have denoted a free Q-semimodule over a set B by $Q(B)$. The existence of free Q-semimodule over a set B can be found in similar way as in [16].

Proposition 2 *Let **Q-Mod** and **Set** be the categories of Q-semimodule and sets, respectively, then $\exists$ an adjunction between them.*

Proof Let $U' :$ ***Q*-Mod** $\to$ **Set**, be a functor forgetting the semimodule structure, and F' be free functor which correspond to U' under which a set X is send to a free Q-semimodule on X. Further, consider $G \in$ ***Q*-Mod**, the adjunction $\psi :$ **Set** $\to$ ***Q*-Mod** sending each $\phi : F'(X) \to G$ (a Q-semimodule map) to a map $\eta : X \to U'(G)$ (a set map), where $\eta = \phi|_X$. However, ***Q*-Mod** $(F'(X), G) \cong$ **Set** $(X, U'(G))$.

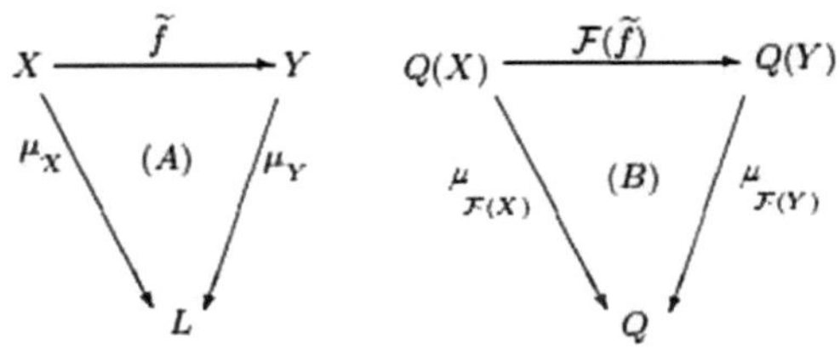

Fig. 1 Diagram for Theorem 1

Definition 9 Let $\mathfrak{M}$ and Q be a Q-semimodule and a quantale, respectively, then $(\mathfrak{M}, \lambda)$, where $\lambda : \mathfrak{M} \to Q$ be a $\boldsymbol{Q}$**-valued map**, is known as $\boldsymbol{Q}$**-Set**.

Remark 1 The object class of the category $\boldsymbol{Q}$-**Set** is the class $\{Q^{\mathfrak{M}}\}$, where $\mathfrak{M}$ is a member of $\{\boldsymbol{Q}\text{-}\mathbf{Mod}\}$, the class of morphisms between objects $Q^{\mathfrak{M}}$ and $Q^{\mathfrak{N}}$ are the maps $\{\widetilde{\varphi} : Q^{\mathfrak{M}} \to Q^{\mathfrak{N}}$, where $\varphi : \mathfrak{M} \to \mathfrak{N}\}$ satisfying for $\lambda \in Q^{\mathfrak{M}}$,

$$\widetilde{\varphi}(\lambda)(n) = \begin{cases} \bigvee \lambda(m) & \text{if } m \in \varphi^{-1}(n) \quad \text{and} \quad \lambda(m) > 0, \\ 0 & \text{otherwise.} \end{cases}$$

Further, $\exists$ a functor $\mathcal{Q} : \boldsymbol{Q}\text{-}\mathbf{Mod} \to \boldsymbol{Q}\text{-}\mathbf{Set}$, known as covariant Q-valued power-set functor sending $\mathfrak{M} \in \boldsymbol{Q}\text{-}\mathbf{Mod}$ to $\mathcal{Q}(\mathfrak{M}) = Q^{\mathfrak{M}}$ and $\varphi : \mathfrak{M} \to \mathfrak{N}$ in $\boldsymbol{Q}\text{-}\mathbf{Mod}$ to $\mathcal{Q}(\varphi) : Q^{\mathfrak{M}} \to Q^{\mathfrak{N}}$ such that $\mathcal{Q}(\varphi)(\lambda)(n) = \bigvee_{\{m:\varphi(m)=n\}} \lambda(m)$, for $\lambda \in Q^{\mathfrak{M}}$.

Theorem 1 *For categories* $\boldsymbol{L}$-**Set** *to* $\boldsymbol{Q}$-**Set**, *there exist maps* $\mathcal{F} : \boldsymbol{L}\text{-}\mathbf{Set} \to \boldsymbol{Q}\text{-}\mathbf{Set}$ *and* $\mathcal{U} : \boldsymbol{Q}\text{-}\mathbf{Set} \to \boldsymbol{L}\text{-}\mathbf{Set}$ *such that* $\mathcal{F}$ *and* $\mathcal{U}$ *are functor and mutually inverse isomorphisms.*

Proof For $\mathcal{F} : \boldsymbol{L}\text{-}\mathbf{Set} \to \boldsymbol{Q}\text{-}\mathbf{Set}$, the commutativity of the Diagrams in Fig. 1 imply that $\mathcal{F}(X, \mu_X)) = (Q(X), \mu_{\mathcal{F}(X)})$, i.e., whenever $(X, \mu_X) \in \boldsymbol{L}\text{-}\mathbf{Set}$, its image $(Q(X), \mu_{\mathcal{F}(X)})$ under $\mathcal{F}$ if a $\boldsymbol{Q}$-**Set** and if $\widetilde{f}$ is an $\boldsymbol{L}$-**Set**-morphism, then $\mathcal{F}(\widetilde{f})$ is a $\boldsymbol{Q}$-**Set**-morphism, i.e., if $\mu_X = \mu_Y \circ \widetilde{f} \in \boldsymbol{L}\text{-}\mathbf{Set}$ morphism, then

$$\begin{aligned} \mathcal{F}(\mu_X) = \mu_{\mathcal{F}(X)} &= \mathcal{F}(\mu_Y \circ \widetilde{f}) \\ &= \mu_{\mathcal{F}(Y)} \circ \mathcal{F}(\widetilde{f}), \end{aligned}$$

or that $\mathcal{F}(\widetilde{f}) \in \boldsymbol{Q}$-**Set**-morphism. It is easy to show that $\mathcal{F}$ preserve identity and composition law. Thus $\mathcal{F}$ is a functor from $\boldsymbol{L}$-**Set** to $\boldsymbol{Q}$-**Set**. Similarly, we can show that $\mathcal{U} : \boldsymbol{Q}\text{-}\mathbf{Set} \to \boldsymbol{L}\text{-}\mathbf{Set}$ is also functor. Finally, $\mathcal{F}$ and $\mathcal{U}$ are mutually inverse as one can easily show that $\mathcal{UF} = I = \mathcal{FU}$. The above arguments lead us to the following.

Theorem 2 *For an adjunction* φ *between categories* $\boldsymbol{L}$-**Set** *and* $\boldsymbol{Q}$-**Set**, *the diagram in Fig. 2 commutes.*

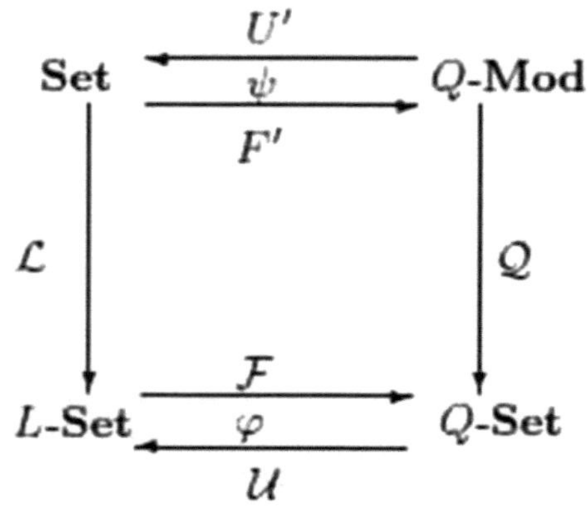

Fig. 2 Diagram for Theorem 2

4 Monoidal Structure of Q-Mod

The concept of Q-semimodule morphism introduced in previous section, is used here to introduce the concept of Q-multi-semimodule morphism which is proved to be the tensor product of Q-semimodule (Proposition 3). We have also shown by Propositions 4 and 5 that the tensor product of Q-semimodule is functorial. Interestingly, it is also shown here that the category ***Q*-Mod** introduced earlier in Sect. 3 is the monoidal category.

Definition 10 Let $\mathfrak{M}_1, \mathfrak{M}_2, \ldots, \mathfrak{M}_n$ and $\mathfrak{N}$ be Q-semimodules. A map $\Psi : \mathfrak{M}_1 \times \mathfrak{M}_2 \times \cdots \times \mathfrak{M}_n \to \mathfrak{N}$ is said to be a ***Q*-multi-semimodule morphism** if it is Q-semimodule morphism in each variable.

Now, we present the method of construction of the tensor product of Q-semimodules over a semiring Q which is commutative.

Suppose $\mathfrak{M}_1, \mathfrak{M}_2, \ldots, \mathfrak{M}_n$ be Q-semimodules and $\mathfrak{N}$ be the free Q-semimodule on $\mathfrak{M}_1 \times \mathfrak{M}_2 \times \cdots \times \mathfrak{M}_n$, and congruence relation $\cong$ on $\mathfrak{N}$ be produced by the equivalences $(\gamma_1, \ldots, \gamma_i +_{\mathfrak{M}_i} \gamma_i', \ldots, \gamma_n) \cong (\gamma_1, \ldots, \gamma_i, \ldots, \gamma_n) +_{\mathfrak{N}} (\gamma_1, \ldots, \gamma_i', \ldots, \gamma_n)$, and $(\gamma_1, \ldots, \alpha\gamma_i, \ldots, \gamma_n) \cong \alpha(\gamma_1, \upsilon, \gamma_i, \ldots, \gamma_n)$, $\forall \alpha \in Q$, $\gamma_i, \gamma_i' \in \mathfrak{M}_i$. Next, suppose $i : \mathfrak{M}_1 \times \mathfrak{M}_2 \times \cdots \times \mathfrak{M}_n \to \mathfrak{N}$ be the canonical injection of $\mathfrak{M}_1 \times \mathfrak{M}_2 \times \cdots \times \mathfrak{M}_n$ into $\mathfrak{N}$ and $\phi = h \circ i$, where $h : \mathfrak{N} \to \mathfrak{N}/\cong$.

Proposition 3 *The map* $\varphi : \mathfrak{M}_1 \times \mathfrak{M}_2 \times \cdots \times \mathfrak{M}_n \to \mathfrak{N}/\cong$ *is a Q-multisemimodule morphism and is a tensor product of* $\mathfrak{M}_1, \mathfrak{M}_2, \ldots, \mathfrak{M}_n$.

Proof Clearly φ is Q-multisemimodule morphism. Let $\mathfrak{K}$ be a Q-semimodule and $\Psi : \mathfrak{M}_1 \times \mathfrak{M}_2 \times \cdots \times \mathfrak{M}_n \to \mathfrak{K}$ be a Q-multisemimodule morphism. By the definition of free Q-semimodule generated by $\mathfrak{M}_1 \times \mathfrak{M}_2 \times \cdots \times \mathfrak{M}_n$, we have an induced Q-semimodule morphism $\mu : \mathfrak{N} \to \mathfrak{K}$ such that the diagram in Fig. 3 commutes.

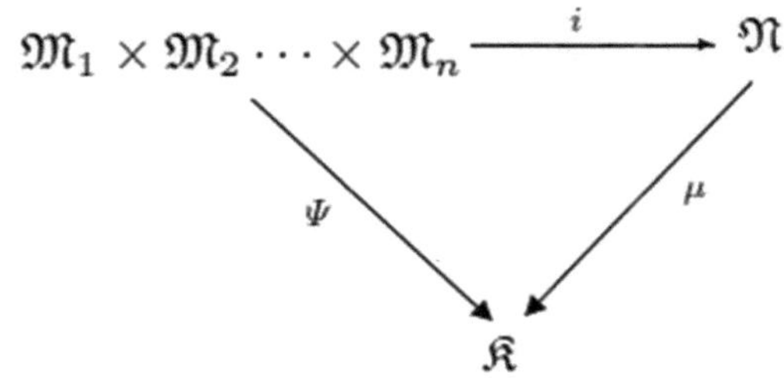

Fig. 3 Diagram for Proposition 3

The kernel of μ denoted $\cong_\mu$, is a congruence relation on $\mathfrak{N}$ given by $k_1 \cong_\mu k_2$ iff $\mu(k_1) = \mu(k_2) \forall k_1, k_2 \in \mathfrak{N}$. Since Ψ is Q-multisemimodule morphism, $k_1 \cong k_2$ implies $k_1 \cong_\mu k_2$, where $\cong$ used here is same as defined in tensor product. Hence it follows that μ can be factored through $\mathfrak{N}/\cong$, and $\exists$ a Q-semimodule morphism $\Psi_* : \mathfrak{N}/\cong \to \mathfrak{K}$ with property that the diagram in Fig. 4 commutes.

The image of φ generates $\mathfrak{N}/\cong$, so Ψ_* is uniquely determined. The module $\mathfrak{N}/\cong$ is denoted by $\mathfrak{M}_1 \otimes_Q \mathfrak{M}_2 \otimes_Q \cdots \otimes_Q \mathfrak{M}_n$. For our convenient we omit the subscript on the $\otimes$ symbol.

Now, we define the tensor product of Q-semimodule morphism.

Proposition 4 *Let $\mathfrak{M}_i$, $\mathfrak{N}_i$ be Q-semimodules, $1 \leq i \leq n$ and let $\Psi_i : \mathfrak{M}_i \to \mathfrak{N}_i$ be Q-semimodule morphisms. Then there is a unique Q-semimodule morphism $\Psi : \mathfrak{M}_1 \otimes \mathfrak{M}_2 \otimes \cdots \otimes \mathfrak{M}_n \to \mathfrak{N}_1 \otimes \mathfrak{N}_2 \otimes \cdots \otimes \mathfrak{N}_n$ such that $\Psi(\gamma_1 \otimes \gamma_2 \otimes \cdots \otimes \gamma_n) = \Psi_1(\gamma_1) \otimes \Psi_2(\gamma_2) \otimes \cdots \otimes \Psi_n(\gamma_n), \forall \gamma_i \in \mathfrak{M}_i$.*

Proof The $\Psi_i's$ induce a map $\Psi_1 \times \Psi_2 \times \cdots \times \Psi_n : \mathfrak{M}_1 \times \mathfrak{M}_2 \times \cdots \times \mathfrak{M}_n \to \mathfrak{N}_1 \otimes \mathfrak{N}_2 \times \cdots \otimes \mathfrak{N}_n$, which is not Q-multisemimodule morphism. It is easy to check that the composition of $\Psi_1 \times \Psi_2 \times \cdots \times \Psi_n$ with the canonical Q-multisemimodule morphism $\phi : \mathfrak{N}_1 \times \mathfrak{N}_2 \times \cdots \times \mathfrak{N}_n \to \mathfrak{N}_1 \otimes \mathfrak{N}_2 \otimes \cdots \otimes \mathfrak{N}_n$ is Q-multisemimodule morphism $(\Psi_1 \times \Psi_2 \times \cdots \times \Psi_n) \circ \phi : \mathfrak{M}_1 \times \mathfrak{M}_2 \times \cdots \times \mathfrak{M}_n \to \mathfrak{N}_1 \otimes \mathfrak{N}_2 \otimes \cdots \otimes \mathfrak{N}_n$ such that $(\Psi_1 \times \Psi_2 \times \cdots \times \Psi_n)(\phi(\gamma_1, \gamma_2, \ldots, \gamma_n)) = \Psi_1(\gamma_1) \otimes \Psi_2(\gamma_2) \otimes \cdots \otimes \Psi_n(\gamma_n)$. By the initiality of the tensor product of the $\mathfrak{M}_i's$ there is a unique Q-semimodule morphism $\Psi : \mathfrak{M}_1 \otimes \mathfrak{M}_2 \otimes \cdots \otimes \mathfrak{M}_n \to \mathfrak{N}_1 \otimes \mathfrak{N}_2 \otimes \cdots \otimes \mathfrak{N}_n$ such that $\Psi(\gamma_1 \otimes \gamma_2 \otimes \cdots \otimes \gamma_n) \mapsto \Psi_1(\gamma_1) \otimes \Psi_2(\gamma_2) \otimes \cdots \otimes \Psi_n(\gamma_n)$. Also, we denote by $T(\Psi_1, \Psi_2, \ldots, \Psi_n)$ or $\Psi_1 \otimes \Psi_2 \otimes \cdots \otimes \Psi_n$, the map Ψ.

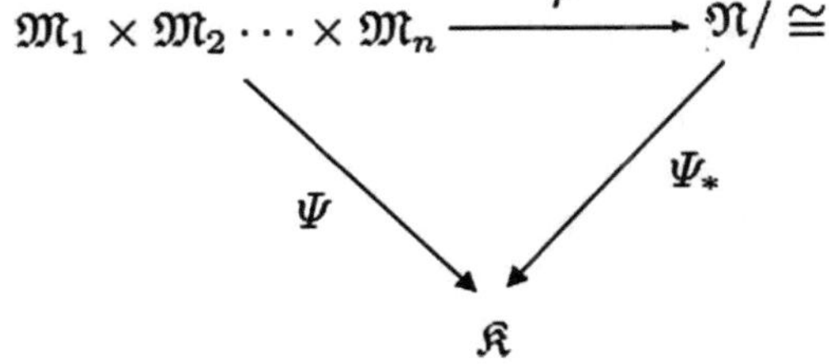

Fig. 4 Diagram for Proposition 3

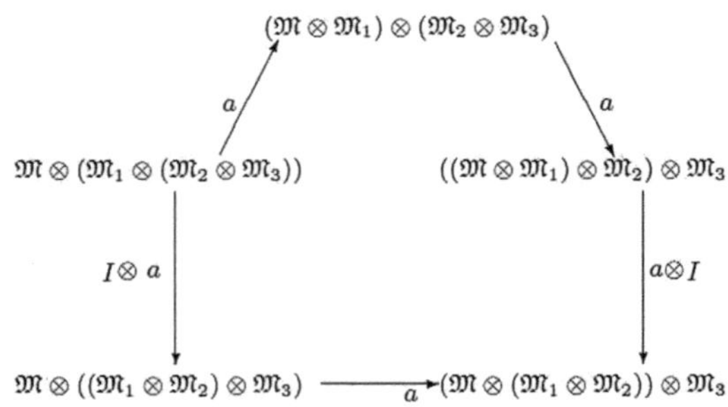

Fig. 5 Diagram for Pentagonal condition

Proposition 5 *Let $\mathfrak{M}_i$, $\mathfrak{N}_i$, $\mathfrak{H}_i$ be Q-semimodules $1 \le i \le n$, $n \in \mathbb{N}$. Also, let $\Psi_i : \mathfrak{M}_i \to \mathfrak{N}_i$ and $b_i : \mathfrak{N}_i \to \mathfrak{H}_i$ be Q-semimodule morphisms. Then $T(\Psi_1 \circ b_1, \Psi_2 \circ b_2, \ldots, \Psi_n \circ b_n) = T(\Psi_1, \Psi_2, \ldots, \Psi_n) \circ T(b_1, b_2, \ldots, b_n)$. Furthermore, $T(1_{\mathfrak{M}_1}, 1_{\mathfrak{M}_2}, \ldots, 1_{\mathfrak{M}_n})$ is the identity function on $\mathfrak{M}_1 \otimes \mathfrak{M}_2 \otimes \cdots \otimes \mathfrak{M}_n$.*

Proof The proof is a straightforward calculation.

Remark 2 From Proposition 4 and 5, it is clear that the tensor product T is functorial.

Proposition 6 *Let $\mathfrak{M}_1, \mathfrak{M}_2, \mathfrak{M}_3$ be Q-semimodules. Then, $\exists$ a unique isomorphism $a = a_{\mathfrak{M}_1\mathfrak{M}_2\mathfrak{M}_3} : (\mathfrak{M}_1 \otimes \mathfrak{M}_2) \otimes \mathfrak{M}_3 \to \mathfrak{M}_1 \otimes (\mathfrak{M}_2 \otimes \mathfrak{M}_3)$ satisfying $a_{\gamma_1\gamma_2\gamma_3}((\gamma_1 \otimes \gamma_2)$
$\otimes \gamma_3) = \gamma_1 \otimes (\gamma_2 \otimes \gamma_3)$, $\forall\ \gamma_i \in \mathfrak{M}_i, 1 \le i \le 3$.*

Proof Since elements of type $(\gamma_1 \otimes \gamma_2) \otimes \gamma_3$ generate the tensor product, the uniqueness of the desired Q-semimodule morphism is obvious. To prove its existence, let $\gamma_1 \in \mathfrak{M}_1$. The map $a_{\gamma_1} : \mathfrak{M}_2 \times \mathfrak{M}_3 \to (\mathfrak{M}_1 \otimes \mathfrak{M}_2) \otimes \mathfrak{M}_3$. Then the map a_{γ_1} is a Q-bisemimodule morphism. Therefore a_{γ_1} defines a Q-semimodule morphism $a^*_{\gamma_1} : \mathfrak{M}_2 \otimes \mathfrak{M}_3 \to (\mathfrak{M}_1 \otimes \mathfrak{M}_2) \otimes \mathfrak{M}_3$ such that $(\gamma_2 \otimes \gamma_3) \mapsto (\gamma_1 \otimes \gamma_2) \otimes \gamma_3$. Now, consider Q-bisemimodule morphism $\mathfrak{M}_1 \times (\mathfrak{M}_2 \otimes \mathfrak{M}_3) \to (\mathfrak{M}_1 \otimes \mathfrak{M}_2) \otimes \mathfrak{M}_3$ such that $(\gamma_1, \gamma_2 \otimes \gamma_3) \mapsto a^*_{\gamma_1}(\gamma_2 \otimes \gamma_3)$. This defines a Q-semimodule morphism $\mathfrak{M}_1 \otimes (\mathfrak{M}_2 \otimes \mathfrak{M}_3) \to (\mathfrak{M}_1 \otimes \mathfrak{M}_2) \otimes \mathfrak{M}_3$. A similar construction yields the inverse map $(\mathfrak{M}_1 \otimes \mathfrak{M}_2) \otimes \mathfrak{M}_3 \to \mathfrak{M}_1 \otimes (\mathfrak{M}_2 \otimes \mathfrak{M}_3)$. Furthermore, the associator a satisfies the pentagonal condition such that $\forall$ objects $\mathfrak{M}, \mathfrak{M}_1, \mathfrak{M}_2, \mathfrak{M}_3 \in$ ***Q*-Mod** the following diagram in Fig. 5 commutes:

Proposition 7 *There exists a unique isomorphism $r : \mathfrak{M} \otimes Q \to \mathfrak{M}$ with $\gamma \otimes \alpha \cong \alpha\gamma$ and $l : Q \otimes \mathfrak{M} \to \mathfrak{M}$ such that $\alpha \otimes \gamma \cong \alpha\gamma$, $\forall\ \alpha \in Q, \gamma \in \mathfrak{M}$.*

Proof Let $\mathfrak{M}$ be a Q-semimodule and $\mathfrak{N}$ be a free Q-semimodule with basis h. We assert that $\mathfrak{M} \otimes \mathfrak{N} \cong \mathfrak{M}$. The map $r' : \mathfrak{M} \times \mathfrak{N} \to \mathfrak{M}$ such that $(\gamma, kh) \mapsto k\gamma$ is Q-bisemimodule morphism from $\mathfrak{M} \times \mathfrak{N} \to \mathfrak{M}$ and hence induce a Q-semimodule morphism $r : \mathfrak{M} \otimes \mathfrak{N} \to \mathfrak{M}$ such that $(\gamma \otimes kh) \mapsto k\gamma$. There is also a Q-semimodule morphism $\Psi : \mathfrak{M} \to \mathfrak{M} \otimes \mathfrak{N}$ such that $\gamma \mapsto \gamma \otimes h$. A simple calculation shows that r and Ψ are inverse to each other. The maps are unique since we

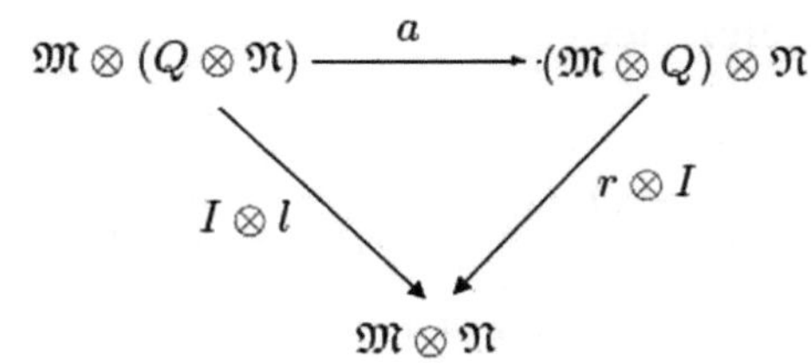

Fig. 6 Diagram for Proposition 7

have specified their actions on generating sets. Since Q can be considered as a free Q-semimodule over itself with basis 1 therefore $\mathfrak{M} \otimes Q \cong \mathfrak{M}$. Similarly $Q \otimes \mathfrak{M} \cong \mathfrak{M}$. Also, r and l make the following diagram in Fig. 6 commutes.

Theorem 3 *The category* ***Q*-Mod** *is a monoidal category.*

Proof As a monoidal category has bifunctor which is associative up to natural isomorphism and which has an object e which is a left and right unit up to a natural isomorphism. Here we consider bifunctor as a tensor product ⊗: ***Q*-Mod** × ***Q*-Mod** → ***Q*-Mod**, the unit object Q is considered as a Q-semimodule over itself. The existence of three natural isomorphism a, r, l (called associator, right unit and left unit) is given in the Propositions 6 and 7.

5 Conclusion

Nowadays, the complete lattices become the important structure to explain many concepts in theoretical computer science. The category theory also plays well for the development of several aspects of computer sciences. In this paper, the complete lattices are used to define quantales which is again applied to develop the idea of quantale-semimodule (Q-semimodule), a free Q-semimodule and Q-set. We have applied the concepts from category theory to introduce the category of ***Q*-Set** and ***Q*-Mod**, and demonstrated the existence of functors and an adjunction between wellknown category ***L*-Set** and category ***Q*-Set**. Further, we have provided the notion of Q-multi-semimodule morphism which is proved to be the tensor product of Q-semimodule (Proposition 3), it is also shown by Propositions 4 and 5 that the tensor product of Q-semimodule is functorial. Finally, we have shown the category ***Q*-Mod** is a monoidal category.

References

1. Abramsky S., Vickers S., Quantales, observational logic and process semantics, *Mathematical Structure in Computer Science*, **3**, 161–227 (1993).
2. Banaschewski B., Nelson E., Tensor products and bimorphisms, *Canadian Mathematical Bulletin*, **19**, 385–402 (1976).
3. Eklund P., Galán M.A., Medina J., Ojeda-Aciego M., Valverde A., Set functors, L-fuzzy set categories and generalized terms, *Computers and Mathematics with Applications*, **43**, 693–705 (2002).
4. Eklund P., Galán M.A., Medina J., Ojeda-Aciego M., Valverde A., Similarities between powersets of terms, *Fuzzy Sets and Systems*, **144**, 213–225 (2004).
5. García J.G., Höhle U., Kubiak T., Tensor products of complete lattices and their application in constructing quantales, *Fuzzy Sets and Systems*, **313**, 43–60 (2017).
6. Goguen J.A., L-fuzzy set, *Journal of Mathematical Analysis and Applications*, **18**, 145–174 (1967).
7. Goguen J.A., Categories of V-set, *Bulletin of American Mathematical Society*, **75**, 622–624 (1969).
8. Golan J.S., Semiring and Their Applications, Springer, Science + Business Media, Dordrecht (1999).
9. Höhle U., On the fundamentals of fuzzy set theory, *Journal of Mathematical Analysis and Application*, **201**, 786–826 (1996).
10. Kruml D., Paseka J., Algebraic and Categorical Aspects of Quantales, in Handbook of Algebra, Vol. **5**, North-Holland, 323–362 (2008).
11. Li Y.M., Zhou M., Li Z., Projective and injective objects in the category of quantales, *Journal of Pure and Applied Algebra*, **176**, 249–258 (2002).
12. Mulvey C.J., Rend. Circolo Mat. Palermo Suppl. Ser.II 12, 99–104 (1986).
13. Priestley H.A., Ordered sets and complete lattices, A primer for computer science, in: R. Backhouse, R. Crole, J. Gibbons (Eds.), Algebraic and Coalgebraic Methods in the Mathematics of Program Construction, in: *Lecture Notes in Computer Science*, **2297**, 21–78 (2002).
14. Rodabaugh S. E., Klement E. P., Höhle U., Applications of Category Theory to Fuzzy Subsets, Kluwer Academic, (1992).
15. Rosenthal K.I., Quantales and Their Applications, Longman Scientific and Technical, London, (1990).
16. Russo C., Quantale modules with applications to logic and image processing, PhD thesis, University of Salerno, Italy, (2007).
17. Sharan S., Tiwari S. P., Yadav Vijay K., Interval Type-2 Fuzzy Rough Sets and Interval Type-2 Fuzzy Closure Spaces, *Iranian Journal of Fuzzy Systems*, **12**, 113–125 (2015).
18. Shmuely Z., The structure of Galois connections, *Pacific Journal of Mathematics*, **54** 209–225 (1974).
19. Shmuely Z., The tensor product of distributive lattices. *Algebra Universalis*, **9**, 281–296 (1979).
20. Solovyov S., Categories of lattice-valued sets as categories of arrows, *Fuzzy Sets and Systems* **157**, 843–854 (2006).
21. Solovyov S., On the category Set(JCPos), *Fuzzy Sets and Systems*, **157**, 459–465 (2006).
22. Solovyov S. A., On the category of Q-Mod, *Algebra universalis*, **58**, 35–58 (2008).
23. Tiwari S. P., Gautam Vinay, Dubey M. K., On fuzzy multiset automata, *Journal of Applied Mathematics and Computing*, **51** 643–657 (2016).
24. Tiwari S. P., Singh Anupam K., Sharan Shambhu, Fuzzy automata based on lattice-ordered monoid and associated topology, *Journal of Uncertain Systems*, **6**, 51–55 (2012).
25. Tiwari S. P., Singh A. K., Sharan S., Yadav, V. K., Bifuzzy core of fuzzy automata, *Iranian Journal of Fuzzy Systems*, **12**, 63–73 (2015).
26. Tiwari S. P., Yadav Vijay K., Dubey M. K., Minimal realization for fuzzy behaviour: A bicategory-theoretic approach, *Journal of Intelligent and Fuzzy Systems*, **30**, 1057–1065 (2016).

27. Tiwari S. P., Yadav Vijay K., Gautam Vinay, On Minimal Fuzzy Realization for a Fuzzy Language: A Categorical Approach, *Multiple-Valued Logic and Soft Computing*, **28**, 361–374 (2017).
28. Tiwari S. P., Yadav Vijay K., Singh A. K., Construction of a minimal realization and monoid for a fuzzy language: a categorical approach, *Journal of Applied Mathematics and Computing*, **47**, 401–415 (2015).
29. Yadav Vijay K., Gautam V., Tiwari S. P., On minimal realization of IF-languages: A categorical approach *Iranian Journal of Fuzzy Systems*, **13**, 19–34 (2016).
30. Zadeh L.A., Fuzzy Sets, *Information and Control*, **8**, 338–353 (1965).

Author Index

C. R. Panigrahi et al. (eds.), *Progress in Advanced Computing and Intelligent Engineering*, Advances in Intelligent Systems and Computing 714,
https://doi.org/10.1007/978-981-13-0224-4

MIX
Papier aus verantwortungsvollen Quellen
Paper from responsible sources
FSC® C105338

If you have any concerns about our products,
you can contact us on
ProductSafety@springernature.com

In case Publisher is established outside the EU,
the EU authorized representative is:
Springer Nature Customer Service Center GmbH
Europaplatz 3, 69115 Heidelberg, Germany

Printed by Libri Plureos GmbH
in Hamburg, Germany